Vertebrates of Florida

Vertebrates of Florida

Identification and Distribution

Henry M. Stevenson

A Florida State University Book

University Presses of Florida
Gainesville

Library of Congress Cataloging in Publication Data

Stevenson, Henry Miller, 1914–
    Vertebrates of Florida.

    "A Florida State University Book."
    Bibliography: p.
    Includes index.
    1. Vertebrates-Florida-Identification.
2. Zoology-Florida.   I.   Title.
QL169.S75   596'.09'759   75–37723
ISBN   0–8130–0437–3

Printed in Florida

Contents

List of Illustrations

Plates

Figures

Maps

Foreword

by WILLIAM B. ROBERTSON, JR.

As if in compensation for its somewhat uninspiring landscape, Florida possesses a remarkable wealth of animals. The great extent of the state, almost 500 miles in both its north–south and east–west dimensions, includes a diverse array of environments from the temperate, vaguely Appalachian, river valleys of the panhandle to the tropical, Antillean, Florida Keys. The variety of native animal life reflects this diversity of living conditions. It also reflects an eventful geologic past when the peninsula was flooded by shallow seas with islands where animals unique to Florida evolved, and where animals that died out elsewhere were able to survive. Unhappily, a growing part of the state's present faunal diversity results from man's uncontrolled itch to improve on nature by importing nonnative animals. In all, *Homo sapiens* now shares Florida with wild populations of some 880 other species of vertebrate animals—freshwater fish, amphibians, reptiles, birds, and mammals.

Dr. Stevenson's book gives, for the first time within one set of covers, the means to identify specimens of all the land and freshwater vertebrates known to occur in Florida and the salient facts about their distribution in the state. It fills the long-standing need for a single reference, up-to-date and authoritative, on the Florida animals of most interest and concern to man. The book is timely in a day when we are beginning to understand that any loss of diversity in the natural world around us cheapens our own lives. Sound information on the identity and distribution of Florida vertebrates is basic to any effort to conserve them.

It is to be hoped that this book will find use in Florida schools, and also among the state's conservation-minded citizens and amateur naturalists, many of whom made important contributions to its contents.

Everglades National Park, 1975.

Preface

There is presently no publication dealing exclusively with all the vertebrates of Florida and the means of identifying them. *Vertebrates of the United States* (Blair et al., 1957) necessarily gave extremely brief treatment of each species and omitted some Florida species, some of which are also not in the 1968 edition. The large number of species involved in that work also compounded the difficulties of writing workable keys for certain groups. Carr and Goin (1955), in their *Guide to the Reptiles, Amphibians, and Fresh-water Fishes of Florida*, provided a means of identifying the great majority of cold-blooded vertebrates (except most marine fishes) then known to Florida, but many additional species have been reported since. *A Key to Florida Birds* (Stevenson, 1960a) featured keys and descriptions for the identification of all birds except a few of accidental occurrence, but this faunal list has also increased in the intervening years. There seems to be no comprehensive treatment (or key) for the mammals of Florida. Thus the present work has been prepared in the belief that a need exists for a single source by which the identity of any specimen of Florida vertebrate may be established.

LIMITATIONS As in the case of its predecessors cited above, this work has omitted marine fishes. Their inclusion would not only have lengthened the work, increased the difficulty of writing and using the keys, and decreased the reliability of the information, but would have introduced an insoluble problem of establishing and using a constant boundary for the state. Potentially, at least, a similar problem may seem to exist for marine birds and mammals, but nearly all of those included have reached the coast occasionally. Only those species believed to have occurred or likely to recur within the present century are included. Although the descriptions of species should prove helpful in identifying both live and preserved specimens, this work is not intended to be a field guide.

GEOGRAPHIC COVERAGE Several species have been included in the keys (though not in the species accounts) that have not been recorded in Florida. In each case the range extends so close to the state line, or is so poorly known, that occurrence in Florida is deemed likely. The inclusion of these extralimital species should make this book more useful in surrounding states. In fact, with the exceptions of salamanders and freshwater fishes, the great majority of species known to occur in both states bordering Florida are included in this book on the basis of Florida records.

VALIDATION OF RECORDS The verification of Florida occurrences of nearly all vertebrates except some birds and cetaceans is based on collected specimens showing locality data. It may seem strange that any other evidence would suffice for birds, but there are practical reasons for these few exceptions. Largely because of public sentiment, most species of birds are protected against collecting without a scientific permit, and several are justifiably protected against collecting under any circumstances. The small percentage of ornithologists who have collecting permits still may not collect on wildlife refuges or in state parks (without special permission) or on private lands (without the owner's permission). Partially compensating for this disadvantage in corroborating records is the fact that experienced field workers, under favorable conditions, can correctly identify all but a few species of Florida birds. There still remains a question as to whether the occurrence of any avian species in a state should rest on any evidence less than a collected specimen or recognizable photograph. Surely if any exceptions are made, the circumstantial evidence in those cases should be overwhelming. As a rule, those vertebrates suspected of Florida occurrence, but not convincingly shown to have occurred here (or introduced but only doubtfully established) will be found in a special Hypothetical List at the end of each section.

DEFINITION OF "FRESHWATER FISHES" Another vexing decision has been that of what species to include as "freshwater" fishes. Ideally, perhaps, the criterion should be a certain maximum concentration of dissolved salts at the site of capture. Unfortunately, however, we do not always have such data, and the matter has been resolved by following most of the designations of "freshwater" or "euryhaline" given by Briggs' (1958) list and the inclusion of fishes collected in the St. Johns River above Jacksonville (Tagatz, 1967).

NOMENCLATURE AND CLASSIFICATION In many instances the correct name to be used for a species, or the higher taxon into which it is placed, is subject to individual opinion. In works of this kind it is desirable to follow a consensus as represented by committees of specialists, and this has been done in most cases in which it was possible. Thus, the names of fishes follow those of the American Fisheries Society (1960), and those of birds the American Ornithologists' Union (1957), except in a few cases indicated by footnotes in the text. A more recent revision of most of the major taxa of bony fishes (Greenwood et al., 1966) is also adopted. The names of amphibians and reptiles are those used by the American Society of Ichthyologists and Herpetologists (1963 et seq.) in their publications to date. For mammals it has been necessary to follow the works of individuals, and Hall and Kelson (1959) provided that standard. In the case of the remaining amphibians and reptiles (and occasionally of recently revised taxa in other groups), I have had to rely on my own judgment, modified by the advice of specialists in the respective fields. In such instances, of course, the responsibility for the names used is mine, not theirs. A more trivial deviation from certain of the above sources has been the hyphenation of compound modifiers in order to insure more uniform treatment of common names for all groups. In other works these names are hyphenated for birds and mammals (for example, Long-nosed Shrew) but not for lower vertebrates (for example, Shortnose Sturgeon).

Few states can boast as large a list of vertebrates as Florida, even when marine fishes are omitted. This text enumerates 880 species, fully accredited when the book was first organized, including 19 addenda following the mammal section of Chapter Three. These 880 species are divided as follows: freshwater fishes, 208; amphibians, 53; reptiles, 98; birds, 428; mammals, 93. Carr and Goin (1955) listed 333 "species" of cold-blooded vertebrates as against 349 listed here, but their count must have included subspecies as well. They claimed 57 species of amphibians for Florida, but only 42 of these forms treated in their book were full species, the rest being subspecies. Complete lists of vertebrates probably do not exist in most states, but such states as Texas and California have more species of birds than Florida, and probably more of all vertebrates combined.

As our knowledge of avian relationships improves it is often necessary to revise our classifications. While the manuscript of this work was being prepared for the printers, the Committee on Classification and Nomenclature of the American Ornithologists' Union published a supplement to the 1957 edition of the A.O.U. *Check-list of North American Birds* (*Auk* 90:411–419). Most of the changes involving species of Florida occurrence have since been made in my text, but there was inadequate time for the extensive changes that would have resulted from the lumping or splitting of species (except for the few such changes I had previously adopted). Accordingly, the other changes of this kind are listed here.

The Great White Heron (*Ardea occidentalis*) becomes a subspecies of the Great Blue Heron (*A. herodias*); it will be known as *A. h. occidentalis*.

The Snow Goose (*Chen hyperborea*) and the Blue Goose (*Chen caerulescens*) are considered color phases of a single species now to be known as the Snow Goose (*Chen caerulescens*).

The species previously known as Traill's Flycatcher (*Empidonax traillii*) is believed to comprise two species: the Alder Flycatcher (*E. alnorum*) and the Willow Flycatcher (*E. traillii*).

Bullock's Oriole (*Icterus bullockii*) is merged with the Baltimore Oriole (*I. galbula*) under the new name of Northern Oriole.

The Ipswich Sparrow (*Passerculus princeps*) is regarded as a subspecies of the Savannah Sparrow (*P. sandwichensis*).

The Dusky Seaside Sparrow (*Ammospiza nigrescens*) and the Cape Sable Sparrow (*A. mirabilis*) are considered subspecies of the widespread Seaside Sparrow (*A. maritima*). They become, respectively, *A. m. nigrescens* and *A. m. mirabilis*.

As both species of the former *Empidonax traillii* have occurred in Florida, the number of species of birds recorded in the state is reduced to 423 by these changes and the total number of vertebrate species to 875.

April, 1974

A later revision of the North American *Check-list* appeared in the fall of 1976 (*Auk*, 93:875–879), necessitating the following changes in the generic and specific names of Florida birds:

Butorides virescens becomes *Butorides striatus*.

The Fulvous and Black-bellied Tree Ducks become, respectively, the Fulvous and Black-bellied Whistling-Ducks.

Ictinia misisippiensis becomes *Ictinia mississippiensis*.

Thalasseus and *Hydroprogne* are merged in the genus *Sterna*, the new names being *Sterna maxima*, *S. sandvicensis*, and *S. caspia*.

Speotyto cunicularia becomes *Athene cunicularia*.

Centurus carolinus becomes *Melanerpes carolinus*.

Dendrocopos is merged in the genus *Picoides*. The new names are *Picoides villosus*, *P. pubescens*, and *P. borealis*.

Telmatodytes palustris becomes *Cistothorus palustris*.

Cassidix major becomes *Quiscalus major*.

Tangavius aeneus becomes *Molothrus aeneus*.

Spinus is merged in the genus *Carduelis*. The new names are *Carduelis pinus* and *C. tristis*.

December, 1976

Acknowledgments

No work of this kind can be entirely original, but if the author is to be more than a compiler he must expect to participate in the research on which the publication is based. To that end I have personally examined representative museum specimens of more than 90 percent of the species included here. As a result, some of the characters used in keys and descriptions are believed to be original in this work. Most of those that are not original were, nevertheless, carefully checked against the specimens available to me. My indebtedness, however, to those who have published previous works on these species is both obvious and enormous.

In a few instances keys prepared by graduate students at Florida State University were generously contributed, and these have not been modified to the point that I could claim credit for them. The key to the Cyprinidae was prepared by Ray Birdsong, that for larval anurans by Sheryl Fanning, and for the Scolopacidae by Horace Loftin.

Specimens critical to the study were provided through gift, loan, or exchange by such institutions as the American Museum of Natural History, Carnegie Museum, Florida State Museum (Gainesville), Miami Seaquarium, National Museum of Canada, Philadelphia Academy of Natural Science, United States National Museum, and museums at the following institutions: Auburn University, University of Illinois, Louisiana State University, University of Michigan, and the University of Oklahoma. In some cases, individuals at these museums identified or confirmed the identification of important specimens.

Various individuals also either provided or helped prepare critical specimens:

Ronald Altig	Richard M. Blaney
J.R. Bailey	Irene Boliek

Roland Brandon
Edwin C. Brown
P. Fairly Chandler
Frank L. Chapman
Ralph Chermok
Clare Cichowski
Allan D. Cruickshank
William E. Duellman
Harold Dundee
George W. Folkerts
O. Earle Frye
Bryan P. Glass
Roy C. Hallman
Frances Hames
Glenn Ivey
H.P. Langridge
Frederick H. Lesser

D. Bruce Means
Burt L. Monroe, Jr.
Russell E. Mumford
John C. Ogden
Mary Ann Olson
Storrs L. Olson
Dennis R. Paulson
Al Pflueger
Allan R. Phillips
Warren M. Pulich
Albert Schwartz
Dorothy E. Snyder
Ernest H. Stevenson
James M. Stevenson
Shirley Whitt
Lovett E. Williams, Jr.

In a number of instances records of birds outside their normal range were documented by photographs. For the privilege of examining these I am indebted to:

William J. Bolte
E.B. Chamberlain
Dorothy Dodd
John B. Edscorn
Myron Elliott
Harold H. Gaither
Samuel A. Grimes
Roy C. Hallman
Lyle S. Hubbard
Curtis L. Kingsbery
Howard P. Langridge

Davies Lazen
James C. McDaniel
Milton G. Nelson
John C. Ogden
Mary Ann Olson
William B. Robertson, Jr.
Alexander Sprunt, IV
Paul Sykes
Marvin L. Wass
Karl Zerbe

My greatest indebtedness, by far, I owe to those specialists who patiently read sections of the manuscript and made numerous, highly valuable comments. The fish sections were read by Ralph W. Yerger, the amphibian and reptile sections by Albert Schwartz, the bird sections by William B. Robertson, Jr., and John C. Ogden, and the mammal sections by James N. Layne and David K. Caldwell (marine species). The section on the preparation of bird skins, from an earlier work of mine, was then examined by Herbert L. Stoddard, Sr. Dr. Yerger has been especially helpful in making available the large fish collection at Florida State University and in providing miscellaneous information about fishes over a period of several years. Oscar T. Owre read the entire manuscript and made numerous valuable suggestions.

Various other specialists provided critical information of a more limited nature in their respective fields.

FISHES
  Glenn H. Clemmer
  Walter Courtenay
  Michael S. Dahlberg
  Carter Gilbert
  Robert Hastings
  Vernon Ogilvie
  Dennis R. Paulson
  William F. Smith-Vaniz
  Camm Swift

AMPHIBIANS AND REPTILES
  Ronald Altig
  Richard M. Blaney
  Steve Christman
  John W. Crenshaw, Jr.

  Herndon G. Dowling
  Hobart Landreth
  D. Bruce Means
  Robert Mount
  Lewis Ober
  Fred Shanholtzer

BIRDS
  Allan D. Cruickshank
  Charles H. Rogers
  Louis A. Stimson

MAMMALS
  W. Wilson Baker
  Richard Laval
  Joseph C. Moore
  Albert Schwartz

For their skilled contributions in preparing the photographs I acknowledge my gratitude to James Capone, Christopher Combs, Rhodes Holliman, and Charlotte Maxwell; also to Miss Maxwell for Figure 7. (Other drawings are my own.)

The publication of this manuscript would not have been possible but for generous financial contributions by Mrs. Bradley Fisk and John Foster. Some of my field work was supported by grants from the U.S. Public Health Service and Florida State University where I was a member of the Biological Sciences faculty over the two decades during which this work was prepared. Finally, but to no less degree, I am grateful to my wife, other members of my family, my students, and other friends who have helped provide time for this work or have encouraged it in various other ways.

Henry M. Stevenson
Tall Timbers Research Station
Route 1, Box 160
Tallahassee, FL 32303

Vertebrates of Florida

Chapter One

Introduction

IDENTIFICATION OF VERTEBRATES

All species included in this book are members of a subphylum (Craniata, or Vertebrata) of the Phylum Chordata. It is obvious that other animals, such as insects, worms, and mollusks, cannot be identified in it. The classical, and easy, definition of a vertebrate as an animal with a vertebral column consisting of consecutive pieces of bone ("backbone") or cartilage is not very helpful unless dissection is practical. In lieu of this evidence, certain external features are either individually or collectively diagnostic. The most important of these are: (1) a postanal tail; (2) usually two pairs of appendages (rarely one pair or none), fins in fishes or limbs bearing digits (usually 5) in most other vertebrates; (3) a body covering of scales, feathers, or hair (all species exceptional as to *both* criteria 2 and 3 show one or more pairs of gill slits). Even by the use of these criteria it is conceivable, though quite unlikely, that a nonvertebrate animal could be erroneously considered a vertebrate and tested in this key, though without success. Given, however, a genuine vertebrate, the investigator must begin with couplet 1 in the Key to Classes (page 9). By looking for paired appendages and examining the gross structure of the mouth, he should decide whether to continue to couplet 2 in the same key, or proceed to the CLASS AGNATHA. In either case he would continue to select the more fitting of two alternative descriptions until he arrived at the name of the family, at which point a Roman numeral would refer him to the location of the key to the species in that family. Finally, when the specimen had been keyed to a species, an Arabic numeral would send him to the descriptive account of that species. Both this account and that of the family will serve as a check on the accuracy of the keying process. Not only should the description of the species agree with that of the animal in question, but the collection data, if known, should agree with the known range of the species. A natural tendency among those unfamiliar with keys is to agree too readily with the first part of a couplet without even reading the second part (or *all* of the first part)—a practice certain to result in many erroneous identifications and wasted time.

In some instances the characteristics used in keying out taxa, especially families, are applicable only to their Florida representatives. In those cases it follows that a specimen from another part of the world could not be keyed even to its proper family. This limitation applies chiefly to taxa represented in Florida by only one or two species (for example, Pycnonotidae).

USE OF SKULLS IN KEYING MAMMALS

Of all vertebrate zoologists, mammalogists alone commonly preserve separately both the skull and the museum skin of the individual specimen. In the great majority of cases, therefore, both of these evidences of the animal's identity are available. This fact may be deemed both a blessing and a curse by the deviser or the user of the key, as it may appear to complicate identification unnecessarily. The truth is that a positive identification by the average person will prove very difficult in many cases when only the skin or the skull is available. By using both, the probability of a correct identification is greatly increased.

The use of these two lines of evidence in other keys familiar to me follows either of two courses. Either there are separate keys for skins and skulls; or a single key is constructed, utilizing both skins and skulls only in the places where that procedure seems most necessary. Unlike other keys, then, the present one endeavors to utilize characters of both the skull and external morphology *throughout*, except in two instances in which no consistent differences could be found. It is thus possible to use the same key with skull only, skin only, or with both. As explained previously, the chances of success are enhanced if both skin and skull are available, but it seems likely that the great majority of species can be identified from only one line of evidence.

TERMINOLOGY

The guiding philosophy in the preparation of this work has been to avoid the use of technical terms whenever feasible, but many are necessary for the sake of precision and conciseness. Those least familiar to the layman will be found in the Glossary (pp. 549) and the remainder in a standard dictionary.

MEASUREMENTS AND COUNTS

There are many cases in which an animal cannot be keyed correctly without the use of special equipment. With small animals (sometimes larger ones) some magnification is necessary. In many cases a hand lens will suffice, but in others a binocular microscope is essential. Measurements and counts often require such magnification, and measuring often requires additionally an inexpensive compass bearing a sharp-pointed pencil. In some cases dividers or calipers are almost a necessity. Even slight inaccuracies in measuring or counting may lead to wrong identifications. All the measurements and counts used frequently in this book are straight-line measurements unless indicated otherwise.

FISHES *Depth*—maximum depth of body, dorsal to ventral, exclusive of fins; *total length*—most anterior part of head to tip of longest rays in caudal fin; *standard length*—tip of upper jaw to hidden base of caudal fin rays; *length of dorsal and anal fins* (the same as "length of fin base" of other authors)—distance along base of fin, from origin to insertion; *length of head*—tip of upper jaw to most posterior edge of operculum; *length of snout*—tip of upper jaw to anterior edge of orbit; *length of caudal peduncle*—oblique distance from insertion of anal fin to hidden middle base of caudal rays; *dorsal and anal fin-ray counts*—in this work the count includes all soft rays, even rudimentary ones, except in the Cyprinidae; in cyprinids, only the branched rays and the one *long* unbranched ray are counted; spines, whether true spines or those consisting of fused, hardened rays, are not included in the count; *lateral-line scale count*—number of scales along lateral line (or, if no lateral line, scales at that level) from edge of operculum to hidden base of caudal rays; also referred to as "number of lateral scales" or of "oblique scale rows"; *count of horizontal scale rows*—the number of horizontal scale rows on one side (sometimes specified as those above or below lateral line), counting obliquely upward from the origin of the anal fin.

AMPHIBIANS *Snout-vent length*—tip of snout to anterior edge of vent (all adult anuran lengths are snout-to-vent); *body length* (tadpoles)—tip of snout to base of tail on lateral axis; *tail length*—posterior edge of vent to tip of tail (except begin at tail base, lateral axis, in tadpoles); *costal-groove count*—number of costal grooves between front and hind limbs, including one each for armpit and groin (extend count to vent in Sirenidae); *costal grooves between appressed limbs*—press forelimb and hind limb on same side against body and toward one another and count costal grooves between tips of longest digits.

REPTILES *Snout-vent length*—see above paragraph; *tail length*—see above paragraph; *count of dorsal scale rows*—count at center of body, going diagonally around body, but omitting enlarged ventral scales;

these rows are *numbered* from the middorsal row downward in lizards, but from the *lowermost* row upward in snakes; *count of ventral plates*—begin with first scale under head that is wider than long and end with last scale before anal plate; *count of subcaudal plates*—begin with first scale behind anal plate and count pairs as single scales (if in 2 rows), continuing to last scale.

BIRDS   *Total length*[1]—tip of upper mandible to tip of longest rectrix; *wing length*—anterior edge of folded wing (wrist joint) to tip of longest primary; *tail length*—base of central rectrices to tip of longest rectrix; *tarsal length*—back of heel joint to lower end of last unmodified tarsal scale, anterior side; *hind-toe length*—dorsal side, from junction with tarsus, usually including hind claw; *culmen length*—length of *exposed* culmen (some workers measure from junction of culmen with skull; bill length *from nostril* is also used); *toe numbering*—begin with hind toe (hallux), continuing from medial to lateral; *primary numbering*—in this work the primaries are numbered from the outermost one inward.

MAMMALS   *Total length*—tip of nose to tip of fleshy part of tail, with specimen lying on its back; *tail length*—base of tail to its fleshy tip (holding tail perpendicular to body in fresh specimens); *hind-foot length*—from back of heel to tip of longest claw; *ear length*—from deepest part of basal notch to fleshy tip of ear; *skull length*—from tip of nasal bones to most posterior part of cranium; *basilar length of skull*—from anterior edge of foramen magnum to anterior edge of hard palate at base of first upper incisors; *tooth counts*—even vestigial teeth are included; magnification may be necessary.

SPECIES ACCOUNTS

As these accounts are intended largely as a check on the correctness of the identification made by keying, the descriptions are not intended to be detailed, although descriptions of "difficult" species are longer. For most species the most important criteria for identification are italicized. Unless of a brownish or blackish color, the color of the unfeathered parts ("soft parts") of birds as it appeared in life is indicated; these parts include the bill, cere, lores and eyelids (in some cases), iris, and feet (tarsus and toes). It should be emphasized that these often change color

---

1. Size is a much more important criterion in birds than in other vertebrates, as there is generally little variation within a species. Even juvenals are almost as large as adults. Probably the most reliable index of size in museum skins is total length. Although total length will vary with tail length, wing length is equally variable. Some preparators tend to minimize the importance of total length and fail to provide this information on the label. In such cases the total length of the museum skin may be assumed to be within 10 percent of the original in most species and within 20 percent in long-necked birds.

in museum skins. The distribution of each species is also given in general terms, including that of each subspecies known to occur in Florida. Unless a species of bird, bat, or cetacean is present throughout the year in Florida, the season of occurrence is also given in general terms. If notable changes in distribution have occurred since the appearance of certain general works from 1954 to 1959 (see the Bibliography of this work), these changes are documented by bibliographic references or the location and catalog numbers of museum specimens. For the latter the following symbols have been used: CM, Carnegie Museum; FSU, Florida State University; RMB, Richard M. Blaney, private collection; TT, Tall Timbers Research Station; UF, University of Florida (Florida State Museum); UM, University of Miami; USF, University of South Florida; NMNS, National Museum of Natural Science.

Relative abundance is indicated for certain birds not frequently occurring in the state. Those considered "accidental" are out of their normal range and have occurred only a few times. "Casual" visitors may occur about once every five to ten years, and a "rare" species as frequently as a few times each year.

The range limits of 68 species are shown on Maps 3–68. The criteria most often employed in selecting these species were: (1) range virtually confined to Florida or to the Southeast; (2) distribution not well known, or misrepresented in some previous publications; (3) recent changes in known range; (4) recent changes in taxonomic status, such as the lumping of two or more nominal species (for example, Yellow-shafted, Red-shafted, and Gilded Flickers). It should be emphasized that the establishment of range limits does not imply that a species occurs in all areas within, even in suitable habitat. Perhaps a few species do so, but the great majority are locally distributed, at least to some extent.

Chapter Two

Key to Vertebrate Classes, Orders, Families, and Species

KEY TO VERTEBRATE CLASSES

1. Mouth provided with jaws, *or* surrounded by black, horny mandibles, *or* tail longer and deeper than body; one or 2 pairs of appendages (fins or limbs) usually present 2
   Mouth without jaws, permanently open and rounded; tail not longer and deeper than body; paired appendages lacking .... AGNATHA, p. 10
2. Paired appendages (if present) finlike, with long, slender, closely paralleled skeletal elements; also median, unpaired fins of similar structure; gill openings present 3
   Paired appendages (if present) limblike, with a single bone in the first segment of each; median fins (if present) without skeletal support; gill openings usually absent 4
3. Gill openings (at least 5) exposed; mouth ventral and crescent-shaped .... CHONDRICHTHYES, p. 10
   Gill openings covered by a usually hardened flap, the operculum; mouth usually terminal, sometimes superior or inferior .... OSTEICHTHYES, p. 11
4. Epidermal derivatives (claws, scales, feathers, hair) entirely lacking; limbs not paddlelike .... AMPHIBIA, p. 42
   Skin covered, at least in part, with scales (plates), feathers, or hair, and digits provided with claws, nails, or hoofs; *or* forelimbs paddlelike and hind limbs lacking 5

5. Skin entirely covered with scales or enlarged plates; hair or feathers never present; limbs sometimes lacking .... REPTILIA, p. 57

 With feathers or hair inserted on some parts of head or body; scales usually localized, if present; at least one pair of limbs present ....
 6

6. Head and body covered with feathers; scales usually present on hind limbs; teeth lacking .... AVES, p. 77

 Partly (usually almost entirely) covered with hair; scales rarely present, but widespread in one group; teeth usually present ......
 .... MAMMALIA, p. 134

CLASS AGNATHA

ORDER PETROMYZONTIFORMES: I. FAMILY PETROMYZONTIDAE, Lampreys.[1]

1. Dorsal fins not united; body mottled; length up to 90 cm ....
 .... Sea Lamprey (*Petromyzon marinus*), sp.1

 Dorsal fin continuous, though sometimes notched; body unicolored; length under 15 cm ....
 .... Southern Brook Lamprey (*Ichthyomyzon gagei*),[2] sp.2

CLASS CHONDRICHTHYES—ORDERS AND FAMILIES

1. Dorsoventrally flattened (or triangular in cross section), with eyes dorsal and gill openings ventral; anal fin lacking ....
 .... RAJIFORMES, 2

 Body terete; gill openings lateral; anal fin present ....
 .... SQUALIFORMES: CARCHARHINIDAE, II

2. With an extended, flattened rostrum and laterally protruding teeth; body triangular in cross section .... PRISTIDAE, III

 Without a saw-shaped snout; body flattened ...... DASYATIDAE, IV

ORDER SQUALIFORMES: II. FAMILY CARCHARHINIDAE, Requiem Sharks.

1. With well-defined labial grooves extending forward from the corners of the mouth (Plate I-1); distance between nostrils scarcely, if any, greater than that from either nostril to tip of snout ....
 Atlantic Sharp-nosed Shark (*Rhizoprionodon terraenovae*), sp.3

 Without well-defined labial grooves at corners of mouth; distance between nostrils at least 30% greater than that from either nostril to tip of snout .... Bull Shark (*Carcharhinus leucas*), sp.4

---

1. The families and orders in the following sections of this book are consecutively numbered, families with Roman numerals and species with Arabic numerals (preceded by "sp.").

2. The occurrence of the Least Brook Lamprey, *Lampetra aepyptera*, in the Yellow River drainage of Alabama, may also be expected in Florida (Smith-Vaniz, 1968).

ORDER RAJIFORMES: III. FAMILY PRISTIDAE, Sawfishes.

1. With more than 22 pairs of teeth ....
.... Small-toothed Sawfish (*Pristis pectinatus*), sp.5
   With fewer than 22 pairs of teeth ....
.... Large-toothed Sawfish (*Pristis perotteti*), sp.6

ORDER RAJIFORMES: IV. FAMILY DASYATIDAE, Sting Rays.

1. Contour of body rhomboidal; width at least 20% greater than
   distance from tip of snout to posterior tip of pectoral fin ......
.... Southern Sting Ray (*Dasyatis sabina*), sp.8
   Posterior contour of body almost semicircular; width about equal to
   length (not exceeding it by more than 10%) ....
.... Atlantic Sting Ray (*Dasyatis americana*), sp.7

CLASS OSTEICHTHYES—ORDERS AND FAMILIES

1. Tail heterocercal (never square or notched); gular plate or snout
   present (Plate I-2) 2
   Tail not heterocercal, *or* snout and gular plate absent 4
2. Caudal fin with prominent, pointed lobes, especially dorsally; mouth
   inferior; body partly covered with rows of plates ....
.... ACIPENSERIFORMES: ACIPENSERIDAE, V
   Caudal fin asymmetrical and rounded; body covered with scales
.... 3
3. Well-developed snout present; gular plate lacking ....
.... LEPISOSTEIFORMES: LEPISOSTEIDAE, VI
   Mouth essentially unmodified; gular plate present ....
.... AMIIFORMES: AMIIDAE, VII
4. Body elongate and covered with plates; elongate snout present; anal
   fin lacking .... GASTEROSTEIFORMES: SYNGNATHIDAE, XXVII
   Body usually not covered with enlarged plates; snout usually absent;
   anal fin present, though sometimes conjoined with caudal 5
5. Body elongate and with no apparent scales; dorsal, caudal, and anal
   fins continuous; pelvic fins wanting ...... ANGUILLIFORMES, 6
   Body not remarkably elongate, usually covered with scales; dorsal,
   caudal, and anal fins separate; pelvic fins present 7
6. Lower jaw protruding beyond upper; scales embedded in skin ......
.... ANGUILLIDAE, X
   Lower jaw shorter than upper; scales lacking ....
.... OPHICHTHIDAE, XI
7. Extremely compressed, with one side of body unpigmented and both
   eyes on the other side .... PLEURONECTIFORMES, 8
   Body not unusually compressed and both sides equally pigmented;
   eyes normally arranged 10

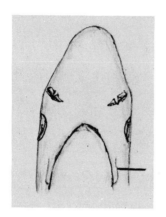

1. Labial grooves of *Rhizoprionodon*

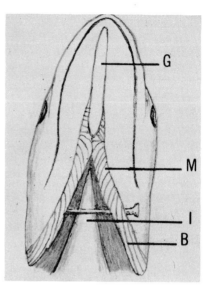

2. Gular plate (G), isthmus (I), branchiostegal rays (B), and gill membrane (M) of *Elops saurus*

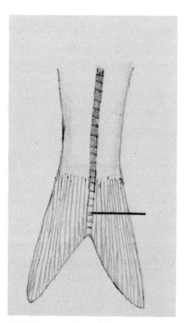

3. Lateral line of *Centropomus undecimalis*

4. Protactile premaxillary of the Gerreidae

PLATE I

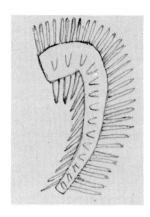

5.
First gill arch of
Lepomis macrochirus

6.
Hybopsis aestivalis,
showing
maxillary barbels
and
inferior mouth

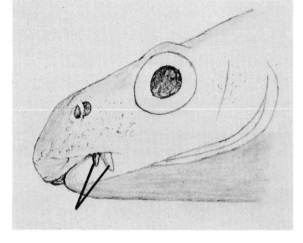

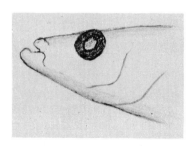

8. Angular lower lip of
Carpiodes velifer

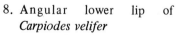

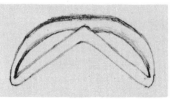

7. Superior mouth of *Fundulus grandis*

FISHES

8. Eyes on right side of body .... SOLEIDAE, XLIII
   Eyes on left side of body 9
9. Standard length fully 3 times width of disk, excluding fins; pectoral
   fins lacking .... CYNOGLOSSIDAE, XLIV
   Standard length much less than 3 times width of disk; one or 2
   pectoral fins present .... BOTHIDAE, XLII
10. Front of head depressed and provided with barbels, usually long;
   body naked or covered with bony plates except on abdomen; strong
   spines in pectoral fins .... SILURIFORMES, 23
   Front of head not strongly depressed; barbels short or lacking; body
   usually scaly; pectoral fins without spines 11
11. First dorsal fin modified as a suckerlike organ; more than 30 rays in
   anal fin .... PERCIFORMES: ECHENEIDAE, XXXII
   First dorsal fin not markedly modified; anal fin with fewer than 30
   rays 12
12. Body long and slender (standard length at least 10 times body depth);
   jaws produced into a long, compressed beak; lateral line ventral ..
   .... ATHERINIFORMES: BELONIDAE, XXIII
   Body usually not extremely long and slender; beak (if present) not
   deeper than wide; lateral line (if present) usually on upper half
   of body 13
13. Mouth of normal size and shape; 2 dorsal fins present; standard length
   at least 6 times body depth; dorsal edge of pectoral fin more than
   halfway up side of body; one spine and at least 16 rays in anal
   fin .... ATHERINIFORMES: ATHERINIDAE, XXVI
   Not with the above combination of characteristics (if one spine in
   anal, then less than 16 rays) 14
14. Lateral line not in evidence; only one dorsal fin present; mouth small
   and superior (Plate I-7); spines lacking; dorsal edge of pectoral fin
   less than half way up side of body; caudal fin not forked or
   notched .... ATHERINIFORMES (part), 15
   Not with the above combination of characteristics (lateral line usually
   present and spines often present; caudal fin usually forked or
   notched) 16
15. Anal fin rounded; third anal ray branched; length of upper jaw (from
   crease) less than width of gape ...... CYPRINODONTIDAE, XXIV
   Anal fin of males slender and its rays long; third anal ray not branched,
   or length of upper jaw greater than width of gape ....
   .... POECILIIDAE, XXV
16. Entire head naked or granular, or spines absent 17
   Head at least partly covered with scales; spines present, but sometimes
   slender and weak 27
17. Gill membranes free from isthmus; at least 4 branchiostegal rays
   present (Plate I-2) 18

Gill membranes united with isthmus; fewer than 4 branchiostegal
rays .... Cypriniformes, 26
18. Gular plate present (Plate I-2) .... Elopiformes, 19
Gular plate absent 20
19. Last ray of dorsal fin not elongate; scales smaller than eye ....
.... Elopidae, VIII
Last ray of dorsal fin greatly elongate; scales larger than eye ....
.... Megalopidae, IX
20. Ventral contour rounded in cross section; dorsal fin on posterior half
of body .... Salmoniformes, 21
Ventral contour angular (ridged) in cross section; dorsal fin about
middle of body .... Clupeiformes, 22
21. Large (length of adults exceeding 15 cm); snout flattened like a duck's
bill; caudal fin forked .... Esocidae, XIV
Small (length of adults about 8 cm); snout blunt; caudal fin rounded
.... Umbridae, XV
22. Mouth inferior; angle of jaws extending posterior to position of eyes;
ventral scales not serrate .... Engraulidae, XIII
Mouth terminal or superior (rarely somewhat inferior); angle of jaws
not extending behind level of eye; ventral scales serrate ....
.... Clupeidae, XII
23. Body covered with bony plates; mouth ventral, with lips disklike
.... Loricariidae, XX
Body naked; mouth usually terminal or nearly so; lips not disklike 24
24. Dorsal and anal fins very long (more than 50% of total length) and
without spines .... Clariidae, XIX
Bases of dorsal and anal fins short, the dorsal with a strong spine
.... 25
25. Nostrils without barbels; total of 4 or 6 barbels present; primarily
marine .... Ariidae, XXI
Posterior nostrils provided with 2 barbels; total of 8 barbels; freshwater fishes .... Ictaluridae, XVIII
26. Mouth ventral, with thickened, fleshy lips; more than 10 dorsal rays;
barbels and spines lacking .... Catostomidae, XVII
Mouth usually terminal, the lips not greatly thickened; not more
than 10 dorsal rays unless dorsal spine or barbels present ....
.... Cyprinidae, XVI
27. Anus far forward (between bases of pelvic fins, or anterior to them);
7 rays in pelvic fin ....
.... Percopsiformes: Aphredoderidae, XXII
Anus normally located; not more than 6 rays in pelvic fin ....
.... Perciformes, 28

28. Pectoral fins at least midway up sides of body; caudal fin deeply forked .... MUGILIDAE, XL

Pectoral fins closer to ventral than to dorsal midline, *or* caudal fin not forked 29

29. Dorsal spines slender and flexible; gill membranes joined to isthmus .... GOBIIDAE, XLI

Dorsal spines stiff; gill membranes usually free from isthmus.... 30

30. Front teeth shaped like human incisors and often protruding .... SPARIDAE, XXXVII

Front teeth not incisorlike 31

31. Scales minute; 2 dorsal fins; anal and second dorsal fins very long (about 35–40% of standard length); peduncle very slender; with 2 free anal spines .... CARANGIDAE, XXXIII

Not with the above combination of characteristics 32

32. Lateral line extending without interruption to end of caudal fin ..... (Plate I-3) 33

Lateral line absent or not extending far onto caudal fin 34

33. Bases of dorsal and anal fins covered with a scaly sheath; lower jaw protruding; one or more spines on operculum .... CENTROPOMIDAE, XXVIII

No scaly sheath covering bases of dorsal and anal fins; lower jaw usually not protruding; no spines on operculum .... SCIAENIDAE, XXXVIII

34. Dorsal fin single 35

With 2 separate dorsal fins 39

35. Lateral line interrupted and displaced near soft dorsal fin; only one pair of nostrils .... CICHLIDAE, XXXIX

Lateral line absent or continuous; 2 pairs of nostrils 36

36. Premaxillary protractile outward and downward from sheath in skull for distance equal to or exceeding diameter of eye (Plate I-4); first dorsal spine less than 20% length of second .... GERREIDAE, XXXV

Premaxillary not more than normally protractile; first dorsal spine more than 20% of second 37

37. With at least one enlarged, fanglike front tooth on each side of the upper jaw .... LUTJANIDAE, XXXIV

Front teeth of upper jaw all of about equal size 38

38. Second anal spine much thicker than third .... POMADASYIDAE, XXXVI

Second anal spine not thicker than the third .... CENTRARCHIDAE, XXX

39. With 3 strongly graduated anal spines .... PERCICHTHYIDAE, XXIX

With less than 3 anal spines .... PERCIDAE, XXXI

ORDER ACIPENSERIFORMES: V. FAMILY ACIPENSERIDAE, Sturgeons.

1. Anterior edge of anal fin directly below that of dorsal fin . . . .
. . . . Short-nosed Sturgeon (*Acipenser brevirostrum*), sp.10
Anterior edge of anal fin well behind that of dorsal fin . . . .
. . . . Atlantic Sturgeon (*Acipenser oxyrhynchus*), sp.9

ORDER LEPISOSTEIFORMES: VI. FAMILY LEPISOSTEIDAE, Gars.

1. Length of snout (measured to angle of jaws) less than that of rest of
head (angle of jaws to back edge of operculum) . . . .
. . . . Alligator Gar (*Lepisosteus spatula*), sp.11
Length of snout greater than that of remainder of head   2
2. Snout length more than twice length of remainder of head . . . .
. . . . Long-nosed Gar (*Lepisosteus osseus*), sp.14
Snout length not more than twice length of rest of head   3
3. Isthmus, at least, scaled . . . .
. . . . Florida Gar (*Lepisosteus platyrhincus*), sp.13
Breast and isthmus not scaled . . . .
. . . . Spotted Gar (*Lepisosteus oculatus*), sp.12

ORDER AMIIFORMES: VII. FAMILY AMIIDAE, Bowfins. One species . . . . . .
. . . . Bowfin (*Amia calva*), sp.15

ORDER ELOPIFORMES: VIII. FAMILY ELOPIDAE, Ladyfishes. One species
in Florida . . . . Ladyfish (*Elops saurus*), sp.16

ORDER ELOPIFORMES: IX. FAMILY MEGALOPIDAE, Tarpons. One species
in Florida . . . . Tarpon (*Megalops atlantica*), sp.17

ORDER ANGUILLIFORMES: X. FAMILY ANGUILLIDAE, Freshwater Eels.
One species in Florida . . . . . . American Eel (*Anguilla rostrata*), sp.18

ORDER ANGUILLIFORMES: XI. FAMILY OPHICHTHIDAE, Snake Eels. One
species in Florida fresh water . . . .
. . . . Speckled Worm Eel (*Myrophis punctatus*), sp.19

ORDER CLUPEIFORMES: XII. FAMILY CLUPEIDAE, Herrings and Shad.

1. Posterior ray of dorsal fin greatly elongate   2
Posterior ray of dorsal fin not elongate   3
2. With 29 to 35 rays in anal fin; mouth almost ventral . . . .
. . . . Gizzard Shad (*Dorosoma cepedianum*), sp.29
With 17 to 25 rays in anal fin; mouth terminal . . . .
. . . . Threadfin Shad (*Dorosoma petenense*), sp.30

3. Visible portion of scales rounded or angular on posterior margin and its width almost equal to its depth 7
   Visible portion of most scales vertically arranged, its depth more than twice its width 4
4. Origin of dorsal fin more anterior than that of pelvics; most scales with smooth edges; maximum length about 15 cm ....
   .... Scaled Sardine (*Harengula pensacolae*), sp.28
   Origin of dorsal fin more posterior than that of pelvics; scales with irregular edges; maximum length of adults much more than 15 cm
   5
5. With fewer than 58 oblique scale rows along one side of body .....
   6
   With more than 58 oblique scale rows per side ....
   .... Yellowfin Shad (*Brevoortia smithi*), sp.27
6. Appressed pectoral fin usually reaching to within 2 scale rows of pelvic base; maximum depth of body near operculum; Gulf of Mexico ...... Large-scaled Menhaden (*Brevoortia patronus*), sp.26
   Appressed pectoral fin falling at least 3 scales short of pelvic base; maximum body depth near tip of pectoral fin; Atlantic Coast
   .... Atlantic Menhaden (*Brevoortia tyrannus*), sp.25
7. Peritoneum dark ..... Blue-backed Herring (*Alosa aestivalis*), sp.22
   Peritoneum not dark 8
8. Teeth present at tip of upper jaw ....
   ....Skipjack Herring (*Alosa chrysochloris*), sp.20
   Teeth entirely lacking 9
9. Lower jaw protruding beyond upper, producing a superior mouth
   .... Hickory Shad (*Alosa mediocris*), sp.21
   Upper and lower jaws almost equally protruding 10
10. With about 60 gill rakers on lower arm of first gill arch (see Plate I-5 for a gill arch); one or more dark spots in a longitudinal row behind operculum ..... American Shad (*Alosa sapidissima*), sp.23
    With about 40 gill rakers on lower arm of first gill arch; only one dark spot behind operculum, sometimes followed by a continuous dark streak .... Alabama Shad (*Alosa alabamae*), sp.24

ORDER CLUPEIFORMES: XIII. FAMILY ENGRAULIDAE, Anchovies.

1. With 25 or more rays in anal fin; silvery stripe indistinct and its width less than diameter of eye ....
   .... Bay Anchovy (*Anchoa mitchilli*), sp.32
   With less than 25 rays in anal fin; silvery stripe conspicuous and its width at least as great as diameter of eye ....
   .... Striped Anchovy (*Anchoa hepsetus*), sp.31

ORDER SALMONIFORMES: XIV. FAMILY ESOCIDAE, Pickerels.

1. Branchiostegal rays 11 to 14 (see Plate I-2 for branchiostegals); scales along lateral line about 105; usually marked with vertical, irregular dark bars ..... Redfin Pickerel (*Esox americanus*), sp.33
   Branchiostegals 14 to 16; scales along lateral line about 125; adults with chainlike or reticulate markings (but young with vertical bars) .... Chain Pickerel (*Esox niger*), sp.34

ORDER SALMONIFORMES: XV. FAMILY UMBRIDAE, Mud Minnows. One species in Florida ...... Eastern Mud Minnow (*Umbra pygmaea*), sp.35

ORDER CYPRINIFORMES: XVI. FAMILY CYPRINIDAE, Minnows and Carps.[1]

1. Dorsal and anal fins with a strong serrated spine; dorsal fin with more than 15 rays .... Carp (*Cyprinus carpio*), sp.58
   Dorsal and anal fins without spines; dorsal fin with less than 15 rays 2
2. One or more pairs of maxillary barbels present (often small or obsolete), *or* scales becoming noticeably smaller anteriad 3
   No maxillary barbels present; scales of uniform size over body 6
3. Lateral line with 40 or more scales and deeply decurved; basicaudal spot usually triangular, with apex pointing anteriad; barbels often hidden in groove above angle of jaw; scales smaller anteriad ....
   .... Creek Chub (*Semotilus atromaculatus*), sp.56
   Lateral line with less than 40 scales and not deeply decurved; basicaudal spot not triangular; barbel at angle of jaw; scales of uniform size over body 4
4. Two pairs of maxillary barbels present (Plate I-6); eyes dorsolateral
   .... Speckled Chub (*Hybopsis aestivalis*), sp. 55
   One pair of maxillary barbels present, though sometimes obscure; eyes lateral 5
5. Lateral stripe broad and dark; a light U-shaped (or V-shaped) mark on snout (yellowish in life) and a light rectangular mark on top of head between the eyes; more than 17 scales (usually about 22) along the dorsum from origin of dorsal fin to posterior margin of head; mouth subterminal; maxillary barbels small and often obscure .... Red-eyed Chub (*Hybopsis harperi*), sp.54
   Lateral stripe narrow and light; no conspicuous light markings on head; less than 17 scales (usually about 13) along the dorsum from origin of dorsal fin to posterior margin of head; mouth inferior; maxillary barbels usually conspicuous ....
   .... Big-eyed Chub (*Hybopsis amblops*), sp.53

---

1. Modified from a key prepared by Ray Birdsong. In addition to the species included here, the Goldfish (*Carassius auratus*) is occasionally released from captivity and may persist in a wild state for short periods.

6. Lower lateral and underside of head divided internally into large cuboidal chambers which are externally visible ....
.... Silver-jawed Minnow (*Ericymba buccata*), sp.52
No large cuboidal chambers in head 7
7. Anal fin with 13 or more rays; lateral line markedly decurved .....
.... Golden Shiner (*Notemigonus crysoleucas*), sp.36
Anal fin with less than 13 rays; lateral line usually not decurved
.... 8
8. Mouth small and oblique (almost vertical), extending half, or less than half, the distance to the eye; dorsal fin typically with 9 rays; scales below lateral line strongly outlined in black ....
.... Pug-nosed Minnow (*Opsopoeodus emiliae*), sp.57
Mouth not small, extending to, or almost to, the eye; dorsal fin typically with 8 rays 9
9. Anal fin with 10 to 12 rays 10
Anal fin with 7 or 8 rays (rarely 9) 14
10. Dorsum finely scaled (22 or more scales between origin of dorsal fin and posterior margin of head) ....
.... Blacktip Shiner (*Notropis atrapiculus*), sp.37
Dorsum coarsely scaled (less than 22 scales between origin of dorsal fin and posterior margin of head) 11
11. Lateral stripe narrow and light, absent or almost absent anteriad; typically with more than 37 scales in lateral line ....
.... Bandfin Shiner (*Notropis zonistius*), sp.38
Lateral stripe broad and dark, extending to or onto the head; typically with fewer than 37 scales in lateral line 12
12. Body along base of anal fin darkly pigmented ....
.... Dusky Shiner (*Notropis cummingsae*), sp.44
Body along base of anal fin not darkly pigmented 13
13. Predorsal streak dark and conspicuous; dorsal fin membranes typically with much dark pigment, concentrated on membranes between first 3 rays in young specimens ....
................Sailfin Shiner (*Notropis hypselopterus*), sp.42
Predorsal streak faint or absent; dorsal fin membranes with little or no dark pigment, usually concentrated along margins of rays if present .... Flagfin Shiner (*Notropis signipinnis*), sp.43
14. Lateral line complete 15
Lateral line incomplete 21
15. Anal fin typically with 7 rays 16
Anal fin typically with 8 rays (occasionally 9) 18
16. Mouth inferior; lateral stripe and basicaudal spot obscure or absent
.... Long-nosed Shiner (*Notropis longirostris*), sp.50
Mouth terminal or subterminal; lateral stripe and basicaudal spot conspicuous 17

17. Scales in first one or 2 rows below anterior portion of lateral line outlined in dark pigment; pigmentation in anal fin concentrated along last 3 or 4 rays; a dark spot on dorsum at origin of dorsal fin .... Weed Shiner (*Notropis texanus*), sp.46
    Scales in first one to 2 rows below anterior lateral line not outlined in dark pigment; pigmentation in anal fin not concentrated along last 3–4 rays, but scattered along margins of all rays; no spot on dorsum at origin of dorsal fin ....
    .... Coastal Shiner (*Notropis petersoni*), sp.47
18. Basicaudal spot large and dark (more than half diameter of orbit .... Black-tailed Shiner (*Notropis venustus*), sp.41
    Basicaudal spot less than half diameter of orbit and often light or obsolete 19
19. Lateral stripe of silvery pigment, continuing onto head; (intestine long and convoluted into many coils) ....
    .... Cypress Minnow (*Hybognathus hayi*), sp.51
    Lateral stripe of dark pigment, becoming faint anteriorly; (intestine essentially straight) 20
20. Anterior lateral-line pores outlined above and below in dark pigment; a decurved, crescent-shaped row of melanophores extending from anterior orbit of eye to premaxillary; dorsal fin pigmentation restricted to membrane around first dorsal ray in young specimens; light pigmentation scattered throughout dorsal membranes in mature specimens ....
    .... Blue-striped Shiner (*Notropis callitaenia*), sp.39
    Anterior lateral-line pores not outlined above and below in dark pigment; no crescent-shaped row of melanophores between orbit and premaxillary; dorsal fin pigmentation restricted to membranes between first 3 dorsal rays in young specimens; mature specimens with pigment concentrated in a dark, horizontal slash through the dorsal fin .... Ohoopee Shiner (*Notropis leedsi*), sp.40
21. No dark pigment in mouth; basicaudal spot round, and its diameter much greater than width of lateral stripe ....
    .... Tail-light Shiner (*Notropis maculatus*), sp.49
    Much dark pigment in mouth (concentrated on oral valves); basicaudal spot not round and little, if any, wider than lateral stripe
    22
22. Basicaudal spot smaller than diameter of orbit of eye; scale rows below lateral stripe without pigmentation; males without blue in life; fins unmodified ....
    .... Iron-colored Shiner (*Notropis chalybaeus*), sp.45
    Basicaudal spot at least as wide as diameter of orbit; scale rows below lateral stripe with dark pigment (heavily pigmented in males, but only outlined in females); mature males (in life) with

a strikingly blue nose and enlarged, darkly pigmented dorsal and anal fins .... Blue-nosed Shiner *(Notropis welaka)*, sp.48

ORDER CYPRINIFORMES: XVII. FAMILY CATOSTOMIDAE, Suckers.

1. Base of dorsal fin very long (at least one-third of standard length), and the anterior rays more than twice as long as the posterior 2
   Dorsal fin of normal proportions (not more than 20% of standard length) and the length of its rays diminishing gradually posteriad 3

2. Lower lip angular, with a nipple or point at its center (Plate I-8); angle of jaws posterior to nostril .... Highfin Carpsucker *(Carpiodes velifer)*, sp.60
   Lower lip curved and without a nipple at center; angle of jaws directly below nostril ..... Quillback *(Carpiodes cyprinus)*, sp.59

3. With longitudinal rows of black spots on sides, sometimes faint; lateral line incomplete .... Spotted Sucker *(Minytrema melanops)*, sp.63
   No longitudinal rows of black spots, though sometimes a *single, interrupted* stripe; lateral line complete or absent 4

4. Lateral line present 5
   Lateral line absent 7

5. Maximum depth more than 25% of standard length; width of lower lip at least as great as diameter of eye .... River Redhorse *(Moxostoma carinatum)*, sp.198H
   Depth not more than 25% of standard length; width of lower lip less than diameter of eye 6

6. Lower lobe of caudal fin evenly colored ....? Redhorse *(Moxostoma sp.)*,[1] sp.61
   Lower lobe of caudal fin with contrasting black and white .... Black-tailed Redhorse *(Moxostoma poecilurum)*, sp.62

7. First ray of dorsal fin at least 20% longer than base of dorsal, producing a concave or straight posterior margin of fin (Plate II-1) .... Sharpfin Chubsucker *(Erimyzon tenuis)*, sp.64
   First ray of dorsal fin about same length as dorsal base (not more than 10% longer); dorsal fin rounded posteriad 8

8. Oblique scale rows along side not more than 38; depth of body at pectoral base less than that at pelvic base .... Lake Chubsucker *(Erimyzon sucetta)*, sp.65
   More than 38 oblique scale rows along side; depth of body at pectoral base at least as great as at pelvic base (except in very

---

1. *Moxostoma duquesnei* of some authors.

small specimens) ....
.... Creek Chubsucker (*Erimyzon oblongus*), sp.66

ORDER SILURIFORMES: XVIII. FAMILY ICTALURIDAE, Freshwater Catfishes.[1]

1. Adipose fin free (posterior end not fused to back or connected to caudal fin) 2
   Adipose fin adnate (fused to back and caudal fin, or separated from caudal fin by a slight or incomplete notch; Plate II-2) 7
2. Tail forked 3
   Tail emarginate or rounded, but never truly forked 4
3. Width of gape greater than length of snout; bony ridge from head to origin of dorsal fin not quite complete; lobes of tail rounded; no distinct dark spots on body; anal rays usually 21 to 24 ....
   .... White Catfish (*Ictalurus catus*), sp.72
   Width of gape equal to or less than length of snout; bony ridge unbroken from head to origin of dorsal fin; lobes of caudal fin usually pointed, with upper lobe longer than lower; few to many dark spots usually present on body; anal rays 20 to 30 ....
   .... Channel Catfish (*Ictalurus punctatus*), sp.71
4. Chin barbels at least partly white, lacking melanophores; body uniform; anal rays 25 or more; length of eye about 10% (in large specimens) to 12% that of head ....
   .... Yellow Bullhead (*Ictalurus natalis*), sp.70
   Chin barbels dark, or at least sprinkled with melanophores; body mottled or spotted; anal rays 24 or fewer; length of eye about 12% (large specimens) or more that of head 5
5. Upper and lower jaws about equal in length; color pattern of irregular dark mottling on light background ....
   .... Brown Bullhead (*Ictalurus nebulosus*), sp.69
   Upper jaw projecting beyond lower; body speckled or uniform 6
6. Fine serrations on pectoral spines; anal rays usually 16 to 20 (to 22 in St. Johns River); pectoral fins without narrow black margins; olive green in life ..... Snail Bullhead (*Ictalurus brunneus*),[2] sp.67
   Strong serrations on pectoral spines; anal rays usually 20 to 24; pectoral (and other) fins with a narrow black margin, and body dark with rounded light spots ....
   .... Spotted Bullhead (*Ictalurus serracanthus*), sp.68
7. Median fins heavily speckled; width of head about 20% of standard length .... Speckled Madtom (*Noturus leptacanthus*), sp.74
   Median fins uniformly brown or black; width of head more than 20% of standard length 8

1. The anal ray counts in this section of the Key include all rudiments.
2. Formerly considered identical with the Flat Bullhead, *Ictalurus platy-cephalus*, which species is now thought not to occur in Florida.

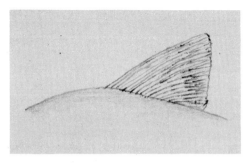

1. Dorsal fin of
   Erimyzon tenuis

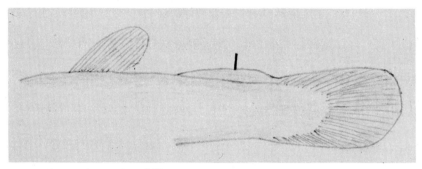

2. Adnate adipose fin of *Noturus*

3.
Enlarged humeral
scale of *Cyprinodon*

PLATE   II

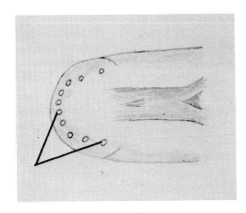

4.
Mandibular pores of
Fundulus grandis
(ventral aspect)

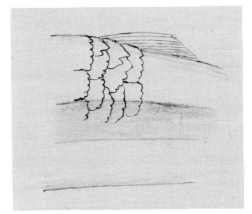

5.
Ragged scale edges of
Membras martinica

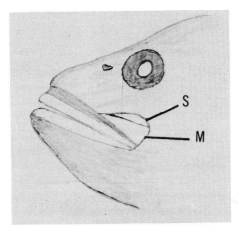

6.
Supramaxilla (S) and
maxilla (M) of
Lepomis gulosus

FISHES

8. Upper jaw protruding beyond lower; with more than 18 anal rays
.... Black Madtom (*Noturus funebris*), sp.75
Upper and lower jaws of about equal length; anal rays 14 to 16 .....
.... Tadpole Madtom (*Noturus gyrinus*), sp.73

ORDER SILURIFORMES: XIX. FAMILY CLARIIDAE, Clariid Catfishes. One
species in Florida .... Walking Catfish (*Clarias batrachus*), sp.76

ORDER SILURIFORMES: XX. FAMILY LORICARIIDAE, Armored Catfishes.
One species in Florida ..... Armored Catfish (*Hypostomus* sp.), sp.860

ORDER SILURIFORMES: XXI. FAMILY ARIIDAE, Sea Catfishes.

1. With long filaments on dorsal and pectoral fins and only 2 barbels on
lower jaw .... Gafftopsail Catfish (*Bagre marinus*), sp.77
No long filaments on fins, but with 4 barbels on lower jaw ....
.... Sea Catfish (*Arius felis*), sp.78

ORDER PERCOPSIFORMES: XXII. FAMILY APHREDODERIDAE, Pirate Perches.
One species in Florida ....
.... Pirate Perch (*Aphredoderus sayanus*), sp.79

ORDER ATHERINIFORMES: XXIII. FAMILY BELONIDAE, Needlefishes.

1. Peduncle keeled laterally and not deeper than wide ....
.... Atlantic Needlefish (*Strongylura marina*), sp.80
Peduncle compressed and without a lateral keel ....
.... Timucu (*Strongylura timucu*), sp.81

ORDER ATHERINIFORMES: XXIV. FAMILY CYPRINODONTIDAE, Killifishes.

1. Standard length less than 3 times body depth 2
Standard length at least 3 times body depth 5
2. Humeral scale greatly enlarged, its depth about twice that of
surrounding scales (Plate II-3); caudal fin square-tipped, its central
rays not longer than adjacent ones ....
.... Sheepshead Minnow (*Cyprinodon variegatus*), sp.98
No enlarged humeral scale; caudal fin somewhat rounded, its central
rays slightly longer than adjacent ones 3
3. Base of dorsal fin at least 25% of standard length; black spots present
on sides (and on dorsal fin of males) ....
.... Flagfish (*Jordanella floridae*), sp.101
Base of dorsal fin less than 25% of standard length; without
conspicuous black spots 4
4. Dorsal fin extending farther anterior than pelvics; body marked
with conspicuous vertical bars ....
.... Diamond Killifish (*Adinia xenica*), sp.85

Dorsal and pelvic fins extending equally far forward; conspicuous vertical bars lacking .... Gold-spotted Killifish (*Floridichthys carpio*), sp.100

5. Humeral scale almost twice depth of nearby scales (Plate II-3) .... Lake Eustis Minnow (*Cyprinodon hubbsi*), sp.99 Scales in humeral region of uniform size 6

6. With an isolated, rounded black spot on peduncle; dorsal rays fewer than 8, *or* lateral scale count less than 30 and dorsal rays 8 .... Pygmy Killifish (*Leptolucania ommata*), sp.84 No rounded black spot on peduncle unless connected to a longi- tudinal stripe; dorsal rays at least 10 if lateral scale count is less than 30 7

7. Eye relatively small, its length less than that of snout; length of head at least 4 times that of eye (or slightly less if total length under 50 mm) .... 8 Length of head less than 4 times that of eye (often more if total length more than 100 mm) 9

8. Snout long, and its profile flattened or concave; length of eye not more than 50% that of snout in adults; lateral scale count about 33 .... Long-nosed Killifish (*Fundulus similis*), sp.90 Profile of snout flattened or convex; length of eye usually more than 50% that of snout in adults; lateral scale count about 36 .... Striped Killifish (*Fundulus majalis*), sp.91

9. With a broad, continuous, dark stripe on each side .... 10 Black stripes usually lacking or numerous; if single, then inter- rupted 11

10. Central caudal rays not longer than adjacent ones; often with blue on fins in life; origin of dorsal fin anterior to that of anal fin .... Bluefin Killifish (*Lucania goodei*), sp.83 Caudal fin rounded, with central rays longest; no blue in fins in life; origin of dorsal fin posterior to that of anal .... Black-spotted Topminnow (*Fundulus olivaceus*), sp.97

11. Scales along side of body 50 or more; about 17 rays in dorsal fin .... Seminole Killifish (*Fundulus seminolis*), sp.89 Lateral scale count less than 40; less than 15 rays in dorsal fin 12

12. With one or 2 longitudinal rows of spots along sides;[1] lateral scale count usually less than 33 .... Salt-marsh Topminnow (*Fundulus jenkinsi*), sp.86 Several, or no, longitudinal streaks or rows of spots along sides; lateral scale count usually more than 33 13

_____

1. Some specimens from Pensacola presently referred to *F. jenkinsi* have no conspicuous markings of any kind.

13. With a dark area around or below the eye and several longitudinal streaks or rows of spots on body (rarely absent; also, vertical bars may be present); caudal fin long and rounded, its length at least twice depth of peduncle 14

No prominently darkened area around eye; no longitudinal streaks or rows of spots; caudal fin not more than slightly rounded, and its length less than twice depth of peduncle 15

14. Body depth at least 20% of standard length; dark area extending below eye for distance almost equal to diameter of eye; longitudinal streaks often broken into rows of spots; scales along side of body 34 to 36 ....
.... Northern Starhead Minnow (*Fundulus notti*), sp.95

Body depth less than 20% of standard length; dark area extending scarcely, if at all, below orbit; longitudinal streaks typically unbroken; scales along sides 32 to 34 ....
.... Southern Starhead Minnow (*Fundulus lineolatus*), sp.96

15. Less than 30 scales along side of body; without conspicuous markings (except males in life); maximum length less than 50 mm
.... Rainwater Killifish (*Lucania parva*), sp.82

More than 30 scales along side of body; usually with dark vertical bars or other conspicuous markings; maximum length 50 mm or more 16

16. With 10 or 11 rays in dorsal fin and usually more than 34 oblique scale rows; maximum length more than 65 mm 17

With 7 to 9 dorsal rays and usually less than 34 oblique rows of scales; maximum length about 65 mm 19

17. With a discrete black spot on dorsal fin (female) *or* some of the dark vertical bars joining one another on sides of body (males); usually with longitudinal rows of black spots on dorsum; no light spots; maximum length about 100 mm ....
.... Marsh Killifish (*Fundulus confluentus*), sp.92

No discrete black spot on dorsal or rows of black spots anterior to it; dark vertical bars not joining unless dorsally; large specimens usually with light spots; maximum length about 150 mm 18

18. With 4 pairs of mandibular pores and at least 18 predorsal scales
.... Mummichog (*Fundulus heteroclitus*) sp.87

With 5 pairs of mandibular pores (Plate II-4) and less than 18 predorsal scales .... Gulf Killifish (*Fundulus grandis*), sp.88

19. Origin of dorsal fin posterior to that of anal fin; usually less than 12 dark vertical bars (none in females); golden flecks on body in life .... Golden Topminnow (*Fundulus chrysotus*), sp.94

Origin of dorsal fin directly above that of anal; usually more than 12 dark vertical bars (though faint anteriad); without gold flecks in life .... Banded Topminnow (*Fundulus cingulatus*), sp.93

ORDER ATHERINIFORMES: XXV. FAMILY POECILIIDAE, Live-bearers.[1]

1. Length of upper jaw (from crease at its base, dorsally) greater than width of gape; a single dark spot at base of caudal rays .... Pike Killifish (*Belonesox belizanus*), sp.106
   Length of upper jaw from crease less than width of gape; no discrete, enlarged dark spot on bases of caudal rays 2
2. With a single dark spot in the dorsal fin (also in anal of females), but caudal fin clear .... Least Killifish (*Heterandria formosa*), sp.104
   With numerous black flecks in dorsal and caudal fins 3
3. Anterior portion of back heavily pigmented; dorsal rays 14 to 16 .... Sailfin Molly (*Poecilia latipinna*), sp.105
   Anterior portion of back only medium dark; dorsal rays 7 to 9 4
4. With longitudinal rows of black flecks on body; 28 to 30 oblique scale rows .... Mangrove Mosquitofish (*Gambusia rhizophorae*), sp.103
   Sides of body unspotted; 30 to 32 oblique scale rows .... Common Mosquitofish (*Gambusia affinis*), sp.102

ORDER ATHERINIFORMES: XXVI. FAMILY ATHERINIDAE, Silversides.

1. With more than 70 scales along sides and more than 20 rays in anal fin; standard length more than 6 times depth .... Brook Silverside (*Labidesthes sicculus*), sp.109
   With less than 50 scales along sides and less than 20 rays in anal fin; standard length about 5 or 6 times depth 2
2. Edges of scales scalloped or ragged (Plate II-5) .... Rough Silverside (*Membras martinica*), sp.108
   Edges of scales smooth .... Tidewater Silverside (*Menidia beryllina*), sp.107

ORDER GASTEROSTEIFORMES: XXVII. FAMILY SYNGNATHIDAE, Pipefishes and Seahorses.

1. Vent and dorsal fin closer to tip of snout than to caudal fin .... Gulf Pipefish (*Syngnathus scovelli*), sp.110
   Vent and dorsal fin closer to caudal fin than to tip of snout .... Opossum Pipefish (*Oostethus lineatus*), sp.111

ORDER PERCIFORMES: XXVIII. FAMILY CENTROPOMIDAE, Snooks.

1. Pelvic fin, when appressed, reaching beyond vent 2

---

1. For species recently established in Florida see the Addenda following the mammal section of Chapter Three.

Appressed pelvic fin falling short of vent in adults or barely reaching
it in young . . . .
. . . . Common Snook (*Centropomus undecimalis*), sp.112

2. Lateral-line scales 80 or more; gill rakers less than 14; anal rays
usually 6 . . . . Little Snook (*Centropomus parallelus*), sp.114
Lateral-line scales 70 or fewer; gill rakers 15 or more; anal rays 7
. . . . Tarpon Snook (*Centropomus pectinatus*), sp.113

ORDER PERCIFORMES: XXIX. FAMILY PERCICHTHYIDAE, Temperate Basses.

1. Second anal spine not more than one-fifth of head length and less
than half length of longest anal ray . . . .
. . . . Striped Bass (*Morone saxatilis*), sp.115
Second anal spine about one-third of head length and at least half
length of longest anal ray . . . .
. . . . White Bass (*Morone chrysops*), sp.116

ORDER PERCIFORMES: XXX. FAMILY CENTRARCHIDAE, Sunfishes.

1. Lateral line absent; length not exceeding 50 mm   2
Lateral line present; adults more than 50 mm long   4

2. Dorsal spines 5; with a dark spot on body near tip of pectoral fin;
males not blue in life; with 7 or more dark vertical bars . . . .
. . . . .Banded Pygmy Sunfish (*Elassoma zonatum*), sp.117
Dorsal spines 3 or 4; no dark spot on side; males with blue in life;
with 5 or fewer dark vertical bars (sometimes none) near posterior
end   3

3. Top of head without scales; tip of mouth light, contrasting with sides
of mouth; pattern blotched anteriad; caudal fin unspotted (although
light or dark) . . . .
. . . . . . Okefenokee Pygmy Sunfish (*Elassoma okefenokee*), sp.118
Top of head scaled; mouth evenly colored; anterior half of body with
little or no contrast; caudal fin spotted, although sometimes only
faintly . . . .
. . . . Everglades Pygmy Sunfish (*Elassoma evergladei*), sp.119

4. With 5 or more spines in anal fin   5
Anal spines only 3   9

5. Caudal fin rounded . . . .
. . . . Mud Sunfish (*Acantharchus pomotis*), sp.137
Caudal fin notched (Fig. 1)   6

6. With 11 or more dorsal spines   7
With not more than 8 dorsal spines   8

7. Anal fin at least 80% as long as dorsal; striped or with longitudinal
rows of spots . . . . Flier (*Centrarchus macropterus*), sp.140

Anal fin much shorter than dorsal; young irregularly blotched, but pattern becoming obscurely dark with age ....
.... Rock Bass (*Ambloplites rupestris*), sp.136

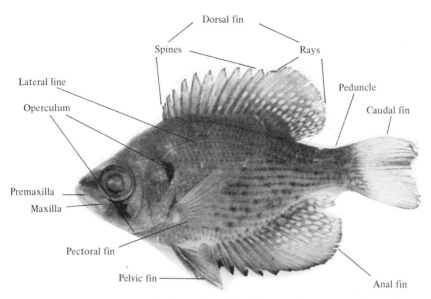

Fig. 1. External morphology of a fish (*Centrarchus macropterus*)

8. Dorsal spines 5 or 6; distance from first dorsal spine to rim of orbits greater than length of dorsal base; with dark vertical bars on sides .... White Crappie (*Pomoxis annularis*), sp.139
   Dorsal spines usually 7 or 8; distance from first dorsal spine to rim of orbit about same as length of dorsal base; with horizontal rows of dark spots on sides ....
   .... Black Crappie (*Pomoxis nigromaculatus*), sp.138
9. Caudal fin rounded; dark vertical bar below eye 10
   Caudal fin notched; no dark vertical bar below eye 12
10. Middle dorsal spines noticeably longer than posterior ones; anterior part of dorsal fin dark ....
   .... Black-banded Sunfish (*Enneacanthus chaetodon*), sp.135
   Middle and posterior dorsal spines of about equal length; front of dorsal fin not darker than remainder 11
11. Not more than 18 scale rows around caudal peduncle; with light spots on sides (blue in life); dark vertical bars lacking, *or* indistinct and not more than 5 in number; spinous portion of pelvic fins very dark ....
   .... Blue-spotted Sunfish (*Enneacanthus gloriosus*), sp.134

With 19 or more scale rows around caudal peduncle; sides marked
with vertical dark bars (usually more than 5); with golden spots
on sides in life; spinous portion of pelvics only medium dark
.... Banded Sunfish (*Enneacanthus obesus*), sp.133
12. Standard length more than 3 times body depth 13
Standard length less than 3 times body depth 16
13. Longest dorsal spine at least twice length of shortest posterior dorsal
spine; membranes of soft dorsal and anal fins not scaled ....
.... Largemouth Bass (*Micropterus salmoides*), sp.123
Longest dorsal spine less than twice length of shortest posterior
dorsal spine; membranes of soft dorsal and anal fins scaly 14
14. With 75 or more scales along lateral line; confined (in Florida) to
Apalachicola River system ....
.... Red-eyed Bass (*Micropterus coosae*), sp.122
Not more than 70 scales along lateral line 15
15. Pattern typically very dark, even ventrally (underparts of adults blue
in life); no longitudinal rows of spots; depth of peduncle about
13% of standard length; about 59 to 63 scales in lateral line;
largely confined to the Suwannee River System ....
.... Suwannee Bass (*Micropterus notius*), sp.120
Underparts light, sometimes silvery; longitudinal rows of spots often
present; depth of peduncle about 11% of standard length; 62 to
70 scales in lateral line; from north Florida northwestward .....
.... Spotted Bass (*Micropterus punctulatus*), sp.121
16. Angle of jaws directly below front of eye; length of supramaxilla
greater than width of maxilla (Plate II-6) ....
.... Warmouth (*Lepomis gulosus*),[1] sp.124
Angle of jaws somewhat anterior to eye; length of supramaxilla
(when present) less than width of maxilla 17
17. Mouth large, with posterior end of maxilla directly below front part
of eye; often with a dark spot in soft dorsal fin ....
.... Green Sunfish (*Lepomis cyanellus*), sp.125
Mouth smaller, posterior end of maxilla rarely extending to (and
never *below*) the eye; most species without a dark spot in soft
dorsal 18
18. Pectoral fin long and pointed, with rays 3 to 6 much longer than
others and, when appressed, almost reaching to origin of first dorsal
fin (length of pectoral usually at least 30% of standard length;
if not, then with a dark spot in soft dorsal fin) 19
Pectoral fin shorter and more rounded, with rays 3 to 6 not the
longest and usually not reaching near origin of first dorsal fin

---

1. A hybrid population between this species and the Bluegill has been found
in Jackson County (Birdsong and Yerger, 1967).

when appressed (length of pectoral usually less than 30% of standard length; usually no dark spot in soft dorsal) 20

19. Dark opercular spot extending backward as a flap and with a light margin (orange red in life; Plate III-1); gill rakers much shorter than longest gill filaments; no dark spot in soft dorsal; usually 22 lateral scale rows ....
.... Red-eared Sunfish (*Lepomis microlophus*), sp.127
Dark opercular spot not extending appreciably backward and without a light margin; gill rakers about as long as longest gill filaments; usually with a dark spot on soft dorsal; usually 24 lateral scale rows .... Bluegill (*Lepomis macrochirus*), sp.132

20. With blackish pigment spots on body, sometimes reddish in life (not merely dusky scales with light margins), *or* eye longer than snout
.... 21
Without blackish spots on body (or reddish spots in life); eye not longer than snout 22

21. Opercular spot entirely dark; body usually heavily spotted with black .... Spotted Sunfish (*Lepomis punctatus*), sp.126
Opercular spot with a light margin; body moderately spotted if at all (spots reddish in life) ....
.... Orange-spotted Sunfish (*Lepomis humilis*), sp.131

22. Length of pectoral fin less than 25% of standard length; 12 rays in pectoral fin; body depth usually more than 50% of standard length .... Dollar Sunfish (*Lepomis marginatus*), sp.129
Length of pectoral fin at least 25% of standard length; usually more than 12 rays in pectoral fin; body depth usually less than 50% of standard length 23

23. Opercular flap usually stiff and dark to posterior edge; eye slightly shorter than snout ....
.... Red-breasted Sunfish (*Lepomis auritus*), sp.128
Opercular flap soft and with a light margin; eye about same length as snout ...... Long-eared Sunfish (*Lepomis megalotis*), sp.130

ORDER PERCIFORMES: XXXI. FAMILY PERCIDAE, Perches and Darters.

1. Caudal fin deeply forked (depth of fork at least half length of shortest caudal rays); northwest Florida only 2
Caudal fin notched, square, or rounded 3

2. Color pattern of dark vertical bars; pelvic fins almost in contact ....
.... Yellow Perch (*Perca flavescens*), sp.141
Color pattern blotched; distance between pelvic fins equal to width of pelvic base .... Sauger (*Stizostedion canadense*), sp.142

1. Light margin of opercular spot in *Lepomis microlophus*

2. Sharply angled profile of *Etheostoma stigmaeum*

PLATE   III

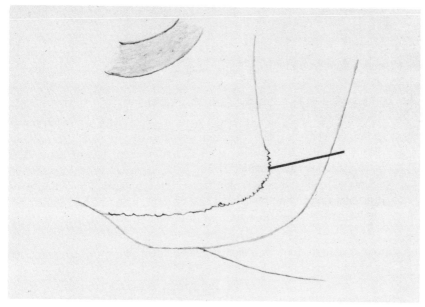

3. Serrate preopercle of *Diapterus*

4. Asymmetric pelvic fins of *Citharichthys*
(ventro-lateral view)

FISHES

3. Very pale, the body virtually devoid of pigment; standard length 7 times body depth; body largely naked ....
.... Naked Sand Darter (*Ammocrypta beani*),[1] sp.146

Body scaly and pigmented, its length usually less than 7 times its depth 4

4. Midline of belly naked or with one or 2 enlarged scales between pelvic fins, and the posterior portion often with a single row of enlarged scales; space between pelvic fins almost equal to pelvic base; lateral line complete; anal and soft dorsal fins of about equal area; caudal fin notched 5

Midline of belly without enlarged scales (sometimes naked); pelvic fins almost in contact; lateral line usually incomplete; area of anal fin somewhat less than that of soft dorsal; caudal fin usually not notched 7

5. Snout decidedly longer than diameter of eye; pattern of discontinuous vertical bars of alternating length ....
.... Logperch (*Percina caprodes*), sp.143

Snout length scarcely, if at all, greater than diameter of eye; blotches on sides confluent 6

6. Distance between eyes equal to diameter of eye ....
.... Black-banded Darter (*Percina nigrofasciata*), sp.144

Distance between eyes less than diameter of eye ....
.... Star-gazing Darter (*Percina uranidea*), sp.145

7. With only one anal spine (but first soft ray sometimes unbranched); 9 anal rays; lateral line complete ....
.... Johnny Darter (*Etheostoma nigrum*), sp.149

With 2 anal spines; usually less than 9 anal rays; lateral line incomplete in most species 8

8. Lateral line almost wanting (up to 7 pored scales) ....
.... Cypress Darter (*Etheostoma proeliare*), sp.156

Lateral line extending at least to first dorsal fin 9

9. Profile of head angling (or curving) abruptly downward at about 45 degrees anterior to eyes (Plate III-2) 10

Profile of head curved normally downward, or sloping at an angle of much less than 45 degrees 13

10. Premaxillary protractile, being separated from snout by a groove; with W-shaped markings often present on sides 11

Premaxillary not protractile, but usually connected to snout by a ridge, the frenum; color pattern various 12

---

1. *A. asprella* was collected in the Escambia River in 1974 (Hal Beecher, personal communication).

11. With a pattern of about 10 dark lateral blotches, tending to form vertical bars (males), *or* a long, free genital papilla present (females) ....
.... Choctawhatchee Darter (*Etheostoma davisoni*), sp.148
With a pattern of smaller, scattered dark spots, some tending to form M's or W's; genital papilla of female inconspicuous and not extending as a free tube ....
.... Speckled Darter (*Etheostoma stigmaeum*), sp.147

12. Pectoral, pelvic, and anal fins clear; anal rays 8 to 10; with 7 or 8 discrete, sometimes M-shaped, dark blotches on sides ....
.... Orange-sided Darter (*Etheostoma* sp.)[1]
Pectoral, pelvic, and anal fins speckled; anal rays 7 to 8; lateral blotches connected to dorsal blotches; extreme west Florida only
.... Harlequin Darter (*Etheostoma histrio*), sp.150

13. Breast completely scaled; lateral line light colored and conspicuous, at least anteriad 14
Breast naked, at least anteriad; lateral line usually not whitish or conspicuous 15

14. With about 10 dark vertical bars; lateral line nearly straight, yellowish in life ....
.... Gold-striped Darter (*Etheostoma parvipinne*), sp.152
With dark blotches of irregular size and distribution; lateral line arched, not yellowish in life ....
.... Swamp Darter (*Etheostoma fusiforme*), sp.155

15. Lateral line virtually complete, with not more than 4 unpored scales; with a marked depression at back of head 16
Lateral line incomplete (7 to 14 unpored scales); no depression at back of head ...... Brown Darter (*Etheostoma edwini*), sp.153

16. Very dark, except midventrally; confined to Choctawhatchee Bay drainage ..... Okaloosa Darter (*Etheostoma okaloosae*), sp.154
With small, rounded spots tending to form longitudinal rows on a light background; Apalachicola River westward ....
.... Gulf Darter (*Etheostoma swaini*), sp.151

ORDER PERCIFORMES: XXXII. FAMILY ECHENEIDAE, Remoras. One species in Florida fresh water ....
.... Sharksucker (*Echeneis naucrates*), sp.157

ORDER PERCIFORMES: XXXIII. FAMILY CARANGIDAE, Jacks and Pompanos. One species in Florida fresh water ....
.... Common Jack (*Caranx hippos*), sp.158

---

1. Undescribed species.

ORDER PERCIFORMES: XXXIV. FAMILY LUTJANIDAE, Snappers.

1. Body depth less than 40% of standard length; more than 47 scales along lateral line; profile of snout usually convex .... Gray Snapper (*Lutjanus griseus*), sp.159

Body depth more than 40% of standard length; less than 47 scales along lateral line; profile of snout concave or straight .... Schoolmaster (*Lutjanus apodus*), sp.160

ORDER PERCIFORMES: XXXV. FAMILY GERREIDAE, Mojarras.

1. Second anal spine obviously thickened; longest dorsal spines about twice length of its longest rays; maximum length 200 mm or more 2

Second anal spine not appreciably thickened; longest dorsal spines and rays about equally long; maximum length about 125 mm .... Spotfin Mojarra (*Eucinostomus argenteus*), sp.161

2. Preopercle serrate ventrally (Plate III-3); second anal spine greatly thickened and projecting far beyond second and subsequent anal rays .... Irish Pompano (*Diapterus olisthostomus*), sp.163

Preopercle not serrate; second anal spine only moderately thickened and not projecting far beyond anal rays .... Yellowfin Mojarra (*Gerres cinereus*), sp.162

ORDER PERCIFORMES: XXXVI. FAMILY POMADASYIDAE, Grunts.

1. Anal fin short, with 6 or 7 soft rays; dorsal fin decidedly notched, its spines thickened .... Burro Grunt (*Pomadasys crocro*), sp.164

Anal fin long, with 11 to 13 rays; dorsal fin low, only slightly notched .... Pigfish (*Orthopristis chrysopterus*), sp.165

ORDER PERCIFORMES: XXXVII. FAMILY SPARIDAE, Porgies.

1. With a pattern of strongly contrasting, black vertical bars; not more than 50 scales along lateral line .... Sheepshead (*Archosargus probatocephalus*),[1] sp.167

Black vertical bars, if present, not strongly contrasting with ground color; more than 50 scales in lateral line 2

2. With a large black spot on peduncle and none above base of pectoral fin; scales along lateral line less than 60 .... Spot-tailed Pinfish (*Diplodus holbrooki*), sp.168

With a dark spot above base of pectoral fin, but no well-defined spot on peduncle; more than 60 scales along sides........... .... Pinfish (*Lagodon rhomboides*), sp.166

1. This widely accepted form of the common name, like that of the Sheepshead Minnow, is improperly spelled.

ORDER PERCIFORMES: XXXVIII. FAMILY SCIAENIDAE, Drums.

1. Snout pointed, with lower jaw protruding slightly beyond upper; rounded black spots on body and fins ....
   .... Spotted Seatrout (*Cynoscion nebulosus*), sp.170
   Snout blunt, with upper jaw protruding slightly beyond lower; mostly without rounded black spots on body 2

2. With barbels on chin 3
   No barbels on chin 4

3. Standard length more than 3 times body depth; more than 25 rays in soft dorsal fin; color pattern mottled or speckled ....
   .... Atlantic Croaker (*Micropogon undulatus*), sp.174
   Standard length less than 3 times body depth; less than 25 rays in soft dorsal; about 5 dark vertical bars in young ....
   .... Black Drum (*Pogonias cromis*), sp.173

4. Central rays of caudal fin shorter than adjacent ones; about 15 oblique bars on sides (yellowish in life) and a dark spot behind operculum; no teeth in lower jaw ....
   .... Spot (*Leiostomus xanthurus*), sp.172
   Central rays of caudal fin longer than adjacent ones; ground color reddish or silvery in life, without oblique yellowish bars; teeth present in both jaws 5

5. Ground color reddish (in life), with one or more black spots at or near base of caudal fin ....
   .... Red Drum (*Sciaenops ocellata*), sp.171
   Ground color silvery, without black spots ....
   .... Silver Perch (*Bairdiella chrysura*), sp.169

ORDER PERCIFORMES: XXXIX. FAMILY CICHLIDAE, Cichlids.[1]

1. Soft dorsal and anal fin scaleless, or scaly only at the base; rays in soft dorsal, 7–12; in anal, 7–11 2
   Soft dorsal and anal fins covered with small scales; 19–21 rays in soft dorsal and 15 or 16 in anal fin ....
   .... Oscar (*Astronotus ocellatus*), sp.176

2. With a large, dark blotch at mid-body on each side and another at base of caudal fin ....
   .... Two-spotted Cichlid (*Cichlasoma bimaculatum*), sp.175
   No large blotches as described above, but often one each on soft dorsal or operculum ....
   .... Black-chinned Tilapia (*Tilapia melanotheron*), sp. 177

---

1. Other species of the Cichlidae, more recently established in Florida, are listed in the Addenda following the mammal section of Chapter Three.

ORDER PERCIFORMES: XL. FAMILY MUGILIDAE, Mullets.

1. With adipose tissue partly covering eye (adults); no dark spot at base of pectoral fin (immatures) 2
   No visible adipose tissue around eye; dark spot at base of pectoral fin .... Mountain Mullet (*Agonostomus monticola*), sp.181
2. Soft dorsal and anal fins sparsely scaled; about 40 oblique scale rows along side; sides striped in adults ....
   .... Striped Mullet (*Mugil cephalus*), sp.178
   Soft dorsal and anal fins fully scaled; less than 40 oblique scale rows; never striped 3
3. Anal fin with 8 rays (and 3 spines); less than 35 oblique scale rows along side .... Fan-tailed Mullet (*Mugil trichodon*), sp.180
   Anal fin with 9 rays; usually more than 35 oblique scale rows along side; St. Johns River drainage (and salt water) ....
   .... White Mullet (*Mugil curema*), sp.179

ORDER PERCIFORMES: XLI. FAMILY GOBIIDAE, Gobies and Sleepers.

1. Pelvic fins united to form suctorial disk 4
   Pelvic fins separate 2
2. Scales relatively large, with less than 40 along lateral line; width of head at eyes less than its depth ....
   .... Fat Sleeper (*Dormitator maculatus*), sp.184
   Scales relatively small, with 50 or more along lateral line; width of head at eyes at least as great as its depth 3
3. Gill openings extending anterior to position of eyes ....
   .... Big-mouth Sleeper (*Gobiomorus dormitor*), sp.183
   Gill openings not extending anterior to position of eyes ....
   .... Spiny-cheeked Sleeper (*Eleotris pisonis*),[1] sp.182
4. Dorsal fins united to each other and to the caudal fin ....
   .... Violet Goby (*Gobioides broussonneti*), sp.198H
   Dorsal fins separate from one another and from caudal fin 5
5. Body naked 6
   Body mostly scaly 7
6. Visible portion of longest ventral (pelvic) rays at least twice as long as distance from their tips to base of anal fin; general coloration light in preservative (irregular, light vertical bars on darker background in life); usually fewer than 13 dorsal and 11 anal rays .....
   .... Robust Goby (*Gobiosoma robustum*), sp.191
   Visible portion of longest ventral rays less than twice as long as distance between them and anal base; general coloration dark in

---

1. Formerly known as *Eleotris amblyopsis*. The record of *Eleotris picta* in the St. Johns River (Tagatz, 1968) is probably in error (Gilbert, personal communication).

preservative; usually with regular, light vertical bars; usually at least 13 dorsal and 11 anal rays ....
.... Naked Goby (*Gobiosoma bosci*), sp.192

7. Heavy-bodied, very dark, with a blunt snout, thick lips, and a fleshy crest on head; fins large, almost hiding midline of body; mouth oblique ...... Crested Goby (*Lophogobius cyprinoides*), sp.190
Various in form and color, but with no crest on head, and mouth not markedly oblique 8

8. With 60 or more oblique scale rows along side; snout length almost 50% of head length ...... River Goby (*Awaous tajasica*), sp.188
With less than 50 oblique scale rows along side; snout about 30% of head length 9

9. Pectoral fins with a dorsal fringe of free rays; length up to 150 mm .... Frillfin Goby (*Bathygobius soporator*), sp.185
Pectoral fins without free rays; length under 100 mm 10

10. Soft dorsal fin long, with 15 or more rays; median fins often edged with blackish ...... Clown Goby (*Microgobius gulosus*), sp.189
Soft dorsal fin with not more than 13 rays; median fins not edged with blackish 11

11. With a dark spot about base of pectoral fin, but no horizontal dark bar on cheek; usually 11 rays in soft dorsal fin and 12 in anal; 30 to 33 oblique scale rows ....
.... Darter Goby (*Gobionellus boleosoma*), sp.186
With no dark humeral spot, but a dusky horizontal bar on cheek; usually 12 rays in soft dorsal and 13 in anal fin; 34 to 36 oblique scale rows ..... Freshwater Goby (*Gobionellus shufeldti*), sp.187

ORDER PLEURONECTIFORMES: XLII. FAMILY BOTHIDAE, Lefteye Flounders.

1. Pelvic fins asymmetrically placed and their bases unequal in length (Plate III-4); lateral line sloping downward above pectoral fin, but not arched .... Bay Whiff (*Citharichthys spilopterus*), sp.193
Pelvic fins symmetrical and their bases equally long; lateral line strongly arched over pectoral fin 2

2. Dark spots on body (if present) indefinite in number and not forming a triangle; distance between eyes about equal to diameter of eye; 85 or more scales in lateral line ....
.... Southern Flounder (*Paralichthys lethostigma*), sp.195
Dark spots on body often 3 and arranged as a triangle; distance between eyes decidedly less than diameter of eye (but approaching it more closely in large specimens); less than 85 scales in lateral line .... Gulf Flounder (*Paralichthys albigutta*), sp.194

ORDER PLEURONECTIFORMES: XLIII. FAMILY SOLEIDAE, Soles.

1. Right pectoral fin present; no dark crossbars ....
.... Lined Sole (*Achirus lineatus*), sp.196

Both pectoral fins lacking; usually with several dark crossbars .....
.... Hogchoker (*Trinectes maculatus*), sp.197

ORDER PLEURONECTIFORMES: XLIV. FAMILY CYNOGLOSSIDAE, Tonguefishes. One species in fresh water in Florida ....
.... Black-cheeked Tonguefish (*Symphurus plagiusa*), sp.198

CLASS AMPHIBIA—ORDERS AND FAMILIES

1. Postanal tail present; forelimbs present; hind limbs (if present)
   2-segmented and not more than 50% longer than forelimbs .....
   .... CAUDATA, 2
   Postanal tail absent in adults (present in larvae, which have no limbs
   or hind limbs only); hind limbs of adults and subadults (tailed)
   consisting of 3 segments (plus toes) and at least twice as long
   as forelimbs .... SALIENTIA, 7
2. Only the forelimbs present .... SIRENIDAE, XLV
   With 2 pairs of limbs 3
3. Eyes much reduced and without lids; skin fully pigmented; limbs
   often reduced in length 4
   Eyes of normal size and with a distinct upper lid, *or* skin devoid of
   pigment; limbs of normal length 5
4. External gills present; length of hind limbs about equal to maximum
   depth of tail .... PROTEIDAE, XLVI
   External gills lacking in adult; limbs vestigial ....
   .... AMPHIUMIDAE, XLVIII
5. Costal grooves lacking or indistinct ...... SALAMANDRIDAE, XLVII
   Costal grooves well developed 6
6. Nasolabial grooves present, though sometimes indistinct (Plate IV-1);
   parasphenoid teeth present (Plate IV-2) ....
   .... PLETHODONTIDAE, L
   Nasolabial grooves and parasphenoid teeth absent ....
   .... AMBYSTOMATIDAE, XLIX
7. Postanal tail present; forelimbs lacking or concealed[1] ....
   .... Tadpoles, p. 52
   Postanal tail lacking; forelimbs well developed 8
8. Parotoid glands present 9
   Parotoid glands lacking 10
9. Outline of parotoid glands rounded; cranial crests lacking ....
   .... PELOBATIDAE, LI
   Outline of parotoid glands elliptical; cranial crests often present (Plate
   IV-3) .... BUFONIDAE, LII

---

1. Tadpoles collected during metamorphosis usually cannot be keyed.

10. Tympanum lacking; toes not webbed; a fold of skin across head behind eyes; thigh not striped .... MICROHYLIDAE, LVI
Tympanum present (often not visible in 2 species in which the thigh
is alternately striped with light and dark); hind toes usually
webbed; no fold of skin across back of head 11
11. Tips of toes expanded into adhesive disks, *or* maximum length less
than 40 mm and webbing reduced 12
Toes without expanded disks; length of adults much more than
40 mm; webs on hind feet well developed .... RANIDAE, LV
12. Skin granular ventrally; pads under toe joints low and rounded ....
.... HYLIDAE, LIV
Skin smooth ventrally; subarticular pads elongate, pointing toward tips
of toes (Plate IV-4) .... LEPTODACTYLIDAE, LIII

ORDER CAUDATA: XLV. FAMILY SIRENIDAE, Sirens.

1. Total length less than 25 cm; stripes on body; 3 toes per foot ....
.... Dwarf Siren (*Pseudobranchus striatus*), sp.201
Total length of adults over 25 cm; no stripes on body; 4 toes per
foot 2
2. Total length up to 100 cm (specimens less than 50 cm long usually
have dorsal fin continuous with caudal); costal grooves usually more
than 34; light markings on body usually conspicuous ....
.... Greater Siren (*Siren lacertina*), sp.199
Total length not over 50 cm (large specimens lack a dorsal fin);
costal grooves usually not more than 34; light markings on body
absent or inconspicuous ....
.... Lesser Siren (*Siren intermedia*), sp.200

ORDER CAUDATA: XLVI. FAMILY PROTEIDAE, Mud Puppies and Waterdogs.[1]

1. With many rounded black spots on a medium dark background ....
.... Gulf Coast Waterdog (*Necturus beyeri*), sp.202
With few or no black spots on a very dark background ....
.... Dwarf Waterdog (*Necturus punctatus*), sp.203

ORDER CAUDATA: XLVII. FAMILY SALAMANDRIDAE, Newts.

1. Usually with light stripe (red in life) down each side of back,
bordered with black ....
.... Striped Newt (*Notophthalmus perstriatus*), sp.205
With no light (or red) stripe down each side of back ....
.... Spotted Newt (*Notophthalmus viridescens*), sp.204

---

1. Conant now refers all Florida specimens to *N. alabamensis*.

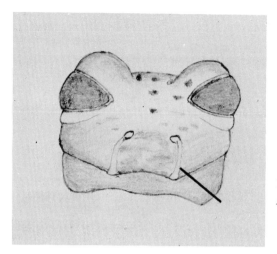

1.
Nasolabial grooves of
Eurycea longicauda

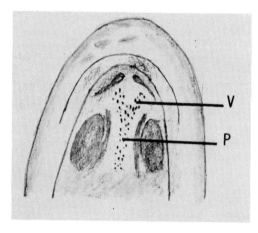

2.
Vomerine (V) and
parasphenoid (P)
teeth of
Eurycea longicauda

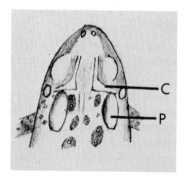

3.
Elliptical parotoid glands (P)
and cranial crests (C) of
Bufo woodhousei

PLATE IV

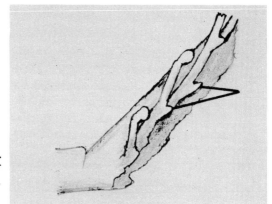

4.
Subarticular pads of
Eleutherodactylus

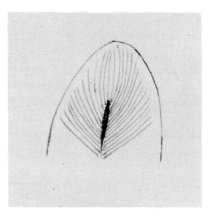

5. Pinnately grooved tongue of
Ambystoma cingulatum

6.
Relative length of fingers in
Rana grylio

AMPHIBIANS

ORDER CAUDATA: XLVIII. FAMILY AMPHIUMIDAE, Amphiumas.

1. Only one toe per foot; total length not exceeding 30 cm; underparts almost as dark as upperparts; appressed forelimb not extending over ear opening ....
.... One-toed Amphiuma (*Amphiuma pholeter*), sp.207
Two-toed; total length of adults up to 1 m; underparts much lighter than upperparts; appressed forelimb extending over ear opening ...... Two-toed Amphiuma (*Amphiuma means*), sp.206

ORDER CAUDATA: XLIX. FAMILY AMBYSTOMATIDAE, Mole Salamanders.

1. Interorbital distance greater than length of snout; usually only 10 costal grooves; usually medium gray above with small dark spots .... Mole Salamander (*Ambystoma talpoideum*), sp.209
Interorbital distance not greater than (about same as) snout length; back usually with contrasting light and dark markings; 11 or more costal grooves 2
2. Grooves on upper surface of tongue pinnately arranged (Plate IV-5); markings on back forming a reticulate pattern ....
.... Flatwoods Salamander (*Ambystoma cingulatum*), sp.208
Grooves on tongue absent or not pinnately arranged (either median groove or oblique grooves lacking); no network of light lines on back 3
3. Usually with 11 costal grooves; with several transverse light blotches on upper surface (rarely lacking, or regularly in immatures) ....
.... Marbled Salamander (*Ambystoma opacum*), sp.210
Usually with 12 or more costal grooves; light markings rounded and scattered 4
4. Adults with light markings (yellowish in life) of irregular size, shape, and distribution, some extending to lower sides; immatures medium gray with small light spots along sides of body, often in a row ....
.... Tiger Salamander (*Ambystoma tigrinum*), sp.211
Adults with rounded light spots of similar size arranged essentially in 2 dorsolateral rows on a dark background; no Florida records .... Spotted Salamander (*Ambystoma maculatum*), sp.250H

ORDER CAUDATA: L. FAMILY PLETHODONTIDAE, Lungless Salamanders.

1. Pale and unpigmented; legs extremely slender; external gills present ...... Georgia Blind Salamander (*Haideotriton wallacei*), sp.214
Normally colored; legs well developed; external gills absent in adults 2
2. With a constriction at base of tail ....
...... Four-toed Salamander (*Hemidactylium scutatum*), sp.213
No constriction at base of tail 3

3. With at least 20 costal grooves (13 between appressed limbs); legs extremely short; unrecorded in Florida ....
...... Red Hills Salamander (*Phaeognathus hubrichti*), sp.250H
Less than 20 costal grooves (less than 12 between appressed limbs); legs of normal length 4
4. Head usually angling abruptly downward; with a conspicuous swelling on each side at back of head; usually a light line from eye to angle of jaws; interorbital distance equal to distance from eye to nostril; no dark middorsal stripe 5
Head scarcely angling downward, if at all; no large swelling behind angle of jaws; usually no light line from eye to angle of jaws; interorbital distance greater than eye-nostril distance, *or* with a black stripe down dorsal midline[1] 7
5. Distal one-fourth of tail slender, its depth not greater than thickness of forelimb and less than that of hind limb; often with a dark, wavy line on each side of back ....
..... Northern Dusky Salamander (*Desmognathus fuscus*), sp.220
Distal fourth of tail compressed and bladelike, its maximum depth exceeding thickness of forelimb and sometimes that of hind limb; no dark, wavy line on each side of back 6
6. Ground color usually blackish above and dark gray below, with numerous light flecks often in a row on lower side; widespread in north Florida ....
Southern Dusky Salamander (*Desmognathus auriculatus*), sp.221
Ground color mainly grayish above and whitish below, with scattered dark flecks on dorsum; extreme northwest corner of Florida .... Seal Salamander (*Desmognathus monticola*), sp.222
7. With 17 or 18 costal grooves (6 or 7 between appressed limbs); with reddish mottling below and usually a middorsal red stripe, often with wavy margins; unrecorded in Florida ....
.... Zigzag Salamander (*Plethodon dorsalis*), sp.250H
Usually fewer than 17 costal grooves; if reddish present, not in form of a dorsal stripe 8
8. Upperparts longitudinally striped; underparts uniformly light, without conspicuous dark spots 9
Upperparts not striped, *or* underparts more or less dark-spotted
.... 10
9. Costal grooves usually 16; only 4 toes on hind foot (rarely 5) .....
.... Dwarf Salamander (*Manculus quadridigitatus*), sp.219

---

1. The Many-lined Salamander (*Stereochilus marginatus*), discovered in Florida in January, 1974 (S. Christman, personal communication), probably will not key beyond this point.

Costal grooves 13 or 14; 5 toes on hind foot ....
.... Two-lined Salamander (*Eurycea bislineata*), sp.218
10. Ground color blackish, with irregular, rounded white spots ....
.... Slimy Salamander (*Plethodon glutinosus*), sp.212
Ground color reddish or yellowish, at least ventrally (light in
preservative) 11
11. Tail more than 50% of total length; with 3 dark stripes down
back; 14 costal grooves ....
.... Long-tailed Salamander (*Eurycea longicauda*), sp.217
Tail less than 50% of total length; pattern of dark spots on reddish
or yellowish background; 15 or more costal grooves 12
12. Length of eye almost equal to snout length; no light line from eye to
nostril ...... Mud Salamander (*Pseudotriton montanus*), sp.216
Length of eye less than snout length; an indistinct, irregular light
line from eye to nostril ....
.... Red Salamander (*Pseudotriton ruber*), sp.215

ORDER SALIENTIA: LI. FAMILY PELOBATIDAE, Spadefoot Toads. One species
in Florida .... Eastern Spadefoot (*Scaphiopus holbrooki*), sp.223

ORDER SALIENTIA: LII. FAMILY BUFONIDAE, Toads.

1. Length up to 150 mm or more; greatest width of parotoid gland
anteriad and much greater than distance between orbits ....
.... Giant Toad (*Bufo marinus*), sp.227
Length usually less than 100 mm; greatest width of parotoid gland
near its center and about equal to distance between orbits 2

2. Greatest diameter of tympanum scarcely half that of eye; parotoid
glands diverging sharply posteriad; maximum length about 30
mm .... Oak Toad (*Bufo quercicus*), sp.226
Greatest diameter of tympanum more than half that of eye; parotoid
glands nearly or quite parallel; length of adults more than 30
mm 3

3. Postorbital ridge connected to parotoid gland only by a spur, or not
at all; warts irregular in size and distribution ....
.... Southern Toad (*Bufo terrestris*), sp.224
Postorbital ridge in direct contact with parotoid gland (Plate IV-3);
a few large warts in each pigment spot ....
.... Woodhouse's Toad (*Bufo woodhousei*), sp.225

ORDER SALIENTIA: LIII. FAMILY LEPTODACTYLIDAE, Tropical Frogs.
One species in Florida ....
...... Greenhouse Frog (*Eleutherodactylus planirostris*), sp.228

1. Disks on fingers more than half diameter of tympanum; distance from angle of jaws to tip of snout less than width of head at angle of jaws; hind toes noticeably webbed (web extending about halfway out on second, third, and fifth toes); no dark longitudinal stripe on back of thigh 2

   Disks on fingers not more than half diameter of tympanum, *or* tympanum not in evidence; length of mouth at least as great as its width; webbing of hind toes vestigial, *or* with a dark longitudinal stripe on back of thigh 10

2. Skin fused to top of skull; disks on fingers about as wide as tympanum; maximum length of adults 100 mm or more ....
   .... Cuban Tree Frog (*Hyla septentrionalis*), sp.236

   Skin movable against top of skull, not fused; disks on fingers not as wide as tympanum; maximum length of adults not more than 80 mm 3

3. Rear surface of thigh marked with light areas (orange, yellow, or greenish in life), often of rounded outline, on dark background; dark mask across eye usually present 4

   Never with the above combination of markings (dark mask absent in species more than 45 mm long) 7

4. With a conspicuous light line from eye to groin and no dark blotches dorsally ..... Pine Barrens Tree Frog (*Hyla andersoni*), sp.870

   No prominent light line laterally; dark blotches present dorsally 5

5. Without a light spot under the eye; usually 2 or more separate dark blotches on back; not warty above; several spots on thigh smaller than tympanum; maximum length about 35 mm ....
   .... Pine-woods Tree Frog (*Hyla femoralis*), sp.230

   Usually with a light spot under each eye; dark dorsal blotches usually conjoined along midline; most spots on thigh large as tympanum; rather warty above; maximum length 50 to 55 mm 6

6. Light areas on thigh and groin orange yellow in life; ground color light gray dorsally; light area below eye diagonally arranged and its width dorsally about half length of eye ....
   .... Southern Gray Tree Frog (*Hyla chrysoscelis*), sp.233

   Light areas on thigh and groin greenish in life; dorsal ground color dark or medium gray; light area below eye trapezoidal, but portion near eye almost as wide as length of eye ....
   .... Bird-voiced Tree Frog (*Hyla avivoca*), sp.234

7. With a discrete light line along margin of upper jaw, but no dark X on back; maximum length usually more than 40 mm (2 of 3 species) 8

No definite light line along edge of upper jaw, but with a dark X on back; maximum length less than 40 mm ....
.... Spring Peeper (*Hyla crucifer*), sp.229

8. Light line of upper jaw continuing straight, distinct, and unbroken along body to near groin ....
.... Green Tree Frog (*Hyla cinerea*), sp.235
Light line never continuing to near groin as described above 9

9. Back marked with numerous large, dark, oval spots and often with light flecks (golden in life; either marking may become obscure in preservative); width of head at angle of jaws at least 40% of snout-vent length; maximum length about 70 to 80 mm ....
.... Barking Tree Frog (*Hyla gratiosa*), sp.231
Color pattern not as described above (black dorsal markings, when present, fewer and smaller); width of head at angle of jaws less than 40% of snout-vent length; maximum length 35 to 40 mm
.... Squirrel Tree Frog (*Hyla squirella*), sp.232

10. With a longitudinal dark bar on rear of thigh and light vertical bars on upper jaw; no dark mask over eye; hind toes provided with webs and digital disks 11
No dark longitudinal bar on thigh or light vertical bars on upper jaw; black mask (sometimes interrupted) running across eye; webs and disks on hind toes usually reduced 12

11. With discrete light and dark stripes on rear of thigh; back and anal region not markedly warty; snout pointed (distance between nostrils less than that from nostril to tip of snout) ....
.... Southern Cricket Frog (*Acris gryllus*), sp.241
Dark thigh stripe with ragged edges, and the light stripe often indefinite; back and anal region warty; snout rounded (distance between nostrils about equal to that from nostril to tip of snout)
.... Northern Cricket Frog (*Acris crepitans*), sp.242

12. Webbing and disks of hind toes moderately developed; tibia more than 50% of snout-vent length; maximum length less than 18 mm .... Little Grass Frog (*Limnaoedus ocularis*), sp.237
Webbing and disks vestigial; tibia not more than 50% of snout-vent length; maximum length more than 20 mm 13

13. Usually with 2 or 3 oval, black, light-bordered spots on each side of body; 2 dark stripes down back faint or not in evidence; interorbital distance less than length of eye ....
.... Ornate Chorus Frog (*Pseudacris ornata*), sp.240
With 3 or more dark stripes or rows of spots arranged longitudinally on back, but no enlarged spots on sides of body; interorbital distance about equal to length of eye 14

14. Distance from armpit to snout greater than that from armpit to groin; snout-vent length usually less than 30 mm; usually no dark

blotch between eyes; ground color usually light gray; warty .....
.... Southern Chorus Frog (*Pseudacris nigrita*), sp.239
Distance from armpit to snout about equal to that from armpit to
groin; snout-vent length often more than 30 mm; usually a dark
blotch between eyes; ground color brownish; not very warty .....
.... Upland Chorus Frog (*Pseudacris triseriata*), sp.238

ORDER SALIENTIA: LV. FAMILY RANIDAE, True Frogs.

1. With conspicuous dorsolateral ridges on body 2
   Dorsolateral ridges vestigial or lacking 5
2. Dorsolateral ridges complete; with elongate pigment spots arranged
   in longitudinal rows on back 3
   Dorsolateral ridges not extending to level of groin; color pattern
   not as described above 4
3. Greatest distance between adjacent pigment spots in one longitudinal
   dorsal row less than width of any large spot; usually a single pigment spot between nostrils; pigment spots in young green (in life),
   and orange present on back of thighs in adults; no Florida records
   .... Pickerel Frog (*Rana palustris*), sp.250H
   Greatest distance between adjacent pigment spots greater than width
   of a large spot; pigment spots not green in life, and no orange on
   back of thigh; no pigment spot between nostrils ....
   .... Leopard Frog (*Rana pipiens*), sp.248
4. Upperparts unspotted and not warty, though sometimes granular
   (north of Florida dark spots smaller than the eye may be present)
   .... Bronze Frog (*Rana clamitans*), sp.247
   Upperparts with dark spots (some about diameter of eye), or with
   warts .... Gopher Frog (*Rana areolata*), sp.249
5. Upperparts marked with 4 light stripes (2 dorsal, 2 lateral); less
   than 75 mm long; Okefenokee Swamp ....
   .... Carpenter Frog (*Rana virgatipes*), sp.246
   Upperparts sometimes with 2 dorsal stripes, but never with 2 additional lateral stripes; adults 100 mm or more in maximum length
   6
6. With conspicuous white spots at edges of mouth; much blackish
   ventrally .... River Frog (*Rana heckscheri*), sp.244
   No white spots at edges of mouth; dark pigment of underparts
   grayish and restricted 7
7. First finger at least as long as second; fourth toe longer than fifth by
   about length of first toe; webs not extending to tips of hind toes;
   back of thigh usually mottled; no light lines on back ....
   .... Bullfrog (*Rana catesbeiana*), sp.243
   First finger slightly shorter than second (Plate IV-6); fourth toe
   longer than fifth by less than length of first toe; webs extending to

tips of hind toes; hind surface of thigh usually with alternating dark and light stripes; with 2 faint, light dorsolateral stripes .....
.... Pig Frog (*Rana grylio*), sp.245

ORDER SALIENTIA: LVI. FAMILY MICROHYLIDAE, Narrow-mouthed Toads.

One species in Florida ....
Eastern Narrow-mouthed Toad (*Gastrophryne carolinensis*), sp.250

KEY TO LARVAL ANURANS[1]

1. Labial teeth, horny beak, and papillae lacking; spiracle in midline near anus; blackish in color and dorsoventrally flattened; eyes lateral; white markings on belly, and a longitudinal white line on tail musculature .... *Gastrophryne carolinensis*
   Horny beak and papillae present; labial teeth usually present; spiracle sinistral; eyes lateral or dorsal (see Figs. 2 and 3) 2

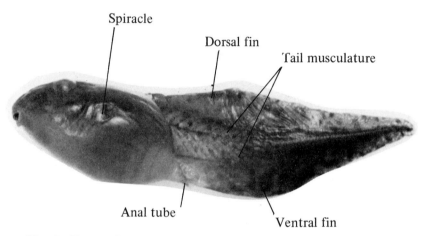

Fig. 2. External morphology of a tadpole (*Rana areolata aesopus*)

2. Labial tooth rows numerous (7 to 12), some incomplete and lateral to horny beak; papillae completely encircling oral disk except for a small space above first upper tooth row ....
   .... *Scaphiopus holbrooki*
   Labial teeth arranged in rows above and below horny beak (1 to 3 above, 2 to 4 below; rarely lacking); papillae always lacking for some distance on upper labium, and often absent under lower tooth rows 3

3. Anal tube opening in midline of ventral tail crest (appearing symmetrical); papillae not extending along entire lower lip; dorsum and venter dark; eyes dorsal, not visible from below 4

---

1. Modified from a key prepared by Sheryl Fanning.

Anal tube opening to right side of ventral crest (appearing asymmetrical); papillae extending across lower labium, *or* eyes lateral; part of venter usually lighter than dorsum 7

4. Upper edge of tail musculature with series of light areas that have dark pigment between .... *Bufo quercicus*
   Upper portion of tail musculature uniformly dark (macroscopically) 5

5. One row of papillae (sometimes incomplete) on sides of labium; spiracle on lateral axis; eyes nearer middorsal line than lateral outline; length of horny beak greater than that of third lower tooth row .... *Bufo woodhousei*
   Two or more complete rows of papillae on sides of labium; spiracle below lateral axis; eyes at least as close to lateral outline as to middorsal line; length of horny beak not greater than that of third lower tooth row 6

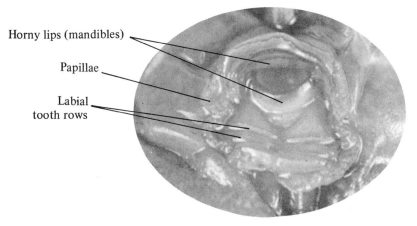

Horny lips (mandibles)

Papillae

Labial
tooth rows

Fig. 3. Mouthparts of a tadpole (*Rana heckscheri*)

6. Internasal distance about 10-15% of body length; gap in second upper tooth row about half as long as horizontal portion of upper horny beak .... *Bufo marinus*
   Internasal distance about 20–25% of body length; gap in second upper tooth row about equal to length of horizontal portion of upper horny beak .... *Bufo terrestris*

7. Eyes lateral in position, usually reaching lateral outline from dorsal aspect; lateral papillae not indented; tooth rows normally developed 8

Eyes dorsal in position, not reaching lateral outline from dorsal aspect; papillae emarginate; tooth rows sometimes lacking or incomplete 23

8. Tooth-row formula typically 2/2 9
   Tooth-row formula typically 2/3 or 2/4 11

9. Dorsal edge of tail musculature with an irregularly margined light area (from dorsal aspect); appressed lips in form of a triangle, with apex pointed upward; tip of tail not black; dorsal half of tail musculature darker than ventral half .... *Hyla crucifer*
   Dorsal edge of tail musculature with solid black stripe or a series of dark spots; appressed lips not in form of triangle; tip of tail often jet black; tail musculature mostly of uniform color; belly with pink iridescence in life 10

10. Spiracle short, not projecting as a free tube; dark pigment on dorsal tail musculature in form of squares or bars; greatest depth of tail about equal to that of body .... *Acris crepitans*
    Spiracle in form of a long projecting tube; dark pigment on base of dorsal tail musculature tending to run together, not forming distinct squares; greatest depth of tail at least 10% greater than that of body .... *Acris gryllus*

11. Tooth-row formula 2/4 .... *Hyla septentrionalis*
    Tooth-row formula 2/3 12

12. Body blackish with prominent light stripes in front of the eyes and light areas behind the eyes; tail musculature dark dorsally with several broad, light saddle marks (all light areas coppery in life); fins appearing clear macroscopically .... *Hyla avivoca*
    Not with the above combination of colors 13

13. Third lower tooth row not longer than horny beak and not much more than half length of first lower tooth; lower labial corner with less than 3 rows of papillae 14
    Third lower tooth row at least as long as horny beak and at least three-fourths length of first lower tooth row; lower labial corner with 3 or 4 rows of papillae 21

14. Papillae absent below part of third lower tooth row ....
    .... *Hyla crucifer*
    Papillae (sometimes small) extending completely across lower labium 15

15. Maximum body length about 9 mm, limb buds appearing at about 5 mm and reaching length of 3 mm at body length of about 8 mm; tail musculature with longitudinal brown stripe bordered above and below by a light stripe; pigment in fins concentrated near edges .... *Limnaoedus ocularis*
    Maximum body length more than 9 mm, limb buds appearing at 7 mm or more, not reaching 3 mm in length before body length is about 10 mm; tail not as described above 16

16. Eye usually closer to spiracular opening than to tip of snout; ventral half of tail musculature not lighter than dorsal half (except rarely near base of tail) 17

Eye equidistant between spiracular opening and tip of snout; ventral half of tail musculature lighter than dorsal half for most of its length 19

17. Pigmented part of lower beak wider (deeper) than its unpigmented part; black saddle spot showing through skin between eyes; deep globose body with wide tail fins; depth at eyes about half of greatest depth of tail, including crest .... *Hyla gratiosa*

Pigmented part of lower beak not wider than its unpigmented part; interorbital area uniformly dark; body more slender; depth at eyes much more than half greatest depth of tail, including fins 18

18. Pigment not concentrated dorsally (though sometimes laterally) on musculature near base of tail; eye decidedly closer to spiracular opening than to tip of snout; maximum body length about 20 mm .... *Hyla cinerea*

Pigment concentrated dorsally on tail musculature near its base; eye almost equidistant between spiracular opening and tip of snout; maximum body length about 15 mm .... *Hyla andersoni*

19. Length of tail about 3 times its depth at deepest point; 2 or more rows of papillae below third lower tooth row; third lower tooth row usually less than half length of first lower tooth row; ventral half of tail musculature largely devoid of pigment anteriad..... 20

Length of tail less than 3 times its depth at deepest point; one row of papillae below third lower tooth row; third lower tooth row at least half length of first lower row; ventral half of tail musculature with some flecks of pigment scattered throughout ....
.... *Pseudacris ornata*

20. With some pigment spots in dorsal fin at least half diameter of eye; dark pigment of dorsal tail musculature gradually merging with the lighter ventral part .... *Pseudacris nigrita*

Pigment spots in dorsal fin minute; dark pigment of dorsal tail musculature abruptly contrasting with white ventrally ....
.... *Pseudacris triseriata*

21. Third lower tooth row almost as long as second (at least 90%); tail fins heavily spotted except clear near musculature ....
.... *Hyla femoralis*

Third lower tooth row not more than 80% as long as second; tail fins usually not as described above 22

22. Pigment often concentrated at margin of tail fins; tail musculature evenly pigmented; usually with reddish in tail in life ....
.... *Hyla chrysoscelis*

Pigment evenly distributed in tail fins, not concentrated at margin; tail musculature often bicolored; no reddish in tail in life .... *Hyla squirella*

23. Dorsal edge of tail musculature a prominent, contrasting black stripe; edge of tail crests also contrastingly black in specimens of more than 35 mm total length; entire tail otherwise clear .... *Rana heckscheri*

   Tail not as described above (not strongly bicolored) 24

24. Pigmented portion of lower beak wide (almost or quite as deep as its unpigmented portion), *or* abdominal wall thin, dark, and transparent at body lengths up to 20 or 25 mm 25

   Pigmented portion of lower beak narrow (not more than half depth of the unpigmented portion); abdominal wall thicker and opaque at body lengths of more than 15 mm 27

25. No concentration of dark pigment at dorsal base of tail musculature; anal tube large, its length almost twice width of spiracle in specimens over 20 mm body length and greater than twice in smaller specimens; nearly all of lower beak pigmented in large specimens; hind limb buds appear at body length of about 25 mm, reaching length of 2 to 5 mm at body length of 28 mm (see Fig. 2) .... *Rana areolata*[1]

   Usually with a concentration of dark pigment at dorsal base of tail musculature; length of anal tube much less than twice width of spiracle in large specimens or barely twice in smaller specimens; pigmented part of lower beak scarcely, if at all, deeper than its unpigmented portion; hind limb buds appear at body length of 11 to 21 mm, reaching length of more than 5 mm at body length of 28 mm or less 26

26. Median space in second upper tooth row at least twice length of either lateral part; no light stripes on back; spots on tail smaller than diameter of eye; pigmented part of lower beak not deeper than that of upper beak; no Florida records ...... *Rana palustris*

   Median space in second upper tooth row less than twice length of either lateral part; often with 2 faint dorsolateral stripes; some dark spots on tail near size of eye; pigmented part of lower beak deeper than that of upper beak; widespread .... *Rana pipiens*

27. Dorsal fin with a longitudinal row of black spots 28

   Pigment in tail fins often sparse, never forming a longitudinal row of spots 29

28. Tooth-row formula 1/2 or 2/2, the second upper row (if present) vestigial (each lateral part about 10% of first upper row); streak in dorsal crest conspicuous and contrasting with surrounding clear

---

1. Many specimens cannot be distinguished from *Rana pipiens*.

area; tail fin edged with dark pigment, continuous or interrupted; streak in tail musculature continuing to tip of tail; at least 2 rows of papillae below lowermost tooth row .... *Rana virgatipes*

Tooth-row formula 2/3, each lateral part of the second upper row more than 10% length of first upper row; streak on dorsal fin faint and not in bold contrast with the surrounding area; dark pigment scattered throughout remainder of tail fin, not concentrated at margins; dark streak in tail musculature not continuing to tip of tail; only one row of papillae below most of lowermost tooth row .... *Rana grylio*

29. Tail crest clear (sometimes with small, sharply defined, scattered, dark dots); basal fourth of tail crests often more opaque than remainder; body light yellow green in life (pale in preservative), sometimes with small dark dots; hind limb buds appearing at about 20 mm body length, not reaching a length of 20 mm until about 30 mm body length .... *Rana catesbeiana*

Tail more heavily pigmented; tail crests not opaque basally; upper parts of body usually dark in life; hind limb buds appearing at about 15 mm body length; maximum body length about 25 mm .... *Rana clamitans*

CLASS REPTILIA—ORDERS AND FAMILIES

1. Eyes, ear openings, and limbs lacking; tail less than twice length of head .... SQUAMATA (SUBORDER AMPHISBAENIA): AMPHISBAENIDAE[1] LXIX

   Eyes always present, ear openings and limbs usually so; tail usually at least twice length of head 2

2. Body covered with a bony or leathery shell, the upper part (carapace) being connected to the lower part (plastron) by a relatively narrow bridge; no true teeth present (see Fig. 4) .... CHELONIA, 3

   Body covered with scales or epidermal plates, but never with a two-part shell; teeth present 9

3. Limbs paddlelike, with toes and claws scarcely in evidence, if at all 4

   Limbs less modified; toes in evidence and at least 3 on each foot provided with claws 5

4. With an external body covering of ridged, leathery skin (but scales present in young); claws lacking ...... DERMOCHELYIDAE, LVII

   Shell covered with epidermal (horny) plates; one or more claws present .... CHELONIIDAE, LXII

---

1. Any specimen of the Family Typhlopidae probably could not be keyed past here. One specimen of *Typhlops lumbricalis* has been collected in Florida (Myers, 1958), but there is little likelihood of an established population here.

5. Body covering a leathery skin; underlying shell flexible; snout long and tubular .... TRIONYCHIDAE, LXIII
   Shell hard and bony, covered with horny plates; no well-developed snout present 6
6. Carapace rough; plastron greatly reduced and cross-shaped; tail more than half length of carapace .... CHELYDRIDAE, LVIII
   Carapace essentially smooth; plastron oval and of normal size; tail less than half length of carapace 7
7. Plastral plates 10 or 11; pectorals not in contact with marginals .... KINOSTERNIDAE, LIX
   Plastral plates 12; pectorals in contact with marginals (or separated only by skin) 8
8. Forelimb at wrist almost as wide as at foot and with laterally flattened claws; toes not webbed; plastron not hinged ....
   .... TESTUDINIDAE, LXI
   Foot much wider than wrist joint and claws slender; toes with some webbing, *or* plastron hinged ....EMYDIDAE, LX
9. Anal slit longitudinal; toes webbed ....
   .... CROCODILIA: CROCODYLIDAE, LXXIII
   Anal slit transverse; toes (if present) not webbed ....
   .... SQUAMATA, 10
10. Ear openings or limbs present .... SUBORDER LACERTILIA,[1] 11
    Ear openings and limbs invariably lacking ....
    .... SUBORDER SERPENTES,[1] 15
11. Without limbs .... ANGUIDAE, LXVI
    Two pairs of limbs present 12
12. Eyelids absent .... GEKKONIDAE, LXV
    Eyelids present 13
13. Scales small (all of nearly equal size except on head), smooth, and rounded .... SCINCIDAE, LXVIII
    Scales keeled or granular, not usually rounded, and those on ventral side often larger than dorsal scales 14
14. With light lines (sometimes incomplete) on back; ventral scales in 8 to 10 longitudinal rows .... TEEIDAE, LXVII
    Without light lines dorsally; ventral scales not in obvious rows .....
    .... IGUANIDAE, LXIV
15. With a pit located between eye and nostril; head wide and triangular; some of subcaudals in one row .... CROTALIDAE, LXXII
    No pit between eye and nostril; head usually scarcely wider than neck; subcaudals in 2 rows 16
16. Pattern of red, yellow, and black rings, with red and yellow in contact; some subcaudals in 2 rows .... ELAPIDAE, LXXI

---

1. The Lacertilia, Serpentes, and Amphisbaenia are given full ordinal rank in the ASIH *Catalogue*.

Pattern never of red, yellow, and black rings with red and yellow in contact; subcaudals in one row . . . . COLUBRIDAE, LXX

ORDER CHELONIA: LVII. FAMILY DERMOCHELYIDAE, Leatherbacks. One species in Florida . . . . . . Leatherback (*Dermochelys coriacea*), sp.251

ORDER CHELONIA: LVIII. FAMILY CHELYDRIDAE, Snapping Turtles.

1. With 2 rows of marginal plates; eyes not clearly visible from dorsal aspect; with many small, rounded scales under tail . . . .
   . . . . . . Alligator Snapping Turtle (*Macroclemys temmincki*), sp.253
   Only one row of marginals; eyes clearly visible from above; with 2 rows of enlarged scales under tail . . . .
   . . . . Common Snapping Turtle (*Chelydra serpentina*), sp.252

ORDER CHELONIA: LIX. FAMILY KINOSTERNIDAE, Mud and Musk Turtles.

1. Pectoral plates quadrangular (length of medial suture almost equal to that of lateral edge); plastron almost immovable 2
   Pectoral plate almost triangular (lateral edge 3 times length of medial suture; Plate V-1); plastron movable at pectoral-abdominal junction 3
2. Barbels confined to chin (Plate V-2); head light (or medium dark) with blackish spots . . . .
   . . . . Loggerhead Musk Turtle (*Sternotherus minor*), sp.255
   Barbels (when present) on skin of throat as well as on chin; head conspicuously striped in young, becoming dark and obsolescent in old age . . . . Stinkpot (*Sternotherus odoratus*), sp.254
3. Carapace with 3 prominent light stripes . . . .
   . . . . Striped Mud Turtle (*Kinosternon bauri*), sp.256
   Carapace uniformly dark . . . .
   . . . . Common Mud Turtle (*Kinosternon subrubrum*), sp.257

ORDER CHELONIA: LX. FAMILY EMYDIDAE, Terrapins.

1. Plastron hinged; webbing between front toes vestigial or absent
   . . . . Box Turtle (*Terrapene carolina*), sp.259
   Plastron not hinged; some webbing between front toes 2
2. Plates of carapace with rounded light centers and (in young) with concentric ridges; second to fourth vertebrals with centrally located, rounded knobs; width of head greater than anteroposterior width of second costal . . . .
   . . . . Diamondback Terrapin (*Malaclemys terrapin*), sp.260
   Plates of carapace without rounded light centers or concentric ridges; second to fourth vertebrals without rounded knobs at their centers; width of head not greater than that of second costal 3

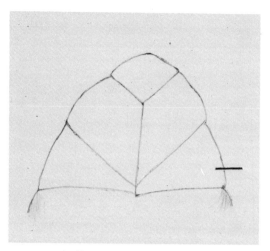

1. Triangular pectoral plates
   of *Kinosternon bauri*

2.
Location of barbels in
Sternotherus minor

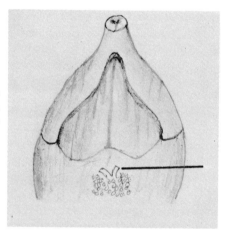

3.
C-shaped marking on
second costal plate
of *Chrysemys concinna*

PLATE   V

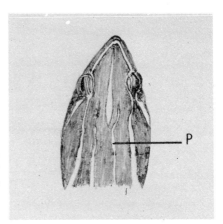

4.
Paramedian stripe (P)
and arrow of
Chrysemys nelsoni

P

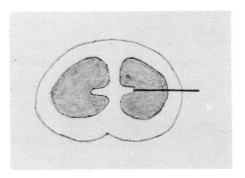

5. Fleshy ridges in nostrils of
Trionyx spiniferus

6.
Second finger of *Hemidactylus*,
showing medially divided
transverse lamellae

REPTILES

3. With rounded light spots on carapace and head; carapace smooth; plastron mostly blackish ....
.... Spotted Turtle (*Clemmys guttata*), sp.258
Without rounded light spots on carapace and head; carapace usually with ridges or striations; plastron scarcely more than 50% blackish, if at all 4

4. Posterior edges of first 3 vertebral plates with high projections; highest point of carapace anterior to middle of body; posterior marginals with pointed projections 5
Posterior edges of vertebral plates not raised; highest point of carapace not anterior to middle of body; posterior marginals without marked V-shaped projections 8

5. Light blotch behind eye (yellowish or greenish in life) as wide as, or wider than, the orbit; posterior marginals single-toothed (second tooth, if present, much smaller); restricted in Florida to Panhandle 6
Light markings behind eye not as wide as orbit; posterior marginals double-toothed in young; not known to occur in Florida 7

6. With U-shaped light markings on anterior costals; plastron uniformly dull; a curved light marking on chin ....
.... Barbour's Map Turtle (*Malaclemys barbouri*), sp.261
With a reticulate pattern of light lines on carapace (often indistinct); plastron yellow with darker pigment along seams; a light longitudinal bar on chin ....
.... Alabama Map Turtle (*Malaclemys pulchra*), sp.262

7. Second and third dorsal spines much enlarged, rounded, and darkened terminally; dark pigment on plastron largely restricted to seams
.....Black-knobbed Map Turtle (*Malaclemys nigrinoda*), sp.348H
Second and third dorsal spines pointed and not greatly enlarged; dark pigment covering about half of plastron ....
.... Mississippi Map Turtle (*Malaclemys kohni*), sp.348H

8. With prominent light and dark vertical stripes on posterior surface of body and thighs; neck and head length combined about equal to length of plastron; pattern of light reticulations on carapace (yellowish in life) ....
.... Chicken Turtle (*Deirochelys reticularia*), sp.269
No light and dark vertical bars on thighs (except *C. scripta*); neck not unusually long; light pattern on carapace not reticulate 9

9. Carapace smooth and often with a median light line; posterior marginals entire; costals and vertebrals arranged in transverse rows (Fla. subsp.) .... Painted Turtle (*Chrysemys picta*), sp.263
Carapace keeled, wrinkled, more or less serrate posteriad, and without a light median line; costals and vertebrals in offset positions 10

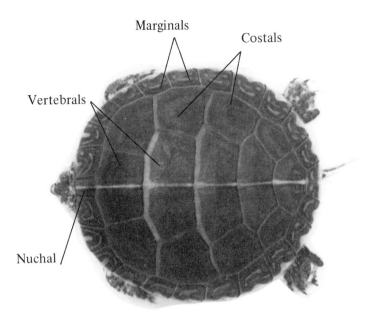

Marginals

Costals

Vertebrals

Nuchal

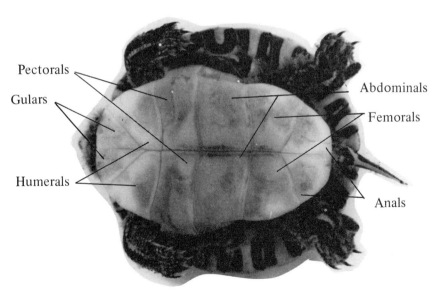

Pectorals

Gulars

Abdominals

Femorals

Humerals

Anals

Fig. 4. Plates on the carapace and plastron of a turtle (*Chrysemys picta*)

10. Cutting edge of upper jaw running essentially straight for more than half its length and with a distinct notch anteriad; with broad, yellow, vertical bars on costals and a large yellow or red patch behind eye (may be absent in old age) ....
.... Yellow-bellied Turtle (*Chrysemys scripta*),[1] sp.264
Cutting edge of upper jaw more curved and distinct median notch sometimes lacking; vertical bars on costals (if present) usually faint or narrow; pattern of narrow yellow or red stripes on sides of head (less than diameter of eye) 11
11. Notch at tip of upper jaw rounded and not bordered by cusps; both jaws smooth or only weakly serrate; no arrow-shaped marking between eyes 12
Notch at tip of upper jaw deep and narrow, bordered on each side by a cusp; lower jaw noticeably serrate; an arrow-shaped marking between eyes pointing toward tip of snout 13
12. With a light C-shaped marking on the posteromedial portion of the second costal scute (Plate V-3); plastron with dark markings
.... River Cooter (*Chrysemys concinna*), sp.265
Without a light C-shaped marking on the second costal scale; plastron unmarked .... Florida Cooter (*Chrysemys floridana*), sp.266
13. Paramedian stripe terminating behind eye (unless shell length under 100 mm; Plate V-4) ....
.... Florida Red-bellied Turtle (*Chrysemys nelsoni*),[2] sp.267
Paramedian stripe continuing anterior to eye ....
....Alabama Red-bellied Turtle (*Chrysemys alabamensis*),[2] sp.268

ORDER CHELONIA: LXI. FAMILY TESTUDINIDAE, Land Tortoises. One species in Florida ....
.... Gopher Tortoise (*Gopherus polyphemus*), sp.270

ORDER CHELONIA: LXII. FAMILY CHELONIIDAE, Sea Turtles.

1. With only 4 costal plates on each side 2
With more than 4 pairs of costal plates 3
2. With 2 pairs of prefrontal scales; lower jaw weakly serrate if at all .... Hawksbill (*Eretmochelys imbricata*), sp.272
Only one pair of prefrontals; lower jaw strongly serrate in adults
.... Green Turtle (*Chelonia mydas*), sp.271

---

1. The turtles in this genus are taxonomically difficult, each species virtually matching the characteristics of the others in its variations. The species assignments (from Conant) should perhaps be regarded as tentative. For use of *Chrysemys* and *Malaclemys* see McDowell, 1964.

2. These two turtles probably should be regarded as conspecific; immatures can be distinguished only with difficulty, if at all, and sympatry has not been amply demonstrated.

3. With 4 enlarged inframarginals (plus smaller ones) ....
.... Atlantic Ridley (*Lepidochelys kempi*), sp.274
With only 3 enlarged inframarginals ....
.... Loggerhead (*Caretta caretta*), sp.273

ORDER CHELONIA: LXIII. FAMILY TRIONYCHIDAE, Softshell Turtles.

1. With a fleshy ridge projecting into each nostril from the medial
side (Plate V-5); tubercles present toward anterior edge of cara-
pace (poorly developed in young) 2
No ridges in nostrils or tubercles on carapace ....
.... Smooth Softshell (*Trionyx muticus*), sp.275
2. Ground color of carapace dark; tubercles flattened and hemispherical
.... Florida Softshell (*Trionyx ferox*), sp.277
Ground color of carapace light; tubercles cone-shaped or spiny
.... Spiny Softshell (*Trionyx spinifer*), sp.276

ORDER SQUAMATA: SUBORDER LACERTILIA: LXIV. FAMILY IGUANIDAE,
Iguanas and allies.[1]

1. Dorsoventrally flattened and with spines on back of head and sides
of body; tail (measured from vent) scarcely, if any, longer than
width of body ....
.... Texas Horned Lizard (*Phrynosoma cornutum*), sp.285
Body roughly terete, its width less than twice its depth; spines
lacking; tail much longer than width of body 2
2. All dorsal scales pointed and of normal size; toes not widened near
tips 6
Most dorsal scales not pointed and those along midline (usually
others also) minute; toes laterally expanded near tip 3
3. Most body scales keeled 4
Most body scales flat, conical, or granular 5
4. Tail terete .... Green Anole (*Anolis carolinensis*), sp.279
Tail laterally compressed, with a strong dorsal keel ....
.... Brown Anole (*Anolis sagrei*), sp.281
5. Most scales flattened and rectangular; a prominent white (or light)
slash on body above shoulder; total length of adults exceeding
300 mm .... Knight Anole (*Anolis equestris*), sp.280
Dorsal scales conical to granular; no white markings on body; total
length less than 150 mm ....
.... Bahaman Bark Anole (*Anolis distichus*), sp.278
6. Tail approximately terete 7
Tail markedly compressed and with a strong dorsal keel ....
.... Curly-tailed Lizard (*Leiocephalus carinatus*), sp.282
7. With a pair of distinct, brown dorsolateral stripes ....
.... Florida Scrub Lizard (*Sceloporus woodi*), sp.284

_____
1. Conant's latest field guide also accredits *Anolis cybotes* to the Miami area.

Dorsolateral stripes, if present, blackish and not clearly delimited
.... Eastern Fence Lizard (*Sceloporus undulatus*), sp.283

ORDER SQUAMATA: SUBORDER LACERTILIA: LXV. FAMILY GEKKONIDAE, Geckos.

1. Transverse lamellae (under toes) divided medially (Plate V-6) 2
Most transverse lamellae not divided medially 3
2. Back warty; ground color very light, with scattered dark markings
.... Mediterranean Gecko (*Hemidactylus turcicus*), sp.288
Back not warty; ground color grayish with scattered light spots
.... Indo-Pacific Gecko (*Hemidactylus garnoti*), sp.289
3. Toes not expanded; tip of tail (unless regenerated) light ....
.... Yellow-headed Gecko (*Gonatodes albogularis*), sp.286
Toes expanded near tip; tip of tail not lighter than other portions
.... 4
4. Total length up to 350 mm; tail with dark bands ....
.... Tokay Gecko (*Gekko gecko*), sp.287
Total length less than 100 mm; tail not dark-banded 5
5. Dorsal scales of normal size (about 2 per mm in adults); maximum
width of head greater than length of snout from back of eye .....
.... Reef Gecko (*Sphaerodactylus notatus*), sp.290
Dorsal scales minute (about 4 per mm in adults); maximum width
of head not greater than length of snout from back of eye 6
6. Scales of back granular and not keeled; color pattern medium light,
with dark stripes on neck (young with dark crossbars and red
tail) .... Ashy Gecko (*Sphaerodactylus cinereus*), sp.292
Scales of back flattened, though sometimes keeled; ground color
dark, with numerous light spots ....
.... Ocellated Gecko (*Sphaerodactylus argus*), sp.291

ORDER SQUAMATA: SUBORDER LACERTILIA: LXVI. FAMILY ANGUIDAE, Glass Lizards and allies.

1. Frontonasal scale (or larger one if 2 are present) less than twice
area of each prefrontal (Plate VI-1); 1½ rows of scales above
lateral fold devoid of pigment from cloaca about halfway to head
.... Island Glass Lizard (*Ophisaurus compressus*), sp.294
Only one frontonasal, its area about twice that of each prefrontal;
less than 1½ rows of unpigmented scales posteriad above lateral
fold 2
2. Light dorsal spots or stripes involving the centers of the scales, except
when forming crossbars; one or 2 dark stripes (or rows of spots)
below lateral fold ....
.... Slender Glass Lizard (*Ophisaurus attenuatus*), sp.295

Light dorsal spots or stripes involving only the edges of the scales; no light bars across back; little or no pigmentation below lateral fold ...... Eastern Glass Lizard (*Ophisaurus ventralis*), sp.293

ORDER SQUAMATA: SUBORDER LACERTILIA: LXVII. FAMILY TEEIDAE, Whiptail Lizards.

1. With 6 light dorsal stripes and 5 or 6 pairs of enlarged sublabials; not more than 110 granular (dorsal and lateral) scale rows; total length less than 300 mm; statewide ....
..... Six-lined Racerunner (*Cnemidophorus sexlineatus*), sp.297
Back not striped; only 3 pairs of enlarged sublabials; 130 or more granular scale rows; up to 500 mm long; known only from Miami .... Colombian Ground Lizard (*Ameiva ameiva*), sp.296

ORDER SQUAMATA: SUBORDER LACERTILIA: LXVIII. FAMILY SCINCIDAE, Skinks.

1. Limbs vestigial, the forelimbs almost wanting ....
.... Sand Skink (*Neoseps reynoldsi*), sp.304
Limbs normally developed 2
2. Lower eyelid with a transparent disk (Plate VI-2); without longitudinal light stripes; maximum snout-vent length less than 2 inches .... Ground Skink (*Scincella lateralis*), sp.298
No transparent area on lower lid; longitudinal light stripes usually present 3
3. Median row of subcaudal scales not noticeably wider than adjacent rows ....
.....Southeastern Five-lined Skink (*Eumeces inexpectatus*), sp.301
Median row of subcaudals decidedly wider than adjacent rows 4
4. No postnasal scales (but 2 loreals); with less than 30 rows of scales at mid-body; maximum snout-vent length less than 75 mm 5
One postnasal and 2 loreals present; often at least 30 scale rows at mid-body; maximum snout-vent length of adults more than 75 mm 6
5. Only one postmental plate; ground color of tail usually dark; 24 to 28 scale rows at mid-body ....
.... Coal Skink (*Eumeces anthracinus*), sp.302
With 2 postmentals (Plate VI-3); tail orange, reddish, or bluish in life (usually light in preservative); only 22 scale rows at mid-body .... Red-tailed Skink (*Eumeces egregius*), sp.303
6. White stripe on side of head involving scales above ear opening; auriculars rounded and not projecting over ear opening; light dorsolateral stripe (when present) on fourth and fifth scale rows; usually with one or no enlarged postlabial plate (Fig. 5) ....
.... Broad-headed Skink (*Eumeces laticeps*), sp.300

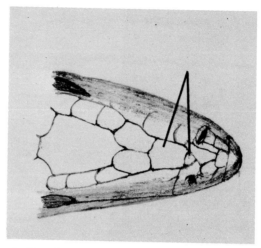

1. Two frontonasals of *Ophisaurus compressus*

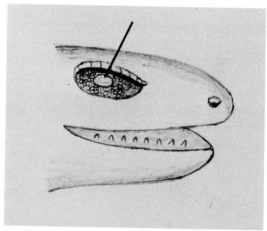

2. Transparent window in lower eyelid of
Scincella lateralis

PLATE VI

68 / VERTEBRATES OF FLORIDA

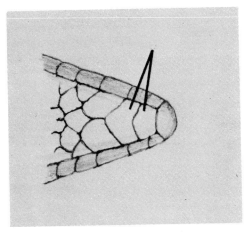

3. Two postmental scales of
   Eumeces egregius

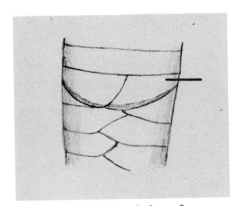

4. Divided anal plate of
   Natrix erythrogaster

REPTILES

White stripe on side of head not involving scales above ear opening; some auriculars spinose and projecting over ear opening; light dorsolateral stripe (when present) on third and fourth scale rows; usually with 2 enlarged postlabial scales (much larger than body scales) ....

ORDER SQUAMATA: SUBORDER AMPHISBAENIA: LXIX. FAMILY AMPHIS-BAENIDAE, Ringed Lizards. One species in Florida ....

ORDER SQUAMATA: SUBORDER SERPENTES: LXX. FAMILY COLUBRIDAE, Colubrid Snakes.

1. Scales at least weakly keeled along middorsal line at middle of body 2
   Dorsal scales smooth, at least at mid-body 22
2. Dorsal color uniformly medium gray or grayish tan, somewhat lighter below; head scarcely wider than neck; preocular lacking; loreal horizontally elongate; maximum length about 300 mm ....
   .... Rough Earth Snake (*Virginia striatula*), sp.319
   Dorsal color pattern not uniform unless in sharp contrast with ventral color; head noticeably wider than neck; preocular usually present and vertically elongate; loreal usually not horizontally elongate, sometimes lacking; maximum length of most species more than 300 mm 3
3. Anal plate single 4
   Anal plate divided obliquely (Plate VI-4) 7
4. Depth of rostral plate greater than its width; tail less than 20% of total length ....
   .... Pine Snake (*Pituophis melanoleucus*), sp.331
   Vertical dimension of rostral less than the horizontal dimension; tail more than 20% of total length 5
5. With 23 or more dorsal scale rows; upperparts plain or with large dark blotches 18
   With only 19 dorsal scale rows; upperparts striped or with small dark blotches 6
6. Tail more than 27% of total length; more than 85 pairs of caudal plates ..... Eastern Ribbon Snake (*Thamnophis sauritus*), sp.318
   Tail less than 27% of total length; less than 85 pairs of caudal plates ...... Common Garter Snake (*Thamnophis sirtalis*), sp.317
7. Rostral plate turned up and pointed 8
   Rostral plate neither turned up nor pointed 9
8. Prefrontal scales at least partly in contact; under side of tail often lighter than belly ....
   ..... Eastern Hog-nosed Snake (*Heterodon platyrhinos*), sp.338

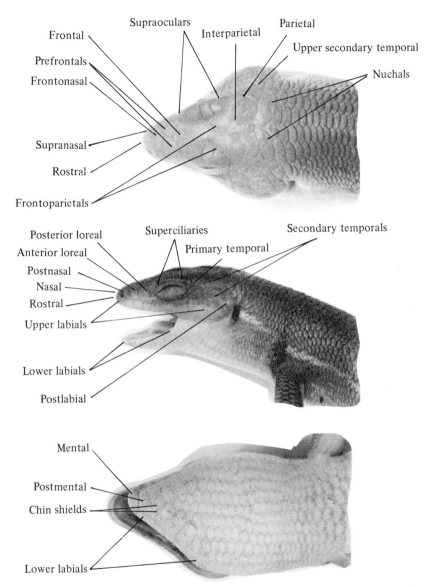

Fig. 5. Specialized scales on the head of a lizard (*Eumeces laticeps*)

Prefrontal scales separated by smaller scales; under side of tail not lighter than belly ....
.... Southern Hog-nosed Snake (*Heterodon simus*), sp.339

9. Dorsal scales only weakly keeled; tail usually not more than 20% of total length 10

Dorsal scales strongly keeled; tail usually more than 20% of total length 12

10. Only 19 scale rows at mid-body ....
.... Striped Swamp Snake (*Regina alleni*), sp.308

At least 25 dorsal scale rows 11

11. With a broad, V-shaped light band on top of head, terminating between the eyes; adult with coppery or pinkish blotches in life; with 2 dark stripes (sometimes broken) under tail of young .....
.... Corn Snake (*Elaphe guttata*), sp.329

Head not as described above, though sometimes with a *narrow* dark line running *through* the eye and crossing the head anterior to the eyes; never with pinkish or coppery marks in life; no dark stripes under tail .... Rat Snake (*Elaphe obsoleta*), sp.330

12. Grass green above (often bluish in preservative) and light below; tail about 35 to 40% of total length ....
.... Rough Green Snake (*Opheodrys aestivus*), sp.327

Never bright green; tail less than 35% of total length 13

13. Posterior nasal large, reaching preoculars 14

Nasal separated from preocular by a loreal scale 16

14. With 17 dorsal scale rows and one preocular ....
.... Brown Snake (*Storeria dekayi*), sp.315

With 15 dorsal scale rows and usually 2 preoculars 15

15. With less than 127 ventral scales, their lateral edges of the same color as sides of body (remainder of ventrals reddish in life, pale in preservative) ....
.... Red-bellied Snake (*Storeria occipitomaculata*), sp.316

With 127 or more ventrals that are uniformly whitish, unless with black spots at lateral edges of some ....
.... Brown Snake (*Storeria dekayi*), sp.315

16. Dark dorsal markings arranged in a chainlike fashion; papillae present on chin of male; not in Florida unless extreme northwestern corner
..... Diamond-backed Water Snake (*Natrix rhombifera*), sp.348H

No chainlike markings on back; no papillae present on chin 17

17. With more than 26 scale rows 18

With 21 to 25 scale rows 19

With 19 scale rows 20

18. Scales on sides of body smooth to weakly keeled; suboculars present; no large dark blotches dorsally ....
.... Green Water Snake (*Natrix cyclopion*), sp.309

All scales on sides of body strongly keeled; suboculars absent; upperparts with large dark blotches ....
.... Brown Water Snake (*Natrix taxispilota*), sp.311
19. Underparts uniformly pale (reddish to yellowish in life); upperparts
uniformly medium dark, except for blackish blotches in young;
more than 140 ventral plates ....
.... Red-bellied Water Snake (*Natrix erythrogaster*), sp.310
Underparts usually varied, rarely uniform in color; upperparts
blotched, banded, or striped; less than 140 ventral plates 21
20. With a light stripe (yellowish in life) on the first and (usually)
second scale row; 4 dark stripes on ventral plates; first dorsal
scale row keeled ....
.... Queen Snake (*Regina septemvittata*), sp.307
No light stripes on dorsal scales, but 3 on ventral plates, alternating
with 2 dark stripes (or rows of spots); first dorsal scale row
smooth .... Glossy Water Snake (*Regina rigida*), sp.306
21. Top of head not blackish; no dark stripe from eye to angle of jaws;
body with many dark crossbars, some toward posterior end being
broken and the 2 parts offset ....
.... Common Water Snake (*Natrix sipedon*), sp.313
Top of head and band from eye to angle of jaws dark or black; crossbars, when present, not broken but continuous (body may be
striped or essentially unicolored) ....
.... Banded Water Snake (*Natrix fasciata*), sp.312
22. With 19 or more scale rows 23
With fewer than 19 scale rows 31
23. Anal plate entire 24
Anal plate with an oblique division (Plate VI-4) 28
24. With 21 to 23 rows of dorsal scales at mid-body; length up to 1 m
or more 25
With only 19 rows of dorsal scales; maximum length less than 1 m
.... 26
25. Dorsal color pattern of black (or dark brown) contrasting with
white, the pattern ringed, blotched, or speckled; *or* pale and almost
devoid of pattern in south Florida ....
.... Common Kingsnake (*Lampropeltis getulus*), sp.333
Without white and with little black except ventrally; brown or gray
predominant (not pale); confined to northwest Florida ....
.... Prairie Kingsnake (*Lampropeltis calligaster*), sp.332
26. Loreal, and sometimes preocular, absent; tail less than 10% of total
length .... Short-tailed Snake (*Stilosoma extenuatum*), sp.335
Loreal and preoculars present (see Fig. 6); tail more than 10% of
total length 27

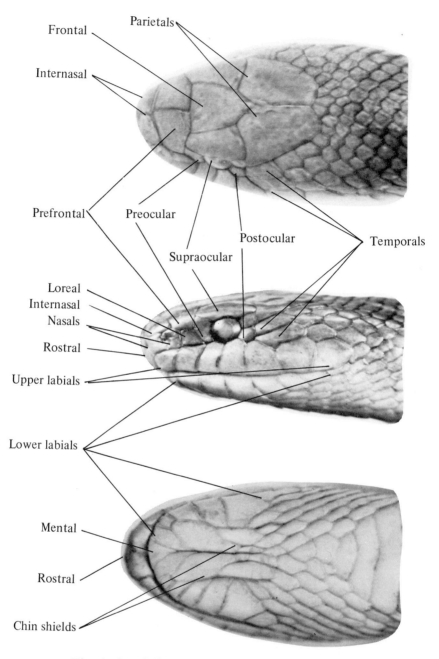

Fig. 6. Specialized scales on the head of a snake
(*Lampropeltis calligaster*)

27. With narrow black rings alternating with narrow and wide light rings (the 2 latter yellowish and red in life); if rings not complete, then some black present ventrally ....
.... Scarlet Kingsnake (*Lampropeltis triangulum*), sp.334
Pattern of transverse blotches or of rings (black, yellowish, and red in life); if rings, then interrupted ventrally so that no black is present there ...... Scarlet Snake (*Cemophora coccinea*), sp.336

28. Preoculars lacking, the loreal reaching the eye; with some red on body in life 30
With one or more preoculars separating the loreal from the eye; ground color sometimes coppery, but never red 29

29. Scale rows more than 19 at mid-body 11
Only 19 scale rows at mid-body ....
.... Striped Swamp Snake (*Regina alleni*), sp.308

30. With a pattern of red bands, interrupted dorsally by black; only one internasal .... Mud Snake (*Farancia abacura*), sp.324
With longitudinal red stripes and 2 internasals ....
.... Rainbow Snake (*Farancia erytrogramma*), sp.323

31. Preoculars lacking, the loreal reaching the eye; small gray to tan snakes, somewhat lighter below 38
One or more preoculars separating the loreal (if present) from the eye; color pattern not uniformly gray or tan dorsally unless in strong contrast with ventral color 32

32. Anal plate single ...... Indigo Snake (*Drymarchon corais*), sp.328
Anal plate divided (Plate VI-4) 33

33. Blackish above, reddish ventrally; total length less than 50 cm .....
.... Black Swamp Snake (*Seminatrix pygaea*), sp.314
Not blackish above and reddish below; sizes various 34

34. Loreal absent; total length less than 38 cm ....
.... Crowned Snake (*Tantilla coronata*),[1] sp.337
Loreal present; length variable 35

35. With 15 rows of scales at mid-body; underparts yellowish and continuing dorsally as a ring around the neck (sometimes incomplete)
.... Ring-necked Snake (*Diadophis punctatus*), sp.322
With 17 rows of scales at mid-body; color pattern not as described above 36

36. Rostral flattened; a contrasting dark line through eye; total length under 50 cm ....
.... Yellow-lipped Snake (*Rhadinaea flavilata*), sp.321

---

1. Telford (1966) has divided these snakes into 3 species (*Tantilla coronata, T. relicta,* and *T. oolitica*) on bases that are not fully convincing.

Rostral angular at midline; no contrasting dark line through eye; total length of adults more than 50 cm 37
37. With 15 scale rows just anterior to vent; upperparts of adults unvarying in color; ground color of young bluish ....
........................Racer (*Coluber constrictor*), sp.325
With 11 to 13 scale rows just anterior to vent; anterior parts of adults darker than remainder; ground color of young light brown .... Coachwhip (*Masticophis flagellum*), sp.326
38. Head slightly wider than neck; nasal plate divided; with 15 scale rows .... Smooth Earth Snake (*Virginia valeriae*), sp.320
Head not wider than neck; nasal plate not divided; only 13 scale rows; no Florida record ....
.... Worm Snake (*Carphophis amoena*), sp.348H

ORDER SQUAMATA: SUBORDER SERPENTES: LXXI. FAMILY ELAPIDAE, Coral Snakes. One species in Florida ....
.... Eastern Coral Snake (*Micrurus fulvius*), sp.340

ORDER SQUAMATA: SUBORDER SERPENTES: LXXII. FAMILY CROTALIDAE, Pit Vipers.

1. With a rattle on end of tail 2
No rattle on tail 4
2. With 9 enlarged plates on top of head, the width of each at least as great as that of orbit ....
.... Pigmy Rattlesnake (*Sistrurus miliarius*), sp.343
Most scales on top of head smaller than orbit 3
3. Upperparts with a brown median band and brown transverse bars; scales in 25 rows or fewer; north Florida only, where generally rare .... Timber Rattlesnake (*Crotalus horridus*),[1] sp.344
Upperparts with dark (light-bordered) diamond-shaped markings (rarely lacking); scales in 29 rows; statewide ....
Eastern Diamondback Rattlesnake (*Crotalus adamanteus*), sp.345
4. With subocculars separating upper labials from eye; loreal present; northwest Florida only ....
.... Copperhead (*Agkistrodon contortrix*), sp.341
With at least one upper labial entering the orbit; loreal absent; common and widespread ....
.... Cottonmouth (*Agkistrodon piscivorus*), sp.342

ORDER CROCODILIA: LXXIII. FAMILY CROCODYLIDAE, Crocodilians.

1. With a curved, bony ridge in front of the eyes; light bands on tail wider than dark bands ....
.... Spectacled Caiman (*Caiman sclerops*), sp.348

---

1. The Florida race, *C. h. atricaudatus,* is called the Canebrake Rattlesnake.

No curved, bony ridge anterior to eyes; light bands on tail not wider
than dark ones, *or* tail not banded 2

2. Fourth tooth of lower jaw exposed in adults; width of head at eyes
usually less than 80% of distance from front of eye to tip of snout;
nostrils inconspicuous, crescentic slits in a single rounded disk ..
.... American Crocodile (*Crocodylus acutus*), sp.346
Fourth tooth of lower jaw not exposed when mouth is shut; width
of head at eyes more than 80% of snout length; nostrils conspicuous, each in a separate disc ....
.... American Alligator (*Alligator mississipiensis*), sp.347

CLASS AVES—ORDERS AND FAMILIES

1. Adapted for life in or around water (toes webbed or lobed, or lower
tibia bare); *or* bill straight, at least 60 mm long, less than 15 mm
deep at base, *and* the 3 outermost primaries less than half width
of other primaries 2
Not adapted for aquatic life (toes neither webbed nor lobed; tibia
feathered to within 10 mm of ankle joint); not with above combination of bill and wing characters 31

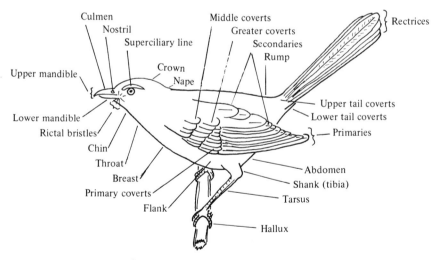

Fig. 7. External features of a bird

2. Front toes (or all 4) fully webbed, though webs sometimes much
incised 3
Webs lacking or vestigial (extending less than halfway along toes to
the claws) 19

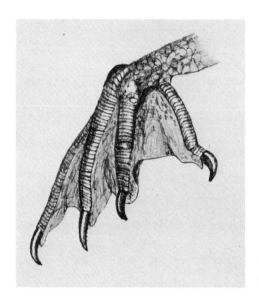

1.
Totipalmate foot of
Phalacrocorax auritus

2. Single nostril tube of
Oceanodroma castro

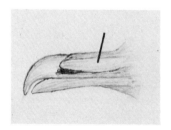

3.
Upper mandible and cere of
Stercorarius parasiticus

PLATE VII

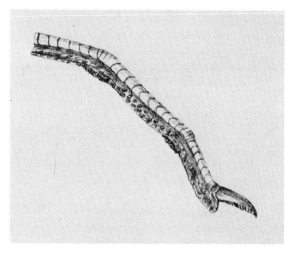

4. Pectinate claw on middle toe
of *Ardea herodias*

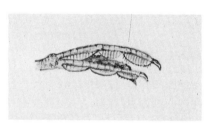

5.
Lobed toes of
Lobipes lobatus

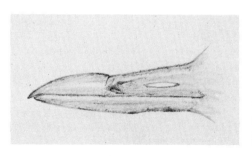

6.
Shape of bill in Charadriidae
(Pluvialis squatarola)

B I R D S

3. Totipalmate (all 4 toes included in web; Plate VII-1) ....
.... PELECANIFORMES, 14
Hind toe not included in web (or hind toe lacking) 4
4. Tomium serrate or lamellate; tip of upper mandible usually provided
with a nail; bill often wider than deep 13
Tomium neither serrate nor lamellate, and depth of bill at least as
great as its width; nail usually lacking 5
5. Bill hooked; nostrils opening through 1 or 2 tubes ....
.... PROCELLARIIFORMES, 6
Bill not as described above (if hooked, then without nostril tubes)
.... 8
6. With 2 parallel nostril tubes; total length more than 230 mm 7
Nostril tube single externally (Plate VII-2); length less than 230 mm
.... HYDROBATIDAE, LXXVIII
7. Hallux entirely lacking; total length more than 70 cm; part of culmen
separating nostril tubes .... DIOMEDEIDAE, LXXVI
Vestigial hallux present; total length less than 70 cm; nostril tubes
connected medially .... PROCELLARIIDAE, LXXVII
8. Tarsus greatly compressed; outer toe the longest one ....
.... GAVIIFORMES: GAVIIDAE, LXXIV
Tarsus not unusually compressed; middle toe at least as long as
outer one .... CHARADRIIFORMES (part), 9
9. Feathering extending on culmen to position of nostrils; hallux lacking
.... ALCIDAE, CVIII
Feathering not reaching nostrils; hallux usually present 10
10. Cere present; sheath of upper mandible in 3 parts (Plate VII-3)
.... STERCORARIIDAE, CV
Cere lacking; sheath of upper mandible in one part 11
11. Bill decidedly compressed unless very short; not upturned 12
Bill approximately round in cross section, very long and slender, and
more or less upcurved .... RECURVIROSTRIDAE (part), CIII
12. Bill moderately compressed and the upper mandible at least as long
as the lower .... LARIDAE, CVI
Bill extremely compressed, with the lower mandible much longer than
the upper .... RYNCHOPIDAE, CVII
13. Legs very long; bill abruptly and strongly decurved ....
...... PHOENICOPTERIFORMES:[1] PHOENICOPTERIDAE, LXXXVIII
Legs not relatively long; bill essentially straight, never abruptly decurved .... ANSERIFORMES: ANATIDAE, LXXXIX
14. Webs incised; tail forked .... FREGATIDAE, LXXXIV
Webs entire; tail rounded or pointed 15
15. Bill depressed; gular pouch immense .... PELECANIDAE, LXXX
Bill not depressed; gular pouch small or lacking 16

---

1. Included under CICONIIFORMES in A.O.U. *Check-list.*

16. Bill strongly hooked .... PHALACROCORACIDAE, LXXXII
    Bill not hooked 17
17. Head small; neck long and slender (head only twice as wide as tarsus)
    .... ANHINGIDAE, LXXXIII
    Size and proportions of head and neck normal (head more than
    twice as wide as tarsus) 18
18. Chin feathered; central rectrices usually elongate ....
    .... PHAETHONTIDAE, LXXIX
    Chin bare; central rectrices not greatly elongate ....
    .... SULIDAE, LXXXI
19. Toes lobed; claws and tarsi greatly flattened ....
    .... PODICIPEDIFORMES: PODICIPEDIDAE, LXXV
    Never with the above combination of characters; claws and tarsi not
    unusually flattened 20
20. Lores bare; legs and bill long; hind toe well developed and at about
    same level as other toes (more than half length of 2nd and 4th
    toes) .... CICONIIFORMES, 21
    Lores with feathers or bristles; hind toe slightly above level of other
    toes and often reduced (less than half length of 2nd and 4th toes
    except in Limpkin and Jaçana) 23
21. Bill long, straight, spearlike; middle claw pectinate (Plate VII-4);
    head mostly feathered .... ARDEIDAE, LXXXV
    Bill not spearlike; middle claw not pectinate 22
22. Bill somewhat compressed; without true nasal grooves; head bare (or
    sparsely feathered in immatures) .... CICONIIDAE, LXXXVI
    Bill cylindrical or spatulate; nasal groove extending nearly to tip of
    bill; only the face bare .... THRESKIORNITHIDAE, LXXXVII
23. Outermost primary usually about as wide as, but much shorter than,
    second; second primary of normal width; no spurs on wings .....
    .... GRUIFORMES, 24
    Outermost primary as long as second, or 3 outermost primaries less
    than half width of others, or with spurs on bend of wing ......
    .... CHARADRIIFORMES (part), 26
24. Outermost primary narrow except at tip; culmen about 80% of tarsal
    length; total length more than 50 cm .... ARAMIDAE, XCVII
    Outermost primary not narrowed; culmen about 50% of tarsal length,
    or total length less than 50 cm 25
25. Total length more than 50 cm; culmen at least 125 mm long; inner
    secondaries longer than primaries, drooping at the ends ....
    .... GRUIDAE, XCVI
    Total length less than 50 cm; culmen less than 100 mm long; inner
    secondaries not particularly elongate .... RALLIDAE, XCVIII
26. Tarsus more than 90 mm long ....
    .... RECURVIROSTRIDAE (part), CIII
    Tarsus less than 90 mm long 27

27. Toes lobed (Plate VII-5) .... PHALAROPODIDAE, CIV
    Toes not lobate 28
28. Hind toe long and its claw even longer; spur on bend of wing .....
    .... JAÇANIDAE, XCIX
    Hind toe reduced or absent; no spur on bend of wing 29
29. Bill greatly compressed, its depth at middle about 2.5 times its
    width .... HAEMATOPODIDAE, C
    Bill not unusually compressed 30
30. Bill long and slender; tip of culmen slightly expanded laterally and
    decurved; tarsus usually scutellate in front; 4 toes present (one
    exception) .... SCOLOPACIDAE, CII
    Bill of moderate length, with a swollen dertrum at tip (except pointed
    and with culmen flattened near base in one species; Plate VII-6);
    only 3 toes, or tarsus reticulate in front ...... CHARADRIIDAE, CI
31. With 3 toes permanently directed forward and their claws approxi-
    mately parallel; facial disk lacking 38
    With only 2 toes permanently directed forward (numbers 2 and 3),
    number 4 being entirely reversed (zygodactyl), partly reversed
    (semizygodactyl), or lacking; never with 3 claws parallel; if semi-
    zygodactyl, then a facial disk often present 32
32. Foot semizygodactyl, the claw of the fourth toe not parallel to any
    other claw (Fig. 8) 33
    Foot zygodactyl, the 2 front claws (or hind claws) essentially parallel
    to one another 36
33. Facial disk present, some of its feathers hiding base of bill; tarsus
    at least sparsely feathered .... STRIGIFORMES, 34
    No facial disk; base of bill exposed; tarsus bare 35
34. Middle claw pectinate (Plate VII-4); total length about 45 cm ....
    .... TYTONIDAE, CXII
    Middle claw not pectinate; length variable ...... STRIGIDAE, CXIII
35. Total length less than 50 cm; depth of bill at base about equal to
    its length .... PSITTACIFORMES: PSITTACIDAE, CX
    Total length more than 50 cm; depth of bill at base not more than
    half its length .... FALCONIFORMES: PANDIONIDAE, XCII
36. Bill chisel-like; tail feathers unusually stiffened, the vanes narrowed
    toward tip .... PICIFORMES: PICIDAE, CXIII
    Bill not chisel-shaped; rectrices not unusually stiffened and vanes
    usually not narrowed toward tip 37
37. Bill long, slender, and gently decurved, or greatly compressed ....
    .... CUCULIFORMES: CUCULIDAE, CXI
    Length of bill about equal to depth of bill at base; bill strongly
    decurved and hooked at tip ....
    .... PSITTACIFORMES: PSITTACIDAE, CX

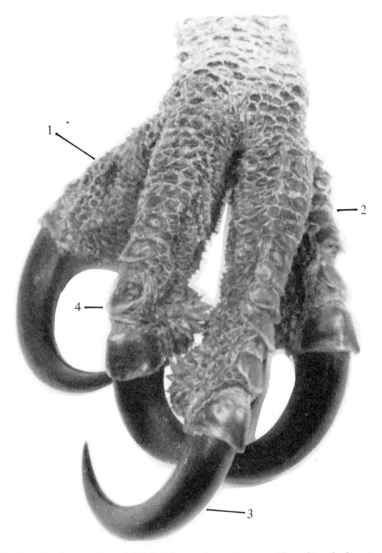

Fig. 8. Semizygodactyl (right) foot of an Osprey (*Pandion haliaetus*)

38. Bill strong, sharply hooked, and with cere at base; tip of bill pointed
    vertically or backward .... FALCONIFORMES (part), 39
    Bill never strongly hooked if cere present; tip of bill usually not
    pointing vertically and never backward 41

39. Nostrils perforate; head naked .... CATHARTIDAE, XC
    Nostrils imperforate; head feathered 40
40. Tarsus at least partly reticulate in front; upper mandible with a tooth
    near tip (weak in one species); nostrils circular (or slitlike and
    sloping downward and forward) (Plate VIII-1) ....
    .... FALCONIDAE, XCIII
    Never with the above combination of bill and tarsal characters .....
    .... ACCIPITRIDAE, XCI
41. Bill reduced and gape enlarged (wider than length of culmen); total
    length over 200 mm ....
    .... CAPRIMULGIFORMES: CAPRIMULGIDAE, CXIV
    Bill not as described above unless total length under 200 mm 42
42. Foot syndactyl (Plate VIII-2); head large; bill long, without dertrum
    or cere .... CORACIIFORMES: ALCEDINIDAE, CXVII
    At least one front toe free at base 43
43. Head relatively small; bill short, slender, and with dertrum and cere
    (bill narrower near middle; Plate VIII-3) ....
    .... COLUMBIFORMES: COLUMBIDAE, CIX
    Head of normal size; dertrum and cere lacking 44
44. Bill length about 10 times its depth, or bill reduced in all dimensions;
    rectrices 10; tarsus not scaly, appearing wrinkled in museum skins
    (may be partly feathered) .... APODIFORMES, 45
    Length of bill less than 10 times its depth; bill much reduced only
    in Hirundinidae; number of rectrices usually greater than 10; tarsus
    scaly 46
45. Bill short and wide; wing length more than 80 mm ....
    .... APODIDAE, CXV
    Bill long and cylindrical; wing length less than 80 mm ....
    .... TROCHILIDAE, CXVI
46. Hind toe relatively small and inserted above level of other toes, its
    claw somewhat smaller than middle claw; wing coverts not
    arranged in 3 distinct ranks .... GALLIFORMES, 47
    Hallux scarcely smaller than (often as large as) second or fourth toe
    and placed at same level; claw of hallux larger than middle claw
    (thicker near base or longer); wing coverts in 3 ranks ....
    .... PASSERIFORMES, 48
47. Head mostly naked; feathers of back subtruncate ....
    .... MELEAGRIDIDAE, XCV
    Head largely or entirely feathered; contour feathers mostly rounded
    or pointed .... PHASIANIDAE, XCIV
48. Tarsus not bilaminate, the line of junction of the scales being medial;
    bill broad, depressed at base, and hooked at tip; rictal bristles
    strong .... TYRANNIDAE, CXIX
    Tarsus bilaminate, the lines of scale junction on the medial and
    lateral sides; other characters variable 49

49. Inner secondaries fully as long as inner primaries; hind claw about twice length of longest front claw 50
    Inner secondaries not elongate; hind claw less than twice as long as longest front claw; tarsus laminiplantar (Plate VIII-4) 51
50. Posterior aspect of tarsus blunt and its scales scutellate; adults with black on face and throat .... ALAUDIDAE, CXX
    Tarsus laminiplantar (Plate VIII-4); no black on face and throat .... MOTACILLIDAE, CXXXI
51. Width of bill at junction of mandibular rami greater than its depth at same point 52
    Width of bill at junction of mandibular rami not greater than its depth at same point 53
52. Outermost primary definitely the longest one; tail notched or forked and without yellow terminal band .... HIRUNDINIDAE, CXXI
    Outermost primary not longer than second one; tail square-tipped and with yellowish terminal band ..... BOMBYCILLIDAE, CXXXII
53. Bill long, straight, and slightly compressed, the culmen not decurved at tip; total length less than 155 mm; upperparts chiefly gray ....
    .... SITTIDAE, CXXIV
    Not with the above combination of characteristics; if culmen long and straight, then at least slightly decurved at tip *and* total length usually more than 155 mm 54
    Culmen slightly decurved, at least near tip, and usually less than 4 times depth of bill at base; length of culmen usually exceeding that of tarsus by more than 10% 54
54. Bill long and straight, its length (about 25 mm) being almost 3 times its depth at base, but culmen slightly decurved near tip; length of tail less than half that of wing; outermost primary minute; some contour feathers of adult pointed (adults black with iridescence) .... STURNIDAE,[1] CXXXIV
    Bill usually not as described above; length of tail usually more than half that of wing; outermost primary more than 15% as long as second (or rudimentary and laterally displaced); contour feathers not pointed 55
55. Tarsus booted (most of anterior portion covered by a single scale, but often with one smaller scale at lower end) 56
    Tarsus scutellate (scales in front large or small, but never with a single scale covering most of its length) 58
56. Small greenish birds (total length less than 130 mm) ....
    .... SYLVIIDAE (part), CXXX
    Larger and without green in plumage 57

---

1. This description does not apply to the Hill Mynah (*Gracula religiosa*), which has recently nested at Boynton Beach.

1.
Toothed bill and
circular nostrils of
Falco peregrinus

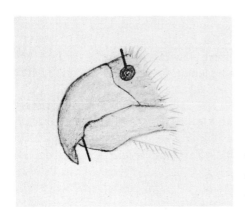

2. Syndactylous foot of
   Megaceryle alcyon

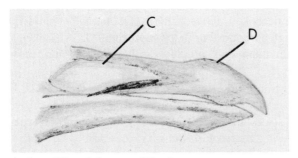

3. Cere (C) and dertrum (D) of
   Zenaida macroura

PLATE VIII

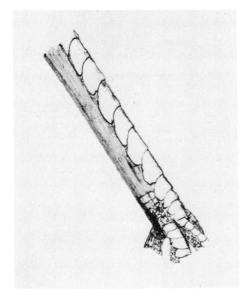

4.
Bilaminate, laminiplantar
tarsus of most passerines
(Corvus)

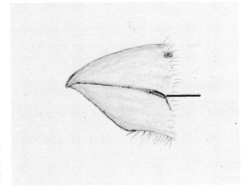

5.
Angled tomium
of finches
(Cardinalis cardinalis)

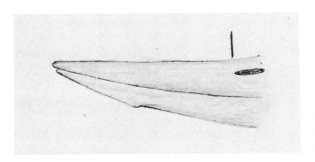

6.
Concave culmen of
Gavia stellata

B I R D S

57. Plumage mostly gray, but with red on vent and below eye of adults; crest present; length of tarsus less than 22 mm .... PYCNONOTIDAE, CXXV

Never gray ventrally; crest lacking; tarsus more than 22 mm long .... TURDIDAE, CXXIX

58. Upper mandible decidedly hooked and with a notch (sometimes a tooth also) near tip; tip of upper mandible pointing almost vertically 59

Upper mandible not definitely hooked and usually without a notice-able notch or a tooth 60

59. Plumage largely gray above, without green or olive; tooth of upper mandible well developed; total length more than 175 mm; tail rounded .... LANIIDAE, CXXXIII

Dorsal plumage partly greenish; tooth on upper mandible weak or lacking; total length less than 175 mm; tail square-tipped .... VIREONIDAE, CXXXV

60. Length of bill from anterior edge of nostril less than twice its depth at same point; exposed portion of outermost obvious primary more than half length of longest primary; rictal bristles present 61

Length of bill from anterior edge of nostril more than twice depth at same point, *or* exposed portion of outermost obvious primary less than half length of longest primary, *or* rictal bristles absent 63

61. Tomium not abruptly downcurved or angled near base, though some-times gently curving throughout .... THRAUPIDAE, CXXXIX

Tomium abruptly downcurved or angled near base (but base may be hidden by bristles or feathers; Plate VIII-5) 62

62. Outermost primary vestigial (often hidden by covert of next primary); with a light brownish area (speculum) near base of outer primaries, immediately distal to primary coverts, *or* nostrils more than 6 mm apart .... PLOCEIDAE, CXL

Outermost primary much more than 50% length of next primary; speculum, if present, white; nostrils less than 6 mm apart .... FRINGILLIDAE, CXLI

63. Bill long, slender, and conspicuously decurved, about as long as tarsus, and about 4 times as long as its depth at base; total length less than 150 mm; wings and tail not barred 64

Bill not conspicuously decurved; if length of bill more than 3 times its depth at base, then shorter than tarsus; total length often more than 150 mm 65

64. Mostly brownish above and white below; tail more than 45 mm in length; rectrices pointed .... CERTHIIDAE, CXXVI

Blackish above and mostly yellow below; tail less than 45 mm in length; rectrices rounded .... COEREBIDAE, CXXXVI

65. Small grayish birds less than 125 mm long; no crest or black hood present .... Sylviidae (in part), CXXX
Total length usually more than 125 mm; if small and gray, then head with black hood or crested 66
66. Exposed portion of outermost primary less than 70% as long as second (the measurement of each taken from the same point— that is, the base of the exposed portion of the medial side of the outermost primary); tail more or less rounded 67
Exposed portion of outermost primary at least 80% as long as the second primary; tail usually square or notched 70
67. Total length more than 175 mm 68
Total length less than 175 mm 69
68. Nostrils largely concealed by bristles .... Corvidae, CXXII
Nostrils fully exposed .... Mimidae, CXXVIII
69. Plumage gray, with crest or black on head; bill subconical ....
.... Paridae, CXXIII
Plumage brown, with blackish bars on wings and tail; bill long, slender, and somewhat decurved ..... Troglodytidae, CXXVII
70. Total length usually more than 150 mm; rictal bristles absent ......
.... Icteridae, CXXXVIII
Total length less than 150 mm, *or* weak rictal bristles present ....
.... Parulidae, CXXXVII

ORDER GAVIIFORMES: LXXIV. Family Gaviidae, Loons.

1. Culmen straight or convex 2
Culmen concave (Plate VIII-6) ....
.... Red-throated Loon (*Gavia stellata*), sp.351
2. Bill more than 45 mm long from nostril and 18 mm deep at same point .... Common Loon (*Gavia immer*), sp.349
Corresponding bill measurements less than 45 and 18 mm ....
.... Arctic Loon (*Gavia arctica*), sp.350

ORDER PODICIPEDIFORMES: LXXV. Family Podicipedidae, Grebes.

1. Total length more than 40 cm 2
Total length less than 40 cm 3
2. Total length more than 53 cm; length of bill from nostril more than 35 mm ....
.... Western Grebe (*Aechmophorus occidentalis*), sp.355
Total length less than 53 cm; length of bill from nostril less than 35 mm .... Red-necked Grebe (*Podiceps grisegena*), sp.352
3. Bill blunt at tip, its depth about half its length ....
.... Pied-billed Grebe (*Podilymbus podiceps*), sp.356
Bill sharp at tip, its depth much less than half its length 4

4. Wing length less than 100 mm; sight records only ....
    ....Least Grebe (*Podiceps dominicus*), sp.769H
    Wing length more than 100 mm 5
5. Depth of bill at base greater than its width; common ....
    .... Horned Grebe (*Podiceps auritus*), sp.353
    Depth of bill at base not greater than its width; rare ....
    .... Eared Grebe (*Podiceps nigricollis*), sp.354

ORDER PROCELLARIIFORMES: LXXVI. FAMILY DIOMEDEIDAE, Albatrosses.

1. Bill mostly yellowish; middle toe with claw about 150 mm long;
    rump dark ....
    .... Wandering Albatross (*Diomedea exulans*), sp.769H
    Bill mostly blackish, with culmen yellowish; middle toe with claw
    about 105 mm long; rump white ....
    ....Yellow-nosed Albatross (*Diomedea chlororhynchos*), sp.769H

ORDER PROCELLARIIFORMES: LXXVII. FAMILY PROCELLARIIDAE,
Shearwaters and Petrels.[1]

1. Underparts dark ...... Sooty Shearwater (*Puffinus griseus*), sp.359
    Underparts mostly white 2
2. Total length 45 cm or more; bill from nostril tube more than 20 mm
    long 3
    Total length less than 45 cm; bill from nostril tube less than 20
    mm long 4
3. Bill dark, its depth anterior to nostril tubes not more than 10 mm;
    top of head dark, abruptly contrasting below eye with white throat
    .... Greater Shearwater (*Puffinus gravis*), sp.358
    Bill yellowish, its depth anterior to nostrils more than 10 mm; ashy
    crown gradually shading to white throat ....
    .... Cory's Shearwater (*Puffinus diomedea*), sp.357
4. Upper tail coverts largely white; forehead white, contrasting with
    dark crown; length of culmen less than 3 times depth of bill at
    base ...... Black-capped Petrel (*Pterodroma hasitata*), sp.362
    Upper tail coverts and entire top of head dark; length of culmen at
    least 3 times depth of bill at base 5
5. Tail strongly graduated, its length more than one-third that of wing
    .... Audubon's Shearwater (*Puffinus lherminieri*), sp.361
    Tail only slightly rounded, its length less than one-third that of
    wing .... Manx Shearwater (*Puffinus puffinus*), sp.360

---

1. A "probable" Bulwer's Petrel (*Bulweria bulwerii*) was seen between Key
West and the Dry Tortugas in May, 1969 (*Wilson Bull.*, 8:198).

ORDER PROCELLARIIFORMES: LXXVIII. FAMILY HYDROBATIDAE, Storm-Petrels.

1. Tail usually more than 78 mm long, forked for about 12 mm .....
.... Leach's Storm-Petrel (*Oceanodroma leucorhoa*), sp.363
Tail usually less than 78 mm long and not forked 2
2. Tarsus more than 28 mm long; upper tail coverts not white with dark tips 3
Tarsus less than 28 mm long; upper tail coverts white with abruptly contrasting dark tips ....
.... Harcourt's Storm-Petrel (*Oceanodroma castro*), sp.364
3. Length of wing more than 160 mm; length of culmen at least 13 mm; usually with some white on belly ....
..... White-bellied Storm-Petrel (*Fregetta grallaria*)[1] sp.769H
Length of wing less than 160 mm; length of culmen less than 13 mm; underparts dark except for flanks and under tail coverts .... Wilson's Storm-Petrel (*Oceanites oceanicus*), sp.365

ORDER PELECANIFORMES: LXXIX. FAMILY PHAETHONTIDAE, Tropic-birds.

1. Bill from nostril more than 35 mm long; shafts of central rectrices partly white above; bill reddish in life ....
.... Red-billed Tropicbird (*Phaethon aethereus*), sp.769H
Bill from nostril less than 35 mm long; shafts of central rectrices entirely dark dorsally; bill yellow or orange ....
.... White-tailed Tropicbird (*Phaethon lepturus*), sp.366

ORDER PELECANIFORMES: LXXX. FAMILY PELECANIDAE, Pelicans.

1. Total length usually more than 140 cm; plumage essentially white, with black primaries ....
.... White Pelican (*Pelecanus erythrorhynchos*), sp.367
Total length usually less than 140 cm; plumage essentially gray, with white on head and neck of adults ....
.... Brown Pelican (*Pelecanus occidentalis*), sp.368

ORDER PELECANIFORMES: LXXXI. FAMILY SULIDAE, Boobies and Gannets.

1. Total length usually more than 86 cm, with a narrow strip of bare skin running from lower mandible along center of otherwise feathered chin .... Gannet (*Morus bassanus*), sp.372
Total length not more than 86 cm; entire lower mandible and chin bare to a point below or behind eye 2

---

1. Record formerly referred to Black-bellied Storm-Petrel (*Fregetta tropica*), but see Palmer (1962, pp. 223–224).

2. Culmen less than 90 mm long; tail more than half as long as wing .... Red-footed Booby (*Sula sula*), sp.371
   Culmen at least 90 mm long; tail less than half length of wing 3
3. Breast and belly white; upperparts at least partly white .... Blue-faced Booby (*Sula dactylatra*), sp.369
   Dark breast, contrasting with white (or light) belly; upperparts entirely brownish ..... Brown Booby (*Sula leucogaster*), sp.370

ORDER PELECANIFORMES: LXXXII. FAMILY PHALACROCORACIDAE, Cormorants.

1. Total length 86 cm or more; rectrices 14 in number; midline of gular sac partly feathered .... Great Cormorant (*Phalacrocorax carbo*), sp.373
   Total length less than 86 cm; rectrices 12; no feathers extending onto gular sac .... Double-crested Cormorant (*Phalacrocorax auritus*), sp.374

ORDER PELECANIFORMES: LXXXIII. FAMILY ANHINGIDAE, Darters. One species in Florida .... Anhinga (*Anhinga anhinga*), sp.375

ORDER PELECANIFORMES: LXXXIV. FAMILY FREGATIDAE, Frigatebirds. One species in Florida .... Magnificent Frigatebird (*Fregata magnificens*), sp.376

ORDER CICONIIFORMES: LXXXV. FAMILY ARDEIDAE, Herons, Egrets, and Bitterns.

1. Plumage essentially white 2
   Plumage various, white constituting less than 50% of surface 7
2. Tarsus and toes greenish or yellowish in life; total length more than 114 cm ...... Great White Heron (*Ardea occidentalis*), sp.377
   Tarsus and toes black in life, *or* total length less than 90 cm 3
3. Total length at least 90 cm ..... Great Egret (*Egretta alba*),[1] sp.383
   Total length less than 90 cm 4
4. Wing length more than 28 cm; basal half of bill often lighter than distal half (flesh-colored in life) .... Reddish Egret (*Egretta rufescens*),[1] sp.382
   Wing less than 28 cm long; bill not flesh-colored at base, though sometimes light blue 5
5. Middle toe (with claw) longer than bill; bill usually yellowish in life .... Cattle Egret (*Bubulcus ibis*), sp.380
   Middle toe with claw shorter than bill; bill never yellow (except in nestlings) 6
6. Plumage entirely white; bill black; lores and toes yellow; tarsi black or dark green .... Snowy Egret (*Egretta thula*),[1] sp.384

Primaries edged with pale slate gray; irregular dark slaty blotches
  sometimes present; tarsi, toes, and bill gray green.............
  ....Little Blue Heron, immature (*Egretta caerulea*),[1] sp.381
7. Total length 105 to 130 cm; head white (or light) with dark
  plumes .... Great Blue Heron (*Ardea herodias*), sp.378
  Total length less than 105 cm 8
8. Depth of bill at base at least 22% length of culmen 9
  Depth of bill at base less than 22% length of culmen 11
9. Depth of bill at base less than 25% length of culmen; rectrices 10;
  fourth toe at least 10% longer than second ....
  ....American Bittern (*Botaurus lentiginosus*), sp.389
  Depth of bill at base at least 25% length of culmen; rectrices 12;
  second and fourth toes about same length 10
10. Unfeathered portion of lower tibia at least 38 mm long ....
  ..... Yellow-crowned Night Heron (*Nyctanassa violacea*), sp.387
  Unfeathered lower tibia less than 38 mm long ....
  .... Black-crowned Night Heron (*Nycticorax nycticorax*), sp.386
11. Wing length more than 280 mm; tarsal length more than 115 mm ..
  .... Reddish Egret (*Egretta rufescens*),[1] sp.382
  Wing length less than 280 mm; tarsal length less than 115 mm ....
  12
12. Length of tarsus more than 58 mm 13
  Length of tarsus less than 58 mm 14
13. Plumage entirely dark, or with scattered patches of white ....
  .... Little Blue Heron, adult (*Egretta caerulea*),[1] sp.381
  Plumage chiefly dark above and light underneath ....
  .... Louisiana Heron (*Egretta tricolor*),[1] sp.385
14. Total length more than 38 cm ....
  .... Green Heron (*Butorides virescens*), sp.379
  Total length less than 38 cm ....
  .... Least Bittern (*Ixobrychus exilis*), sp.388

ORDER CICONIIFORMES: LXXXVI. FAMILY CICONIIDAE, Storks. One species in Florida .... Wood Stork (*Mycteria americana*), sp.390

ORDER CICONIIFORMES: LXXXVII. FAMILY THRESKIORNITHIDAE, Ibises
and Spoonbills.

1. Bill cylindrical and decurved 2
  Bill spatulate at tip, not decurved ....
  .... Roseate Spoonbill (*Ajaia ajaja*), sp.395

---

1. These seven species are listed under four other generic names in the A.O.U.
Check-list, but see Dickerman and Parkes, 1968.

2. Belly white or scarlet; claws noticeably curved 3
   Belly not white or scarlet; claws almost straight 4
3. Total length 71 cm or more; plumage of adults scarlet with black wing tips; bill, head, and neck of immatures uniformly dark ....
   .... Scarlet Ibis (*Eudocimus ruber*), sp.394
   Total length less than 71 cm; plumage of adults mostly white with black wing tips; immatures with bill partly reddish and head and neck streaked .... White Ibis (*Eudocimus albus*), sp.393
4. Feathers behind eye of adult white; lores (bare) reddish..........
   .... White-faced Ibis (*Plegadis chihi*),[1] sp.392
   Feathers behind eye of adult dark; lores (bare) black or greenish
   .... Glossy Ibis (*Plegadis falcinellus*),[1] sp.391

ORDER PHOENICOPTERIFORMES: LXXXVIII. FAMILY PHOENICOPTERIDAE, Flamingoes. One species in Florida ....
.... American Flamingo (*Phoenicopterus ruber*), sp. 396

ORDER ANSERIFORMES: LXXXIX. FAMILY ANATIDAE, Swans, Geese, and Ducks.

1. Total length more than 112 cm; length of tarsus about 100 mm ....
   2
   Total length less than 112 cm; length of tarsus much less than 100 mm 3
2. Tail rounded; length of culmen more than 95 mm; feral, but rare in Florida .... Whistling Swan (*Olor columbianus*), sp.397
   Tail cuneate; length of culmen less than 95 mm; domestic ....
   ....Mute Swan (*Cygnus olor*), sp.769H
3. Greatest depth of bill at least half length of culmen and never less than 23 mm 4
   Greatest depth of bill less than half length of culmen, *or* less than 23 mm 12
4. Total length usually 585 mm or more; tarsus longer than middle toe without claw 5
   Total length 585 mm or less; tarsus shorter than middle toe without claw 9
5. Head and neck largely black, but with some white laterally 6
   Head and neck white (sometimes stained with rust) 7
   Head and neck brownish gray 8
6. Large white cheek patch present; wings and back brownish ......
   .... Canada Goose (*Branta canadensis*), sp.398
   White patch smaller and on side of neck; wings and back fuscous
   .... Brant (*Branta bernicla*), sp.399

1. Young of the Glossy and White-faced Ibis are probably indistinguishable, and perhaps the 2 populations should be considered conspecific.

7. Body white or ashy gray .. Snow Goose (*Chen hyperborea*), sp.401
   Body dark .... Blue Goose, adult (*Chen caerulescens*),[1] sp.402
8. Tomia of upper and lower mandibles separated at middle of bill by
   a space of about 5 mm; legs pink in life ....
   .... Blue Goose, immature (*Chen caerulescens*),[1] sp.402
   Tomia of the 2 mandibles almost flush; legs yellow or orange in life
   .... White-fronted Goose (*Anser albifrons*), sp.400
9. With a white patch (speculum) on wing 10
   No white patch on wing 11
10. Plumage chiefly dark ....
    .... White-winged Scoter (*Melanitta deglandi*), sp.428
    Considerable white in plumage, at least ventrally ....
    ....Common Goldeneye (*Bucephala clangula*), sp.422
11. Rectrices 14; 2 white or light patches on head ....
    .... Surf Scoter (*Melanitta perspicillata*), sp.429
    Rectrices 16; no light patches on head of male, but one on female
    .... Black Scoter (*Melanitta nigra*), sp.430
12. Tarsus at least 50 mm long 45
    Tarsus less than 50 mm long 13
13. Edges of mandibles serrate 14
    Edges of mandibles lamellate (Plate IX-1) 16
14. Total length less than 51 cm; serrations on bill vertical ....
    .... Hooded Merganser (*Lophodytes cucullatus*), sp.433
    Total length usually 51 cm or more; serrations on bill projecting
    backward 15
15. Feathering on culmen extending anterior to that on sides of upper
    mandible ...... Common Merganser (*Mergus merganser*), sp.434
    Feathering on culmen not extending beyond that on sides of upper
    mandible .... Red-breasted Merganser (*Mergus serrator*), sp.435
16. Rectrices graduated, stiffened, and the central ones more than half
    length of wing 17
    Rectrices scarcely, or less than, half length of wing 18
17. Nail of bill reduced and curving under the upper mandible (Plate
    IX-2); outer toe at least as long as middle toe ....
    .... Ruddy Duck (*Oxyura jamaicensis*), sp.431
    Nail well developed; outer toe shorter than middle toe ....
    .... Masked Duck (*Oxyura dominica*), sp.432
18. Hind toe with a well-developed lobe, making its greatest width more
    than 6 mm 19
    Lobe on hind toe poorly developed; greatest width of hind toe not
    more than 6 mm 28

---

1. Now regarded as a color phase of the Snow Goose.

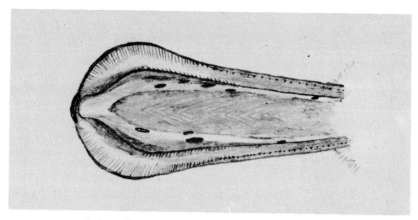

1. Lamellate bill of most ducks *(Anas clypeata)*

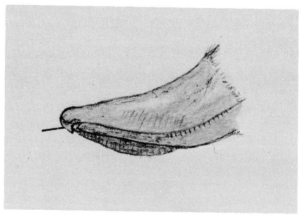

2. Retrorse nail of *Oxyura jamaicensis*

PLATE   IX

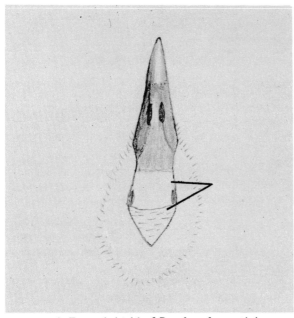

3. Frontal shield of *Porphyrula martinica*

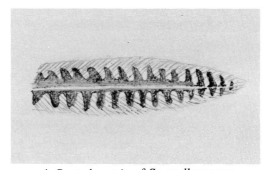

4. Central rectrix of *Sturnella magna*

B I R D S

19. Culmen definitely longer than tarsus 20
    Culmen about same length as, or shorter than, tarsus 24
20. Culmen more than 50 mm long ....
    .... Canvasback (*Aythya valisineria*), sp.419
    Culmen not more than 50 mm long 21
21. With a white or light band across the upper mandible near its tip,
    sometimes indistinct 22
    No light band across tip of upper mandible 23
22. Wing more than 205 mm long; band across upper mandible usually
    indistinct .... Redhead (*Aythya americana*), sp.417
    Wing not more than 205 mm long; band across upper mandible
    distinct .... Ring-necked Duck (*Aythya collaris*), sp.418
23. Least width of upper mandible not less than 21 mm; width of nail
    not less than 7.0 mm ....
    .... Greater Scaup (*Aythya marila*), sp.420
    Least width of upper mandible not more than 20 mm; width of nail
    not more than 6.5 mm ....
    .... Lesser Scaup (*Aythya affinis*), sp.421
24. Central rectrices elongate and pointed (at least 7 mm longer than
    any others); under tail coverts white ....
    .... Oldsquaw (*Clangula hyemalis*), sp.424
    Central rectrices scarcely longer than adjacent ones and not pointed;
    under tail coverts not pure white 25
25. Wing length more than 250 mm 26
    Wing length less than 225 mm 27
26. Feathering on culmen (but not on side of bill) extending to about
    level of nostrils ...... King Eider (*Somateria spectabilis*), sp.427
    Feathering on side of bill (but not on culmen) extending to about
    level of nostrils ....
    .... Common Eider (*Somateria mollissima*), sp.426
27. Breast dark; rectrices 14 ....
    .... Harlequin Duck (*Histrionicus histrionicus*), sp.425
    Breast whitish; rectrices 16 ....
    .... Bufflehead (*Bucephala albeola*), sp.423
28. Ventral side of head and neck white; base of bill red; speculum
    green with buffy ochraceous border; tail buffy ....
    .... Bahama Duck (*Anas bahamensis*), sp.409
    Color pattern never as described above 29
29. Length of culmen at least 46 mm 30
    Length of culmen less than 46 mm 37
30. Greatest width of bill more than 25 mm (Plate IX-1) ....
    .... Northern Shoveler (*Anas clypeata*), sp.415
    Greatest width of bill less than 25 mm 31

31. Secondaries with considerable white 32
    Secondaries narrowly tipped with white or not at all 36
32. Lamellae crowded (about 1 per mm on upper mandible) ....
    .... Gadwall (*Anas strepera*), sp.407
    Lamellae less crowded (less than 1 per mm on upper mandible)
    33
33. Head more or less solidly colored with green or chocolate brown ..
    34
    Head brownish streaked 35
34. Head metallic green; central rectrices recurved ....
    .... Mallard, ♂ (*Anas platyrhynchos*), sp.404
    Head rich brown; central rectrices very long and straight ....
    .... Pintail, ♂ (*Anas acuta*), sp.408
35. Belly white .... Pintail, ♀ (*Anas acuta*),[1] sp.408
    Belly mottled .... Mallard, ♀ (*Anas platyrhynchos*),[1] sp.404
36. Longer feathers of back with longitudinal buffy bands in vanes; chin
    evenly buffy .... Mottled Duck (*Anas fulvigula*), sp.406
    Only the edges of the back feathers buffy; chin streaked with
    dusky .... Black Duck (*Anas rubripes*), sp.405
37. Total length 432 mm or more; wing more than 200 mm long
    38
    Total length 432 mm or less; wing less than 200 mm long 43
38. Greatest width of bill less than 45% length of culmen ....
    .... Gadwall (*Anas strepera*), sp.407
    Greatest width of bill at least 45% length of culmen 39
39. Rectrices rounded .... Wood Duck (*Aix sponsa*), sp.416
    Rectrices pointed 40
40. With a white or buffy patch on top of head and green or reddish
    on sides of head 41
    Head brownish streaked 42
41. With greenish on sides of head ....
    .... American Wigeon, ♂ (*Anas americana*), sp.414
    With reddish on sides of head ....
    .... European Wigeon, ♂ (*Anas penelope*), sp.413
42. Head and neck with fuscous streaks on whitish background ......
    .... American Wigeon, ♀ (*Anas americana*), sp.414
    Dark streaks of head and neck on cinnamon or buffy background ..
    .... European Wigeon, ♀ (*Anas penelope*), sp.413
43. With green but no light blue in wing ....
    .... Green-winged Teal (*Anas crecca*), sp.410
    With a large chalky blue patch on wing 44

---

1. Immature males and those in or near eclipse plumage may key out here.

44. Greatest width of bill near its tip; male with rich maroon on head
and body; no white on head; length of bill from nostril usually
more than 34 mm ....
.... Cinnamon Teal (*Anas cyanoptera*),[1] sp.412
Base of bill at least as wide as any portion; male without maroon,
but with a white crescent in front of eye; length of bill from
nostril usually less than 34 mm ....
.... Blue-winged Teal (*Anas discors*),[1] sp.411
45. Underparts buffy brown; statewide ....
.... Fulvous Tree Duck (*Dendrocygna bicolor*), sp.403
Belly mostly black; Miami area ....
.. Black-bellied Tree Duck (*Dendrocygna autumnalis*), sp.769H

ORDER FALCONIFORMES: XC. FAMILY CATHARTIDAE, New World Vultures.

1. Wing length 455 mm or more; nostrils large and broad (about 6 mm
wide) .... Turkey Vulture (*Cathartes aura*), sp.436
Wing length less than 455 mm; nostrils narrow (about 3 mm
wide) .... Black Vulture (*Coragyps atratus*), sp.437

ORDER FALCONIFORMES: XCI. FAMILY ACCIPITRIDAE, Kites, Hawks, and
Eagles.[2]

1. Wing more than 455 mm long 2
Wing not more than 455 mm long 3
2. Entire tarsus feathered...Golden Eagle (*Aquila chrysaetos*), sp.451
Lower tarsus bare ....
.... Bald Eagle (*Haliaeetus leucocephalus*), sp.452
3. Tarsus less than 44 mm long 4
Tarsus more than 44 mm long 6
4. Tail very long and deeply forked ....
.... Swallow-tailed Kite (*Elanoides forficatus*), sp.439
Tail moderately long and not forked 5
5. Underparts white ...... White-tailed Kite (*Elanus caeruleus*), sp.438
Underparts dark or streaked ....
.... Mississippi Kite (*Ictinia misisippiensis*), sp.440
6. Wing length at least 65% of total length 7
Wing length less than 65% of total length 12
7. Culmen greatly decurved, its length (from cere) at least 25 mm, and
more than 50% of tarsal length ....
.... Everglade Kite (*Rostrhamus sociabilis*), sp.441
Culmen from cere less than 40% length of tarsus 8

---

1. Females of these 2 species are very difficult to distinguish.

2. Harris' Hawk (*Parabuteo unicinctus*) is omitted in the belief that Florida
records represent escaped individuals.

8. Tarsus feathered to toes ....
...............Rough-legged Hawk (*Buteo lagopus*), sp.450
Tarsus partly bare 9
9. Length of tarsus less than 20% that of wing ....
.... Swainson's Hawk (*Buteo swainsoni*), sp.448
Length of tarsus at least 20% that of wing 10
10. Four outermost primaries indented abruptly about 100 mm from
tip; wing length usually more than 330 mm ....
.... Red-tailed Hawk (*Buteo jamaicensis*), sp.445
Only the 3 outermost primaries emarginate; wing length less than
330 mm 11
11. No white feathers at base of bill; wing length usually less than 290
mm; less than one-third of tarsus feathered ....
.............Broad-winged Hawk (*Buteo platypterus*), sp.447
Usually with white between eye and base of bill; wing length at
least 290 mm; at least upper third of tarsus feathered. ....
.... Short-tailed Hawk (*Buteo brachyurus*), sp.449
12. Upper tail coverts entirely white ....
.... Marsh Hawk (*Circus cyaneus*), sp.453
Upper tail coverts not entirely white 13
13. Tail at least 70% of wing length, so that folded wing usually does not
reach halfway down length of tail 14
Tail less than 70% of wing length; tip of folded wing usually extend-
ing beyond midpoint of tail ....
.... Red-shouldered Hawk (*Buteo lineatus*), sp.446
14. Tail tip rounded, the central rectrices at least 20 mm longer than the
outer ones 15
Tail tip square (all rectrices about same length) ....
.............Sharp-shinned Hawk (*Accipiter striatus*), sp.443
15. Wing length at least 300 mm ....
.......................Goshawk (*Accipiter gentilis*), sp.442
Wing length less than 300 mm ....
.... Cooper's Hawk (*Accipiter cooperii*), sp.444

ORDER FALCONIFORMES: XCII. FAMILY PANDIONIDAE, Ospreys. One
species .... Osprey (*Pandion haliaetus*), sp.454

ORDER FALCONIFORMES: XCIII. FAMILY FALCONIDAE, Falcons and Caracaras.

1. Nostrils linear, slanted upward and backward; upper mandible not
distinctly toothed; total length more than 50 cm; tail with 12 or
more dark bars .... Caracara (*Caracara cheriway*), sp.455
Nostrils circular; upper mandible with a distinct tooth near its tip;
total length not more than 50 cm; tail with less than 10 dark
bars 2

2. Total length more than 355 mm; wing length more than 250 mm
   .... Peregrine Falcon (*Falco peregrinus*), sp.456
   Total length less than 355 mm; wing length less than 250 mm 3
3. Sides of head longitudinally streaked; back brownish or bluish .....
   .... Merlin (*Falco columbarius*), sp.457
   Sides of head with vertical black bars; back reddish ....
   .... American Kestrel (*Falco sparverius*), sp.458

ORDER GALLIFORMES: XCIV. FAMILY PHASIANIDAE, Pheasants and Quail.

1. Tail long and graduated, making total length more than 50 cm ....
   .... Ring-necked Pheasant (*Phasianus colchicus*), sp.769H
   Tail rather short; total length less than 50 cm 2
2. Tarsus spurred; total length more than 28 cm ....
   .... Black Francolin (*Francolinus francolinus*), sp.769H
   Spurs lacking; total length less than 28 cm ....
   .... Bobwhite (*Colinus virginianus*), sp.459

ORDER GALLIFORMES: XCV. FAMILY MELEAGRIDIDAE, Turkeys. One
species in Florida .... Turkey (*Meleagris gallopavo*), sp.460

ORDER GRUIFORMES: XCVI. FAMILY GRUIDAE, Cranes.

1. Plumage largely white; depth of bill at nostrils more than 20 mm;
   no recent records ...... Whooping Crane (*Grus americana*), sp.461
   Plumage largely gray, only the cheeks sometimes whitish; depth of
   bill at nostrils less than 20 mm; widespread ....
   .... Sandhill Crane (*Grus canadensis*), sp.462

ORDER GRUIFORMES: XCVII. FAMILY ARAMIDAE, Limpkins. One species
.... Limpkin (*Aramus guarauna*), sp.463

ORDER GRUIFORMES: XCVIII. FAMILY RALLIDAE, Rails, Gallinules, and
Coots.

1. Frontal shield present (Plate IX-3) 2
   Frontal shield lacking 4
2. Toes lobed; back blackish ....
   .... American Coot (*Fulica americana*),[1] sp.472
   Toes not lobed; back brownish or greenish 3
3. Back brownish; hind claw less than 14 mm long ....
   ....Common Gallinule (*Gallinula chloropus*), sp.471
   Back greenish; hind claw more than 14 mm long ....
   .... Purple Gallinule (*Porphyrula martinica*), sp.470

1. Records of the Caribbean Coot (*Fulica caribaea*) occurred and a specimen was taken on the lower east coast in the spring of 1974 (Bolte, *American
Birds,* 28:730).

4. Culmen at least as long as tarsus 5
Culmen much shorter than tarsus 7

5. Total length more than 280 mm; bill from nostril more than 30 mm
long 6
Total length less than 280 mm; bill from nostril less than 30 mm
long .... Virginia Rail (*Rallus limicola*), sp.466

6. More gray than brown on sides of head; bend of wing plain brown
.... Clapper Rail (*Rallus longirostris*),[1] sp.465
Brown predominating on sides of head; bend of wing reddish brown
.... King Rail (*Rallus elegans*),[1] sp.464

7. Wing more than 100 mm long 9
Wing less than 100 mm long 8

8. With a patch of white in the secondaries; underparts buffy brown
.... Yellow Rail (*Coturnicops noveboracensis*), sp.468
No white in secondaries; underparts smoky gray ....
.... Black Rail (*Laterallus jamaicensis*), sp.469

9. Total length less than 300 mm; back spotted with black and white
.... Sora (*Porzana carolina*), sp.467
Total length more than 300 mm; back unspotted ....
.... Gray-necked Wood Rail (*Aramides cajanea*), sp.769H

ORDER CHARADRIIFORMES: XCIX. FAMILY JAÇANIDAE, Jaçanas. One
species in Florida .... Jaçana (*Jacana spinosa*), sp.769H

ORDER CHARADRIIFORMES: C. FAMILY HAEMATOPODIDAE, Oystercatchers.
One species in Florida ....
.... American Oystercatcher (*Haematopus palliatus*), sp.473

ORDER CHARADRIIFORMES: CI. FAMILY CHARADRIIDAE, Plovers and Turnstones.

1. Culmen slightly depressed at middle of bill, with a swollen dertrum
beyond (see Plate VII-1); scales on front of tarsus reticulate
(Plate VII-6) 2
Bill pointed; dertrum not apparent; scales on front of tarsus largely
scutellate (see Plate VIII-4) ....
.... Ruddy Turnstone (*Arenaria interpres*),[2] sp. 483

2. With a vestigial hallux 3
Hallux lacking 5

---

1. These two rails interbreed frequently and by some are considered conspecific.
2. This species and the Surfbird have been transferred to the Family Scolopacidae.

3. Long crest present; back greenish; one sight record ....
.... Eurasian Lapwing (*Vanellus vanellus*),[1] sp.769H
Crest lacking; back not greenish 4
4. Length of culmen at least 25 mm, and that of tarsus at least 40 mm
.... Black-bellied Plover (*Pluvialis squatarola*), sp.481
Length of culmen and tarsus, respectively, less than 25 and 40 mm
.... Surfbird (*Aphriza virgata*), sp.482
5. Bill from nostril at least 12 mm long 6
Bill from nostril less than 12 mm long 9
6. With one or 2 dark bands across breast and throat 7
Black of underparts (if present) not forming a band across breast or
throat 8
7. With 2 black bands across breast and throat (only one in downy
young) .... Killdeer (*Charadrius vociferus*), sp.478
With one black or dark band across throat ....
.... Wilson's Plover (*Charadrius wilsonia*), sp.477
8. Middle toe without claw more than 20 mm long ....
.... American Golden Plover (*Pluvialis dominica*), sp.480
Middle toe without claw not more than 20 mm long ....
.... Mountain Plover (*Charadrius montana*), sp.479
9. Culmen about as long as middle toe without claw; back light ......
.... Snowy Plover (*Charadrius alexandrinus*), sp.476
Culmen much shorter than middle toe without claw; back gray brown
or darker 10
10. Back medium to dark brown ....
.... Semipalmated Plover (*Charadrius semipalmatus*), sp.474
Back light brownish gray ....
.... Piping Plover (*Charadrius melodus*), sp.475

ORDER CHARADRIIFORMES: CII. FAMILY SCOLOPACIDAE, Sandpipers and
allies.

1. Bill from nostril at least 40 mm long 2
Bill from nostril less than 40 mm long 12
2. Bill strongly decurved 3
Bill not markedly decurved 5
3. Bill from nostril less than 92 mm long; underparts partly whitish ..
.... 4
Bill from nostril more than 92 mm long; underparts entirely cinnamon or buffy ....
.... Long-billed Curlew (*Numenius americanus*), sp.486

---

1. Records of the Southern Lapwing (*Belonopterus chilensis*) in Florida are
thought to be based on escaped birds.

4. Wing more than 220 mm long; bill from nostril more than 65 mm long; primaries barred ....
.... Whimbrel (*Numenius phaeopus*), sp.487
Wing less than 220 mm long; bill from nostril less than 65 mm long; primaries not barred ....
.... Eskimo Curlew (*Numenius borealis*), sp.769H
5. Bill from nostril more than 70 mm long 6
Bill from nostril less than 70 mm long 7
6. Tarsal length less than 65 mm; tail black, with a white base and tip .... Hudsonian Godwit (*Limosa haemastica*), sp.511
Tarsus more than 65 mm long; tail barred ....
.... Marbled Godwit (*Limosa fedoa*),[1] sp.510
7. Length of wing 175 mm or more 8
Wing length less than 175 mm 9
8. Rectrices barred with black; wings spotted or barred with white ..
.... Greater Yellowlegs (*Tringa melanoleuca*), sp.491
Rectrices not barred; wings with large white patch ....
.... Willet (*Catoptrophorus semipalmatus*), sp.493
9. Three outermost primaries less than 50% as wide as other primaries
.... American Woodcock (*Philohela minor*), sp.484
All primaries about same width 10
10. Upper rump black with narrow white bands ....
.... Common Snipe (*Capella gallinago*), sp.485
Upper rump white 11
11. Tarsal length usually less than 27% of wing length; total length usually over 275 mm; bill from nostril often more than 60 mm. long[2] ....
.... Long-billed Dowitcher (*Limnodromus scolopaceus*), sp.509
Tarsus usually more than 27% of wing length; total length usually less than 275 mm; bill from nostril often less than 60 mm long[2] ....
.... Short-billed Dowitcher (*Limnodromus griseus*), sp.508
12. Wing length 150 mm or greater 13
Wing length less than 150 mm 16
13. Tarsus shorter than 44 mm ....
.... Red Knot (*Calidris canutus*), sp.494
Tarsus 44 mm or longer 14

---

1. A Bar-tailed Godwit, *Limosa lapponica* (sp. 871), appeared at Cape Canaveral in October, 1970, remained for several months, and was photographed many times. It differs in having the tarsus about 20 to 25% of wing length, versus 30% or more in *L. fedoa*.
2. Dowitchers for which this measurement falls between 50 and 63 mm cannot be safely identified to species without expert assistance, though there is little overlap between *male* Short-billed and *female* Long-billed.

14. Outer primaries barred black and white ....
    .... Upland Sandpiper (*Bartramia longicauda*), sp.488
    Outer primaries uniformly dark 15
15. Wing more than 5 times length of bill from nostril; tail· not barred
    .... Ruff (*Philomachus pugnax*), sp.506
    Wing less than 5 times length of bill from nostril; tail barred ......
    .... Lesser Yellowlegs (*Tringa flavipes*), sp.492
16. Length of bill from nostril at least 30 mm, *or* culmen markedly
    decurved 17
    Bill from nostril less than 30 mm long and not more than slightly
    decurved 19
17. Bill markedly decurved and its length from nostril not more than
    30 mm 18
    Bill essentially straight (culmen slightly drooped near tip) and its
    length from nostril more than 30 mm ....
    .... Stilt Sandpiper (*Micropalama himantopus*), sp.507
18. Rump entirely white ....
    .... Curlew Sandpiper (*Calidris ferruginea*), sp.769H
    Middle of rump dark .... Dunlin (*Calidris alpina*), sp.501
19. Tarsus about twice length of bill from nostril ....
    ....Buff-breasted Sandpiper (*Tryngites subruficollis*), sp.505
    Tarsus less than twice length of bill from nostril 20
20. Hind toe absent .... Sanderling (*Calidris alba*), sp.504
    Hind toe present 21
21. Tail barred black and white ....
    .... Solitary Sandpiper (*Tringa solitaria*), sp.490
    Tail barred brown and white, or not barred 22
22. Third and fourth toes webbed at base 23
    Third and fourth toes not webbed 24
23. Bill from nostril not more than 18% length of wing ....
    .... Semipalmated Sandpiper (*Calidris pusillus*), sp.502
    Bill from nostril at least 18% length of wing ....
    .... Western Sandpiper (*Calidris mauri*), sp.503
24. Wing less than 110 mm long 25
    Wing more than 110 mm long 26
25. Back marked with rounded dark spots ....
    .... Least Sandpiper (*Calidris minutilla*), sp.500
    Dark markings on upperparts in form of streaks or crossbars ......
    .... Spotted Sandpiper (*Actitis macularia*), sp.489
26. Upper tail coverts chiefly white, forming a conspicuous white patch
    .... White-rumped Sandpiper (*Calidris fuscicollis*), sp.498
    Upper tail coverts mostly dark 27
27. Culmen more than 10% longer than tarsus ....
    .... Purple Sandpiper (*Calidris maritima*), sp.495
    Culmen about equal to or less than tarsus 28

28. Tarsus at least 24 mm long; bill drooped at tip 29
    Tarsus less than 24 mm long; bill straight ....
    .... Baird's Sandpiper (*Calidris bairdii*), sp.499
29. Culmen shorter than tarsus; shafts of primaries white for part of
    length ...... Sharp-tailed Sandpiper (*Calidris acuminata*), sp.496
    Culmen longer than tarsus; only the outermost primary shaft white ..
    .... Pectoral Sandpiper (*Calidris melanotos*), sp.497

ORDER CHARADRIIFORMES: CIII. FAMILY RECURVIROSTRIDAE, Avocets
and Stilts.

1. Culmen more than 75 mm long and markedly upcurved; small hallux
   present ..... American Avocet (*Recurvirostra americana*), sp.512
   Culmen less than 75 mm long and not more than slightly upcurved;
   hallux lacking ....
   .... Black-necked Stilt (*Himantopus mexicanus*), sp.513

ORDER CHARADRIIFORMES: CIV. FAMILY PHALAROPODIDAE, Phalaropes.

1. Bill widened near tip and its length from nostril less than 5 times
   depth at same point ....
   .... Red Phalarope (*Phalaropus fulicarius*), sp.514
   Bill not widened near tip and its length from nostril more than 5
   times depth at same point 2
2. Total length more than 205 mm; bill from nostril more than 25 mm
   long .... Wilson's Phalarope (*Steganopus tricolor*), sp.515
   Total length not more than 205 mm; bill from nostril less than 25
   mm long .... Northern Phalarope (*Lobipes lobatus*), sp.516

ORDER CHARADRIIFORMES: CV. FAMILY STERCORARIIDAE, Jaegers and
Skuas.

1. Wing more than 38 cm long and with a white spot in primaries;
   length of middle toe without claw more than 50 mm ....
   .... Skua (*Catharacta skua*), sp.769H
   Wing less than 38 cm long and with no white spot on primaries;
   length of middle toe without claw less than 50 mm 2
2. Elongate central rectrices rounded at tips; culmen more than 36 mm
   long ...... Pomarine Jaeger (*Stercorarius pomarinus*), sp.517
   Elongate central rectrices pointed at tips; culmen not more than 36
   mm long 3
3. Shafts of all primaries white; tarsus entirely black ....
   .... Parasitic Jaeger (*Stercorarius parasiticus*), sp.518
   Only the 2 or 3 outermost primaries with white shafts; tarsus not
   entirely black (blue in life) ....
   .... Long-tailed Jaeger (*Stercorarius longicaudus*), sp.519

1. Upper mandible hooked at tip, and nostrils halfway out on bill (Larinae) 2

   Bill spear-shaped, without a noticeable hook at tip of upper mandible; nostrils closer to base than to tip of bill (Sterninae) 14

2. Outer rectrices slightly longer than middle ones; hallux rudimentary or absent 3

   Outer rectrices never the longest; hallux without claw usually about 20% length of second toe without claw 4

3. Tail notched; wing length more than 290 mm; bill from nostril more than 16 mm long ....
   .... Black-legged Kittiwake (*Rissa tridactyla*), sp.529

   Tail forked; wing length less than 290 mm; bill from nostril less than 16 mm long .... Sabine's Gull (*Xema sabini*), sp.530

4. Total length 455 mm or more; wing length more than 330 mm 5

   Total length less than 455 mm; wing length not more than 330 mm
   10

5. Primaries light or white 6

   Primaries mostly dark 7

6. Bill from nostril more than 20 mm long; tarsus 63 mm or more in length .... Glaucous Gull (*Larus hyperboreus*), sp.520

   Bill from nostril less than 20 mm long, and tarsus less than 63 mm long .... Iceland Gull (*Larus glaucoides*), sp.521

7. Total length 685 mm or more; wing length more than 445 mm ....
   .... Great Black-backed Gull (*Larus marinus*), sp.522

   Total length less than 685 mm; wing length usually less than 445 mm
   8

8. Total length 535 to 660 mm; wing length 405 to 445 mm 9

   Total length not more than 505 mm; wing length less than 405 mm
   .... Ring-billed Gull (*Larus delawarensis*), sp.524

9. Mantle of adult black or dark gray and the legs usually yellowish ..
   .... Lesser Black-backed Gull (*Larus fuscus*),[1] sp.872

   Mantle of adult light or medium gray and the legs pinkish ....
   .... Herring Gull (*Larus argentatus*),[1] sp.523

10. Total length 380 mm or more; wing length more than 305 mm ....
    .... Laughing Gull (*Larus atricilla*), sp.525

    Total length 380 mm or less; wing length less than 305 mm 11

11. Wing length usually less than 265 mm; tarsus not more than 38 mm long; inner edge of outermost primary white, except near tip ....
    12

    Wing length more than 265 mm; tarsus more than 38 mm long; inner edge of outermost primary dark 13

---

1. First-year birds of these two species are said to be indistinguishable.

12. Wing length more than 240 mm; culmen more than 27 mm long ..
.... Bonaparte's Gull (*Larus philadelphia*), sp.527
Wing length less than 240 mm; culmen less than 27 mm long ......
.... Little Gull (*Larus minutus*), sp.528
13. Outer primaries chiefly white; secondaries not broadly tipped with white; mantle light gray (brownish in immature); upper tail coverts and basal portion of rectrices white ....
.... Black-headed Gull (*Larus ridibundus*), sp.873
Outer primaries mostly dark, but with black band contrasting with white in adults; secondaries broadly tipped with white in contrast to the *dark* gray mantle; upper tail coverts white, contrasting with pearl gray of rectrices basally ....
.... Franklin's Gull (*Larus pipixcan*), sp.526
14. Total length 455 mm or more; wing length more than 330 mm ....
.... 15
Total length usually 330 to 430 mm; wing length 230 to 315 mm (note *third* choice) 16
Total length less than 330 mm, and wing length less than 230 mm ..
.... 25
15. Bill red or orange red, its greatest depth usually more than 19 mm; tail forked for less than 50 mm ....
.... Caspian Tern (*Hydroprogne caspia*), sp.540
Bill orange or yellow, its depth never exceeding 19 mm; tail forked for more than 50 mm ....
.... Royal Tern (*Thalasseus maximus*), sp.538
16. Plumage dark ventrally and whitish on crown; tail not forked 17
Plumage never dark ventrally if crown white; tail forked 18
17. Length of bill from nostril less than 4 times depth of bill at same point; plumage mostly dark brownish gray ....
.... Brown Noddy (*Anous stolidus*), sp.542
Length of bill from nostril more than 4 times depth of bill at same point; plumage mostly gray black ....
.... Black Noddy (*Anous tenuirostris*), sp.543
18. Fork in tail less than 50 mm deep; greatest depth of bill at least 30% of its length ....
.... Gull-billed Tern (*Gelochelidon nilotica*), sp.531
Fork in tail usually more than 50 mm deep; greatest depth of bill less than 30% of its length 19
19. Mantle very dark 20
Mantle pearl gray (or black-spotted in immatures) 21
20. Nape dark; total length of adult at least 380 mm and wing length about 305 mm .... Sooty Tern (*Sterna fuscata*), sp.535
Nape at least partly white; total length 380 mm or less; wing length about 265 mm ...... Bridled Tern (*Sterna anaethetus*), sp.536

21. Bill black with yellow tip (may be mostly yellow or entirely black in immatures); head crested; culmen more than 48 mm long in adults ...... Sandwich Tern (*Thalasseus sandvicensis*), sp.539
    Bill not lighter toward tip than toward base, and not mostly yellowish unless tip dark; head not crested; culmen less than 48 mm long 22
22. Outermost tail feather entirely white .... Roseate Tern (*Sterna dougallii*), sp.534
    Only one side of outermost tail feather entirely whitish 23
23. Inner web of outermost tail feather whitish, outer web gray 24
    Outer web of outermost rectrix entirely white, inner web dark toward tip .... Forster's Tern (*Sterna forsteri*), sp.532
24. Tarsus more than two-thirds length of bill from nostril .... Common Tern (*Sterna hirundo*), sp.533
    Tarsus less than two-thirds length of bill from nostril .... Arctic Tern (*Sterna paradisaea*), sp.769H
25. Some dark pigment ventrally, at least on sides of breast; wing length 190 mm or more ...... Black Tern (*Chlidonias niger*), sp.541
    No dark pigment ventrally; wing less than 190 mm long .... Least Tern (*Sterna albifrons*), sp.537

ORDER CHARADRIIFORMES: CVII. FAMILY RYNCHOPIDAE, Skimmers. One species in Florida ...... Black Skimmer (*Rynchops niger*), sp.544

ORDER CHARADRIIFORMES: CVIII. FAMILY ALCIDAE, Auks, Murres, and Puffins.
1. Total length more than 30 cm; bill decidedly compressed 2
   Total length less than 30 cm; width of bill almost as great as its depth .... Dovekie (*Alle alle*), sp.546
2. Depth of bill at nostrils about 3 times its width at same point; central rectrices pointed .... Razorbill (*Alca torda*), sp.545
   Depth of bill not more than twice its width; rectrices rounded .... Common Murre (*Uria aalge*), sp.769H

ORDER COLUMBIFORMES: CIX. FAMILY COLUMBIDAE, Pigeons and Doves.[1]
1. Tarsus partly feathered 2
   Tarsus entirely bare 5
2. Tail long and graduated .... Passenger Pigeon (*Ectopistes migratorius*), sp.769H
   Tail moderate and square-tipped 3

---

1. Two Band-tailed Pigeons (*Columba fasciata*) reported in Sarasota (Letson, 1968) seem more likely to have escaped from captivity than to have wandered so far from their usual range.

3. Crown entirely white to smoky ....
.... White-crowned Pigeon (*Columba leucocephala*), sp.547
Crown usually dark, but sometimes brown and white 4
4. Nape purplish with scalelike, reddish brown edgings on each feather
.... Scaly-naped Pigeon (*Columba squamosa*), sp.548
Feathers of nape unicolored, though sometimes with metallic reflec-
tions; semidomestic ...... Rock Dove (*Columba livia*), sp.549
5. Wing length over 125 mm 7
Wing length less than 125 mm 6
6. Abdomen crossbarred; tail more than 75 mm long ....
.... Inca Dove (*Scardafella inca*), sp.555
Abdomen not crossbarred; tail less than 75 mm long ....
.... Ground Dove (*Columbina passerina*), sp.554
7. Tail dark with whitish tip 8
Tail not tipped with whitish 11
8. Tail strongly graduated, the central rectrices extending at least 25
mm beyond the lateral rectrices.............................
.... Mourning Dove (*Zenaida macroura*), sp.552
Tail square or rounded, but not graduated 9
9. Coloration light, with dark collar on nape ....
.... Ringed Turtle Dove (*Streptopelia risoria*), sp.553
Coloration reddish to olive brown; no dark collar on nape 10
10. Reddish brown; white on wings confined to flight feathers ....
.... Zenaida Dove (*Zenaida aurita*), sp.550
Olive brown (lighter underneath); with a white stripe extending
from bend of wing onto secondaries ....
.... White-winged Dove (*Zenaida asiatica*), sp.551
11. Crown chalky blue; scales on front of tarsus hexagonal, vertically
elongate ....
.. Blue-headed Quail-Dove (*Starnoenas cyanocephala*), sp.769H
Crown not blue; front of tarsus scutellate 12
12. Back highly iridescent; tail length more than 80 mm ....
.... Key West Quail-Dove (*Geotrygon chrysia*), sp.556
Back only slightly iridescent; tail length less than 80 mm ....
.... Ruddy Quail-Dove (*Geotrygon montana*), sp.557

ORDER PSITTACIFORMES: CX. FAMILY PSITTACIDAE, Parrots and allies.

1. Total length 380 mm or more; bill (or at least upper mandible) red
.... Ring-necked Parakeet (*Psittacula krameri*), sp.769H
Total length less than 380 mm; no red on bill 2
2. Forehead and breast gray ....
.... Monk Parakeet (*Myiopsitta monachus*), sp.769H
Forehead and breast not gray 3

3. With rounded black spots on throat (or plumage entirely white) ..
.... Budgerigar (*Melopsittacus undulatus*), sp.559
No rounded black spots on throat; plumage not white 4
4. Upper wing coverts largely brownish; with an orange spot on chin ..
...... Orange-chinned Parakeet (*Brotogeris jugularis*), sp.769H
Upper wing coverts not brownish; no orange on chin unless on other
parts of head as well 5
5. Upper wing coverts largely yellow, sometimes partly white ....
...... Canary-winged Parakeet (*Brotogeris versicolurus*), sp.874
Upper wing coverts neither true yellow nor white 6
6. Cere and nostrils entirely concealed by dense feathering; culmen
rounded .... Carolina Parakeet (*Conuropsis carolinensis*), sp.558
Cere partly naked, the nostrils exposed; culmen flattened 7
7. Plumage entirely green to yellowish green ....
.... Green Parakeet (*Aratinga holochlora*), sp.769H
With some yellow or orange and some blue in plumage ....
.... Orange-fronted Parakeet (*Aratinga canicularis*), sp.769H

ORDER CUCULIFORMES: CXI. FAMILY CUCULIDAE, Cuckoos and Anis.

1. Plumage black; bill at least 13 mm deep at base 2
Plumage brownish above, white or cinnamon below; bill less than
13 mm deep at base 3
2. Greatest depth of bill less than 20 mm; usually with grooves on
upper mandible ....
.... Groove-billed Ani (*Crotophaga sulcirostris*), sp.564
Greatest depth of bill at least 20 mm; upper mandible without grooves
.... Smooth-billed Ani (*Crotophaga ani*), sp.563
3. Underparts cinnamon or buffy; dark stripe through eye ....
.... Mangrove Cuckoo (*Coccyzus minor*), sp.560
Underparts whitish; no dark stripe through eye 4
4. Lower mandible yellow toward base; cinnamon present on remiges
.... Yellow-billed Cuckoo (*Coccyzus americanus*), sp.561
Lower mandible entirely dark; no true cinnamon in wings, but some-
times reddish brown ....
...... Black-billed Cuckoo (*Coccyzus erythropthalmus*), sp.562

ORDER STRIGIFORMES: CXII. FAMILY TYTONIDAE, Barn Owls. One species in Florida .... Barn Owl (*Tyto alba*), sp.565

ORDER STRIGIFORMES: CXIII. FAMILY STRIGIDAE, Typical Owls.

1. Toes covered with thick, white feathers which cover bases of claws;
wing length usually more than 410 mm; ear tufts not in evidence
.... Snowy Owl (*Nyctea scandiaca*), sp.769H
Feathering of toes sparser, usually buffy, and not covering bases of
claws; ear tufts present if wing length as much as 410 mm 2

2. Total length more than 455 mm; wing length at least 330 mm
   3
   Total length less than 435 mm; wing length less than 330 mm
   4
3. Ear tufts present; dark crossbars on abdomen ....
   .... Great Horned Owl (*Bubo virginianus*), sp.567
   Ear tufts lacking; longitudinal streaks on abdomen ....
   .... Barred Owl (*Strix varia*), sp.569
4. Total length more than 300 mm; wing length more than 250 mm
   5
   Total length less than 300 mm; wing length less than 250 mm
   6
5. Ear tufts long (more than 25 mm); tarsus less than 38 mm long
   .... Long-eared Owl (*Asio otus*), sp.570
   Ear tufts rudimentary (often not in evidence); tarsus more than
   38 mm long .... Short-eared Owl (*Asio flammeus*), sp.571
6. Ear tufts present .... Screech Owl (*Otus asio*), sp.566
   Ear tufts lacking 7
7. With rounded white spots on breast; tarsus more than 38 mm long
   .... Burrowing Owl (*Speotyto cunicularia*), sp.568
   No rounded white spots on breast; tarsus less than 38 mm long....
   .... Saw-whet Owl (*Aegolius acadicus*), sp.572

ORDER CAPRIMULGIFORMES: CXIV. FAMILY CAPRIMULGIDAE, Nightjars.

1. Rictal bristles inconspicuous; white patch present on primaries 2
   Rictal bristles long (more than 20 mm); no white in wings 4
2. Spot on outermost primary white and at least 70 mm from tip 3
   Spot on outermost primary often buffy and not more than 60 mm
   from tip .... Lesser Nighthawk (*Chordeiles acutipennis*), sp.769H
3. Tail more than 100 mm long; underparts with white background, but
   sometimes buffy in female ....
   .... Common Nighthawk (*Chordeiles minor*), sp.575
   Tail less than 100 mm long; underparts with buffy background in
   both sexes ....
   .... Antillean Nighthawk (*Chordeiles gundlachii*),[1] sp.576
4. Rictal bristles pinnate toward base; wing length more than 175 mm
   .... Chuck-will's-widow (*Caprimulgus carolinensis*), sp.573
   No lateral extensions on rictal bristles; wing length less than 175 mm
   .... Whip-poor-will (*Caprimulgus vociferus*), sp.574

ORDER APODIFORMES: CXV. FAMILY APODIDAE, Swifts.

1. Tail spiny-tipped; plumage entirely dusky; widespread ....
   .... Chimney Swift (*Chaetura pelagica*), sp.577

---

1. Sometimes considered a subspecies of the Common Nighthawk.

Tail not spiny-tipped; underparts whitish with a dark breast band; accidental at Key West . . . .
. . . . Antillean Palm Swift (*Tachornis phoenicobia*), sp.875

ORDER APODIFORMES: CXVI. FAMILY TROCHILIDAE, Hummingbirds.

1. Tail noticeably forked   2
   Tail not noticeably forked   3
2. Underparts mostly green . . . .
   . . . . Cuban Emerald (*Chlorostilbon ricordii*), sp.578
   Underparts chiefly brownish and white . . . .
   . . . . Bahama Woodstar ♂ (*Calliphlox evelynae*), sp.875
3. Outer rectrices much shorter than middle ones; with some brownish pigment in tail   4
   Outer rectrices about same length as central tail feathers; no brown in tail   5
4. All rectrices tipped with white; throat spotted . . . .
   . . . . Bahama Woodstar, ♀ (*Calliphlox evelynae*), sp.876
   Middle rectrices (or all) not tipped with white; male with a reddish gorget . . . . Rufous Hummingbird (*Selasphorus rufus*), sp.581
5. Rectrices metallic blue below; width of bill at base at least 3 mm . . .
   . . . . . Broad-billed Hummingbird (*Cynanthus latirostris*), sp.769H
   Rectrices not metallic blue; width of bill at base less than 3 mm
     6
6. With an iridescent gorget   7
   Gorget lacking; throat whitish with dark spots   8
7. Gorget metallic red . . . .
   Ruby-throated Hummingbird, ♂ (*Archilochus colubris*), sp.579
   Gorget metallic purplish . . . .
   Black-chinned Hummingbird, ♂ (*Archilochus alexandri*), sp.580
8. Central tail feathers usually not longer than outermost ones . . . .
   Ruby-throated Hummingbird, ♀ (*Archilochus colubris*), sp.579
   Central rectrices longer than outer pair . . . .
   Black-chinned Hummingbird, ♀ (*Archilochus alexandri*), sp.580

ORDER CORACIIFORMES: CXVII. FAMILY ALCEDINIDAE, Kingfishers. One species in Florida . . . .
. . . . Belted Kingfisher (*Megaceryle alcyon*), sp.582

ORDER PICIFORMES: CXVIII. FAMILY PICIDAE, Woodpeckers.

1. Only 3 toes present, 2 directed forward and one backward . . . .
   Black-backed Three-toed Woodpecker (*Picoides arcticus*), sp.769H
   With 2 front and 2 hind toes   2
2. Total length 375 mm or more; underparts blackish   3
   Total length less than 375 mm; underparts not blackish   4
3. Folded wing largely white; bill whitish (almost extinct) . . . .
   . . . . . Ivory-billed Woodpecker (*Campephilus principalis*), sp.591

With little white visible on folded wing; upper mandible dark (rather common) .... Pileated Woodpecker (*Dryocopus pileatus*), sp.584

4. Total length more than 280 mm; with round black spots underneath .... Common Flicker (*Colaptes auratus*),[1] sp.583

   Total length less than 280 mm; no round black spots underneath
   5

5. With crossbars on back 6

   Without crossbars on back 7

6. Total length 230 mm or more; bill from nostril 20 mm or more; with considerable red on crown of adults ....
   .... Red-bellied Woodpecker (*Centurus carolinus*), sp.585

   Total length less than 230 mm; bill from nostril less than 20 mm; with little or no red on crown ....
   ..... Red-cockaded Woodpecker (*Dendrocopos borealis*), sp.590

7. Total length not more than 175 mm; wing length less than 115 mm; black bars on outer rectrices ....
   .... Downy Woodpecker (*Dendrocopos pubescens*), sp.589

   Total length more than 175 mm; wing length more than 115 mm; no black bars on outer rectrices 8

8. With a large white patch on the secondaries, but without rounded spots ....
   Red-headed Woodpecker (*Melanerpes erythrocephalus*), sp.586

   White on remiges in form of rounded spots 9

9. Black patch present on upper breast ....
   .... Yellow-bellied Sapsucker (*Sphyrapicus varius*), sp.587

   Underparts entirely light gray ....
   .... Hairy Woodpecker (*Dendrocopos villosus*), sp.588

ORDER PASSERIFORMES: CXIX. FAMILY TYRANNIDAE, Tyrant Flycatchers.[2]

1. Tail long and deeply forked, making total length 250 mm or more ..
   2

   Tail of moderate length, not noticeably forked; total length less than 250 mm 3

2. Outer rectrices mostly white; flanks pinkish; top of head pale gray ...
   .... Scissor-tailed Flycatcher (*Muscivora forficata*), sp.595

   Outer rectrices mostly blackish; flanks white; top of head black with yellow crown ....
   .... Fork-tailed Flycatcher (*Muscivora tyrannus*), sp.769H

3. Tail blackish, with broad white band across tip ....
   .... Eastern Kingbird (*Tyrannus tyrannus*), sp.592

---

1. Included here are *Colaptes* "*cafer*" and *C.* "*chrysoides.*"
2. Also see Loggerhead Kingbird in Hypothetical List (769H).

Tail not black with white tip 4
4. With a broad white stripe over the eye ....
.... Greater Kiskadee (*Pitangus sulfuratus*), sp.769H
No white stripe over eye 5
5. Bill from nostril at least as long as tarsus 6
Bill from nostril shorter than tarsus 7
6. Upperparts and sides of breast dark gray; wing tip reaching to
within 30 mm of tail tip ....
.... Olive-sided Flycatcher (*Nuttallornis borealis*), sp.605
Upperparts light gray, underparts almost white; tip of folded wing
about 60 mm short of tail tip ....
.... Gray Kingbird (*Tyrannus dominicensis*), sp.593
7. Abdomen tawny and never streaked ....
.... Say's Phoebe (*Sayornis saya*), sp.769H
Abdomen sometimes yellowish or reddish, and sometimes streaked,
but never tawny 8
8. Total length 200 mm or more 9
Total length 175 mm or less 13
9. With cinnamon or rufous brown in remiges and rectrices 10
Wing and tail feathers blackish 12
10. Cinnamon of outer rectrix separated from shaft by gray ....
...... Wied's Crested Flycatcher (*Myiarchus tyrannulus*), sp.597
Cinnamon of outermost tail feather contacting shaft 11
11. Throat and breast medium gray; belly sulphur yellow; back olive
brown; width of bill at nostril more than 50% length of bill from
nostril .... Great Crested Flycatcher (*Myiarchus crinitus*), sp.596
Throat and breast pale gray; belly pale yellow; back grayish; width
of bill at nostril not more than 50% length of bill from nostril ..
...... Ash-throated Flycatcher (*Myiarchus cinerascens*), sp.598
12. Entire outer web of outermost tail feather white ....
.... Western Kingbird (*Tyrannus verticalis*), sp.594
Outermost rectrix only narrowly edged with white ....
.... Tropical Kingbird (*Tyrannus melancholicus*), sp.769H
13. Underparts reddish or streaked ....
.... Vermilion Flycatcher (*Pyrocephalus rubinus*), sp.606
Underparts never reddish or streaked 14
14. Without wing bars 15
With 2 distinct wing bars 16
15. Throat and breast whitish ....
.... Eastern Phoebe (*Sayornis phoebe*), sp.599
Throat and breast blackish ....
.... Black Phoebe (*Sayornis nigricans*), sp.769H
16. Length of wing about 6 times that of tarsus; eye-ring inconspicuous or
lacking ...... Eastern Wood Pewee (*Contopus virens*), sp.604

Wing length not more than 5 times tarsal length; eye-ring conspicuous
.... 17

17. Distance from tip of longest secondary to that of longest primary ("wing tip") greater than length of tarsus; crown greenish (but feathers with lighter edges in immatures) ....
.... Acadian Flycatcher (*Empidonax virescens*),[1] sp.601
Length of wing tip less than tarsal length; crown usually not greenish
.... 18

18. Upperparts chiefly greenish; underparts almost entirely yellow ....
.... Yellow-bellied Flycatcher (*Empidonax flaviventris*),[1] sp.600
Upperparts chiefly brownish to olivaceous, but never greenish; underparts white (not more than slightly yellowish) 19

19. Total length usually more than 140 mm; length of culmen usually 10 mm or more; width of bill at base usually more than 6 mm; all rectrices about equal in length ....
.... Traill's Flycatcher (*Empidonax traillii*),[1] sp.602
Total length usually less than 140 mm; length of culmen usually less than 10 mm; width of bill at base usually less than 6 mm; tail usually notched ....
.... Least Flycatcher (*Empidonax minimus*),[1] sp.603

ORDER PASSERIFORMES: CXX. FAMILY ALAUDIDAE, Larks. One species in Florida .... Horned Lark (*Eremophila alpestris*), sp.607

ORDER PASSERIFORMES: CXXI. FAMILY HIRUNDINIDAE, Swallows.

1. Wing length more than 125 mm 2
Wing length less than 125 mm 4

2. Length of tail about 50% of total length; belly and under tail coverts dark in female ..... Southern Martin (*Progne modesta*),[2] sp.617
Length of tail about 40% of total length; belly and under tail coverts whitish in female 3

3. With concealed white on abdomen (male), or with creamy white contrasting abruptly with color of breast (female) ....
.... Cuban Martin (*Progne dominicensis*),[2] sp.616
Underparts entirely glossy black (male), or dull white on the abdomen, gradually merging with drab gray breast (female) ....
.... Purple Martin (*Progne subis*),[2] sp.615

4. Tail forked for at least 14 mm 5
Tail forked for less than 14 mm, if at all 6

---

1. Distinguishing these small flycatchers without extensive comparisons is not always possible.

2. Martins may be very difficult to identify and possibly represent a single species (see Peters, 1960); only the Purple Martin is regularly found in Florida; the Cuban Martin is *Progne cryptoleuca* of the A.O.U. *Check-list.*

5. Underparts and forehead chiefly pale pinkish buff or cinnamon brown
.... Barn Swallow (*Hirundo rustica*), sp.612
Underparts chiefly, or entirely, white; forehead violet blue or dull
brown .... Bahama Swallow (*Callichelidon cyaneoviridis*), sp.608
6. Nostrils partly covered by membranes 7
Nostrils not operculate 9
7. Tarsus entirely bare; breast band indistinct or lacking 8
Tarsus feathered near hind toe; with distinct dark band on breast
.... Bank Swallow (*Riparia riparia*), sp.610
8. Ear coverts dark ...... Tree Swallow (*Iridoprocne bicolor*), sp.609
Ear coverts partly white ....
.... Violet-green Swallow (*Tachycineta thalassina*), sp.769H
9. Plumage grayish brown and white; outermost primary usually pro-
vided with minute hooks ....
...... Rough-winged Swallow (*Stelgidopteryx ruficollis*), sp.611
With some reddish brown in plumage; outermost primary smooth-
edged 10
10. Chin and throat largely blackish or brown; rump light (buffy) brown
.... Cliff Swallow (*Petrochelidon pyrrhonota*), sp.613
Chin and throat buffy or pale cinnamon; rump dark (reddish)
brown .... Cave Swallow (*Petrochelidon fulva*), sp.614

ORDER PASSERIFORMES: CXXII. FAMILY CORVIDAE, Crows, Jays, and
Magpies.

1. Plumage largely black and white; tail 225 to 300 mm long and
graduated; accidental ....
.... Black-billed Magpie (*Pica pica*), sp.769H
Plumage not black and white; tail less than 225 mm long (not grad-
uated if more than 150 mm long) 2
2. Plumage blackish; total length more than 355 mm 3
Plumage chiefly gray and blue; total length less than 355 mm 4
3. Tarsus more than 50 mm long ....
.... Common Crow (*Corvus brachyrhynchos*), sp.620
Tarsus not more than 50 mm long ....
.... Fish Crow (*Corvus ossifragus*), sp.621
4. Crest present; with black and white on secondaries ....
.... Blue Jay (*Cyanocitta cristata*), sp.618
Crest lacking; secondaries without true black or white ....
.... Scrub Jay (*Aphelocoma coerulescens*), sp.619

ORDER PASSERIFORMES: CXXIII. FAMILY PARIDAE, Titmice.[1]

1. Crest present; throat light gray ....
.... Tufted Titmouse (*Parus bicolor*), sp.623

1. The likelihood of a natural occurrence of the Black-crested Titmouse in
Florida (See Sprunt, 1954, p. 501) is considered too remote to include the
species in the Hypothetical List.

Crest wanting; throat black ....
.... Carolina Chickadee (*Parus carolinensis*), sp.622

ORDER PASSERIFORMES: CXXIV. FAMILY SITTIDAE, Nuthatches.
1. Underparts more or less reddish; with a white line over the eye ....
.... Red-breasted Nuthatch (*Sitta canadensis*), sp.625
Underparts grayish, with no trace of reddish (may be brown on flanks); no white line over eye 2
2. Total length over 125 mm; cap black ....
.... White-breasted Nuthatch (*Sitta carolinensis*), sp. 624
Total length less than 125 mm; cap brown ....
.... Brown-headed Nuthatch (*Sitta pusilla*), sp.626

ORDER PASSERIFORMES: CXXV. FAMILY PYCNONOTIDAE, Bulbuls. One species in Florida ....
.... Red-whiskered Bulbul (*Pycnonotus jocosus*), sp.627

ORDER PASSERIFORMES: CXXVI. FAMILY CERTHIIDAE, Creepers. One species in Florida ...... Brown Creeper (*Certhia familiaris*), sp.628

ORDER PASSERIFORMES: CXXVII. FAMILY TROGLODYTIDAE, Wrens.
1. With a distinct white line over the eye 2
White line over eye indistinct or lacking 4
2. With white streaks on upper back ....
...... Long-billed Marsh Wren (*Telmatodytes palustris*), sp.633
Without white streaks on back 3
3. Outer tail feathers partly whitish ....
.... Bewick's Wren (*Thryomanes bewickii*), sp.631
No white in tail ....
.... Carolina Wren (*Thryothorus ludovicianus*), sp.632
4. Feathers on belly with black markings ....
.... Winter Wren (*Troglodytes troglodytes*), sp.630
No black on underparts 5
5. Crown uniformly brown ....
.... House Wren (*Troglodytes aedon*), sp.629
Crown streaked with black ....
.... Short-billed Marsh Wren (*Cistothorus platensis*), sp.634

ORDER PASSERIFORMES: CXXVIII. FAMILY MIMIDAE, Mimic Thrushes.
1. Upperparts chiefly light gray, but with *large* white patches in the tail; underparts whitish, without dark spots in adults ....
.... Mockingbird (*Mimus polyglottos*), sp.635
Color pattern not as described above 2

2. Underparts white, with rows of elongate black markings 3
Underparts uniformly dark gray, except for chestnut under tail coverts
.... Gray Catbird (*Dumetella carolinensis*), sp.636
3. Upperparts dark brown or reddish brown 4
Upperparts gray or brownish gray 5
4. Lower mandible light toward base; gonys approximately straight;
ground color of underparts tinged with buff; statewide ....
.... Brown Thrasher (*Toxostoma rufum*), sp.637
Lower mandible dark; gonys markedly decurved; ground color of
underparts white; doubtfully recorded in Florida ....
.... Long-billed Thrasher (*Toxostoma longirostre*), sp.769H
5. Bill long and strongly decurved (length from nostril at least 18 mm);
total length more than 250 mm ....
.... Curve-billed Thrasher (*Toxostoma curvirostre*), sp.638
Bill shorter and almost straight (length from nostril less than 15 mm);
total length less than 250 mm ....
.... Sage Thrasher (*Oreoscoptes montanus*), sp.639

ORDER PASSERIFORMES: CXXIX. FAMILY TURDIDAE, Thrushes.

1. Total length 230 mm or more; wing length more than 115 mm;
upperparts gray ....
.... American Robin (*Turdus migratorius*), sp.640
Total length less than 230 mm; wing length not more than 115 mm;
upperparts not truly gray (brownish gray in *Oenanthe*) 2
2. Upperparts chiefly blue ....
.... Eastern Bluebird (*Sialia sialis*), sp.646
No blue in upperparts 3
3. With large white patches on rectrices ....
.... Wheatear (*Oenanthe oenanthe*), sp.647
No white in rectrices 4
4. Ground color of underparts white, with large dark spots extending
back on flanks to a point near the legs ....
.... Wood Thrush (*Hylocichla mustelina*), sp.641
Underparts partly creamy or buffy in background; dark spots not
extending to legs 5
5. With reddish brown in upperparts 6
Upperparts entirely dull olive brown 7
6. Tail less reddish than remainder of upperparts ....
.... Veery (*Catharus fuscescens*), sp.645
Tail and rump more reddish than remaining upperparts ....
.... Hermit Thrush (*Catharus guttatus*), sp.642
7. With extensive buffy coloring on underparts anteriad, including chin
and eye-ring ....
.... Swainson's Thrush (*Catharus ustulatus*), sp.643

Buffy coloring reduced or lacking, never including chin or eye-ring
.... Gray-cheeked Thrush (*Catharus minimus*), sp.644

ORDER PASSERIFORMES: CXXX. FAMILY SYLVIIDAE, Gnatcatchers, Kinglets, and Old-World Warblers.

1. Upperparts chiefly gray; outer rectrices white ....
.... Blue-gray Gnatcatcher (*Polioptila caerulea*), sp.648
Upperparts greenish; all rectrices dark 2
2. With yellow in crown ....
.... Golden-crowned Kinglet (*Regulus satrapa*), sp.649
Crown same color as upper back, or with concealed red spot ......
.... Ruby-crowned Kinglet (*Regulus calendula*), sp.650

ORDER PASSERIFORMES: CXXXI. FAMILY MOTACILLIDAE, Pipits and
Wagtails.

1. Upperparts essentially unicolored; underparts extensively buffy (cinnamon in spring); hind toe plus claw not more than 21 mm long;
tarsus and toes dark ....
.... Water Pipit (*Anthus spinoletta*), sp.651
Feathers of upperparts dark with light edgings, with head and back
of streaked appearance; buffy underparts anteriad only; hind toe
plus claw at least 23 mm long; tarsus and toes light ....
.... Sprague's Pipit (*Anthus spragueii*), sp.652

ORDER PASSERIFORMES: CXXXII. FAMILY BOMBYCILLIDAE, Waxwings.
One species in Florida ....
.... Cedar Waxwing (*Bombycilla cedrorum*), sp.653

ORDER PASSERIFORMES: CXXXIII. FAMILY LANIIDAE, Shrikes. One
species in Florida ....
.... Loggerhead Shrike (*Lanius ludovicianus*), sp.654

ORDER PASSERIFORMES: CXXXIV. FAMILY STURNIDAE, Starlings.

1. Entire bill decurved; with bare skin on head and a white patch on
wing .... Hill Mynah (*Gracula religiosa*), sp.796H
Only the culmen decurved; no bare skin on head nor white patch on
wing .... Starling (*Sturnus vulgaris*), sp.655

ORDER PASSERIFORMES: CXXXV. FAMILY VIREONIDAE, Vireos.

1. White or light wing bars and eye-ring present 2
Wing bars and eye-ring lacking 6
2. Eye-ring yellow 3
Eye-ring white 5
3. Lower back olive or greenish; head and upper back olive to gray ..
.... 4

Lower back gray; head and upper back greenish ....
.... Yellow-throated Vireo (*Vireo flavifrons*), sp.658
4. Length of bill from nostril more than 8 mm; depth of bill at nostril
   at least 5 mm; bill brownish gray; center of abdomen at least
   partly yellowish; accidental in Florida ....
   .... Thick-billed Vireo (*Vireo crassirostris*), sp.769H
   Length of bill from nostril not more than 8 mm; depth of bill at
   nostril less than 5 mm; bill bluish gray; center of abdomen white;
   common .... White-eyed Vireo (*Vireo griseus*), sp.656
5. Outermost primary almost 25 mm long (at least as long as tarsus);
   wing less than 64 mm long .... Bell's Vireo (*Vireo bellii*), sp.657
   Outermost primary about 12 mm long and much shorter than tarsus;
   wing more than 64 mm long ....
   .... Solitary Vireo (*Vireo solitarius*), sp.659
6. Outermost primary minute (not more than half length of the
   second) .... Warbling Vireo (*Vireo gilvus*), sp.664
   Outermost primary almost as long as second 7
7. Center of breast yellow; bill from nostril not more than 8 mm long
   .... Philadelphia Vireo (*Vireo philadelphicus*), sp.663
   Center of breast white; bill from nostril more than 8 mm long 8
8. With a dark malar stripe ....
   .... Black-whiskered Vireo (*Vireo altiloquus*), sp.660
   No dark streak below eye 9
9. Under tail coverts decidedly yellow; wing length less than 50% of
   total length; accidental ....
   .... Yellow-green Vireo (*Vireo flavoviridis*),[1] sp.661
   Under tail coverts white or merely tinged with yellow; wing length
   more than 50% of total length; common ....
   .... Red-eyed Vireo (*Vireo olivaceus*),[1] sp.662

ORDER PASSERIFORMES: CXXXVI. FAMILY COEREBIDAE, Honeycreepers.
One species in Florida ...... Bananaquit (*Coereba flaveola*), sp.665

ORDER PASSERIFORMES: CXXXVII. FAMILY PARULIDAE, Wood Warblers.

1. With white, yellow, or orange bars at least 1 mm wide across wing
   coverts or one similar patch on primaries 2
   No light or orange markings as described above on wings 28
2. Markings on wing yellow or orange 3
   Markings on wing white or whitish (sometimes yellowish white in
   Dendroica pensylvanica) 5
3. With yellow or orange patches on sides of breast, on wing, and on
   each side of tail ....
   .... American Redstart (*Setophaga ruticilla*), sp.706

_____

1. These two vireos are probably conspecific.

With 2 yellow wing bars, but no distinct large patches of contrasting
yellow or orange on sides of breast or tail 4
4. With dark gray or black on head ....
...... Golden-winged Warbler (*Vermivora chrysoptera*),[1] sp.670
Head chiefly yellow, without dark gray or black ....
.... Yellow Warbler (*Dendroica petechia*), sp.677
5. Head chiefly yellow, with narrow black line through eye ....
.... Blue-winged Warbler (*Vermivora pinus*),[1] sp.671
Head pattern various, but not as described above 6
6. Back blue (tinged with green in immature) or gray, with a greenish
patch 7
Back never blue or gray if a distinct greenish patch is present 8
7. Upperparts chiefly blue (tinged with greenish in young) ....
.... Northern Parula (*Parula americana*), sp.676
Upperparts chiefly gray ....
.... Sutton's Warbler (*"Dendroica potomac"*),[2] sp.769H
8. Rump yellow or greenish yellow, in conjunction with a dark upper
back 9
Rump not yellow or greenish unless upper back also predominantly
greenish (little contrast between back and rump) 12
9. With a yellowish spot (faint in female) on side of neck ....
.... Cape May Warbler (*Dendroica tigrina*), sp.679
Sides of neck gray or black 10
10. Underparts mostly yellow, with black streaking ....
.... Magnolia Warbler (*Dendroica magnolia*), sp.678
Yellow of underparts confined to patches on sides of upper breast
.... 11
11. Throat white ...... Myrtle Warbler (*Dendroica coronata*),[3] sp.681
Throat yellow ....
.... Audubon's Warbler (*Dendroica auduboni*),[3] sp.769H
12. Color pattern essentially gray, black, and white, with virtually no
green, yellow, or blue 13
Color pattern not entirely of gray, black, and white 15
13. Throat mostly black, sometimes suffused, but never streaked with
white ....
.... Black-throated Gray Warbler (*Dendroica nigrescens*), sp.682
Throat white or streaked black and white 14

1. The frequent interbreeding of these 2 warblers (north of Florida) produces
2 hybrids, one with whitish underparts (Brewster's Warbler) and one with
yellow underparts (Lawrence's Warbler). Each is rarer than the parent species,
thus extremely rare in Florida.

2. Probably a hybrid between the Parula and Yellow-throated Warblers.
3. These 2 birds are now merged under the common name of Yellow-rumped
Warbler.

14. Top of head entirely black ....
.... Blackpoll Warbler, ♂ (*Dendroica striata*), sp.690
   Top of head with white central stripe ....
.... Black-and-white Warbler (*Mniotilta varia*), sp.666
15. With a single white patch on the primaries, its greatest width more
   than 5 mm ....
.. Black-throated Blue Warbler (*Dendroica caerulescens*), sp.680
   Wing bars usually 2 in number, not located on primaries, and their
   greatest width less than 5 mm 16
16. Back dark gray, unstreaked; throat yellow, bordered by black ......
.... Yellow-throated Warbler (*Dendroica dominica*), sp.687
   Never with a combination of dark gray back and yellow throat
   bordered by black 17
17. Underparts extensively deep yellow or orange 18
   Underparts not more than faintly yellow 22
18. Side of head with 2 dark streaks ....
.... Prairie Warbler (*Dendroica discolor*), sp.693
   Side of head without clearly defined dark streaks 19
19. Ground color of back greenish to olive brown 20
   Ground color of back blackish 21
20. With a dark patch on side of head, bordered above and below with
   yellow; accidental ....
.... Townsend's Warbler (*Dendroica townsendi*), sp.769H
   No dark, yellow-bordered patch on side of head; common ......
.... Pine Warbler (*Dendroica pinus*), sp.691
21. With a yellow or orange stripe over the eye ....
.... Blackburnian Warbler (*Dendroica fusca*), sp.686
   With an incomplete white eye-ring, but no stripe over the eye ....
.... Kirtland's Warbler (*Dendroica kirtlandii*), sp.692
22. Side of head bright yellow, in contrast to dark green or black crown
   23
   Side of head not bright yellow 24
23. Some white present on 4 outermost rectrices; adults partly black
   above ....
...... Golden-cheeked Warbler (*Dendroica chrysoparia*), sp.684
   Only the 3 outermost rectrices with white; adults mostly greenish
   above ....
...... Black-throated Green Warbler (*Dendroica virens*), sp.683
24. Crown green or yellowish; underparts chiefly whitish, with brown on
   sides of adults ....
...... Chestnut-sided Warbler (*Dendroica pensylvanica*), sp.688
   Crown not greenish unless underparts also greenish 25

25. Total length 125 mm or less; wing length not more than 69 mm ..
.... Cerulean Warbler (*Dendroica cerulea*), sp.685
Total length 125 mm or more; wing length more than 69 mm 26
26. Devoid of streaks above and below ....
...... Pine Warbler, ♀ and immature (*Dendroica pinus*), sp.691
Back or breast at least faintly streaked 27
27. Under tail coverts buffy; legs dark; crown chestnut or greenish, not
noticeably streaked ....
.... Bay-breasted Warbler (*Dendroica castanea*), sp.689
Under tail coverts usually white; legs usually light (pinkish in life);
crown grayish with blackish streaks ....
Blackpoll Warbler, ♀ and immature (*Dendroica striata*), sp.690
28. Some of underparts plainly streaked 29
Underparts very indistinctly streaked or not at all 33
29. Upperparts brownish or crown golden brown; underparts extensively and boldly streaked 30
Upperparts not truly brownish nor crown golden brown; streaking
less distinct or confined to breast regions 32
30. Crown golden brown, bordered by black ....
.... Ovenbird (*Seiurus aurocapillus*), sp.695
Crown entirely dark brown 31
31. Chin unstreaked for longitudinal distance of 20 mm or more; flanks
with buffy ground color; superciliary line usually white ......
.... Louisiana Waterthrush (*Seiurus motacilla*), sp.697
Chin unstreaked for distance of less than 20 mm (if at all); ground
color of flanks not buffy, though underparts sometimes yellowish;
superciliary line usually buffy or whitish ....
.... Northern Waterthrush (*Seiurus noveboracensis*), sp.696
32. Back essentially slate gray; eye-ring present ....
.... Canada Warbler (*Wilsonia canadensis*), sp.705
Back olive green to olive gray; superciliary line present ....
.... Palm Warbler (*Dendroica palmarum*), sp.694
33. Back brownish ....
.... Swainson's Warbler (*Limnothlypis swainsonii*), sp.668
Back greenish 34
34. Under tail coverts yellow or greenish yellow 35
Under tail coverts not yellowish, usually white 44
35. With a black mask, complete or incomplete, running across the eye,
contrasting with head color above and below eye 36
No black mask across eye, but lore black in one species 37
36. Abdomen rich yellow; mask interrupted by yellow around eye .....
.... Kentucky Warbler (*Oporornis formosus*), sp.698

Abdomen not entirely yellow; mask not interrupted at eye ......
.... Common Yellowthroat, ♂ (*Geothlypis trichas*),[1] sp.701

37. With a complete white ring around the eye 38
    White eye-ring incomplete or lacking 39
38. Throat yellowish; top of head grayish, often with a chestnut patch
    on crown; wing length less than 64 mm ....
    .... Nashville Warbler (*Vermivora ruficapilla*), sp.675
    Throat gray or buffy; top of head greenish; wing length more than
    64 mm .... Connecticut Warbler (*Oporornis agilis*), sp.699
39. Throat and upper breast dingy white or greenish yellow, with faint
    dusky streaks ....
    .... Orange-crowned Warbler (*Vermivora celata*), sp.674
    Throat and breast gray, black, buffy, or yellow, and never even
    faintly streaked 40
40. Outer rectrices largely white ....
    .... Hooded Warbler (*Wilsonia citrina*), sp.703
    No white in rectrices 41
41. Length of tarsus less than 18 mm ....
    .... Wilson's Warbler (*Wilsonia pusilla*), sp.704
    Length of tarsus at least 18 mm 42
42. Outermost rectrices shorter than middle ones by less than 5 mm ....
    Black-throated Blue Warbler, immature (*Dendroica caerulescens*),
    sp.680
    Outermost rectrices shorter than middle ones by more than 5 mm
    43
43. Wing length exceeding that of tail by more than 15%; belly yellow;
    eye-ring obvious; throat and breast usually gray to black, but
    sometimes buffy ....
    .... Mourning Warbler (*Oporornis philadelphia*), sp.700
    Wing length exceeding that of tail by less than 15%, if at all; belly
    not truly yellow; eye-ring usually lacking or indistinct; throat often
    yellowish ....
    Common Yellowthroat, ♀ and immature (*Geothlypis trichas*),
    sp.701
44. Total length more than 165 mm; wings comparatively short and tail
    long (wing falling at least 50 mm short of tail tip) ....
    .... Yellow-breasted Chat (*Icteria virens*), sp.702
    Total length less than 165 mm; distance from tip of wing to tip of
    tail much less than 50 mm 45

---

1. A bird believed to be a Bahama Yellowthroat (*Geothlypis rostrata*) was
mist-netted, but escaped, near the lower east coast in the fall of 1968 (Paul
Sykes, personal communication).

45. Throat and breast orange yellow; under tail coverts very long, reaching to within 13 mm of tail tip ....
.... Prothonotary Warbler (*Protonotaria citrea*), sp.667
No trace of orange on throat or breast; distance from tip of under tail coverts to tip of tail more than 13 mm 46
46. Crown buffy, bordered by black stripes on each side ....
.... Worm-eating Warbler (*Helmitheros vermivorus*), sp.669
No stripes of buff and black on top of head 47
47. Forehead yellow, contrasting with black or gray on top of head; yellow patch on bend of wing ....
.... Bachman's Warbler (*Vermivora bachmanii*), sp.672
Forehead and crown unicolored; no yellow on bend of wing ......
.... Tennessee Warbler (*Vermivora peregrina*), sp.673

ORDER PASSERIFORMES: CXXXVIII. FAMILY ICTERIDAE, Blackbirds, Orioles, and allies.

1. All rectrices pointed ....
.... Bobolink (*Dolichonyx oryzivorus*), sp.707
Some rectrices square-tipped or rounded 2
2. Throat yellow, contrasting with a dark belly ....'........
Yellow-headed Blackbird (*Xanthocephalus xanthocephalus*), sp.710
Throat not yellow unless belly also yellow or light 3
3. With true yellow or orange on under side of body 4
Under side of body sometimes buffy, but never deep yellow or orange
.... 9
4. Top of head with buffy and blackish stripes; yellow of underparts interrupted on breast by a blackish V-shaped mark 5
Top of head not distinctly striped; no definite black V on breast 6
5. Cheek usually grayish; feathers of back tipped with medium brown; black bars of central rectrices usually connected to one another along the shafts (Plate IX-4) ....
.... Eastern Meadowlark (*Sturnella magna*), sp.708
Cheek usually buffy; feathers of back tipped (and edged) with whitish; some black bars of central rectrices usually separate from one another .... Western Meadowlark (*Sturnella neglecta*), sp.709
6. Crown orange or yellowish orange; wing length usually more than 100 mm ..... Spotted-breasted Oriole (*Icterus pectoralis*), sp.714
Crown black to olive; wing length usually less than 100 mm 7
7. Underparts mostly dark orange brown or yellow; wing less than 85 mm .... Orchard Oriole (*Icterus spurius*), sp.713
Underparts with some true orange, or with the belly whitish; wing length more than 85 mm 8

8. With a contrasting black streak through the eye (♂), or with a whitish belly (♀) ....
.... Bullock's Oriole (*Icterus bullockii*),[1] sp.716
Top and sides of head entirely dark; belly more or less orange ....
.... Baltimore Oriole (*Icterus galbula*),[1] sp.715
9. Tail feathers strongly graduated, the distance from the tip of the shortest to that of the longest being 18 mm or more 10
Rectrices of less varying length, the greatest difference being 13 mm or less 11
10. Total length more than 350 mm (♂), *or* throat and breast light brown (♀) or grayish (immature) ....
.... Boat-tailed Grackle (*Cassidix major*), sp.719
Total length less than 350 mm; plumage glossy, iridescent black ..
.... Common Grackle (*Quiscalus quiscula*), sp.720
11. Depth of bill at nostrils at least 70% length of bill from same point
.... Brown-headed Cowbird (*Molothrus ater*), sp.721
Depth of bill at nostrils less than 70% length of bill from same point 12
12. Plumage black, with red and buff (or tawny) on bend of wing; *or* with dark streaks ventrally 13
Plumage never streaked; no red, buff, or tawny on bend of wing 14
13. Bend of wing with red and buff (♂), or heavily streaked ventrally (♀); widespread and common ....
.... Red-winged Blackbird (*Agelaius phoeniceus*), sp.711
Bend of wing tawny; accidental on Keys ....
...... Tawny-shouldered Blackbird (*Agelaius humeralis*), sp.712
14. Depth of bill at nostrils less than 60% length of bill from nostril; feathers of nape and upper back not forming a ruff 15
Depth of bill at nostrils at least 60% length of bill from same point; accidental ...... Bronzed Cowbird (*Tangavius aeneus*), sp.722
15. Depth of bill at nostrils less than 50% length of bill from nostril; male black, without purplish gloss on head; feather tips often rusty; female gray without iridescence on lower back ....
.... Rusty Blackbird (*Euphagus carolinus*), sp.717
Depth of bill at nostrils at least 50% of bill length from same point; male black, with purplish reflections on head; female gray, with lower back iridescent; no rust on feathers in either sex ....
.... Brewer's Blackbird (*Euphagus cyanocephalus*), sp.718

ORDER PASSERIFORMES: CXXXIX. FAMILY THRAUPIDAE, Tanagers.

1. Color predominantly light grayish blue ....
.... Blue-gray Tanager (*Thraupis virens*), sp.723

1. These 2 birds are now merged under the common name of Northern Oriole (A.O.U., 1973).

Without blue in plumage 2
2. Under tail coverts white; bill dark ....
.... Stripe-headed Tanager (*Spindalis zena*), sp.724
Under tail coverts yellow or red; bill light or medium dark 3
3. With 2 white or yellowish wing bars ....
.... Western Tanager (*Piranga ludoviciana*), sp.725
Wing bars lacking (or one rarely present) 4
4. Bill from nostril less than 13 mm long; upper mandible toothed near
base .... Scarlet Tanager (*Piranga olivacea*), sp.726
Bill from nostril more than 13 mm long; upper mandible not
distinctly toothed ...... Summer Tanager (*Piranga rubra*), sp.727

ORDER PASSERIFORMES: CXL. FAMILY PLOCEIDAE, Weaver Finches.

1. Plumage largely brownish (but a black patch on breast of male);
nostrils less than 6 mm apart; statewide ....
.... House Sparrow (*Passer domesticus*), sp.728
Plumage chiefly gray, black, and white; nostrils more than 6 mm
apart; Miami area ...... Java Sparrow (*Padda oryzivora*), sp.729

ORDER PASSERIFORMES: CXLI. FAMILY FRINGILLIDAE, Finches, Grosbeaks, Buntings, and Sparrows.[1]

1. Mandibles crossed near tip 2
Mandibles not crossed 3
2. Two white wing bars present ....
.... White-winged Crossbill (*Loxia leucoptera*), sp.769H
Wing bars lacking ...... Red Crossbill (*Loxia curvirostra*), sp.734
3. Hind claw about half length of tarsus 4
Hind claw not more than 40% of tarsal length 6
4. With some secondaries and wing coverts white ....
.... Snow Bunting (*Plectrophenax nivalis*), sp.769
All secondaries and wing coverts largely dark 5
5. Lateral rectrices entirely white except near tip ....
.... Chestnut-collared Longspur (*Calcarius ornatus*), sp.768
Lateral rectrices partly dusky at base ....
.... Lapland Longspur (*Calcarius lapponicus*), sp.767
6. Combined length of tail, tarsus, and hallux (with claw) greater than
length of wing 7
Combined length of tail, tarsus, and hallux less than length of wing
.... 49
7. Red crest present; bill heavy, usually orange to reddish; total length
more than 175 mm .... Cardinal (*Cardinalis cardinalis*), sp.735
Head not crested; crown never red; bill neither orange nor red; total
length usually not more than 175 mm 8

_____

1. A Cuban Bullfinch (*Melopyrrha nigra*) seen in Miami was probably an
escaped cage bird.

8. Plumage mostly slate gray, with the abdomen and lateral rectrices white 9
   Never slate gray with white outer tail feathers 10
9. Two white wing bars present on coverts; doubtfully recorded ......
   .... White-winged Junco (*Junco aikeni*),[1] sp.769H
   No wing bars; regular in winter ....
   .... Slate-colored Junco (*Junco hyemalis*), sp.756
10. With extensive and conspicuous white areas on rectrices, the greatest dimension of which is not less than 3 mm 11
    No true white on tail feathers unless as narrow edgings 16
11. Outer web of outermost tail feather entirely white (except at extreme base); other rectrices mostly dark ....
    .... Vesper Sparrow (*Pooecetes gramineus*), sp.753
    Outer web of outer rectrices partly dark; considerable white in other rectrices 12
12. Depth of bill at nostril at least 12 mm; with a large white patch on primaries 13
    Depth of bill at nostril less than 12 mm; no white patch on primaries
    14
13. Breast rose-colored ....
    .....Rose-breasted Grosbeak, ♂ (*Pheucticus ludovicianus*), sp.736
    Breast brownish ....
    Black-headed Grosbeak, ♂ (*Pheucticus melanocephalus*), sp.737
14. Tail longer than wing; back virtually unicolored, but rarely streaked
    .... Rufous-sided Towhee (*Pipilo erythrophthalmus*), sp.743
    Tail shorter than wing; back streaked 15
15. With chestnut markings on head; longest primaries at least 14 mm longer than longest secondaries ....
    .... Lark Sparrow (*Chondestes grammacus*), sp.754
    No chestnut markings on head; wing tip not more than 12 mm long ...... Lark Bunting, ♀ (*Calamospiza melanocorys*), sp.744
16. With contrasting yellow or red on bend of wing (sometimes concealed)
    17
    Bend of wing without yellow unless remainder of wing yellowish or greenish 29
17. Wing length more than 90 mm; depth of bill at nostril more than 12 mm 18
    Wing length less than 90 mm; depth of bill at nostril less than 12 mm 19

---

1. Now merged with *J. hyemalis,* the enlarged species to be called the Dark-eyed Junco (A.O.U., 1973).

18. Breast streaked; bend of wing pink (immature ♂) or yellow (♀)
..... Rose-breasted Grosbeak (*Pheucticus ludovicianus*),[1] sp.736
Breast evenly brown; bend of wing yellow ....
.. Black-headed Grosbeak (*Pheucticus melanocephalus*),[1] sp.737
19. Depth of bill at nostril less than 60% of its length from nostril 20
Depth of bill at nostril more than 60% its length from nostril 23
20. Breast buffy, with narrow streaks; length of bill from nostril less than
10 mm .... Sharp-tailed Sparrow (*Ammospiza caudacuta*), sp.749
Breast dark gray or broadly streaked with black; length of bill from
nostril at least 10 mm 21
21. Upperparts essentially black; with distinct black stripes on white
background ventrally ....
.... Dusky Seaside Sparrow (*Ammospiza nigrescens*), sp.751
Upperparts not entirely blackish; ventral ground color mostly grayish
or buffy (creamy in juvenal), sometimes with dusky streaking ..
.... 22
22. Upperparts blackish mixed with brown ....
.... Seaside Sparrow (*Ammospiza maritima*), sp.750
Upperparts with a greenish cast ....
.... Cape Sable Sparrow (*Ammospiza mirabilis*), sp.752
23. Rectrices narrow (less than 5 mm) and pointed at tips 24
Rectrices wider (more than 5 mm) and not markedly pointed at
tips 25
24. Head and neck largely buffy olive; outer rectrices about 10 mm
shorter than longest ones ....
.... Henslow's Sparrow (*Ammodramus henslowii*), sp.747
No olive on head or neck; outer rectrices about as long as any ....
.... Grasshopper Sparrow (*Ammodramus savannarum*), sp.746
25. With some yellow on breast; bill from nostril more than 9 mm long
.... Dickcissel (*Spiza americana*), sp.742
Breast not marked with yellow; bill from nostril usually not more
than 9 mm long 26
26. Underparts decidedly streaked; outer rectrices about as long as any
.... 27
Underparts unstreaked (or faintly streaked in juvenal); outermost
rectrices at least 4 mm shorter than longest ones 28
27. Total length usually less than 150 mm and wing length not more than
73 mm; pale edges to longer tertials not more than 2 mm wide;
throat more or less streaked; statewide ....
.... Savannah Sparrow (*Passerculus sandwichensis*), sp.745
Total length at least 150 mm and wing length at least 73 mm; pale
edges of longer tertials more than 2 mm wide; throat unstreaked
except on sides; northeast coast only ....
.... Ipswich Sparrow (*Passerculus princeps*), sp.877

1. Hybrids between these two grosbeaks may not fit either description.

28. With a white throat patch; outermost tail feathers about 4 to 6 mm
    shorter than longest ones ....
    .... White-throated Sparrow (*Zonotrichia albicollis*), sp.762
    Throat and breast buffy; outermost rectrices at least 10 mm shorter
    than longest ones ....
    .... Bachman's Sparrow (*Aimophila aestivalis*), sp.755
29. Total length usually less than 120 mm; wing length less than 60 mm;
    back olive or greenish 30
    Total length more than 120 mm; wing length more than 60 mm; *or*
    back not olive or greenish 31
30. With some yellow on head; wing length less than 52 mm ....
    .... Melodious Grassquit (*Tiaris canora*), sp.769H
    No yellow on head; wing length more than 52 mm ....
    ....Black-faced Grassquit (*Tiaris bicolor*), sp.741
31. Culmen slightly concave; rectrices and ventral spotting bright reddish
    brown; wing length more than 75 mm ....
    .... Fox Sparrow (*Passerella iliaca*), sp.763
    Culmen usually convex, never noticeably concave; rectrices and
    ventral spotting (when present) not bright reddish brown; wing
    length usually less than 75 mm if brown ventral spots present ....
    32
32. Chin (often much of body) black; depth of bill at nostrils more than
    80% length of bill from nostril ....
    .... Lark Bunting, ♂ (*Calamospiza melanocorys*), sp.744
    Chin not black, *or* depth of bill at nostrils less than 80% length of
    bill from nostril 33
33. Tail notched, the central rectrices either shorter than the longest ones
    or else narrowed toward tip and appearing to curve outward 34
    Tail square-tipped or rounded, the central rectrices about as long
    as any and not narrowed or pointed terminally 42
34. Rectrices narrow (greatest width not more than 5 mm) and their
    tips pointed; wing length less than 57 mm ....
    ....Le Conte's Sparrow (*Ammospiza leconteii*), sp.748
    Rectrices wider (greatest width at least 6 mm) and their tips more
    or less rounded; wing length more than 57 mm 35
35. Length of tail at least 80% that of wing; adults streaked with brown
    dorsally but not ventrally 36
    Length of tail less than 80% that of wing; brown streaking present
    below if upperparts streaked 37
36. With a single dark spot in center of breast; 2 white wing bars
    present .... Tree Sparrow (*Spizella arborea*), sp.769H
    Breast unspotted (adults) or streaked (juvenals); wing bars less
    conspicuous, not truly white 38
37. With a pale buffy band across breast; remainder of underparts
    white .... Clay-colored Sparrow (*Spizella pallida*), sp.758
    Underparts not white with buffy breast-band 39

38. Top of head rusty brown (gray and not conspicuously streaked in juvenals); sides of breast also rusty brown in adults ....
.... Field Sparrow (*Spizella pusilla*), sp.759
Crown chestnut red (or conspicuously streaked in juvenals); no rusty brown on sides of breast ....
.... Chipping Sparrow (*Spizella passerina*), sp.757

39. Upperparts streaked 40
Upperparts unstreaked 41

40. Total length usually less than 150 mm and wing length not more than 73 mm; pale edges to longer tertials not more than 2 mm wide; throat more or less streaked; statewide ....
.... Savannah Sparrow (*Passerculus sandwichensis*), sp.745
Total length at least 150 mm and wing length at least 73 mm; pale edges of tertials more than 2 mm wide; throat unstreaked except on sides; northeast coast only ....
.... Ipswich Sparrow (*Passerculus princeps*), sp.877

41. Underparts largely red or yellowish ....
.... Painted Bunting (*Passerina ciris*), sp.740
Underparts partly blue or light gray brown (faintly streaked in juvenals) .... Indigo Bunting (*Passerina cyanea*), sp.739

42. Upperparts unstreaked, brightly colored in male; length of wing exceeding that of tail by more than 25% 43
Upperparts streaked, never brightly colored; length of wing exceeding that of tail by less than 25% 44

43. Total length more than 150 mm; depth of bill at nostrils at least 10 mm .... Blue Grosbeak (*Guiraca caerulea*), sp.738
Total length less than 150 mm; depth of bill at nostrils less than 10 mm 41

44. Underparts decidedly streaked, *or* greater coverts bright reddish brown; wing length less than 75 mm 45
Underparts not more than faintly streaked unless wing length more than 75 mm 47

45. Breast smoky gray, lacking definite streaks; greater coverts (and crown of adults) bright reddish brown ....
.... Swamp Sparrow (*Melospiza georgiana*), sp.765
Breast white or buffy, markedly streaked; greater coverts and crown not bright reddish brown 46

46. Breast white, heavily streaked or spotted with brown ....
.... Song Sparrow (*Melospiza melodia*), sp.766
Breast with a transverse buffy band, finely streaked with blackish ..
.... Lincoln's Sparrow (*Melospiza lincolnii*), sp.764

47. Breast heavily marked with black; crown entirely black or black-spotted .... Harris' Sparrow (*Zonotrichia querula*), sp.760

No true black on breast; black on crown lacking or arranged as stripes 48

48. With a yellow spot in front of eye, merging into the white superciliary line; throat patch contrastingly light or white ....
.... White-throated Sparrow (*Zonotrichia albicollis*), sp.762

No yellow in front of eye; throat patch not contrastingly light ......
...... White-crowned Sparrow (*Zonotrichia leucophrys*), sp.761

49. With some yellow on body; wings and tail conspicuously black and white 50

No yellow on body; wings and tail not conspicuously black and white 51

50. Total length more than 175 mm; depth of bill at nostrils more than 10 mm ..... Evening Grosbeak (*Hesperiphona vespertina*), sp.730

Total length less than 175 mm; depth of bill at nostrils less than 10 mm .... American Goldfinch (*Spinus tristis*), sp.733

51. Length of tarsus less than 15 mm; plumage streaked, with yellow in wings and tail .... Pine Siskin (*Spinus pinus*), sp.732

Length of tarsus more than 15 mm; no yellow in plumage ....
.... Purple Finch (*Carpodacus purpureus*), sp.731

CLASS MAMMALIA—ORDERS AND FAMILIES

1. With 2 pairs of limbs; external nares opening forward; teeth usually heterodont 7

Only the forelimbs (flippers) present; external nares opening upward; teeth homodont or absent, *or* incisors rudimentary, usually concealed, and separated from other teeth by a wide diastema 2

2. Tail finlike and notched; eyes lateral; teeth peglike or absent ......
.... CETACEA, 3

Tail paddle-shaped; eyes directed forward; teeth molariform (incisors sometimes visible in young) .... SIRENIA: TRICHECHIDAE, CLXV

3. Teeth absent; baleen (whalebone) present in upper jaw (Fig. 10); with 2 blowholes 4

Teeth present in upper or lower jaw (rarely below gum line); baleen plates lacking; only one blowhole present 5

4. No dorsal fin; outer surface of throat smooth; baleen plates long and slender (to more than 1 m); length of rostrum about 4 times width of its unflared base .... BALAENIDAE, CLXX

Dorsal fin present; outer surface of throat grooved; baleen plates short (less than 1 m long); rostral length about 3 times width of unflared base of rostrum .... BALAENOPTERIDAE, CLXIX

5. With one or 2 functional teeth on each side of lower jaw; lacrimal and jugal bones separate; upper and lower jaws about equally long; nasal opening median; throat grooved ....
.... HYPEROODONTIDAE, CLXVIII

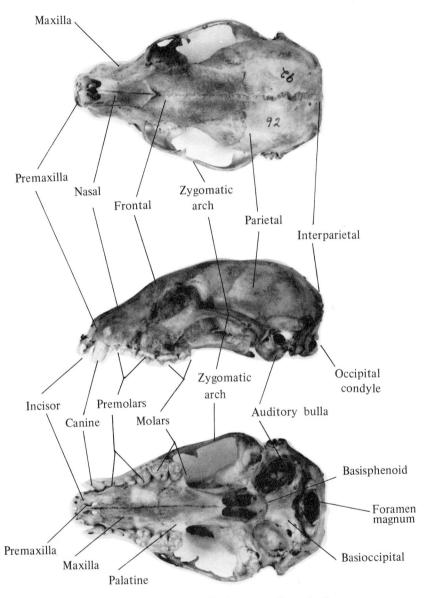

Fig. 9. Bones of a typical mammalian skull

Fig. 10. Portion of a baleen plate (Family Balaenopteridae)

Lower jaw with 2 or more functional teeth on each side; lacrimal and jugal bones not distinctly separate; relative length of the 2 jaws often unequal; nasal opening sinistral; throat smooth 6

6. Upper and lower jaws about equal in length; teeth in upper jaw well developed, *or* not more than 7 pairs in lower jaw ....
.... DELPHINIDAE, CLXVI
Lower jaw ending short of upper jaw; teeth in upper jaw rudimentary or lacking; at least 9 pairs of teeth in lower jaw ....
.... PHYSETERIDAE, CLXVII

7. Upperparts covered with bony plates; teeth homodont ....
.... EDENTATA: DASYPODIDAE, CLI
Covering of bony plates lacking; teeth heterodont 8

8. Forelimbs modified as wings; skull less than 30 mm long and canines longer than incisors; usually with a wide U-shaped notch separating upper incisors of right and left sides ...... CHIROPTERA, 9
Forelimbs not winglike; skull more than 30 mm long, *or* canines shorter than incisors (sometimes lacking); no U-shaped notch between upper incisors 11

9. With 3 bony phalanges in third finger and a leaflike projection on the nose (or flaps on chin); tail not in evidence; palatal notch lacking; skull more than 26 mm long ....
.... PHYLLOSTOMATIDAE, CXLV
Third phalanx of third finger mostly cartilaginous (Plate X-2) and no leaflike projection on the nose or flaps on chin; tail usually obvious; skull less than 26 mm long; palatal notch present (one exception) 10

10. About half of tail extending behind interfemoral membrane; tragus lacking; palatal notch narrowing anteriad to form an oval outline (Plate X-1), *or* notch lacking .... MOLOSSIDAE, CXLVII

Less than half of tail extending beyond interfemoral membrane; tragus present; palatal notch about equally wide anteriad and posteriad .... VESPERTILIONIDAE, CXLVI

11. Limbs flipperlike; canines longer than other teeth; only 4 lower incisors .... PINNIPEDIA: PHOCIDAE, CLXIV

Limbs not flipperlike; if canines longer than other teeth, then 6 or more lower incisors present 12

12. Digits modified to form hoofs; with no upper incisors, *or* with 3 pairs and with upper canines projecting laterally ....
.... ARTIODACTYLA, 13

Digits provided with nails or claws; one or more pairs of upper incisors present; canines not projecting laterally 14

13. Antlers present for part of the year, at least in males; body hair normal; upper incisors lacking .... CERVIDAE, CLXXII

Antlers and horns lacking; hair sparse and stiff; with 3 pairs of upper incisors .... SUIDAE, CLXXI

14. With some of the digits bearing nails rather than claws; orbits directed forward and surrounded by a complete bony ring ....
.... PRIMATES, 15

Digits provided with claws (nails lacking); orbits placed laterally and not usually surrounded by bony rings 17

15. Tail present; incisors decidedly farther anteriad than forehead 16

Tail lacking; incisors almost directly below forehead ....
.... HOMINIDAE, CL

16. With 3 premolars (and 3 molars) in each jaw; tail much longer than head and body .... CEBIDAE, CXLVIII

Only 2 premolars in each jaw; tail not longer than head and body
.... CERCOPITHECIDAE, CXLIX

17. Ear longer than tail and its length about twice its width; canines lacking; incisors like curved chisel blades, with a smaller second pair immediately behind upper pair ....
.... LAGOMORPHA: LEPORIDAE, CLII

Length of ear usually less than twice its width; if longer than tail, then canines well developed; no secondary, displaced incisors 18

18. Ear scarcely visible, if at all; eyes reduced or absent; canines shorter than incisors .... INSECTIVORA, 19

Ear normal, sometimes short though plainly visible; canines lacking, *or* longer than other teeth 20

19. Front feet larger than hind; eyes apparently lacking; teeth whitish; zygomatic arch complete .... TALPIDAE, CXLIV

Front feet normally developed; eyes externally visible; teeth tipped with pigment; zygomatic arch incomplete .... SORICIDAE, CXLIII

20. Tail scaly and only sparsely haired; total length of adults more than 51 cm; first toe of hind foot clawless and opposable; with 10 upper incisors .... MARSUPIALIA: DIDELPHIDAE, CXLII

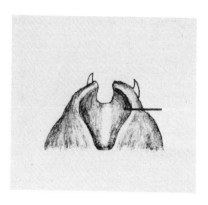

1.
Oval palatal notch of
Tadarida brasiliensis

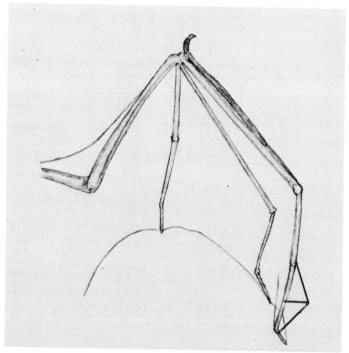

2. Two phalanges in third finger of *Tadarida brasiliensis*

PLATE X

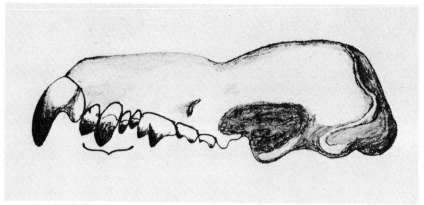

3. Skull of *Blarina,* showing four unicuspids visible laterally
and no zygomatic arch

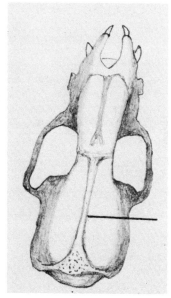

4.
Midsagittal crest of
Myotis sodalis

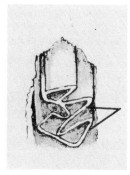

5.
Ridged grinding surface
of upper molar in
Neofiber

MAMMALS

Tail usually hairy (if scaly, then total length usually less than 51 cm); first toe of hind foot usually clawed, never opposable; never as many as 10 upper incisors 21

21. Guard hairs usually well developed; tail never sparsely haired; total length usually more than 305 mm (Florida); with 6 unmodified upper incisors; canines well developed .... CARNIVORA, 22

Guard hairs usually absent or poorly developed; tail often sparsely haired, especially in larger species (more than 500 mm long; squirrels only are both large and bushy-tailed); only one pair each of upper and lower incisors, these being long and chisel-shaped; canines lacking .... RODENTIA, 26

22. Sole of foot bare from contact with ground; teeth 34 to 40 in number, *or* anterior end of nasal bone only halfway between orbit and incisors 23

Bare areas only on the toes; teeth less than 34, *or* anterior end of nasal closer to incisors than to orbits 25

23. Tail shorter than head; teeth 42, including 3 lower molars on each side; nasal bone terminating about halfway between orbits and upper incisors .... URSIDAE, CLX

Tail longer than head; teeth more than 40, with only 2 molars on each side of lower jaw; end of nasal much closer to incisors than to orbits 24

24. Teeth 40, with 2 upper molars per side; tail banded ....
.... PROCYONIDAE, CLXI
Teeth 34 to 36, with one upper molar per side; tail not banded ....
.... MUSTELIDAE, CLXII

25. Teeth 42, with 7 cheek (molariform) teeth on each side of lower jaw; tail long and bushy (hairs 60 to 70 mm long) ....
.... CANIDAE, CLIX
Teeth 28 or 30, with only 3 cheek teeth per side; tail short, or short-haired (hairs 30 to 40 mm long in long-tailed forms) ....
.... FELIDAE, CLXIII

26. Tail about as long as head; cheek pouches present; claws of forelimbs enlarged; front of upper incisors grooved; length of skull about 25 mm or more .... GEOMYIDAE, CLIV

Tail longer than head; cheek pouches absent; claws of forelimbs not enlarged; upper incisors not grooved unless skull less than 20 mm long 27

27. Hind feet webbed; head-body length usually over 405 mm; tail not laterally compressed; grinding surface of molars flat with transverse ridges 28

If head-body length more than 405 mm, then hind foot not webbed, *or* tail laterally compressed (head-body length *usually* less than 405 mm); grinding surface of molars not flat, but usually with

low cusps; if ridges present, then some not arranged transversely
.... 29
28. Tail shaped like the blade of a paddle; infraorbital foramen much
smaller than foramen magnum .... CASTORIDAE, CLV
Tail rounded in cross section; infraorbital foramen about same size
as foramen magnum .... CAPROMYIDAE, CLVIII
29. Tail sparsely haired, *or* with short hairs; teeth 16 in number 30
Tail bushy (hairs 16 to 20 mm long); teeth 20 to 22 ....
.... SCIURIDAE, CLIII
30. Cusps of upper molars in 3 longitudinal rows (Plate XI-3); upper
surface of tail virtually naked in large species (head-body length
more than 100 mm); in smaller species (*Mus*), dorsal fur grayish
and somewhat grizzled, ventral fur not pure white, and tail at
least 95% of head-body length .... MURIDAE,[1] CLVII
If present, cusps of upper molars never in more than 2 longitudinal
rows; if cusps lacking, then with ridges of enamel forming triangular or curving patterns on upper molars (Plate X-5); upper surface of tail obviously, but sparsely, haired in larger species; in
smaller species, dorsal fur not grizzled, ventral fur pure white
unless tail length less than 95% of head-body length ....
.... CRICETIDAE,[1] CLVI

INFRACLASS METATHERIA

ORDER MARSUPIALIA: CXLII. FAMILY DIDELPHIDAE, Opossums. One
species in Florida .... Opossum (*Didelphis marsupialis*), sp.770

INFRACLASS EUTHERIA

ORDER INSECTIVORA: CXLIII. FAMILY SORICIDAE, Shrews.

1. Length of tail at least 40% of total length; first 4 upper unicuspids
about equal in size, the fifth not visible from lateral aspect; northern Florida only ....
.... Southeastern Shrew (*Sorex longirostris*), sp.771
Tail less than 30% of total length; first 2 upper unicuspids much
larger than others (but process of incisor *also* resembles a unicuspid); throughout mainland 2
2. Total length of adults usually over 85 mm (head and body over 70
mm); with 5 upper unicuspids, 4 of which are visible from lateral
aspect (Plate X-3); color slaty gray ....
.... Short-tailed Shrew (*Blarina brevicauda*), sp.772
Total length usually less than 85 mm (head-body under 70 mm); only
4 upper unicuspids, 3 of which are easily visible from lateral aspect,
the fourth being displaced medially; color brownish gray ....
.... Least Shrew (*Cryptotis parva*), sp.773

---

1. These groups should probably be combined under Muridae.

ORDER INSECTIVORA: CXLIV. FAMILY TALPIDAE, Moles.

1. With a ring of fleshy projections on snout; external nares slanted
   upward; unrecorded in Florida ....
   .... Star-nosed Mole (*Condylura cristata*), sp.859H
   No fleshy projections around end of snout; external nares opening
   forward; widespread ....
   .... Eastern Mole (*Scalopus aquaticus*), sp.774

ORDER CHIROPTERA: CXLV. FAMILY PHYLLOSTOMATIDAE, Leaf-nosed
Bats. One species in Florida (accidental at Key West) ....
   ...... Jamaican Fruit-eating Bat (*Artibeus jamaicensis*),[1] sp.775

ORDER CHIROPTERA: CXLVI. FAMILY VESPERTILIONIDAE, Twilight Bats.

1. Total length not more than 90 mm; body hairs dark at base, light
   at center, medium dark at tip (general tone light to medium dark
   above); wing membrane clear near forearm (reddish in life);
   teeth 34 ...... Eastern Pipistrelle (*Pipistrellus subflavus*), sp.780
   Total length often more than 90 mm (if less, then wing membrane
   usually not contrastingly clear near forearm, and dorsal fur dark);
   body hair seldom, if ever, tricolored; with more or fewer than
   34 teeth 2
2. Ear relatively short (10–19 mm) but slender; tragus slender and
   erect (width less than half length); teeth 38 (6 cheek teeth on each
   side of each jaw) 3
   Ear relatively long (over 20 mm), *or* width of tragus at least half
   its length; with 36 teeth or fewer (less than 6 molariform teeth per
   jaw) 6
3. Third metacarpal longer than second; skull with no well-developed
   midsagittal crest, *or* its greatest length at least 15.5 mm[2] 4
   Third metacarpal not definitely longer than second; skull less than
   15.5 mm long and with a well-developed midsagittal crest (Plate
   X-4)[2] 5
4. Length of forearm at least 40 mm; length of ear less than 16 mm;
   dorsal hairs unicolored; end of wing membrane attached to distal
   end of tarsus; sagittal crest well developed; greatest length of
   skull 15.5 mm or more ....
   .... Gray Myotis (*Myotis griscescens*), sp.777

---

1. Although this record is not accepted in some works (for example, Hall
and Kelson, 1959), other species in the family inhabit the West Indies and
may occasionally reach Florida.

2. Probably the skulls, and perhaps the skins also, of *Myotis* bats cannot be
safely distinguished without extensive comparisons.

Length of forearm slightly less than 40 mm; length of ear at least 16 mm; dorsal hairs lighter toward tip; end of wing membrane attached to base of first toe; sagittal crest absent, or present only posteriad; skull less than 15.5 mm long .... Keen's Myotis (*Myotis keeni*), sp.778

5. Dorsal hairs about 5 mm long and not blackish basally; hairs on toes reaching almost or quite to ends of claws; length of foot about 10 mm; width of cranium usually more than 7.2 mm .... Southeastern Myotis (*Myotis austroriparius*), sp.776
Dorsal hairs about 10 mm long and blackish toward base; hairs on toes falling decidedly short of tips of claws; length of foot about 7 mm; width of cranium usually less than 7.2 mm .... Indiana Myotis (*Myotis sodalis*), sp.779

6. Ear more than 20 mm long; teeth 36 .... Eastern Lump-nosed Bat (*Plecotus rafinesquii*), sp.787
Ear less than 20 mm long; teeth fewer than 36 7

7. Ear about 15 mm long and not more than half as wide at midpoint, somewhat narrowed toward tip; dark brown dorsally; upper incisors 4 (2 greatly reduced) .... Big Brown Bat (*Eptesicus fuscus*), sp.781
Ear shorter or more rounded (width about equal to length if as long as 15 mm; dorsal color various; only 2 upper incisors 8

8. Interfemoral membrane densely furred; teeth 32 9
Interfemoral membrane naked or furred only at base; teeth 30 11

9. Total length more than 120 mm; ear black-rimmed; pelage dull brown basally, whitish toward tip; greatest length of skull more than 15.5 mm .... Hoary Bat (*Lasiurus cinereus*), sp.784
Total length less than 120 mm; ear not black-rimmed; pelage reddish or yellowish brown; skull less than 15.5 mm long 10

10. Ground color yellowish basally, reddish toward tips of hairs, and often silvery at tips[1] .... Red Bat (*Lasiurus borealis*), sp.782
Ground color entirely mahogany brown (except when hairs tipped with silver)[1] .... Seminole Bat (*Lasiurus seminolus*), sp.783

11. Total length more than 110 mm; color yellowish brown; upper incisor touching canine at base; third lower incisor smaller than first .... Yellow Bat (*Lasiurus intermedius*), sp.785
Total length less than 110 mm; color medium to dark brown; upper incisor not touching canine at base; third lower incisor about same size as first (all incisors minute) .... Evening Bat (*Nycticeius humeralis*), sp.786

---

1. The skulls of these 2 bats probably cannot be differentiated, and they may represent merely color phases of a single species.

ORDER CHIROPTERA: CXLVII. FAMILY MOLOSSIDAE, Free-tailed Bats.

1. Bases of ears conjoined anteriad; palatal notch lacking; skull more than 20 mm long; Miami only ....
.... Wagner's Mastiff Bat (*Eumops glaucinus*), sp.789
Bases of ears entirely separate; palatal notch present (Plate X-1); skull less than 20 mm long; statewide ....
.... Brazilian Free-tailed Bat (*Tadarida brasiliensis*), sp. 788

ORDER PRIMATES: CXLVIII. FAMILY CEBIDAE, New-World Monkeys. One species in Florida .... Squirrel Monkey (*Saimiri* sp.), sp.859H

ORDER PRIMATES: CXLIX. FAMILY CERCOPITHECIDAE, Old-World Monkeys. One species in Florida ....
.... Rhesus Monkey (*Macaca mulatta*), sp.790

ORDER PRIMATES: CL. FAMILY HOMINIDAE, Man. One species in Florida
.... Man (*Homo sapiens*), sp.791

ORDER EDENTATA: CLI. FAMILY DASYPODIDAE, Armadillos. One species in Florida ....
.... Nine-banded Armadillo (*Dasypus novemcinctus*), sp.792

ORDER LAGOMORPHA: CLII. FAMILY LEPORIDAE, Rabbits and Hares.

1. Length of ear from notch less than 100 mm; no dorsal black stripe on tail; with a discrete interparietal bone (dorsal to the occipital; Plate XI-1) 2
Length of ear from notch 100 mm or more; with a dorsal black stripe on lower back and tail; no discrete interparietal bone (fused with parietals); lower east coast only ....
.... Black-tailed Jack Rabbit (*Lepus californicus*), sp.795
2. Upper side of foot whitish; posterior extension of supraorbital process mostly free from cranium ....
.... Eastern Cottontail (*Sylvilagus floridanus*), sp.794
No white on top of foot; posterior extension of supraorbital process mostly fused to braincase (Plate XI-1) 3
3. Tail white ventrally and usually more than 50 mm long; greatest length of skull more than 8.0 times length of bulla; not recorded in Florida, but likely in northwest corner ....
.... Swamp Rabbit (*Sylvilagus aquaticus*), sp.859H
Tail grayish ventrally and usually less than 50 mm long; skull length usually less than 8 times length of bulla; statewide ....
.... Marsh Rabbit (*Sylvilagus palustris*), sp.793

ORDER RODENTIA: CLIII. FAMILY SCIURIDAE, Squirrels.

1. Length of head and body more than 200 mm in adults; length of skull more than 50 mm 2

Head-body length less than 200 mm; skull length less than 50 mm
.... 4

2. Total length usually less than 56 cm; upperparts, including nose, mostly gray or black; with 5 molars (one vestigial) on each side of upper jaw 3
Total length more than 56 cm in Florida adults; upperparts variously colored, but nose partly white; with 4 unmodified molars on upper jaw .... Fox Squirrel (*Sciurus niger*), sp.799

3. Tail length not more than 240 mm; underparts whitish; length of postorbital process less than twice its width at base; statewide ....
.... Gray Squirrel (*Sciurus carolinensis*), sp.797
Tail length usually more than 240 mm; entirely black, or underparts reddish brown; length of postorbital process about twice its width at base; restricted to Elliott Key ....
.... Red-bellied Squirrel (*Sciurus aureogaster*), sp.798

4. With a fold of loose skin on side of body; pelage mostly gray; 5 upper molars per side; statewide ....
.... Southern Flying Squirrel (*Glaucomys volans*), sp.800
No fold of loose skin on sides; chiefly brownish, with longitudinal white dorsal stripes; 4 upper molars per side; Yellow River drainage only .... Eastern Chipmunk (*Tamias striatus*), sp.796

ORDER RODENTIA: CLIV. FAMILY GEOMYIDAE, Pocket Gophers. One species in Florida ....
.... Southeastern Pocket Gopher (*Geomys pinetis*), sp.801

ORDER RODENTIA: CLV. FAMILY CASTORIDAE, Beavers. One species in Florida .... American Beaver (*Castor canadensis*), sp.802

ORDER RODENTIA: CLVI. FAMILY CRICETIDAE, New-World Rats and Mice.

1. Tail less than 25% of total length, *or* second and third hind toes webbed at base; enamel of molars arranged as 2 series of triangular or V-shaped ridges alternating in position (Plate X-5) 2
Tail more than 25% of total length; hind toes not webbed; enamel of molars in the form of cusps or ridges rarely triangular, but not alternating in position 4

2. Upperparts not reddish; toes webbed at base; length of tail more than 40 mm; length of skull about 45 mm; molars partly blackish
.... 3
Upperparts with a reddish tinge; toes not webbed at base; tail length less than 30 mm; length of skull about 20 to 25 mm; molars whitish .... Pine Vole (*Microtus pinetorum*), sp.811

3. Total length less than 400 mm; tail round in cross section; midline of skull depressed between front of orbits; widespread ....
.... Florida Water Rat (*Neofiber alleni*), sp.812

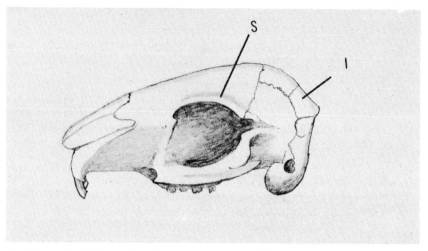

1. Interparietal (I) and fused supraorbital ridge (S) of
   Sylvilagus palustris

2. Five plantar pads of
   Peromyscus floridanus

PLATE XI

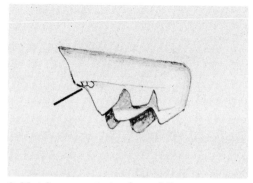

3. Notches on anterior face of first
   upper molar in *Rattus norvegicus*

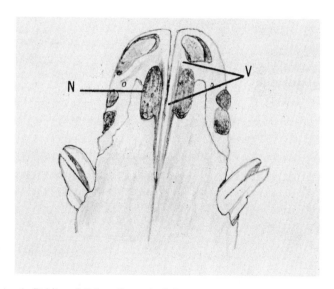

4. Palate of *Odocoileus virginianus*,
   showing vomer (V) dividing nasal cavity (N)

MAMMALS

Total length of adults more than 400 mm; tail laterally compressed; midline of skull ridged between orbits; no Florida records . . . . Muskrat (*Ondatra zibethicus*), sp.859H

4. Hairs of back about 14 mm long; enamel of molars in the form of ridges of almost unvarying height; basilar length of skull usually more than 25 mm 5

  Dorsal hairs much less than 14 mm long; enamel of molars arranged as 2 parallel rows of rounded cusps, connected only by lower ridges; basilar length of skull usually less than 25 mm 6

5. Length of head and body usually more than 170 mm; back rather uniform brownish; basilar length of skull more than 30 mm; molars blackish distally . . . .
  . . . . Eastern Wood Rat (*Neotoma floridana*), sp.810

  Length of head and body usually less than 170 mm; dorsal fur grizzled; enamel ridges of molars not blackish; basilar length of skull less than 30 mm . . . .
  . . . . Hispid Cotton Rat (*Sigmodon hispidus*), sp.809

6. Length of head and body less than 70 mm; belly hairs uniformly gray; upper incisors with a conspicuous median groove; basilar length of skull less than 16 mm . . . .
  . . . . . . Eastern Harvest Mouse (*Reithrodontomys humulis*), sp.804

  Head-body length more than 70 mm; belly hairs bicolored (dark basally, white distally); upper incisors not grooved; basilar length of skull more than 16 mm 7

7. Tail sparsely haired, appearing scaly; length of hind foot 28 mm or more; supraorbital ridge present; basilar length of skull more than 22 mm . . . . Rice Rat (*Oryzomys palustris*), sp.803

  Tail not so sparsely haired, not appearing scaly above; length of hind foot less than 28 mm; supraorbital ridge absent; basilar skull length less than 22 mm 8

8. With 5 plantar pads (one behind fifth toe; Plate XI-2); hind foot more than 24 mm long; ear more than 20 mm long; basilar length of skull more than 20 mm . . . .
  . . . . Florida Deer Mouse (*Peromyscus floridanus*), sp.807

  Usually with 6 plantar pads (2 behind fifth toe), but sometimes with only 5; hind foot usually not more than 24 mm long; ear less than 20 mm long; basilar length of skull usually not exceeding 20 mm 9

9. Upperparts (sides only of immatures) ochraceous to golden brown; ears ochraceous or brownish; posterior palatal foramina at level of posterior part of second molars . . . .
  . . . . Golden Mouse (*Ochrotomys nuttalli*), sp.808

  Upperparts brownish gray to reddish brown (never golden brown to ochraceous); ears mostly dusky; posterior palatal foramina near anterior edge of second molars 10

10. Tail usually less than 65% of head-body length; length of head and body usually less than 92 mm; length of hind foot less than 20 mm; greatest length of skull about 24 mm in adults . . . .
. . . . Old-field Mouse (*Peromyscus polionotus*), sp.805
Tail usually more than 65% of head-body length; length of head and body usually more than 92 mm; length of hind foot at least 20 mm; greatest length of skull about 28 mm in adults . . . .
. . . . Cotton Mouse (*Peromyscus gossypinus*), sp.806

ORDER RODENTIA: CLVII. FAMILY MURIDAE, Old-World Rats and Mice.

1. Total length less than 200 mm; length (antero-posterior) of first molar greater than that of second and third combined . . . .
. . . . House Mouse (*Mus musculus*), sp.815
Total length of adults more than 200 mm; length of first molar not more than combined length of second and third 2
2. Tail longer than head and body; no distinct notches on anterior surface of first molar; greatest distance between temporal ridges much greater than length of parietal bone . . . .
. . . . Black Rat (*Rattus rattus*), sp.813
Tail shorter than head and body; with distinct notches on anterior surface of first molar (Plate XI-3); greatest distance between temporal ridges not greater than length of parietal bone . . . .
. . . . Norway Rat (*Rattus norvegicus*), sp.814

ORDER RODENTIA: CLVIII. FAMILY CAPROMYIDAE. Hutias and Nutrias.
One species in Florida . . . . Nutria (*Myocastor coypus*), sp.816

ORDER CARNIVORA: CLIX. FAMILY CANIDAE, Foxes, Coyotes, and Wolves.[1]

1. Length of tail (to fleshy tip) less than half that of head and body; postorbital processes convex dorsally; basilar length of skull more than 145 mm . . . . Coyote (*Canis latrans*), sp.817
Tail more than half length of head and body; postorbital processes concave dorsally; basilar length of skull less than 145 mm 2
2. Tail provided with stiff black hairs down dorsal midline; no black on feet; lower jaw incised posteriad; statewide . . . .
. . . . Gray Fox (*Urocyon cinereoargenteus*), sp.819
Hair on tail soft and not black on dorsal midline; anterior (upper) surface of feet black; lower jaw not incised posteriad; largely confined to Panhandle.. . . . . . . Red Fox (*Vulpes vulpes*), sp.818

---

1. The Red Wolf (*Canis rufus*), which formerly occurred in Florida, is here considered a race of the Coyote (*Canis latrans*), but the relationships of the wolves and coyotes are complicated by interbreeding among themselves and with the domestic dog (*Canis familiaris*). The native range of *C. latrans* has also been obscured by introductions.

ORDER CARNIVORA: CLX. FAMILY URSIDAE, Bears. One species in Florida .... Black Bear (*Ursus americanus*), sp.820

ORDER CARNIVORA: CLXI. FAMILY PROCYONIDAE, Raccoons. One species in Florida .... Raccoon (*Procyon lotor*), sp.821

ORDER CARNIVORA: CLXII. FAMILY MUSTELIDAE, Mustelids.

1. Toes webbed; tail abnormally thick (diameter at base about 30 mm or more in adults); 5 molariform teeth in each side of upper jaw .... River Otter (*Lutra canadensis*), sp.822
   Toes not webbed; tail normal; only 4 molariform teeth in upper jaw 2

2. Tail bushy; pelage black and white (rarely black only); posterior edge of bony palate at level of last molar 3
   Tail hairs of moderate length; pelage brown above; posterior edge of bony palate well behind last molar 4

3. With one pair of dorsal white to buffy stripes or a single large white or buffy patch (sometimes absent); hind foot more than 55 mm long; frontal bone curving abruptly downward anteriad ....
   .... Striped Skunk (*Mephitis mephitis*), sp.826
   With 2 pairs of broken white stripes dorsally; hind foot less than 55 mm long; frontal gradually sloping downward anteriad ....
   .... Spotted Skunk (*Spilogale putorius*), sp.825

4. Pelage dark underneath; skull length more than 55 mm; length of auditory bulla less than that of upper jaw tooth row and about equal to that of orbit (ventral aspect) ....
   .... Mink (*Mustela vison*), sp.824
   Pelage yellowish to white below; skull length less than 55 mm; auditory bulla about same length as upper jaw tooth row and greater than that of orbit ....
   .... Long-tailed Weasel (*Mustela frenata*), sp.823

ORDER CARNIVORA: CLXIII. FAMILY FELIDAE, Cats.[1]

1. Tail more than half length of head and body; 4 pairs of molariform teeth, the first and third of greatly reduced size 2
   Tail short, much less than half length of head and body; only 3 pairs of upper cheek teeth, the third greatly reduced ....
   .... Bobcat (*Lynx rufus*), sp.829

2. Total length of adult at least 150 cm; skull length more than 140 mm .... Panther (*Felis concolor*), sp.827

---

1. Records of the Jaguar (*Felis onca*) and Ocelot (*Felis pardalis*) in Florida doubtless pertain to escaped individuals (Layne, 1969).

Total length less than 125 cm; skull length less than 140 mm .....
.... Jaguarundi (*Felis yagouaroundi*), sp.828

ORDER PINNIPEDIA: CLXIV. FAMILY PHOCIDAE, Hair Seals.[1]

1. Males with an inflatable pouch on snout; only one pair of lower
   incisors; premaxilla not in contact with nasal ....
   .... Hooded Seal (*Cystophora cristata*), sp.832
   Pouch lacking; with 2 pairs of lower incisors; premaxilla touching
   nasal 2
2. With 3 pairs of upper incisors; all nails well developed ....
   .... Harbor Seal (*Phoca vitulina*), sp.830
   Only 2 pairs of upper incisors; nails of hind foot vestigial ....
   .... West Indian Seal (*Monachus tropicalis*), sp.831

ORDER SIRENIA: CLXV. FAMILY TRICHECHIDAE, Manatees. One species
in Florida .... Manatee (*Trichechus manatus*), sp.833

ORDER CETACEA: CLXVI. FAMILY DELPHINIDAE, Dolphins.

1. Head with a conspicuous beak; width of rostrum at base not more
   than half its length; teeth numerous (not less than 20 in each
   row) 2
   Beak lacking; width of rostrum more than half its length; teeth less
   numerous (less than 20 per row)[2] 8
2. Slope of beak and head gradual throughout; lower jaws united for
   more than 25% length of mandibular ramus; crowns of teeth
   provided with grooves; less than 30 teeth per row ....
   .... Rough-toothed Dolphin (*Steno bredanensis*), sp.834
   Profile indented at base of beak; lower jaws united for less than
   25% length of ramus; crowns of teeth smooth, even if cone-shaped;
   usually more than 30 teeth per row 3
3. With a black ring around eye; a deep groove on each side of bony
   palate .... Common Dolphin (*Delphinus delphis*), sp.839
   No black ring around eye; palatal grooves lacking 4
4. Beak only 2 or 3% of total length; with less than 30 teeth per row,
   each about 10 mm in diameter; common, widespread ....
   ...... Atlantic Bottle-nosed Dolphin (*Tursiops truncatus*), sp.840
   Beak making up more than 5% of total length; more than 30 teeth
   per row, each much less than 10 mm in diameter 5

---

1. Records of California Sea Lions (*Zalophus californianus*, Family Otariidae)
are surely based on escapes; see Gunter, 1968.

2. The latter half of this couplet does not fully apply to the Atlantic Porpoise
(*Phocoena phocoena*), which is likely to stray as far south as Florida.

5. At least 44 teeth per upper row; with 72 or more vertebrae; snout more than 8% of total length, *or* sides of body with conspicuous longitudinal streaks 6

Usually less than 44 teeth per upper row; vertebrae less than 72; snout not more than 8% of total length; no conspicuous longitudinal streaks on sides of body 7

6. Rostral length more than 60% that of skull; flipper less than 15% of total length; no conspicuous longitudinal streaks on sides of body ...... Long-beaked Dolphin (*Stenella longirostris*), sp.836

Rostral length less than 60% that of skull; flipper at least 15% of total length; with a thin black streak from eye to anus and 2 from eye to flipper ...... Gray's Dolphin (*Stenella coeruleoalba*), sp.838

7. Greatest skull width of adults more than 192 mm; flipper length (from anterior insertion) about one-fifth of total length; dorsal color purplish gray, mottled with whitish ....

.... Spotted Dolphin (*Stenella plagiodon*), sp.837

Greatest skull width less than 192 mm; length of flipper about one-sixth of total length; dorsal color blackish ....

.... Cuvier's Dolphin (*Stenella frontalis*), sp.835

8. Anterior end of head truncate in profile; width of rostrum at base at least 80% of its length 9

Anterior aspect of head rounded; width of rostrum at base not more than 70% of its length 10

9. Usually with more than 7 pairs of teeth each on upper and lower jaws; width of rostrum at midpoint almost as great as its length; snout black; length of flipper about 20% of total length ....

Short-finned Pilot Whale (*Globicephala macrorhyncha*), sp.844

Usually no teeth on upper jaw and less than 7 pairs on lower jaw; width of rostrum at midpoint barely more than half its length; snout mostly white; length of flipper less than 20% of total length .... Risso's Dolphin (*Grampus griseus*), sp.843

10. Total length to 6 m or less; flipper slender, its width less than half its length (from anterior insertion); pterygoids in contact 11

Total length to 10 m; flipper large, broad, and rounded, its width about half its length; pterygoids not in contact ....

.... Killer Whale (*Orcinus orca*), sp.842

11. Total length up to 550 cm; color mostly dark; vertebrae about 50 .... False Killer Whale (*Pseudorca crassidens*), sp.841

Total length not more than 300 cm; much white ventrally; vertebrae about 70 .... Pygmy Killer Whale (*Feresa attenuata*), sp.878

ORDER CETACEA: CLXVII. FAMILY PHYSETERIDAE, Sperm Whales.

1. Dorsal fin rudimentary and incomplete; zygomatic arch complete .... Sperm Whale (*Physeter catodon*), sp.845

Dorsal fin well developed, falcate; zygomatic arch interrupted 2
2. Dorsal fin more than halfway back from anterior tip of snout; no maxillary teeth; with a ventral keel on the mandibular symphysis . . . . Pygmy Sperm Whale (*Kogia breviceps*), sp.846
Dorsal fin halfway back on body; usually 1 to 3 pairs of maxillary teeth; no ventral keel on the mandibular symphysis . . . .
. . . . Dwarf Sperm Whale (*Kogia simus*), sp.879

ORDER CETACEA: CLXVIII. FAMILY HYPEROODONTIDAE, Beaked Whales.

1. Mouth extending less than halfway back to eye; nasals forming crest of skull and almost obscuring narial openings from dorsal view; one pair of cone-shaped teeth near tip of lower jaw . . . .
. . . . Goose-beaked Whale (*Ziphius cavirostris*), sp.849
Angle of jaws almost even with eye; nasals sunk between upper (posterior) ends of premaxilla; teeth of lower jaw compressed
. . . . 2
2. Dorsal fin about 60% of distance from anterior to posterior tip (that is, to middle of flukes); functional teeth posterior to mandibular symphysis . . . .
. . . . . Blainville's Beaked Whale (*Mesoplodon densirostris*), sp.880
Dorsal fin about 70% distance from anterior to posterior tip; functional teeth anterior to posterior edge of mandibular symphysis
. . . . 3
3. Teeth located at tip of lower jaw; flipper less than 10% of body length? . . . .
. . . . True's Beaked Whale (*Mesoplodon mirus*), sp.848
Teeth located near posterior edge of mandibular symphysis; flipper more than 10% length of body? . . . .
. . . . . . Antillean Beaked Whale (*Mesoplodon europaeus*), sp.847

ORDER CETACEA: CLXIX. FAMILY BALAENOPTERIDAE, Fin-backed Whales.[1]

1. Length of flipper about one-third of total length; throat grooves 130 to 200 mm apart; sides of rostrum convex anteriad and skull widely flared at rostral base; width of rostrum just anterior to flared portion about 50 to 55% width of flared portion . . . .
. . . . Humpback Whale (*Megaptera novaeangliae*), sp.854
Flipper length less than one-sixth of total length; throat grooves only 50 to 75 mm apart; sides of rostrum mostly straight and skull not so widely flared at rostral base; width of rostral base 65 to 70% that of flared portion 2

1. The Blue Whale (*Balaenoptera musculus*), the largest known animal, probably reaches Florida waters occasionally (Caldwell, personal communication).

2. With a white patch on outer (upper) surface of flipper; baleen yellowish white, and longest plates more than 305 mm; total length of adults less than 11 m ....
    .... Little Piked Whale (*Balaenoptera acutorostrata*), sp.850
   No white patch on outer surface of flipper; baleen not entirely yellowish white, and longest plates more than 305 mm; maximum length more than 11 m 3
3. Total length up to 24.5 m; right side of head lighter than left side; greatest length of baleen plates (exclusive of bristles) about 91 cm; dorsal fin located at 75% of distance from snout to tip of flukes .... Finback Whale (*Balaenoptera physalus*), sp.853
   Total length less than 21.5 m; right and left sides of head of same color; greatest length of baleen plates not more than 76 cm; dorsal fin not more than 70% back from tip of snout 4
4. Ventral grooves ending near mid-body; length of baleen plates about 73 cm and their bristles fine and white; head not ridged; front edge of nasals straight on lateral sides ....
    .... Sei Whale (*Balaenoptera borealis*), sp.851
   Ventral grooves extending back to umbilicus; baleen plates shorter than 73 cm and their bristles coarse and gray; external ridges sometimes present on sides of head; front edge of nasals bent forward on lateral sides ....
    .... Bryde's Whale (*Balaenoptera edeni*), sp.852

ORDER CETACEA: CLXX. FAMILY BALAENIDAE, Whalebone Whales. One species in Florida ....
    .... Atlantic Right Whale (*Eubalaena glacialis*), sp.855

ORDER ARTIODACTYLA: CLXXI. FAMILY SUIDAE, Pigs. One species in Florida .... Pig (*Sus scrofa*), sp.856

ORDER ARTIODACTYLA: CLXXII. FAMILY CERVIDAE, Deer.

1. Branches (tines) of antlers coming separately from a central trunk; tail longer than ear; adults unspotted; posterior portion of nasal cavity separated by the vomer (Plate XI-4); statewide ....
    .... White-tailed Deer (*Odocoileus virginianus*), sp.859
   Antlers dichotomously branched, though only 3-pronged; tail shorter than ear, *or* adults spotted; posterior portion of nasal cavity not divided by vomer; restricted to northeast Florida or to St. Vincent Island 2
2. Central (second) prong of antler the longest one; white spots present at all ages; upper canines lacking; found only east of the St. Johns River .... Axis Deer (*Axis axis*), sp.858
   Third prong of antler at least as long as second; upper canine present, but rudimentary; unspotted; found only on St. Vincent Island
    .... Sambar Deer (*Cervus unicolor*), sp.857

Chapter Three

Species Accounts

CLASS AGNATHA: JAWLESS VERTEBRATES

ORDER PETROMYZONTIFORMES: I. FAMILY PETROMYZONTIDAE, Lampreys. Long, slender, eel-shaped animals with *no scales or paired fins*. The "mouth" (buccal funnel) is *permanently wide open and beset with numerous horny teeth*. There are 7 pairs of gill openings and a single nasal pit.

1. Sea Lamprey, *Petromyzon marinus* L.

IDENTIFICATION: An elongate, terete, eel-like animal without scales or paired fins. It differs from the Southern Brook Lamprey in its *mottled, brownish color pattern, 2 separate dorsal fins* (the second conjoined to the caudal), and its larger size (up to 1 m in total length).

DISTRIBUTION AND VARIATION: Atlantic Coast of North America and Europe, southward to the St. Johns River, Florida; also the Great Lakes. Only *P. m. marinus* occurs in Florida. Anadromous.

2. Southern Brook Lamprey, *Ichthyomyzon gagei* Hubbs and Trautman

IDENTIFICATION: Similar to the Sea Lamprey in its eel-like form, but much smaller (less than 255 mm long). The *dorsal fin is single* and conjoined to the caudal, and the *body a uniform grayish color*. The buccal funnel is relatively small.

DISTRIBUTION: Chiefly in lower Mississippi River drainage, but in Florida streams eastward to Leon County; also western Georgia and most of Alabama.

CLASS CHONDRICHTHYES: CARTILAGINOUS FISHES

ORDER SQUALIFORMES: II. FAMILY CARCHARHINIDAE, Requiem Sharks. A group of rough-skinned, terete fishes with *5 pairs of exposed gill slits*, a *ventral, crescent-shaped mouth*, and a strongly *heterocercal tail*.

3. Atlantic Sharp-nosed Shark, *Rhizoprionodon terraenovae* (Richardson)[1]

IDENTIFICATION: This shark is very similar to the Bull Shark, but has a *longer snout* (see Key), *labial grooves* at the corners of the mouth, and a more *terete body*. The general color is grayish. Total length up to 1 m.

DISTRIBUTION: Tropical and subtropical Atlantic Ocean, straying northward to Massachusetts; thought to enter mouths of Florida rivers (Carr and Goin, 1955).

4. Bull Shark, *Carcharhinus leucas* (Valenciennes)

IDENTIFICATION: Similar to the Sharp-nosed Shark, but body deeper, snout shorter, and mouth *without noticeable labial grooves*. The color pattern (grayish above, whitish below) is not distinctive. Total length up to 3 m.

DISTRIBUTION: New York to Brazil; thought to enter Florida rivers (Carr and Goin, 1955).

ORDER RAJIFORMES: III. FAMILY PRISTIDAE, Sawfishes. Body *triangular in cross section* (flattened ventrally), but otherwise sharkshaped. The *greatly produced, flattened snout, beset laterally with teeth*, separates them from all other Florida fishes. It is also edged with small cirri. Scales uniformly small and flattened. Maximum length about 6 m.

5. Small-toothed Sawfish, *Pristis pectinata* Latham

IDENTIFICATION: A large sawfish with *24 or more teeth on each side*, a dorsal fin originating directly above the pelvics, and *no ventral lobe on the caudal fin*. The saw is about 25 percent of its total length (which may reach 5.5 m). Coloration dark above, lighter below.

DISTRIBUTION: Atlantic Ocean from the Mediterranean Sea to the Equator of West Africa, and from North Carolina (rarely Chesapeake Bay) and the Gulf of Mexico to Brazil. Euryhaline (Bigelow and Schroeder, 1953).

---

1. See Springer, 1964.

6. Large-toothed Sawfish, *Pristis perotteti* Müller and Henle

IDENTIFICATION: A large sawfish with *not more than 20 teeth* on each side, the origin of the dorsal fin anterior to that of the pectoral, and a definite *ventral lobe on the caudal fin*. Saw 20 to 22 percent of total length. Coloration and size as in *P. pectinata*.

DISTRIBUTION: Gulf of Mexico and Atlantic Ocean, from Texas and Louisiana (rarely southern Florida) to Brazil. Euryhaline (Bigelow and Schroeder, 1953).

ORDER RAJIFORMES: IV. FAMILY DASYATIDAE, Sting Rays. *Naked, dorsoventrally flattened fishes*, with a *long, whiplike tail*. Total length up to about 120 cm. Mouth ventral and crescent-shaped. With a prominent spine on dorsal surface of tail.

7. Atlantic Sting Ray, *Dasyatis americana* Hildebrand and Schroeder[1]

IDENTIFICATION: Rays are *extremely flattened dorsoventrally, thus disk-shaped*. This species differs from the Southern Sting Ray in having a more *rhomboidal body outline* and a broader snout (see Key). Coloration brownish gray above and paler below. Maximum width about 150 cm.

DISTRIBUTION: Atlantic Ocean and Gulf of Mexico from New Jersey to Brazil. Enters St. Johns River.

8. Southern Sting Ray, *Dasyatis sabina* (Lesueur)

IDENTIFICATION: Similar in body form, color, and size to the Atlantic Sting Ray, but *outline more circular* (see Key) and snout more angular. Maximum width about 51 cm.

DISTRIBUTION: Gulf of Mexico to Maryland, entering Florida rivers.

CLASS OSTEICHTHYES: BONY FISHES

ORDER ACIPENSERIFORMES: V. FAMILY ACIPENSERIDAE, Sturgeons. Elongate, terete fishes, with a *pronounced snout* and a markedly *heterocercal tail*. Mouth ventral. Body largely covered with *5 rows of bony plates*. Small barbels under snout.

9. Atlantic Sturgeon, *Acipenser oxyrhynchus* Mitchill

IDENTIFICATION: (See description of family.) This sturgeon differs from the Short-nosed Sturgeon in the *position of its anal fin*, the anterior

---

1. *Dasyatis* "*hastata*" (now *D. centroura*), listed by Carr and Goin (1955), is not known to occur in Florida fresh water.

edge of which *is posterior to that of the dorsal fin*. It also reaches larger size, up to 250 or 300 cm (rarely 550 cm in North) in total length. Color grayish or brownish.

DISTRIBUTION AND VARIATION: Gulf of Mexico and Atlantic Ocean, northward to Quebec, entering large rivers. Represented by *A. o. desotoi* in the Gulf of Mexico and by *A. o. oxyrhynchus* in the Atlantic Ocean.

10. Short-nosed Sturgeon, *Acipenser brevirostrum* Lesueur

IDENTIFICATION: Similar to the Atlantic Sturgeon, but smaller (less than 100 cm long) and with the *anterior edge of the anal fin directly below that of the dorsal fin*.

DISTRIBUTION: Atlantic Ocean, from Massachusetts to the St. Johns River, Florida (Kilby et al., 1959). Anadromous.

ORDER LEPISOSTEIFORMES: VI. FAMILY LEPISOSTEIDAE, Gars. With their *terete body form*, *ganoid scales*, *long snouts*, sharp and *prominent teeth*, and rounded, *heterocercal caudal fins*, these are most distinctive fishes.

11. Alligator Gar, *Lepisosteus spatula* Lacépède

IDENTIFICATION: This largest species of gar in Florida has a *proportionately short, wide snout* (see Key). It also differs in having the teeth of the inner row of the upper jaw as wide as those of the outer row. Color an undistinctive grayish or brownish, with little variation over the body. Total length to 250 or 280 cm.

DISTRIBUTION: Ohio and Missouri Rivers southward to the Gulf Coast and northern Mexico, ranging eastward in Florida to the Choctawhatchee River (Bigelow et al., 1963).

12. Spotted Gar, *Lepisosteus oculatus* (Winchell)[1]

IDENTIFICATION: Smaller than the Alligator Gar, and with a *proportionately longer, more narrow snout*. With *prominent dark spots* on a lighter, olive green background, it differs from the Florida Gar in having the *underparts dark with the anterior portion naked*. Young usually have a *longitudinal dark stripe* on the side of the body. Like the Florida and Long-nosed Gars, this gar has the teeth in the inner row of the upper jaw more slender than those in the outer row. Maximum length 60 to 75 cm.

DISTRIBUTION: Lakes Michigan and Erie, Mississippi River drain-

---

1. This fish and the Florida Gar may be conspecific.

age, and the Gulf of Mexico drainage westward to Texas and eastward to the Apalachicola River in Florida and the Flint River in Georgia.

13. Florida Gar, *Lepisosteus platyrhincus* DeKay

IDENTIFICATION: Very similar to the Spotted Gar, but *underparts lighter and completely scaled*. Dark stripe down side of young. Total length up to 75 cm, rarely 120 cm (Bigelow et al., 1963).

DISTRIBUTION: Southern and eastern Georgia throughout Florida Peninsula, ranging westward to the Ochlockonee River (Bigelow et al., 1963; Map 3).

14. Long-nosed Gar, *Lepisosteus osseus* (L.)

IDENTIFICATION: Differs from other gars in its *longer, more slender snout* (more than twice remaining head length). Young have a black lateral line, but are unspotted. Adults not conspicuously spotted and reach larger size (up to 150 cm long).

DISTRIBUTION AND VARIATION: Atlantic drainage from Quebec to Florida; also Great Lakes to Gulf Coast and northern Mexico. The form throughout Florida has been named *L. o. osseus* (Hubbs and Allen, 1943; also see Bigelow et al., 1963).

ORDER AMIIFORMES: VII. FAMILY AMIIDAE, Bowfins. Heavy-bodied fishes with a *depressed head and wide mouth. Head covered with bony plates dorsally and a gular plate ventrally*. Length of dorsal fin about half of total length. Caudal fin decidedly heterocercal in young, less so in larger specimens, when appearing rounded-homocercal. Male has a *dark spot at base of caudal fin*. Scales cycloid. Lateral line present. Color dark with little or no pattern. Total length up to 75 cm. Monotypic.

15. Bowfin, *Amia calva* L.

IDENTIFICATION: See description of family.

DISTRIBUTION: Most of the eastern United States, including all of mainland Florida, westward to the Great Plains; absent from much of New England and at high elevations.

ORDER ELOPIFORMES: VIII. FAMILY ELOPIDAE, Ladyfishes. Strongly tapered at both ends, with the *peduncle relatively slender. Gular plate* and lateral line *present*. Pelvic fins abdominal. Caudal fin forked. No spines. Scales smaller than eye.

16. Ladyfish, *Elops saurus* L.

IDENTIFICATION: See description of family. The body form of the Ladyfish resembles that of a Tarpon, but is more slender (body depth

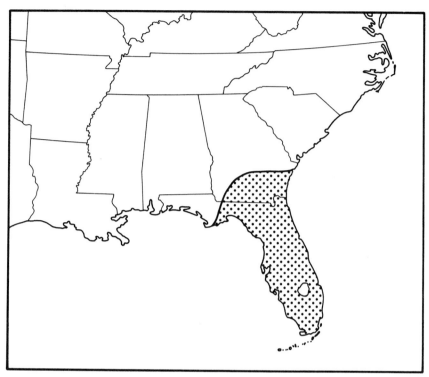

Map 3. Distribution of the Florida Gar (*Lepisosteus platyrhincus*)

about 15 percent of standard length). *No long posterior process on dorsal fin*. Dorsal and anal fins retractable into sheaths. Scales small (120 along lateral line). Upper jaw projects beyond lower. Adipose eyelid well developed. Color silvery, darker above. Total length up to 1 m.

DISTRIBUTION: Tropical seas. In western Atlantic Ocean, ranging northward to Bermuda and North Carolina (rarely Massachusetts), entering Florida rivers and reaching Lake Okeechobee.

ORDER ELOPIFORMES: IX. FAMILY MEGALOPIDAE, Tarpons. Body form similar to that in the Elopidae but size larger, scales larger, and dorsal fin with a *long posterior-projecting ray*. Lower jaw projects beyond upper. Gular plate and lateral line present.

17. Tarpon, *Megalops atlantica* Valenciennes

IDENTIFICATION: Very large, silvery fishes with large rounded scales (larger than eye), about 45 along lateral line. Dorsal fin with a long posterior projection. Mouth superior. Anal fin much longer than in the Ladyfish. Total length up to 250 cm and weight to 130 to 150 kg, but specimens taken in fresh water are much smaller.

DISTRIBUTION: Gulf of Mexico and Atlantic Ocean from Nova Scotia to Brazil, entering Florida rivers and canals.

ORDER ANGUILLIFORMES: X. FAMILY ANGUILLIDAE, Freshwater Eels. Eels are among the most distinctive of fishes, with *long, cylindrical, snakelike bodies and no apparent scales*. The single median fin is very long, representing *continuous dorsal, caudal, and anal fins*. The opercular opening is generally reduced.

18. American Eel, *Anguilla rostrata* (Lesueur)

IDENTIFICATION: (See family description.) This species, like most eels, has no pelvic fins and is of a yellowish brown color. Lower jaw protrudes beyond upper. Maximum length about 150 cm.

DISTRIBUTION: Eastern North America (west to Alberta), Central America, and northern South America, including all of Florida; catadromous. May be conspecific with the European Eel.

ORDER ANGUILLIFORMES: XI. FAMILY OPHICHTHIDAE, Snake Eels. Small, extremely elongate, worm-shaped eels with a dorsal fin running most of the length of the body, or lacking. Lower jaw shorter than upper.

19. Speckled Worm Eel, *Myrophis punctatus* Lütken

IDENTIFICATION: (See description of family.) Dull olive above, lighter below, with upperparts and dorsal and anal fins showing small black dots. Maximum length about 330 mm.

DISTRIBUTION: The West Indies and the Gulf Coast northward to North Carolina. One record in the lower Escambia River of Florida (Bailey, Winn, and Smith, 1954).

ORDER CLUPEIFORMES: XII. FAMILY CLUPEIDAE, Herrings and Shad.[1] Compressed, strongly tapered fishes, with slender peduncles and deeply forked caudal fins. They differ from the Elopiformes in *lacking the gular plate and lateral line. Ventral scales serrate or forming a ridge*.

20. Skipjack Herring, *Alosa chrysochloris* (Rafinesque)

IDENTIFICATION: A compressed fish with a deeply forked caudal fin and a triangular dorsal fin. Top of head essentially straight in profile. *Lower jaw projecting beyond upper*. Teeth present at tip of upper jaw.

---

1. The Alewife (*Alosa pseudoharengus*), once attributed to Florida (Carr and Goin, 1955), does not occur here (Yerger, personal communication).

Color dark above, silvery below, with golden iridescence on sides in life. Total length up to 355 mm.

DISTRIBUTION: Southern Great Lakes southward to the Gulf of Mexico, ranging eastward to the Apalachicola River system. Euryhaline.

21. Hickory Shad, *Alosa mediocris* (Mitchill)

IDENTIFICATION: This shad resembles the Skipjack Herring in its *protruding lower jaw*, but differs in that *adults usually are without teeth* on jaws or tongue. Profile of head not indented. The anal fin has 21 rays and the dorsal 15. Coloration bluish above, silvery below, with faint longitudinal stripes; also a row of dark spots extending backward from operculum. Maximum length about 600 mm.

DISTRIBUTION: Atlantic Coast, Nova Scotia to Florida, entering streams.

22. Blue-backed Herring, *Alosa aestivalis* (Mitchill)

IDENTIFICATION: A shad that differs from others in the genus in having a *black peritoneum* (as opposed to white or light). Lower jaw protrudes beyond upper. Lower lobe of caudal fin slightly longer than upper. Like *A. sapidissima*, it has a dark spot behind the operculum. The faint longitudinal stripes are not entirely diagnostic. The silvery color and the bluish back are shared with other shads. Maximum length about 305 mm.

DISTRIBUTION: Atlantic Coast from Nova Scotia to Florida, entering streams.

23. American Shad, *Alosa sapidissima* (Wilson)

IDENTIFICATION: As in certain other shads, there are many more rays in anal fin than in dorsal (about 20 versus 14), but the peritoneum is light as in most shads. Longitudinal scale count high (60 or more). Silvery spot below eye deeper than long. Teeth lacking. About 60 gill rakers on lower part of first gill arch. *Dark spot behind operculum, often followed by fainter spots in a longitudinal row.* Mouth superior. Length up to 600 or 750 mm.

DISTRIBUTION: Atlantic Coast from Newfoundland to Florida, entering streams.

24. Alabama Shad, *Alosa alabamae* Jordan and Evermann

IDENTIFICATION: Differs from *A. sapidissima* in having *only about 40 gill rakers in lower arm of first gill arch* and never having a longitudinal row of spots behind operculum. Less than 60 oblique scale rows along side of body. About 20 rays in anal fin and 15 in dorsal. Silvery patch on cheek deeper than long. Maximum length 450 mm or more.

DISTRIBUTION: Gulf of Mexico drainage (coastal plain) southward to the Suwannee River.

25. Atlantic Menhaden, *Brevoortia tyrannus* (Latrobe)

IDENTIFICATION: The menhadens are similar to the shads (*Alosa*), but differ in the *dimensions of the scales* (greater in vertical than in horizontal distance) and in their larger mouth. The Atlantic Menhaden differs from others in the genus in having a *lower lateral-scale count* (usually under 56) and in the *numerous dark spots* present in most medium-sized or large specimens. About 36 pairs of modified scales anterior to dorsal fin and 30 to 32 on ventral midline. Upperparts dark bluish green, remainder of ground color brassy yellowish. Maximum length about 300 mm.

DISTRIBUTION: Atlantic Coast from New Brunswick to Palm Beach County, Florida; euryhaline (Bigelow et al., 1963).

26. Large-scaled Menhaden, *Brevoortia patronus* Goode

IDENTIFICATION: Resembles the Atlantic Menhaden in size of scales, but differs in form and length of pectoral fin (see Key). Also there are essentially *2 rows of scales in the sheath subtending the dorsal fin* as against one row under the anterior half of the dorsal fin in *B. tyrannus*. About 40 to 45 oblique scale rows, and *less than 32 pairs of modified scales anterior to dorsal fin*. Coloration similar to that of the Atlantic Menhaden, but typically with fewer black spots (1–2 rows or a single spot) along sides. Maximum length about 250 mm. Hybridizes with *B. smithi*.

DISTRIBUTION: Gulf Coast from Lee County, Florida, to Yucatan; euryhaline (Suttkus, 1958; Turner, 1971).

27. Yellowfin Shad, *Brevoortia smithi* Hildebrand

IDENTIFICATION: Resembling some shads of the genus *Alosa* in having a single dark spot behind the operculum; otherwise mostly silvery. Lateral scale count usually more than 60. About *42 pairs of modified scales anterior to dorsal fin*. Maximum length 300 to 350 mm. Hybridizes with *B. patronus*.

DISTRIBUTION: Atlantic and Gulf coasts from North Carolina to the Bahamas and the Mississippi River Delta; euryhaline (Suttkus, 1958; Turner, 1971).

28. Scaled Sardine, *Harengula pensacolae* Goode and Bean

IDENTIFICATION: A small, slender, deep-bodied fish with large eyes and *scales that are more deep than wide*. The relatively smooth margins of the scales and the *more anterior position of the dorsal fin* distinguish it from a young menhaden, and the lack of an extended last dorsal ray from the smaller shads (*Dorosoma*). Color silvery, darker above, and often with a dark spot just behind the operculum. Maximum length about 150 mm.

DISTRIBUTION: Atlantic and Gulf Coasts from northern Florida to Brazil, entering fresh water in the St. Johns River (Tagatz, 1967).

29. Gizzard Shad, *Dorosoma cepedianum* (Lesueur)

IDENTIFICATION: The 2 species of *Dorosoma* differ from other shads in the *long posterior process of the dorsal fin* and the large number of anal fin rays (20 or more). This species differs from the Threadfin Shad in having an *inferior mouth* and an anal ray count of about 30. There are more than 50 oblique scale rows. Adults have no dark spot behind the operculum. Color silvery, darker above. Up to 300 mm in length, rarely 380.

DISTRIBUTION: Freshwater streams through most of eastern United States, westward through Mississippi River drainage and to northeastern Mexico; in Florida it ranges throughout the mainland. Sometimes enters brackish, or even salt, water.

30. Threadfin Shad, *Dorosoma petenense* (Günther)

IDENTIFICATION: Like the Gizzard Shad in having the last dorsal ray greatly elongate, but differs in its *terminal mouth* and *smaller number of anal rays* (not more than 25) and of oblique scale rows (less than 45). Dark above, silvery on sides and underparts. Dark spot behind operculum present throughout life. Never more than 205 mm in total length.

DISTRIBUTION: Gulf of Mexico drainage from North Carolina and Oklahoma southward to southern Florida and British Honduras; introduced in California, Arizona, eastern Georgia, and elsewhere.

ORDER CLUPEIFORMES: XIII. FAMILY ENGRAULIDAE, Anchovies. Generally smaller than the Clupeidae, more slender, and with a *relatively larger eye*. Snout produced beyond mouth. *Mouth large*, with angle of jaws extending posterior to position of eyes. Silvery lateral stripe present. *Ventral scales not serrate, nor forming a ridge*.

31. Striped Anchovy, *Anchoa hepsetus* (L.)

IDENTIFICATION: *Stripe more prominent* than in the Bay Anchovy, its width as great as diameter of eye. Also with *fewer anal rays* (less than 25). Total length up to 150 mm.

DISTRIBUTION: Atlantic and Gulf Coasts from Massachusetts to Brazil, entering streams and canals in Florida.

32. Bay Anchovy, *Anchoa mitchilli* (Valenciennes)

IDENTIFICATION: Differs from the Striped Anchovy in having a *more narrow stripe*, *at least 25 rays in the anal fin*, and in its smaller size (not more than 100 mm long).

DISTRIBUTION AND VARIATION: Atlantic and Gulf Coasts from Maine to Yucatan; enters fresh water in Florida (*A. m. diaphana*).

ORDER SALMONIFORMES: XIV. FAMILY ESOCIDAE, Pickerels. Long, slender, almost terete fishes with a *depressed, somewhat upturned snout*, and a *dorsal fin placed far back on the body*. There is a lateral line, but no gular plate. Scales small, usually numbering more than 100 oblique rows along side of body. Caudal fin moderately forked. Peduncle not markedly slender.

33. Redfin Pickerel, *Esox americanus* Gmelin

IDENTIFICATION: Aside from the usual reddish color of the fins, the Redfin differs from the only other Florida pickerel in its shorter snout, *dark bars on sides* of adults, and its *lower number of oblique scale rows along the side* (110 or fewer). Also there are only 11 to 13 branchiostegal rays. Young are difficult to identify. Total length up to 300 mm.

DISTRIBUTION AND VARIATION: Most of the eastern United States (and southern Canada), from New Hampshire, Ontario, southern Michigan, and eastern Oklahoma southward to Texas and Florida. Typical *E. a. americanus* occurs only in the northeastern corner of Florida, but intergrades with *E. a. vermiculatus* throughout the rest of the state south to Lake Okeechobee (Crossman, 1966).

34. Chain Pickerel, *Esox niger* Lesueur

IDENTIFICATION: Except when very young, this pickerel differs markedly from the Redfin in its *reticulate pattern of dark markings*. Other differences are its longer snout, *higher lateral-scale count* (about 125), and greater number of branchiostegal rays (14 to 16). Maximum length 75 to 90 cm.

DISTRIBUTION: Southeastern Canada and eastern Great Lakes southward (east of Appalachians) to south Florida (*Fla. Scientist* 37:117); also in the Mississippi River Valley from Missouri, western Tennessee, and northern Alabama to Arkansas, eastern Texas, and the Gulf Coast.

ORDER SALMONIFORMES: XV. FAMILY UMBRIDAE, Mud Minnows. Small, heavy-bodied fishes with a sloping profile and a body form similar to a killifish's (*Fundulus*). Base of dorsal fin rather long, and caudal fin rounded. Lateral line absent.

35. Eastern Mud Minnow, *Umbra pygmaea* (DeKay)

IDENTIFICATION: A heavy-bodied little fish vaguely resembling a dark *Gambusia* in its rounded caudal fin, but with a deeper head and a dark vertical bar (sometimes interrupted) at base of tail. Total length up to 100 mm. (Some individuals show faint longitudinal streaks.)

DISTRIBUTION: Sluggish or calm bodies of water from New York to northeastern Florida (Aucilla River eastward).

ORDER CYPRINIFORMES: XVI. FAMILY CYPRINIDAE, Minnows and Carps. Although this order includes fishes of varying body form and scalation, the Family Cyprinidae is more homogeneous. The mouth is retatively small and completely without teeth. The gill membranes are united to the isthmus. The pelvic fins are abdominal, and the caudal fin deeply forked. Most are of small size and frequently show a *dark lateral stripe*.

36. Golden Shiner, *Notemigonus crysoleucas* (Mitchill)

IDENTIFICATION: Shape rather carplike, but profile of head concave, lower jaw protruding, and dorsal fin short (8 rays). Rays in anal fin (14–17) more than in other shiners. *Lateral line strongly decurved*, with about 3 scale rows below it. *A fleshy keel on belly*. Color silvery with a reddish or golden tint, especially on fins. Usually small, but rarely up to 300 mm long.

DISTRIBUTION: Eastern United States into southern Canada; introduced west of the Rocky Mountains.

37. Blacktip Shiner, *Notropis atrapiculus* (Snelson)[1]

IDENTIFICATION: The shiners (*Notropis*) resemble the chubs in body form, but lack barbels and are not so heavy-bodied as the species of *Semotilus*. They differ from *Notemigonus* in having no ventral ridge

---

1. *Bull. Fla. State Mus.*, 17:1–92.

and usually less than 13 anal rays. Typically the dorsal fin has less than 9 rays. Color usually silvery, often iridescent, and with the black lateral stripe often conspicuous. The Blacktip Shiner differs from other species of the genus in its combination of *10 or more anal rays and 22 or more scales between the dorsal fin and the head*. The front of the dorsal fin is placed well behind that of the pelvics. The rusty dorsal coloration and reddish fins are also unusual. Maximum length about 50 mm.

DISTRIBUTION: Western Georgia, southeastern Alabama, and northwestern Florida (Yellow and Choctawhatchee River drainages).

38. Bandfin Shiner, *Notropis zonistius* (Jordan)

IDENTIFICATION: A large shiner with a *diagonal black bar on the dorsal fin, 2 light spots at the base of the otherwise reddish caudal fin* (obscure in preservative), and a faint or incomplete lateral stripe. Body depth more than 25 percent of standard length. Eye large. Maximum length about 90 mm.

DISTRIBUTION: The Chattahoochee-Flint River system of Alabama, Georgia, and Florida (south to Chattahoochee); also the Tallapoosa River system of Alabama (Smith-Vaniz, 1968) and the upper Savannah River system (Gilbert, 1964) (Map 4).

39. Blue-striped Shiner, *Notropis callitaenia* Bailey and Gibbs

IDENTIFICATION: Rather pale, bluish above and silvery below; *incomplete lateral stripe blue in life*. Head narrow and pointed. Dorsal scales outlined with dusky; dorsal fin mostly clear. Total length up to 90 mm. (Also see Key.)

DISTRIBUTION: Northwestern Florida, southern Alabama, and southern Georgia (Apalachicola River system; see Smith-Vaniz, 1968).

40. Ohoopee Shiner, *Notropis leedsi* Fowler

IDENTIFICATION: Very similar to *N. callitaenia*, but *lateral stripe gray* in life *and more nearly complete. Dorsal fin dusky. Breast naked* (scaled in *N. callitaenia*). Maximum length up to 75 mm. (Also see Key.)

DISTRIBUTION: Southern Georgia and part of northern Florida (Leon, Gadsden, and Liberty Counties, and in the Alapaha River; Yerger and Suttkus, 1962).

41. Black-tailed Shiner, *Notropis venustus* (Girard)

IDENTIFICATION: A large, sometimes pale shiner with *considerable black on the posterior portion of the dorsal fin* and a *prominent basicaudal spot*. Usually 8 rays in anal fin. Lateral stripe complete, but

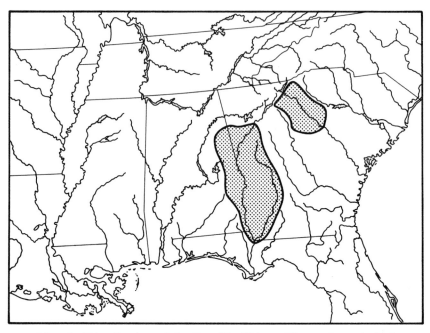

Map 4. Distribution of the Bandfin Shiner (*Notropis zonistius*)

sometimes faint anteriad. Brownish above, silvery below. Sometimes up to 125 mm or more in length.

DISTRIBUTION AND VARIATION: Illinois and Tennessee southward to Texas, the Gulf Coast, and Florida (east to the Suwannee River; *N. v. cercostigma*, Gibbs, 1957).

42. Sailfin Shiner, *Notropis hypselopterus* (Günther)

IDENTIFICATION: One of the *most highly colored of the Florida shiners*. Rusty reddish above, paler below. *Lateral stripe black* (iridescent blue in life), *very broad*, and complete, *bordered above by a light band, reddish in life*. Dorsal fin high, almost vertical behind, and with a dark blotch. With a dark streak running from head to dorsal fin. *Two light spots* (reddish in life) *at base of caudal fin*. Anal fin with 11 rays. Not more than 35 scales in lateral line. Length up to 65 mm.

DISTRIBUTION: From Hillsborough County, Florida, northward into southern Georgia and Alabama, and westward to southern Louisiana; *N. stonei* of South Carolina may be conspecific.

43. Flagfin Shiner, *Notropis signipinnis* Bailey and Suttkus

IDENTIFICATION: Similar to the Sailfin Shiner, but *predorsal streak* of that species *faint or lacking, little or no dark pigment in dorsal fin*, and no reddish color on body in life, but fins often with reddish spots. Basicaudal spots yellowish in life. Not more than 35 scales along lateral line. Maximum length about 65 mm.

DISTRIBUTION: Gulf Coastal Plain from Louisiana to southeastern Alabama and western Florida (east to Chipola River and Bay County).

44. Dusky Shiner, *Notropis cummingsae* Myers

IDENTIFICATION: Similar to the Flagfin and Sailfin Shiners in its large number of anal rays, but *lateral stripe narrower and extending to snout*, and diffuse anteriad. Also eye larger (diameter greater than length of snout). More than 37 scales in lateral line, and no light spots at caudal base. General color silvery, but darker above and with no reddish in life. Maximum length about 75 mm.

DISTRIBUTION: Coastal Plain and Piedmont from North Carolina to eastern Alabama and northern Florida (west to Washington County and south to Marion County; Map 5).

45. Iron-colored Shiner, *Notropis chalybaeus* (Cope)

IDENTIFICATION: Lateral line incomplete, its scale count about 33. Not more than 8 rays (rarely 7) in anal fin. *Considerable black pigment present around anus, caudal peduncle, base of anal fin, and inside mouth.* Lateral stripe dark. Basicaudal spot small. Mouth small and oblique. Dark above (reddish brown in life), light below. A light U-shaped marking on snout as in *Hybopsis harperi*, but dorsal coloration lighter and no barbels at angle of jaws. Maximum length about 65 mm.

DISTRIBUTION: Atlantic and Gulf Coastal Plains from southeastern Connecticut to eastern and south-central Texas and up Mississippi River drainage to vicinity of Lake Michigan (Map 6; Swift, 1970).

46. Weed Shiner, *Notropis texanus* (Girard)

IDENTIFICATION: With *7 anal rays, the last few darkened.* Scales below anterior portion of lateral line outlined with dark pigment. Dusky olive above, silvery below. Fins reddish at base in life. Total length up to 75 mm.

DISTRIBUTION: Southwestern Michigan and central Minnesota through central Iowa and western Indiana to central and southern Texas and the Gulf Coast, and eastward to central Georgia and northern Florida (to Hamilton County) (Map 7; Swift, 1970).

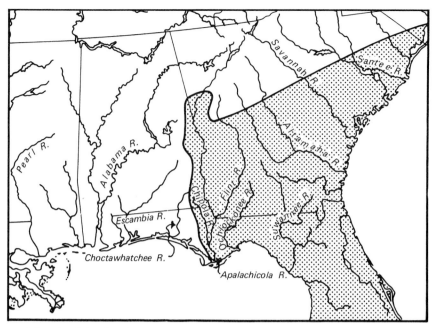

Map 5. Distribution of the Dusky Shiner (*Notropis cummingsae*)

47. Coastal Shiner, *Notropis petersoni* Fowler

IDENTIFICATION: Similar to *N. texanus*, but without dark pigment outlining scales just below lateral line. Also pigment not concentrated in any part of anal fin. Like that species, it has enlarged scales (not more that 15 present) between head and dorsal fin and no conspicuous basicaudal spot. Eye relatively large. Dark above, silvery below. Maximum length about 65 mm.

DISTRIBUTION: Southern Mississippi and Alabama through the Coastal Plain to North Carolina (Map 8; Swift, 1970).

48. Blue-nosed Shiner, *Notropis welaka* Evermann and Kendall

IDENTIFICATION: Females are similar to the Iron-colored Shiner, but the basicaudal spot is larger (at least as great in diameter as the orbit) and the scales below the lateral stripe are pigmented. Light U-shaped mark on snout. Adult males are extremely distinctive, with *hypertrophied dorsal and anal fins* (the former very dark) and a *blue nose in life*. Total length 50 to 65 mm.

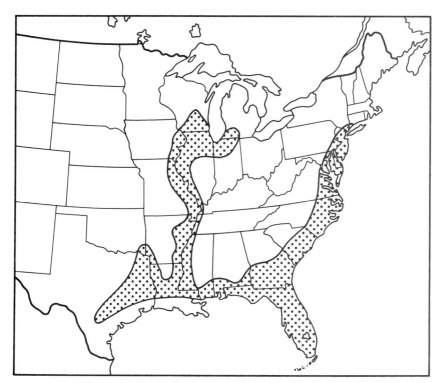

Map 6. Distribution of the Iron-colored Shiner (*Notropis chalybaeus*)

DISTRIBUTION: From Sanford, Florida, locally northward and westward near the coast to the Pearl River, Louisiana (Yerger and Suttkus, 1962).

49. Tail-light Shiner, *Notropis maculatus* (Hay)

IDENTIFICATION: Long, slender, and graceful, with a long, deeply forked caudal fin and a concave posterior margin to the dorsal fin. *Basicaudal spot conspicuous, rounded, and wider than lateral stripe.* Also dark pigment on back and around base of anal fin. Lateral line incomplete. *Often reddish in life.* Total length up to 75 mm or more.

DISTRIBUTION: Southeastern Oklahoma, Missouri, Kentucky and southeastern North Carolina (at low elevations) southward to eastern Texas, the Gulf Coast, and southern Florida (Collier and Palm Beach Counties; *Fla. Scientist* 37:117).

50. Long-nosed Shiner, *Notropis longirostris* (Hay)

IDENTIFICATION: *Length of snout much greater than diameter of eye; mouth inferior.* Lateral stripe faint. Only about 12 scales from head to dorsal fin. Entire coloration pale. Maximum length 50 to 65 mm.

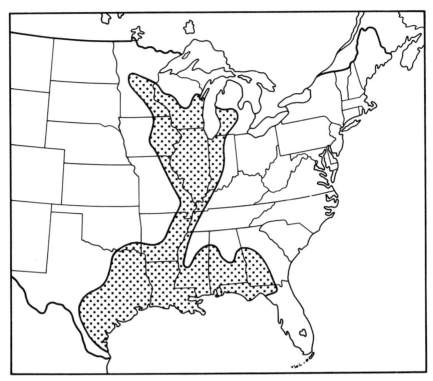

Map 7. Distribution of the Weed Shiner (*Notropis texanus*)

DISTRIBUTION: Coastal Plain from Louisiana to western Florida and southeastern Georgia (Apalachicola River system).

51. Cypress Minnow, *Hybognathus hayi* Jordan

IDENTIFICATION: Internal differences separating this genus from the shiners (*Notropis*) are the *long, highly coiled intestine* and the *black lining of the body cavity*. Both are devoid of barbels. The Cypress Minnow never has more than 8 rays in the dorsal or anal fins. The lateral line is complete, and the lateral stripe is conspicuous (but *silvery*), continuing onto the head. The basicaudal spot is lacking or faint. Total length to 75 mm or more.

DISTRIBUTION: Indiana, Illinois, Missouri, Arkansas, and northeastern Texas southward to Alabama and Mississippi; also found in the Escambia River of Florida (Bailey, Winn, and Smith, 1954).

52. Silver-jawed Minnow, *Ericymba buccata* Cope

IDENTIFICATION: A large, pale minnow with externally visible chambers in the head. Dorsal and anal fins with 8 rays. Lateral line

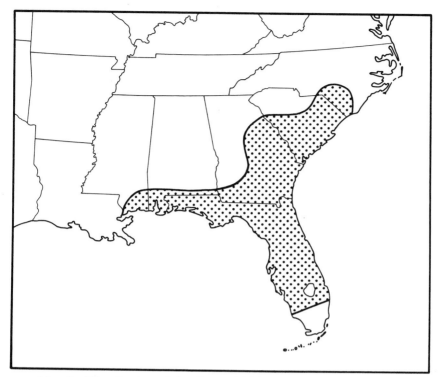

Map 8. Distribution of the Coastal Shiner (*Notropis petersoni*)

almost straight, and lateral stripe present, but inconspicuous. Mouth subterminal, and length of snout slightly greater than diameter of eye. Basicaudal spot lacking. Maximum length 100 to 125 mm.

DISTRIBUTION: West Virginia, western Pennsylvania, southern Michigan, and southeastern Missouri southward to western Florida and Arkansas, except *absent* in southwestern Kentucky, Tennessee, Virginia, the Carolinas, southern Mississippi and Alabama, and parts of northern and western Georgia; eastward in Florida to Gadsden County and the Apalachicola River (*Amer. Midland Nat.*, 89:145–155).

53. Big-eyed Chub, *Hybopsis amblops* (Rafinesque)[1]

IDENTIFICATION: Chubs differ from shiners (*Notropis*) in having one or more *small barbels at the corner of the mouth*. Those in the genus *Hybopsis* differ from the Creek Chub (*Semotilus*) in having fewer than 40 scales along the lateral line. The Big-eyed Chub has an *inferior mouth*, no U-shaped marking on the head, and only one pair of barbels.

1. The taxonomy of the Florida population is open to question, and it may later be re-named *Hybopsis winchelli* (Glenn Clemmer, personal communication).

The *lateral stripe is narrow and faint* and does not terminate in a basi-caudal spot. General color dark above and light below. *Diameter of eye about equal to length of snout and less than distance between eyes.* Maximum length about 75 mm.

DISTRIBUTION: New York and southern Michigan to Oklahoma, Louisiana, and Florida (Panhandle east to Ochlockonee River).

54. Red-eyed Chub, *Hybopsis harperi* (Fowler)

IDENTIFICATION: Similar to the Big-eyed Chub, but with a *smaller eye* (partly red in life), and a more distinct black stripe. The ground color is reddish, the barbels less conspicuous, and there is a *light U-shaped marking on top of the head* (yellowish in life). The basicaudal spot is distinct, and the mouth subterminal. Diameter of eye less than length of snout and less than distance between eyes. Total length up to 50 mm.

DISTRIBUTION: Northern Florida (south to Lake and Hillsborough Counties; Lee, 1969) and the southern parts of Georgia and Alabama.

55. Speckled Chub, *Hybopsis aestivalis* (Girard)

IDENTIFICATION: Differs from others members of its genus in its combination of a *faint* lateral stripe, its *2 pairs of longer barbels* (equal to diameter of pupil), and its blunter, deeper snout, making the *mouth decidedly inferior*. (Populations outside Florida may have only one pair of barbels.) *Eyes more dorsal* and closer together. Total length up to 65 mm.

DISTRIBUTION: Lower Mississippi River drainage and eastern Oklahoma southward to Texas, the Gulf Coast, and western Florida (east to Holmes County).

56. Creek Chub, *Semotilus atromaculatus* (Mitchill)

IDENTIFICATION: Larger and more heavy-set than shiners, and with a *minute barbel at the corner of the mouth*. Mouth relatively large. *Scales smaller toward anterior end of body* than toward posterior end. Bluish above, light below, with a dark lateral stripe, fading in old age; males reddish in spring. *Young have a dark spot at origin of dorsal fin.* Maximum length 175 to 200 mm.

DISTRIBUTION: Small creeks from southeastern Canada and the northern United States (west to Montana) southward to New Mexico and the Florida Panhandle; Florida records extend eastward to Gadsden and Bay Counties.

57. Pug-nosed Minnow, *Opsopoeodus emiliae* Hay[1]

IDENTIFICATION: Resembles shiners (*Notropis*) except for the *small, almost vertical mouth*, the *blunt snout*, and the *higher number* (9) *of dorsal rays*. Lateral stripe rather faint and narrow, and lateral scales outlined with dark pigment. Male has tubercles on snout and chin in spring. General coloration pale. Total length not more than 50 mm.

DISTRIBUTION: Eastern United States, ranging throughout the Florida mainland.

58. Carp, *Cyprinus carpio* L.

IDENTIFICATION: Differs from shiners in its larger size and greater depth (about one-third of standard length); from suckers in its smaller lips; and from both groups in the presence of *well-developed barbels around the mouth*, of single *spines in the dorsal and anal fins*, and of a *long dorsal fin* (at least 20 rays). Head pointed, almost flat in profile. Scales large (about 35 to 40 along lateral line). Color variable. Maximum length about 1 m.

DISTRIBUTION: Asia (native); Europe and the United States (introduced), including northern Florida (Apalachicola River; Spring Creek, Wakulla County; possibly Choctawhatchee River).

ORDER CYPRINIFORMES: XVII. FAMILY CATOSTOMIDAE, Suckers. All species known to Florida have *thick, fleshy lips* surrounding the *ventral mouth*. Gill membranes joined to the isthmus. Most suckers are considerable larger than most members of the Cyprinidae.

59. Quillback, *Carpiodes cyprinus* (Lesueur)

IDENTIFICATION: In the 2 carpsuckers (*Carpiodes*) the *dorsal fin base is almost half the standard length and its anterior rays greatly prolongated*; also a lateral line is present. In the present species the *lower lip is curved*, not coming to a rounded point at the center. The maxillary extends back to the position of the nostril. Color silvery, darker above, with faint spots in longitudinal rows. Total length up to 300 mm.

DISTRIBUTION: Southern Manitoba and most of the Great Lakes southward to Kansas, western Georgia, and western Florida (Apalachicola River westward); also from the St. Lawrence River system to North Carolina.

60. Highfin Carpsucker, *Carpiodes velifer* (Rafinesque)

IDENTIFICATION: Similar to the Quillback, but with the *lower lip angular* and coming to a rounded point at the center; also the mouth

---

1. Included in *Notropis* in the American Fisheries Society bulletin (1970).

extends farther back than the position of the nostril. Silvery with faint longitudinal stripes. Total length up to 225 mm or more.

DISTRIBUTION: Illinois, Indiana, and Ohio southward to Mississippi, Alabama, southern Georgia (?), and western Florida; also Nebraska to Oklahoma.

61. ? Redhorse, *Moxostoma* sp.[1]

IDENTIFICATION: More slender than the River Redhorse (just north of Florida) and the *lower lip not as thick as diameter of eye*; also with more scales in lateral line (usually at least 44). Usually 10 rays in pelvic fin. No concentration of pigment on any part of caudal fin. Maximum length at least 300 mm.

DISTRIBUTION: Endemic in Apalachicola River.

62. Black-tailed Redhorse, *Moxostoma poecilurum* (Jordan)

IDENTIFICATION: Similar to the preceding species, but *lower lobe of caudal fin slender, longer than upper lobe, and with dark pigment concentrated toward fork*; upper lobe reddish in life. Less than 44 scales along lateral line. General color brownish, except fins reddish in life. Maximum length about 455 mm.

DISTRIBUTION: Coastal Plain from eastern Texas to northern Georgia and western Florida (eastward to Holmes County).

63. Spotted Sucker, *Minytrema melanops* (Rafinesque)

IDENTIFICATION: *Lower lip not unusually thickened.* Dorsal fin moderate in depth and length. Lateral line present. General coloration pale, but *with longitudinal rows of dark spots along sides.* Occasionally up to 600 mm in length.

DISTRIBUTION: Lowlands from Maryland, Pennsylvania, and Minnesota southward to eastern Texas and Florida (to Suwannee River system).

64. Sharpfin Chubsucker, *Erimyzon tenuis* (Agassiz)

IDENTIFICATION: This genus consists of small suckers with only moderately thickened lips and *no lateral line*. The young are easily distinguished from most other suckers by their black lateral stripe, but can be confused with certain shiners. Adults are yellowish or reddish in life and lack the lateral stripe. The 3 species differ from one another in the *shape of the dorsal fin*, the *posterior edge being almost vertical in*

---

1. Formerly considered a population of *Moxostoma duquesnei*, which species is no longer accredited to Florida.

this species. The Sharpfin also has a higher number of oblique scale rows (40 or more). Its color in life is more yellowish than that of *E. sucetta*, and its head is more pointed. Large specimens are often dark above, becoming *abruptly light below*. Maximum length about 300 mm.

DISTRIBUTION: Coastal Plain from Louisiana to eastern Alabama and western Florida (at least to Holmes County).

65. Lake Chubsucker, *Erimyzon sucetta* (Lacépède)

IDENTIFICATION: This species is distinguished from *E. tenuis* by the rounded hind edge of its dorsal fin and from *E. oblongus* by its proportions and count of lateral scales (see Key). Immatures are marked with a *broad, dark lateral stripe*; as this feature becomes less distinct in adults, the ground color becomes *reddish*. Maximum length about 300 mm.

DISTRIBUTION: Great Lakes and New York southward to eastern Texas and Florida, except in the mountains.

66. Creek Chubsucker, *Erimyzon oblongus* (Mitchill)

IDENTIFICATION: Similar to *E. sucetta*, but scales smaller (more than 38 along side), and not reddish in life. Except in small specimens, the *back is noticeably humped* anterior to the dorsal fin. Gill rakers shorter than in *E. sucetta*. Maximum length about 250 mm.

DISTRIBUTION: From Minnesota, New Brunswick, and Nova Scotia southward to Texas, southern Arkansas, southern Mississippi, Florida (Escambia River near Cantonment; Bailey, Winn, and Smith, 1954), and Georgia, except not in the mountains.

ORDER SILURIFORMES: XVIII. FAMILY ICTALURIDAE, Freshwater Catfishes. The catfishes are distinctively different from all other fishes, with their *naked bodies*, strong *dorsal and pectoral spines*, wide, *depressed head*, and *barbels* on the head. Those in the family Ictaluridae differ from the sea catfishes in having *barbels at the nostrils*.

67. Snail Bullhead, *Ictalurus brunneus* (Jordan)

IDENTIFICATION: Bullheads differ from other large catfishes in having the *tail only slightly forked*, if at all. This species is unusually flat-headed, with an *inferior mouth* and a dull, olivaceous color pattern, often uniform, but occasionally blotched or spotted. The color pattern approaches that of *I. nebulosus*, and the body form that of *I. serracanthus*. The anal fin typically has only 17 to 20 rays (19–22 in St. Johns River). Maximum length about 25 cm.

DISTRIBUTION: Middle elevations of the Carolinas and Georgia (reaching Coastal Plain locally) southward via the Chattahoochee-

Apalachicola River system through southeastern Alabama into northern Florida; also the upper St. Johns River of Florida (Yerger and Relyea, 1968; Map 9).

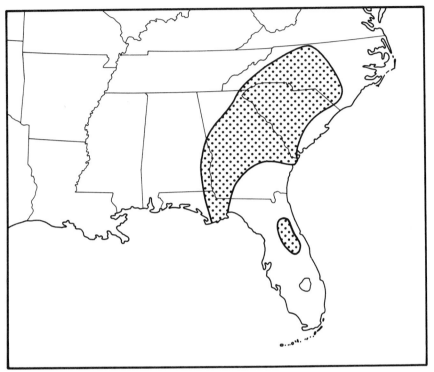

Map 9. Distribution of the Snail Bullhead (*Ictalurus brunneus*)

68. Spotted Bullhead, *Ictalurus serracanthus* Yerger and Relyea

IDENTIFICATION: A small, flat-headed bullhead with *light, rounded spots* (yellowish in life) *on a dark background*. The only catfish in Florida with a similar body form (*I. brunneus*) differs markedly in color, in having smaller serrations on the pectoral spines, and in its anal ray count (*serracanthus* has about 20 to 23). Also distinctive is the *narrow dark margin on each fin*, including the pectorals, in *serracanthus*. Maximum length about 23 cm.

DISTRIBUTION: Econfina Creek (Bay County) to the Suwannee River, Florida, and into extreme southeastern Alabama and southern Georgia (Yerger and Relyea, 1968).

69. Brown Bullhead, *Ictalurus nebulosus* (Lesueur)

IDENTIFICATION: This species, unlike the White and Channel Catfishes, has a *notched* rather than deeply forked *tail*. Its anal ray count (20 to 26) will separate it from other bullheads (notched tails) except for the flat-headed forms (*I. brunneus* and *I. serracanthus*). It differs from both of these in having a *terminal mouth*, as well as in color pattern. The upperparts and sides are *brownish or blackish with dark mottling*, the underparts lighter, and the *chin barbels dark* (see *I. natalis*). Maximum length about 46 cm.

DISTRIBUTION: Southern Canada, the Great Lakes, and North Dakota southward to southern Florida and Mexico; introduced in the Pacific States, Czechoslovakia and elsewhere.

70. Yellow Bullhead, *Ictalurus natalis* (Lesueur)

IDENTIFICATION: Resembling the Brown Bullhead in shape of tail and its *terminal mouth*, but differs in having *more than 24 rays in the anal fin* and *whitish chin barbels*. It is a rather robust fish with a small eye (see Key). Color in life yellowish brown above and pale yellowish below, but sometimes much darker. Maximum length about 38 to 45 cm.

DISTRIBUTION: Southeastern Canada and the eastern United States, ranging throughout the Florida mainland; westward to the Great Plains, but introduced in the Pacific States.

71. Channel Catfish, *Ictalurus punctatus* (Rafinesque)

IDENTIFICATION: A relatively slender catfish with a deeply forked tail, its *2 lobes about equal in length. A continuous bony ridge* extending from the skull to the base of the dorsal fin may be felt just beneath the skin. The *adipose fin* is free posteriad and its origin *is closer to the caudal than to the dorsal*. The ground color is lighter than in *I. catus*, bluish above, fading to white underneath, with *small dark spots on the sides* (except in old age). *Not more than 24 anal rays*. Maximum length about 90 to 110 cm.

DISTRIBUTION: Virtually throughout the United States (partly by introductions), ranging southward throughout the Florida mainland (*Fla. Scientist* 37:118).

72. White Catfish, *Ictalurus catus* (L.)

IDENTIFICATION: Similar to the Channel Catfish, but more robust and *without discrete spots on sides*. Upper lobe of caudal fin somewhat longer than lower. The bony ridge from the skull to the origin of the dorsal fin is incomplete. The *origin of the adipose fin is about equally close to the caudal and dorsal fins*. Including rudiments, there are *at least 25 anal rays*. The common name is misleading, as this species is quite dark (blue gray) above, but whitish underneath. Total length up to 60 cm.

DISTRIBUTION: Coastal Plain from New York to southern Alabama and Florida (to Lake Okeechobee and the Escambia River); locally introduced in the West.

73. Tadpole Madtom, *Noturus gyrinus* (Mitchill)

IDENTIFICATION: Madtoms differ from other catfishes in having the posterior end of the *adipose fin adnate* (connected to the back or the caudal fin, as well as in their smaller size (rarely exceeding 15 cm in Florida). The Tadpole Madtom is heavy-set, with the *upper and lower jaws protruding about equally*. There are only *14 to 16 anal rays*. The spine in the first dorsal fin is about *half the height of that fin*. Chiefly dark brownish, the fins never margined with whitish. Maximum length about 10 cm.

DISTRIBUTION: North Dakota and the Great Lakes southward to the Gulf of Mexico and South Florida chiefly at elevations of less than 300 m (Taylor, 1969; *Fla. Scientist* 37:118).

74. Speckled Madtom, *Noturus leptacanthus* Jordan

IDENTIFICATION: More slender than other madtoms. *Yellowish* brown with *darker spots on body and median fins. Mouth inferior*. First dorsal spine less than half height of that fin. About 14 rays in anal fin. Maximum length about 10 cm.

DISTRIBUTION: Northeastern Mississippi, northern Alabama and Georgia, northwestern and southern South Carolina southward to southeastern Louisiana, the Gulf Coast, and Lee County, Florida (Taylor, 1969; *Fla. Scientist* 37:118).

75. Black Madtom, *Noturus funebris* Gilbert and Swain

IDENTIFICATION: Body heavy-set, as in the Tadpole Madtom, but *upper jaw protrudes beyond the lower*, the *anal fin contains more than 20 rays*, and the fins are margined with whitish. General coloration dark. Maximum length about 16 to 17 cm.

DISTRIBUTION: From Jackson and Bay Counties, Florida, northwestern Georgia (Dahlberg and Scott, 1971), and central Alabama to southeastern Louisiana (Map 10; Taylor, 1969).

ORDER SILURIFORMES: XIX. FAMILY CLARIIDAE, Clariid Catfishes. Long, slender catfishes with 4 pairs of barbels and no scales. *Dorsal and anal fins very long* (to 50 percent of total length), extending backward to, and sometimes joining, the caudal fin. *Dorsal spine lacking*, and pectoral spine relatively weak. Accessory respiratory organs present (Sterba, 1962).

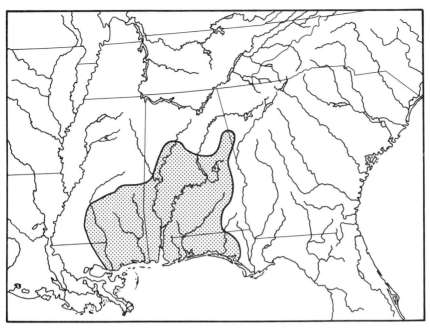

Map 10. Distribution of the Black Madtom (*Noturus funebris*)

76. Walking Catfish, *Clarias batrachus* (L.)

IDENTIFICATION: Upperparts dark, often with a greenish cast; underparts pale brown to reddish. Males with black spots in the dorsal fin. Albinism frequent. Maximum length over 50 cm. (Also see description of family.)

DISTRIBUTION: Ceylon and eastern India to the Malay Archipelago (Sterba, 1962). Introduced into southeastern Florida in 1968; from Lake Okeechobee southeastward by 1974 (*Fla. Scientist* 37:118); appeared in Tampa area in spring of 1969 (Herb Allen, *Tampa Tribune*).

ORDER SILURIFORMES: XX. FAMILY LORICARIIDAE, Armored Catfishes. Small catfishes covered with rows of rough bony plates. Mouth small, with thickened lips. Length usually less than 300 mm. Florida's species appears in the Hypothetical List (198H).

ORDER SILURIFORMES: XXI. FAMILY ARIIDAE: Sea Catfishes. Similar to the Ictaluridae, but with *only 4 or 6 barbels*, having none at the nostrils.

77. Gafftopsail Catfish, *Bagre marinus* (Mitchill)

IDENTIFICATION: A marine catfish with *only 4 barbels*, those at the angle of the jaws being exceptionally long. *The pectoral and dorsal fins have long filaments*. Color bluish above, silvery below. Total length up to 60 cm.

DISTRIBUTION: Atlantic and Gulf Coasts from Massachusetts to Panama, entering Florida streams.

78. Sea Catfish, *Arius felis* (L.)

IDENTIFICATION: Differs from *Bagre* in having 6 *barbels* (an extra pair on the lower jaw), none remarkably long, and in the absence of long filaments on any fin. Maximum length 38 to 45 cm.

DISTRIBUTION: Atlantic and Gulf Coasts from Massachusetts to Texas, entering Florida streams.

ORDER PERCOPSIFORMES: XXII. FAMILY APHREDODERIDAE, Pirate Perches. Heavy bodied and dark brown, with the *anus in an anterior position* (at least as far forward as the pelvic fins). Seven rays in pelvic fins. Caudal fin square-tipped. Males iridescent in spring. Total length up to 125 mm. Monotypic.

79. Pirate Perch, *Aphredoderus sayanus* (Gilliams)

IDENTIFICATION: As given under the family.

DISTRIBUTION: Lowlands from southeastern Minnesota and New York southward to Texas and Florida, ranging to Lake Okeechobee (*Fla. Scientist* 37:119).

ORDER ATHERINIFORMES: XXIII. FAMILY BELONIDAE, Needlefishes.[1] With their extremely *slender body form, elongate and slender beak, and posterior dorsal fin*, the needlefishes might be confused only with gars or pipefishes. The presence of a *homocercal, notched caudal fin* will readily separate them from both groups. Cycloid scales are present, but often difficult to see without magnification. The pelvic fins are much reduced.

80. Atlantic Needlefish, *Strongylura marina* (Walbaum)

IDENTIFICATION: (See description of family.) This species is greenish above and paler below, with about 300 scales in the lateral line. The *peduncle is keeled laterally and its width about equal to its depth*. Maximum length about 120 cm, but averaging much smaller.

DISTRIBUTION: Typically marine, but entering mouths of rivers (occasionally far upstream) on Atlantic and Gulf Coasts from Maine to Texas.

1. *Strongylura notata* has also been reported in fresh water in south Florida (*Fla. Scientist* 37:119).

81. Timucu, *Strongylura timucu* (Walbaum)

IDENTIFICATION: Similar to the Atlantic Needlefish, but with a *compressed peduncle, not laterally keeled*; also a lower lateral-scale count (about 225), and a maximum length of about 45 cm (Breder, 1948).

DISTRIBUTION: From the northern Gulf of Mexico to Rio de Janeiro, Brazil. Euryhaline (Briggs, 1958).

ORDER ATHERINIFORMES: XXIV. FAMILY CYPRINODONTIDAE, Killifishes.[1] A large assemblage of small, often slender fishes with *small, protractile mouths* that are usually *superior in position*. The *caudal fin is never truly forked*, but either rounded or truncate. Lateral line absent or poorly developed. Pelvic fins abdominal. Fins lacking spines. Killifishes are most similar to the Poeciliidae, but their *branched third anal ray* provides an almost infallible distinction. Both groups show marked sexual dimorphism. Size small, with few Florida species exceeding a length of 150 mm.

82. Rainwater Killifish, *Lucania parva* (Baird)

IDENTIFICATION: As this is a small, nondescript fish, its identification is largely a matter of eliminating other plain-colored species. The general coloration is grayish, but males have some orange in the dorsal fin (in life) and a black spot near its anterior base. In both sexes the *scales are margined with darker pigment*, producing a cross-hatched effect. Rarely there is a faint, dusky lateral stripe. Similar in coloration, the mosquitofishes (*Gambusia*) are more slender, have dark specks in the caudal fin, and the males have a prolongation of the anal fin (gonopodium). Female Golden Topminnows (*Fundulus chrysotus*) in preservation are also similar in color, but have a higher lateral scale count (over 30), a more rounded caudal fin, and a more posterior dorsal fin (at least as far back as the anal); in life they show gold flecks on the sides. In the present species the caudal fin is rounded-truncate. Maximum length 50 mm or less.

DISTRIBUTION: Streams along Atlantic and Gulf Coastal Plain from Massachusetts through Florida (except Georgia; Dahlberg and Scott, 1971) to Mexico, inland to New Mexico, usually preferring salt or brackish water; also widely introduced in the West (Hubbs and Miller, 1965).

83. Bluefin Killifish, *Lucania goodei* Jordan

IDENTIFICATION: One of the few killifishes with a *broad, black, longitudinal stripe*. Ground color olivaceous. Bases of dorsal and anal

---

1. The basis for the inclusion of *Rivulus marmoratus* as a *freshwater* species in Florida (Amer. Fisheries Soc. bulletin, 1970) is not known to me.

fins black in male. Fins of males with red and blue in breeding season. The only other killifish in Florida with a black lateral stripe (*Fundulus olivaceus*) differs from the Bluefin in its rounded caudal fin. Length up to 35 or 40 mm.

DISTRIBUTION: Coastal Georgia, southeastern Alabama, and Florida, west to Wakulla and Washington Counties.

84. Pygmy Killifish, *Leptolucania ommata* (Jordan)

IDENTIFICATION: One of Florida's smallest fishes, not exceeding 25 mm in length. Ground color greenish, with an *ocellus at base of caudal fin* (dark spot surrounded by light color). In addition, males have dark vertical bars on the body, and *females a dark spot near the origin of the anal fin*. The Least Killifish (*Heterandria*), also tiny, has a similar spot on the peduncle, but an additional one in the dorsal fin.

DISTRIBUTION: Swamps of Florida and the extreme southern parts of Georgia, Alabama, and Mississippi.

85. Diamond Killifish, *Adinia xenica* (Jordan and Gilbert)

IDENTIFICATION: The common name refers to the roughly *diamond-shaped body outline*, due in part to the straight profile from head to dorsal fin. The effect of this shape is to produce a *pointed snout*. Unlike some other killifishes with dark vertical bars, *Adinia* has the origin of the *dorsal fin decidedly anterior to* that of *the anal*. Caudal fin truncate. Lateral scale count about 25. Body rather deep and compressed, silvery green in life, with light flecks in median fins and a dark spot at anterior base of dorsal fin. The *dark bars are wider than the light ones and faint anteriad*. Length up to 50 mm.

DISTRIBUTION: Brackish water and the mouths of streams from Monroe County, Florida, to Aransas Bay, Texas (Hastings, 1967; Map 11).

86. Salt-marsh Topminnow, *Fundulus jenkinsi* (Evermann)

IDENTIFICATION: The 1 or 2 longitudinal rows of large, dark spots on the body tend to separate this species from all other Florida Killifishes. Occasionally these spots may unite to form indistinct vertical bars or may be lacking. In the latter case, the dusky-edged scales make the color pattern appear crosshatched, as in *Fundulus seminolis* and *Lucania parva*. Maximum length about 60 mm.

DISTRIBUTION: Coastal strip from Texas to Escambia Bay, Florida; euryhaline.

Map 11. Distribution of the Diamond Killifish (*Adinia xenica*)

87. Mummichog, *Fundulus heteroclitus* (L.)

IDENTIFICATION: The numerous species in this genus are distinguished, in part, by having more than 30 oblique scale rows, minute pores on the mandible, and conical teeth. They are also more slender and less compressed than those in the genera *Adinia*, *Cyprinodon*, and *Jordanella*. The Mummichog is *more heavy-set* than most members of the genus, but bears a strong resemblance to the Gulf Killifish. The dorsal fin is set slightly farther forward than the anal (or directly above it in small specimens). The anal fin has a short base, its greatest length slightly less than half that of its longest rays. There are *4 pairs of mandibular pores*. The color in life is dark green above, whitish or yellowish below; the light spots (males) on this dark background distinguish them from most other killifishes (see also *F. grandis* and *chrysotus*). Males invariably and females sometimes show dark vertical bars. Rather large, ranging up to 150 mm in length.

DISTRIBUTION: Salt, brackish, and fresh water from eastern Canada to Volusia County, Florida (Relyea, personal communication).

88. Gulf Killifish, *Fundulus grandis* Baird and Girard

IDENTIFICATION: Similar in color pattern and size to *heteroclitus*, but differs in having *shorter dorsal and anal fins* and a *less rounded caudal fin*. Anal fin rounded, its base slightly more than half length of its longest rays. Origin of dorsal fin slightly anterior to that of anal fin (or directly above in small specimens). *Five pairs of mandibular pores* (Plate II-4). The lateral scale count *averages* lower than in the Mummichog, but counts of 35 and 36 have been recorded in both species. Maximum length about 150 mm.

DISTRIBUTION: Gulf Coast from Marco Island, Florida, to northeastern Mexico, and on Atlantic Coast from Lake Worth to Nassau County, Florida; salt, brackish, and fresh water in Florida (Map 12). The taxonomic status of the Keys population is conjectural (Relyea, personal communication).

Map 12. Distribution of the Gulf Killifish (*Fundulus grandis*)

89. Seminole Killifish, *Fundulus seminolis* Girard

IDENTIFICATION: Of undistinguished color pattern, this killifish is best identified by its *high lateral-scale count* (50 or more). The dorsal and anal fins are unusually long, containing 17 and 13 rays, respectively.

Body proportions long and slender. Ground color rather pale olive green, lighter below. The dusky margin of each scale may produce a *crosshatched appearance*. This color pattern is somewhat like that of *Gambusia affinis* (except in those individuals with numerous, faint vertical bars), but the caudal fin is less rounded and the peduncle thicker. Occasional individuals display small black spots, however. Both the diameter of the eye and the distance between the eyes are less than the snout length. Maximum length about 150 mm.

DISTRIBUTION: Peninsular Florida, from Lee and Martin Counties northward to Franklin and Duval Counties (Tagatz, 1967; Map 13).[1]

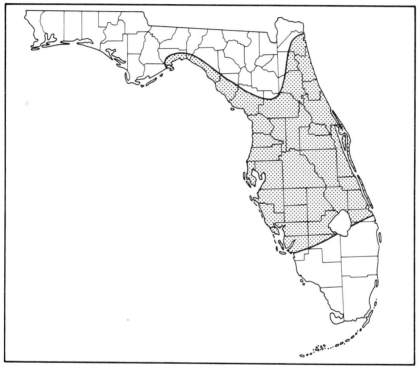

Map 13. Distribution of the Seminole Killifish (*Fundulus seminolis*)

90. Long-nosed Killifish, *Fundulus similis* (Baird and Girard)

IDENTIFICATION: Distinguished chiefly by its *long snout*, about twice the length of the eye (to only 70 percent longer in small specimens). Relatively few scales along sides (about 33). Both sexes have 10 or more dark vertical bars, typically more narrow than the interspaces. Color otherwise usually similar to that of *F. seminolis*, except males in

_____

1. Recently reported south to Monroe County (*Fla. Scientist* 37:120).

breeding season show contrasting black and white. Maximum length about 150 mm.

DISTRIBUTION: Gulf Coast from Marco Island, Florida, to Texas in salt, brackish, and fresh water. The taxonomic status of populations on the Keys and lower east coast remains to be determined (Relyea, personal communication).

91. Striped Killifish, *Fundulus majalis* (Walbaum)

IDENTIFICATION: The only killifish in which the female has *one*, *2*, *or 3 narrow* longitudinal stripes, continuous or broken. The *single* stripe in *F. olivaceus* is much wider, and the narrow stripes in females of *F. lineolatus* are more numerous. Males of this species have numerous dark vertical bars like several other killifishes, but both sexes have small eyes (diameter scarcely more than half of snout length, or one-sixth of head length). This same proportion will separate females from *F. jenkinsi*. Both in lateral-scale count (about 36) and in longitudinal rows of scales on side of body (about 13) the number is higher than in *F. similis*. About 18 predorsal scales. Length up to 150 mm, rarely 200.

DISTRIBUTION: Brackish water of Atlantic Coast from Massachusetts to St. Johns (possibly Palm Beach) County, Florida, entering fresh water here (Relyea, personal communication).

92. Marsh Killifish, *Fundulus confluentus* Goode and Bean

IDENTIFICATION: Smaller than most foregoing species (maximum length about 100 mm). Dark pigment may take pattern of vertical bars (male), a dark spot on the dorsal fin, or dark blotches on the body (female). Typically the male's dark bars are more narrow than the interspaces, as in *F. majalis* and *F. similis*. There are about 36 oblique and 15 longitudinal scale rows on each side. Head length about *4 times diameter of eye*. Origin of dorsal fin directly above that of anal or slightly anterior to it. There are more predorsal scales (18 to 26) than in most similar species. About 7 to 9 anal rays.

DISTRIBUTION AND VARIATION: Brackish and fresh water of Atlantic Coast, Maryland to Texas. *F. c. confluentus* ranges throughout Florida to Apalachicola Bay; westward there is intergradation with *F. c. pulvereus* (Relyea, personal communication).

93. Banded Topminnow, *Fundulus cingulatus* Valenciennes

IDENTIFICATION: A small killifish restricted to fresh water. Both sexes have about 12 dark vertical bars and *reddish fins in life*. *Origin of dorsal fin directly above that of anal*. About 10 longitudinal scale rows per side. Anal fin with about 9 rays. Eye moderate, and snout blunt. The caudal fin is clear in preserved specimens and is rounded. Dorsal color olivaceous, ventral color orange in life. In preserved specimens vertical bars are often in evidence only posteriad. Maximum length about 65 mm.

DISTRIBUTION: Lower Coastal Plain, from New Jersey to Florida and southern Alabama; southward in Florida to Palm Beach and Collier Counties (*Fla. Scientist* 37:119).

94. Golden Topminnow, *Fundulus chrysotus* (Günther)

IDENTIFICATION: Structurally similar to *F. cingulatus*, but easily distinguished in life by the *golden flecks* on the sides. Preserved specimens of the two species are very similar, but in *F. chrysotus* the origin of the dorsal fin is slightly *posterior* to that of the anal, and the number of dark vertical bars in males is typically less than 11. The anal fin may have one more ray (10), and the number of longitudinal scale rows is usually 12. The eye is large and the snout pointed. Preserved specimens often show vertical rows of spots in the caudal fin, which is truncate-rounded. Females are not barred, thus bear some resemblance to *F. seminolis*, *Gambusia affinis*, or females of *Lucania parva*.

DISTRIBUTION: Missouri and Kentucky southward to eastern Texas, the Gulf Coast, and South Carolina, in fresh water, including the entire Florida mainland.

95. Northern Starhead Minnow, *Fundulus notti* (Agassiz)

IDENTIFICATION: A distinctively colored little fish with a *dark spot under the eye, a light one on top of the head, and several longitudinal rows of dark spots* (reddish in life) on an olive brown background. Dark bars in males. There are about 35 oblique scale rows on each side. Maximum length about 65 mm. (Also see Key.)

DISTRIBUTION: Iowa and Ohio southward to Texas and Florida (south and east to Ochlockonee River, perhaps into southwestern Georgia).

96. Southern Starhead Minnow, *Fundulus lineolatus* (Agassiz)

IDENTIFICATION: Similar to *F. notti* in size, form, and in having the dark spot under the eye and the light spot on the head, but differing in color pattern and more slender (body depth less than 25 percent of standard length). Males have prominent *vertical bars* and females *continuous* longitudinal stripes. (Also see Key.)

DISTRIBUTION: Florida Peninsula, from the Ochlockonee River to Palm Beach County (Rivas, 1966); also north in the Coastal Plain to southeastern Virginia (Map 14).

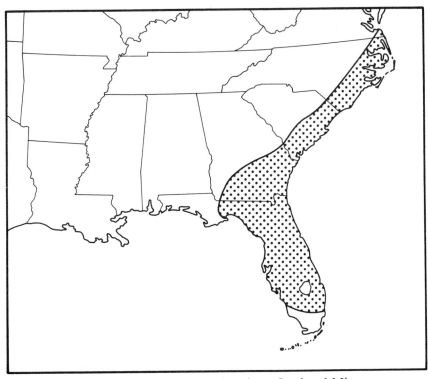

Map 14. Distribution of the Southern Starhead Minnow
(*Fundulus lineolatus*)

97. Black-spotted Topminnow, *Fundulus olivaceus* (Storer)

IDENTIFICATION: The long, slender body proportions and shape of the caudal fin are almost identical to those in *F. notti* and *F. lineolatus*, which it also resembles in having a light spot on the crown. But *F. olivaceus*, unlike others in its genus, *has one broad, dark, lateral stripe*, continuing onto the caudal fin. (Females of *F. majalis* may have one *narrow* stripe.) Ground color olive brown above, lighter below, with small dark spots on the body and dorsal fin. Maximum length about 90 mm.

DISTRIBUTION: Oklahoma, Missouri, and Illinois southward to eastern Texas, western Georgia, and western Florida (east to Chattahoochee River, but not near coast).

98. Sheepshead Minnow, *Cyprinodon variegatus* Lacépède

IDENTIFICATION: Florida members of this genus are *deep-bodied* (depth about one-third of standard length) and differ from other killifishes of similar proportions in their *enlarged humeral scale*. Caudal fin square-tipped and edged with black basally. Compared with the only other member of the genus in Florida (*C. hubbsi*), this species has only 9 or 10 anal rays and is larger (up to 75 mm long). Coloration olive green or olive gray, the dark pigment often forming vertical bars in females, and the males with orange or salmon on belly and fins in life. The young show a black spot on the dorsal fin.

DISTRIBUTION: Salt, brackish, and fresh water of Atlantic and Gulf Coastal Plain from Massachusetts through Florida to the Rio Grande. Throughout Florida mainland, mainly near the coast.

99. Lake Eustis Minnow, *Cyprinodon hubbsi* Carr

IDENTIFICATION: Smaller than the Sheepshead Minnow (about 25 mm in length), and *body less deep* (not more than one-third of standard length). It also has *more anal rays* (11) than that species, but is similar to it in color pattern.

DISTRIBUTION: Marion County to Palm Beach County, Florida, in lakes and ditches (Christensen, 1965; Map 15).

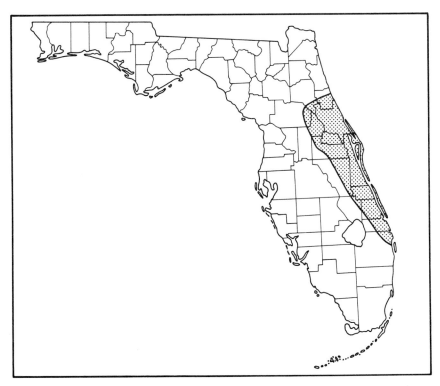

Map 15. Distribution of the Lake Eustis Minnow (*Cyprinodon hubbsi*)

100. Gold-spotted Killifish, *Floridichthys carpio* (Günther)

IDENTIFICATION: Proportions as in the Sheepshead Minnow (*Cyprinodon variegatus*), but snout much more pointed. It lacks that species' enlarged humeral scale. Live adults can be distinguished by their *yellowish spots*, and males show orange in the dorsal and anal fins. The pattern of *dark vertical bars*, *more narrow than the interspaces*, may become obscure with age. Caudal fin not black basally, but with small black specks in vertical rows in young. Maximum length about 75 mm.

DISTRIBUTION AND VARIATION: *F. c. carpio* inhabits brackish water (occasionally fresh) from Key West to Cape San Blas (Gulf County), Florida (Map 16); another subspecies occurs in Yucatan.

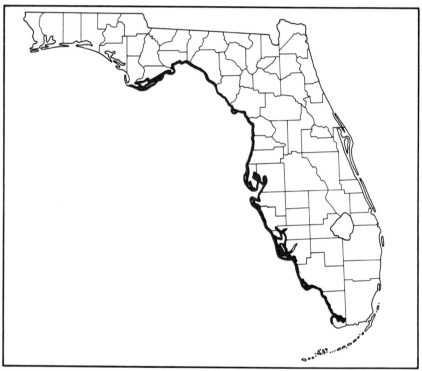

Map 16. Distribution of the Gold-spotted Killifish
(*Floridichthys carpio carpio*)

101. Flagfish, *Jordanella floridae* (Goode and Bean)

IDENTIFICATION: Proportions as in *Floridichthys* and *Cyprinodon*, but easily distinguished by a *large dark spot centrally located on each side*. Another dark spot often present in dorsal fin, as well as *golden flecks on sides in life*. Maximum length about 50 mm.

DISTRIBUTION: Florida, from Franklin, Leon, and Duval Counties southward (Tagatz, 1967; Map 17).

ORDER ATHERINIFORMES: XXV. FAMILY POECILIIDAE, Live-bearers. Similar to the killifishes (Cyprinodontidae) in general body form and mouth structure, but more pot-bellied than most. *Caudal fin rounded* to almost truncate. Males are more easily separated from killifishes by their *modified anal fin* (gonopodium). The most reliable distinction between the two families, however, is the branched third anal ray of the killifishes.

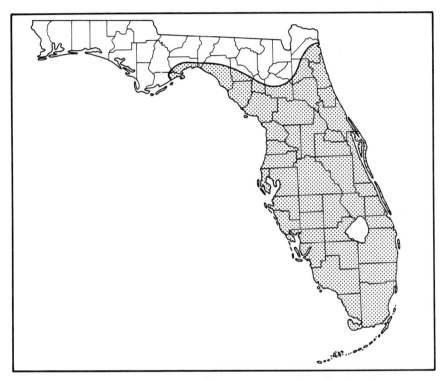

Map 17. Distribution of the Flagfish (*Jordanella floridae*)

102. Common Mosquitofish, *Gambusia affinis* (Baird and Girard)

IDENTIFICATION: Typically light olive gray above and paler below, but occasionally melanistic (black-blotched). The darker scale edges make it similar to *Fundulus seminolis*, and to females of *F. chrysotus* (preserved) and *Lucania parva*, but the Mosquitofish has *vertical rows of dark specks in the caudal fin*. Pregnant females are further distinguished by their *swollen bellies*, visibly darkened within, and males by their *gonopodium*. The dorsal fin has less than 10 rays. Total length up to 65 mm in females, about 50 mm in males. Melanism frequent.

DISTRIBUTION AND VARIATION: Lowlands of southern Illinois, Indiana, and New Jersey southward to Mexico and Florida (throughout); widely introduced elsewhere. *G. a. holbrooki* ranges over most of Florida, but is replaced in the extreme western part by *G. a. affinis* (Rosen and Bailey, 1963).

103. Mangrove Mosquitofish,[1] *Gambusia rhizophorae* Rivas[2]

IDENTIFICATION: Differs from *G. affinis* (and resembles *Poecilia*) in its *longitudinal rows of dark spots*. Unlike the molly, however, it has a slender peduncle and no more than 9 rays in the dorsal fin. The number of oblique scale rows (28 to 30) is lower than in the Common Mosquitofish.

DISTRIBUTION: Extreme southeastern Florida, including the Keys, and northwestern Cuba; euryhaline (Rivas, 1969).

104. Least Killifish, *Heterandria formosa* Agassiz

IDENTIFICATION: A tiny fish *not exceeding 25 mm* in length. Somewhat stockier than *Gambusia* and without spots in caudal fin. Dark olivaceous, with an *irregular dark lateral stripe, expanding on peduncle*. Dorsal fin almost directly above anal and *with a dark spot*. *Females* have an additional *dark spot in the anal fin*.

DISTRIBUTION: Coastal Plain from North Carolina to southeastern Alabama and southern Louisiana, including all of Florida mainland (Map 18).

105. Sailfin Molly, *Poecilia latipinna* (Lesueur)[3]

IDENTIFICATION: Suggestive of a large, heavy-set *Gambusia*, but with *longitudinal rows of dark spots* and a *long dorsal fin* (14 to 16 rays). Ground color olive green, but iridescent bluish in breeding males, which also show black dashes in the dorsal fin. Melanism frequent. Total length up to 75 mm.

DISTRIBUTION: Atlantic and Gulf Coasts from South Carolina through Florida and Mexico to Yucatan; in fresh, brackish, or salt water throughout state, mostly near the coast.

106. Pike Killifish, *Belonesox belizanus* Kner

IDENTIFICATION: A large topminnow with a *pickerel-like mouth* and a *dark spot at the caudal base*. One or more longitudinal rows of dark

1. The describer did not provide a common name.
2. "*Gambusia* species" of Carr and Goin (1955); later lumped with *Gambusia punctata*, of Cuba.
3. For change of generic name, see Rosen and Bailey, 1963.

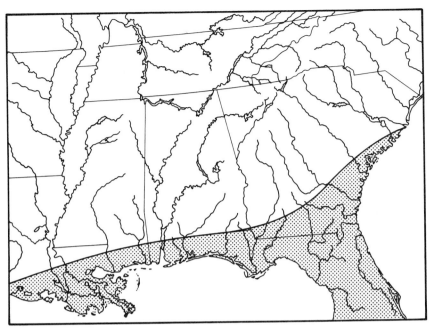

Map 18. Distribution of the Least Killifish (*Heterandria formosa*)

spots on body. Origin of anal fin anterior to that of dorsal. Eye reaching to dorsal contour of head. *Snout long.* Females may reach a length of 175 mm, males about 100 mm.

DISTRIBUTION AND VARIATION: Native to streams entering the Gulf of Mexico and Caribbean Sea from Veracruz to Nicaragua. *B. b. maxillosus* is established in the ditches and canals south of Miami (Rosen and Bailey, 1963).

ORDER ATHERINIFORMES: XXVI. FAMILY ATHERINIDAE, Silversides. Small, *very slender*, delicate fishes with *superior mouths*, 2 widely separated dorsal fins, and *long anal fins*. All are greenish above, lighter below, and have a *dark band along the sides*, appearing silvery at certain angles. The *eyes are relatively large*. Very *slender, delicate spines* are present—4 or 5 in the first dorsal fin, and 1 each in the anal and second dorsal. Total length not greater than 125 mm.

107. Tidewater Silverside, *Menidia beryllina* (Cope)

IDENTIFICATION: Differs from other silversides in its combination of *smooth-edged scales*, the *shorter anal fin* (less than 20 rays), and the *number of oblique scale rows* (less than 40). The standard length is about 6 times body depth, and the snout is rounded in dorsal aspect. Total length up to 115 mm.

DISTRIBUTION: Along the Atlantic and Gulf Coasts from Massachusetts to Veracruz, penetrating rivers and canals to Lake Okeechobee.

108. Rough Silverside, *Membras martinica* (Valenciennes)

IDENTIFICATION: Markedly different from other silversides in the *ragged or scalloped edges of the scales*. In proportions and length of anal fin, very similar to *Menidia*, but with more scales along the sides (45 to 48). Snout rounded in dorsal aspect. Total length up to 100 mm.

DISTRIBUTION: Atlantic Coast from New York to Mexico, entering mouths of rivers.

109. Brook Silverside, *Labidesthes sicculus* (Cope)

IDENTIFICATION: An especially elongate little fish with a form suggestive of a Barracuda (standard length about 8 times body depth). Accordingly, it has *more oblique scale rows along the sides* (at least 70) and *more rays in the anal fin* (22–24) than any other silverside. The edges of the scales are smooth, and the *snout more narrow* than in the 2 preceding species, almost pointed in dorsal aspect. Total length up to 100 mm.

DISTRIBUTION: Low elevations from Minnesota, Wisconsin, lower Michigan, Ontario, and the St. Lawrence River southward to central Oklahoma and the Gulf Coast from Texas to Florida (throughout mainland).

ORDER GASTEROSTEIFORMES: XXVII. FAMILY SYNGNATHIDAE, Pipefishes and Seahorses. With their elongate body, *covered with enlarged bony plates*, their *tube-shaped snout*, vestigial anal fin, and lack of a pelvic fin, the members of this family are about as unmistakable as an animal can be. The peduncle is attenuated into a prehensile tail, terminating in pipefishes in a small caudal fin. The more stout-bodied seahorses (*Hippocampus*) have not been reported from fresh water in Florida. In pipefishes the dorsal fin extends onto the peduncle. Most species do not exceed a length of 200 mm.

110. Gulf Pipefish, *Syngnathus scovelli* (Evermann and Kendall)

IDENTIFICATION: (See description of family.) Pipefishes in this genus differ from those in the genus *Oostethus* in that the *pouch of the male is located under the tail*. In the species *S. scovelli* there are *3 body rings and 5 caudal rings below the dorsal fin*, a total of *18 rings around the body*, *31 to 34 around the tail*, and *30 to 34 rays in the dorsal fin*.

Both the *vent and the dorsal fin are closer to the mouth than to the caudal fin*. Color brownish. Maximum length about 125 mm.

DISTRIBUTION: Northeastern Florida around the coast to Corpus Christi, Texas, entering Florida rivers.

111. Opossum Pipefish, *Oostethus lineatus* (Valenciennes)

IDENTIFICATION: The *egg pouch* of males *is located on the abdomen*, and the *vent and dorsal fin are closer to the caudal fin than to the mouth*. The body contains about 20 rings and the *tail about 25*. Yellowish brown with 5 or 6 dark crossbars under the snout. Maximum length about 95 mm.

DISTRIBUTION: South Carolina and the Gulf of Mexico southward to Brazil (Rio de Janeiro) and Africa; ("euryhaline": Briggs, 1958).

ORDER PERCIFORMES: XXVIII. FAMILY CENTROPOMIDAE, Snooks.[1] The members of this order are so diverse as to defy easy definition. All have *spines in the dorsal fin*, however, and all Florida species in the *anal fin as well*. The scales are usually ctenoid. Snooks (Centropomidae) are long, streamlined fishes with a markedly *protruding lower jaw*, a *deeply forked tail*, and 2 *dorsal fins*. The *lateral line* is complete and *extends onto the caudal fin*, and the bases of the dorsal and anal fins are covered by scaly sheaths. All have a narrow, dark lateral streak, sometimes accompanied by a series of horizontal V's, and the *first anal ray is dark*. The anal fin has fewer rays (6 or 7) than in most other members of the order. Body depth is about one-fourth of standard length, and the profile of the head is concave.

112. Common Snook, *Centropomus undecimalis* (Bloch)

IDENTIFICATION: (See description of family.) This species differs from other snooks in having a shorter pelvic fin, *which does not reach beyond the anus when appressed*. Also the body is relatively slender (not more than one-fourth of standard length). The counts of 11 rays in the dorsal fin and 6 in the anal usually differ from those of other snooks. Its *maximum weight of about 22 kg* is at least 10 times that of other snooks. There are about 67 to 75 scales along the lateral line. The second anal spine, when appressed, does not reach the base of the caudal fin. Color pattern as in other species: olive or bluish above, silvery below, and with a narrow dark stripe along the lateral line. Total length up to 120 cm (Rivas, 1962).

DISTRIBUTION: Florida (rarely to North Carolina) and southern Texas to Brazil and the West Indies, ascending streams and canals in the Florida Peninsula; ranges north on Gulf Coast to Franklin County.

---

1. *Centropomus ensiferus* also occurs in Dade County canals (Rivas, 1962).

113. Tarpon Snook, *Centropomus pectinatus* Poey

IDENTIFICATION: Resembling *C. undecimalis* in its color pattern, elongate body, and protruding jaw, but differing in its smaller size (up to about 50 cm in total length) and more compressed body. From small specimens of *C. undecimalis* it may be distinguished by its anal and dorsal ray counts (7 and 10 respectively), its *longer pelvic fin* (reaching beyond anus when appressed), and its lower lateral scale count (not more than 67). Also the second anal spine reaches at least to the base of the caudal fin. The *number of* well-developed *gillrakers* on the lower limb of the first arch is higher than in the other 2 snooks (15 or more) (Rivas, 1962).

DISTRIBUTION: Southern Florida (Caloosahatchee River and Biscayne Bay) to Brazil, entering freshwater canals in Florida (Rivas, 1962).

114. Little Snook, *Centropomus parallelus* Poey

IDENTIFICATION: Like the Tarpon Snook, this fish is usually less slender than *C. undecimalis* (depth more than one-fourth of standard length), but otherwise all 3 species are very similar in body form and coloration. The combination of 10 rays in the dorsal fin and usually 6 in the anal tends to distinguish this species from the other 2, but the most reliable difference is the *large number of scales in the lateral line* (80 or more). The appressed anal spine barely reaches the level of the caudal base. In having less than 13 well-developed gill rakers on the lower limb of the first gill arch it differs from *C. pectinatus*, and in its smaller size (not more than 46 cm long) from *C. undecimalis* (Rivas, 1962).

DISTRIBUTION: Southern Florida (Lake Okeechobee) and eastern Mexico southward to South America (Rivas, 1962).

ORDER PERCIFORMES: XXIX. FAMILY PERCICHTHYIDAE, Temperate Basses. These are slender, slightly compressed fishes, somewhat deeper than snooks (depth usually more than one-fourth of standard length) and of a pale ground color. Freshwater species have *several narrow dark stripes along the sides*. They show a combination of a moderately forked caudal fin, *2 dorsal fins*, and thoracic pelvic fins. The lateral line does not extend onto the caudal fin, and the *3 anal spines are strongly graduated*. Profile of head concave.

115. Striped Bass, *Morone saxatilis* (Walbaum)

IDENTIFICATION: This species differs from the next most noticeably in having the *second and third anal spines relatively short* (see Key). Typically it also has a *more slender body* (depth less than one-third of standard length). Maximum weight 45 kg or more. (Also see description of family.)

DISTRIBUTION: Atlantic and Gulf coastal plain, New Brunswick to Mississippi; penetrates certain Florida rivers for great distances. Introduced on Pacific Coast. Anadromous.

116. White Bass, *Morone chrysops* (Rafinesque)

IDENTIFICATION: Similar to the Striped Bass in having several narrow dark stripes on a pale background, but differs in having the *second anal spine longer* (one-third of head length). Furthermore, its body *depth is usually more than one-third of standard length*. Smaller (to 45 cm and 1.5 or 2 kg).

DISTRIBUTION: Minnesota, southern Great Lakes, and the St. Lawrence River southward to Texas, Georgia, and western Florida (Kilby et al., 1959).

ORDER PERCIFORMES: XXX. FAMILY CENTRARCHIDAE, Sunfishes. Deep-bodied, markedly compressed freshwater fishes, with *strong spines in the dorsal and anal fins*. There is *only one dorsal fin* (sometimes deeply notched). With few exceptions, the caudal fin is notched or forked, and a lateral line present, but not extending onto the caudal fin.

117. Banded Pygmy Sunfish, *Elassoma zonatum* Jordan

IDENTIFICATION: The pygmy sunfishes (*Elassoma*) differ from others in this family in their small size and *absence of a lateral line*. The Banded Pygmy Sunfish differs from the Okefenokee and Everglades Pygmy Sunfishes in having *more dorsal spines* (5), *more oblique scale rows* (35 or more), and *more longitudinal scale rows* (about 19). Color dusky greenish with several *vertical dark bars*, the males with little or no iridescence. It may be distinguished from other pygmy sunfishes also by the *dark spot on the body near the tip of the pectoral fin*. Maximum length about 65 mm.

DISTRIBUTION: Low elevations from Illinois and North Carolina southward to eastern Texas, the Gulf Coast, and northern Florida (to Marion County).

118. Okefenokee Pygmy Sunfish, *Elassoma okefenokee* Böhlke

IDENTIFICATION: A tiny, dark sunfish with a square-tipped tail, *3 or 4 dorsal spines*, *less than 35 oblique scale rows*, and *no dark spot on the body* near tip of pectoral fin. It resembles the Everglades Pygmy Sunfish in that males of both species are brilliantly iridescent blue in life, but differs in having *no scales on the head*, *contrasting light and dark color on the mouth*, and a *blotched color pattern* except near the caudal peduncle. Maximum length about 30 mm.

DISTRIBUTION: Florida (south to Dixie and Gilchrist Counties, west to Okaloosa County), northward to southwestern and central Georgia (Böhlke, 1956; Smith-Vaniz, 1968) (Map 19).

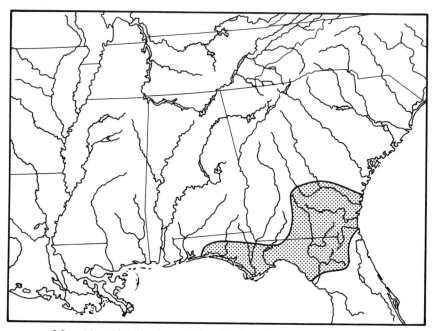

Map 19. Distribution of the Okefenokee Pygmy Sunfish
(*Elassoma okefenokee*)

119. Everglades Pygmy Sunfish, *Elassoma evergladei* Jordan

IDENTIFICATION: Most similar to *E. okefenokee*, but with a *scaly head*, a *uniformly colored mouth*, and a pattern of dark vertical bars *posteriad* (but *no dark blotches*). As in that species, the males are brilliantly blue in life. Maximum length about 30 mm.

DISTRIBUTION: Atlantic and Gulf Coastal Plain from North (?) and South Carolina to Mobile Bay, Alabama (Smith-Vaniz, 1968); south in Florida to tip of mainland.

120. Suwannee Bass, *Micropterus notius* Bailey and Hubbs

IDENTIFICATION: All species of bass (*Micropterus*) differ from other sunfishes in their proportions (depth less than one-third of standard length) and in having a *definite notch between the spinous and soft parts of the dorsal fin*. In this species the *notch is shallow* and the color pattern unusually dark. The *depth of the peduncle is about 13 percent of standard length*, and there are about 60 scales in the lateral line. Up to 300 mm or more in length.

DISTRIBUTION: Suwannee and Ochlockonee River drainages (Hellier, 1967; Yerger, personal communication; Map 20).

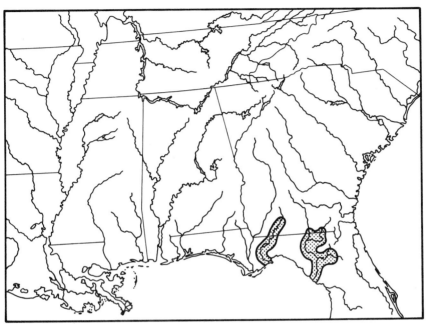

Map 20. Distribution of the Suwannee Bass (*Micropterus notius*)

121. Spotted Bass, *Micropterus punctulatus* (Rafinesque)

IDENTIFICATION: Proportions as in the Suwannee Bass, but even more slender (depth of peduncle about 11 percent of standard length) and usually with a higher lateral-scale count (62–70). The color is much lighter, especially ventrally. Typically longitudinal rows of spots are present. Younger specimens show the irregular dark lateral stripe of *M. salmoides*, but terminating in a *more conspicuous spot at the caudal base*. Maximum length about 45 cm.

DISTRIBUTION: Kansas and the Ohio River system south to eastern Texas, western and southern Georgia (Dahlberg and Scott, 1971), and western Florida (east to Gadsden County).

122. Red-eyed Bass, *Micropterus coosae* Hubbs and Bailey[1]

IDENTIFICATION: Similar to the Suwannee Bass and Spotted Bass, but with *smaller scales* (75 or more along lateral line). Proportions simi-

---

1. "*Micropterus* species" of Carr and Goin (1955).

lar to those of *M. punctulatus* (depth of peduncle 11 to 12 percent of standard length). Median fins and body with numerous dark spots, those on body almost coalescing to form *longitudinal streaks*; pattern less marked in old age. Maximum length about 40 cm.

DISTRIBUTION: Georgia, Alabama, Tennessee (?), and the Chipola and Apalachicola Rivers of Florida (Kilby et al., 1959).

123. Largemouth Bass, *Micropterus salmoides* (Lacépède)

IDENTIFICATION: The largest member of its genus in Florida, this bass is also distinguished by the *deep notch between the spinous and soft parts of the dorsal fin*, and the *absence of scales on the membranes of the dorsal and anal fins*. (Also see Key.) There are about 60 to 65 scales along the lateral line, and the depth of the caudal peduncle is about 12 percent of standard length. Color of adults uniformly dark; young with a row of irregular blotches along the side, typically coalescent, but *not* terminating in a conspicuous dark spot at caudal base. Reaches a length of at least 50 cm and a weight of more than 6 kg (15 lbs).

DISTRIBUTION AND VARIATION: Southeastern Canada and the Great Lakes to Mexico and Florida (throughout the mainland); introduced elsewhere. *M. s. salmoides* occurs in northern and northwestern Florida, and *M. s. floridanus* in the Peninsula.

124. Warmouth, *Lepomis gulosus* (Cuvier)

IDENTIFICATION: An unusually dark-colored, large-mouthed sunfish, less deep-bodied than some (depth usually less than half of standard length). The *mouth extends to a point below the eye*, and the supramaxilla is well developed (see Key). Color dark olive brown with faint darker blotches and several *alternating light and dark stripes on operculum*. There is a tendency toward *one dark spot per scale*. The pattern becomes more obscure in large specimens, but (as in *Pomoxis* and *Centrarchus*) there are always light spots in the anal and dorsal fins. Like *Lepomis microlophus*, it may show an *orange spot on the opercular flap*. Maximum length 200 to 250 mm.

DISTRIBUTION: Throughout eastern United States except the Dakotas, Minnesota, most of Wisconsin, New England, and the mountains.

125. Green Sunfish, *Lepomis cyanellus* Rafinesque

IDENTIFICATION: Rather elongate (body depth less than 50 percent of standard length) and with a larger mouth than most members of the genus (maxilla reaching a point below pupil of eye). With the Bluegill it often shares a *dark spot at the posterior base of the dorsal fin*, but its proportions and rounded pectoral fin distinguish it from that species. Some individuals have dark spots on the body or a large one on the anal fin. The gill rakers are unusually long and slender (length about 6 times width at base). Profile of head *flat to concave*. Ground color iridescent greenish. Maximum length about 150 mm.

DISTRIBUTION: Colorado, Minnesota, southern Ontario, and New York southward (except in the mountains) to New Mexico, northeastern Mexico, Louisiana, Mississippi, Alabama, Georgia, and northwest Florida (Lake Seminole, where probably introduced; Kilby et al., 1959). Widely introduced elsewhere.

126. Spotted Sunfish, *Lepomis punctatus* (Valenciennes)

IDENTIFICATION: Proportions as in most members of this genus (body depth about half of standard length). The pattern of *small dark spots on a dark olivaceous background*, and a *dark opercular flap without a light margin*, set this species apart from other sunfishes. Maximum length about 175 mm.

DISTRIBUTION AND VARIATION: Southern Indiana and Illinois southward to the Gulf Coast, including all of mainland Florida (*Fla. Scientist* 37:122); up the Atlantic Coastal Plain to North Carolina; also in the western United States east to central Texas, Arkansas, and Missouri. The Florida race, *L. p. punctatus*, intergrades with *L. p. miniatus* in west Florida.

127. Red-eared Sunfish, *Lepomis microlophus* (Günther)

IDENTIFICATION: In life, a light colored sunfish with a *bright orange red margin on the opercular flap* (light in preservative). Preserved specimens are more difficult to identify. The very short gill rakers then are helpful (length not more than twice width), as is the *long, pointed pectoral fin* (at least one-third of standard length). Body depth less than half of standard length. Maximum length about 250 mm.

DISTRIBUTION: Missouri and Indiana southward to the Gulf Coast (Texas to Florida), ranging throughout the Florida mainland.

128. Red-breasted Sunfish, *Lepomis auritus* (L.)

IDENTIFICATION: The *reddish or yellowish underparts* distinguish this sunfish from most in life, sometimes along with small dark spots as in *L. punctatus*. In preservative it appears almost uniformly pale. Then the *long, narrow opercular flap* (except in small specimens), *with* an incomplete or *no light margin*, is a good clue to its identity. The pectoral fin is moderately rounded, the gill rakers short and wide. Body depth less than half of standard length, except in very large specimens. Maximum length about 200 mm.

DISTRIBUTION: Near Atlantic Coast from New Brunswick to Florida (to Hillsborough County) and near the Gulf Coast to Texas (introduced); also up certain river systems to Oklahoma, western and northern Alabama, and northern Georgia (introduced; Dahlberg and Scott, 1971).

129. Dollar Sunfish, *Lepomis marginatus* (Holbrook)

IDENTIFICATION: A small sunfish with orange on belly and sides of head, *blue green stripes on sides of head*, and a *green margin on the opercular flap* (light in preservative). Deep-bodied, the *body depth of adults being more than half of standard length*. The pectoral fin is relatively short and rounded. Young may be barred like young Bluegills, but lack the dark spot on the dorsal fin. *Gill rakers short and wide*. Maximum length about 125 mm, rarely more.

DISTRIBUTION: Low elevations from southeastern Oklahoma, Arkansas, Tennessee, and North Carolina southward to Texas and Florida (nearly, if not quite, to tip of mainland).

130. Long-eared Sunfish, *Lepomis megalotis* (Rafinesque)

IDENTIFICATION: Compared with other Florida sunfishes, the Long-ear has a blunter snout, almost *pug-nosed profile* (length of eye same as length of snout), and an *opercular flap with a bright blue margin in life* (pale in preservative). This flap becomes exceptionally long in old age; its width is usually as great as the diameter of the eye, and its length even greater. The gill rakers are short, their length not more than twice their width. The dorsal ground color is bluish, the ventral orange, and there are *often orange spots on the body*. The blue streaks on the head and operculum suggest the Dollar Sunfish, but preserved specimens may be uniformly plain. Depth of body less than half standard length. Maximum length 150 to 200 mm.

DISTRIBUTION AND VARIATION: Iowa, Minnesota, southeastern Canada, and western Pennsylvania southward (except in mountains) to eastern Mexico, South Carolina, western Georgia, and the Gulf Coast. In Florida, *L. m. megalotis* is confined to the Panhandle (Holmes County westward).

131. Orange-spotted Sunfish, *Lepomis humilis* (Girard)

IDENTIFICATION: Somewhat more elongate than most members of the genus, and with the *dorsal profile of the head rather concave*. The color in life suggests both *L. microlophus* (orange red margin of opercular flap) and *L. megalotis* (*orange spots on body*). Preserved specimens can be separated from most species by the *long, slender gill rakers* and lower count of lateral-line scales (less than 40). Maximum length about 100 mm.

DISTRIBUTION: North Dakota and western Ohio southward to Texas, Alabama, western Georgia, and northern Florida (Lake Seminole; Kilby et al., 1959).

132. Bluegill, *Lepomis macrochirus* Rafinesque

IDENTIFICATION: This popular sunfish may be distinguished by its combination of *long, slender gill rakers* and *dark, often double vertical bars* on the body (young) or *dark spot posteriad in the dorsal fin* (adult). At all ages there is a *dark blue opercular spot* (with no light margin) and a long, pointed pectoral fin. The ground color is light in young, dark in large specimens, which may have coppery red underparts and fins. The body depth is less than half of standard length except in the largest specimens. Maximum length 250 to 300 mm.

DISTRIBUTION: Great Lakes and St. Lawrence drainage southward to Mexico, all of mainland Florida, and Georgia, except in the mountains and northeastern states.

133. Banded Sunfish, *Enneacanthus obesus* (Girard)

IDENTIFICATION: Sunfishes in this genus are distinguishable by their *rounded caudal fin* and *emarginate operculum*. They are smaller and proportionately deeper than most species. In this species and the next the dorsal fin is of uniform coloration, and the last 5 dorsal spines are of uniform length. *E. obesus* differs from *E. gloriosus* in having *more scale rows (19–22) around the caudal peduncle*, usually more than 5 dark vertical bars, in the *color of its pelvic fins*, and in its proportions (see Key). There are no light spots in the caudal fin, but *golden spots on its sides* help to distinguish it in life. Maximum length about 75 mm.

DISTRIBUTION: Coastal Plain from southern New Hampshire to Florida (to central parts and westward to Escambia County).

134. Blue-spotted Sunfish, *Enneacanthus gloriosus* (Holbrook)

IDENTIFICATION: Strongly similar to the Banded Sunfish (*E. obesus*), but with a *more slender peduncle* (15 to 18 longitudinal scale rows), not more than 5 dark vertical bars (often lacking or indistinct), the *anterior part of the pelvic fins very dark*, and more slender proportions (see Key). In life it is marked by *light blue to green spots on body and fins*, usually including the caudal. Maximum length about 75 mm.

DISTRIBUTION: Mostly near Atlantic and Gulf coasts, from New York to Alabama, ranging to tip of Florida Peninsula and westward to Mobile Bay drainage (Smith-Vaniz, 1968).

135. Black-banded Sunfish, *Enneacanthus chaetodon* (Baird)

IDENTIFICATION: Easily distinguished from other members of its genus by having the *middle dorsal spines longer than the others* and the *anterior tip of the dorsal fin dark*. Its dark vertical bars resemble those of *E. obesus*, but are more widely spaced and often fewer in number. The peduncle is slender, as in *E. gloriosus* (17 to 18 scale rows). Typically it has one more dorsal spine (10) than others in this genus. Maximum length about 75 mm.

DISTRIBUTION AND VARIATION: Coastal Plain from New Jersey to Florida (where *E. c. elizabethae* inhabits small lakes in the Peninsula southward to Marion County).

136. Rock Bass, *Ambloplites rupestris* (Rafinesque)

IDENTIFICATION: Differs from most sunfishes in its *greater number of spines in the dorsal fin* (11) *and anal fin* (5 to 7). The other 2 species that may show this combination are the Mud Sunfish (*Acantharchus*) and the Flier (*Centrarchus*). From the former the Rock Bass may be separated by its ctenoid scales and *notched caudal fin*, and from the Flier by its *shorter anal fin* (about 50 percent length of dorsal fin). The mouth is relatively large (maxillary extending under eye). General coloration dark olive brown, with irregular darker mottling. Maximum length about 300 mm, though probably less in Florida.

DISTRIBUTION AND VARIATION: Manitoba, Great Lakes, and St. Lawrence River southward (locally) to Louisiana, Georgia, and western Florida (east to Holmes County). The Florida race is *A. r. ariommus*.

137. Mud Sunfish, *Acantharchus pomotis* (Baird)

IDENTIFICATION: The only large sunfish with a *rounded caudal fin*. The numbers of spines in the dorsal fin (12) and anal fin (5 or 6) are unusually high. In each case the *spines are much shorter than the corresponding rays*. There are about *4 longitudinal dark stripes on the head*, continuing (sometimes less distinctly) onto the body, the ground color of which is a dull olivaceous. Maximum length about 150 mm.

DISTRIBUTION: Lowlands from New York to Florida, southward to Alachua County and westward to Leon County (FSU 16687) (Map 21).

138. Black Crappie, *Pomoxis nigromaculatus* (Lesueur)

IDENTIFICATION: In having the *dorsal and anal fins of similar size and shape*, the crappies resemble the Flier, but they differ in having

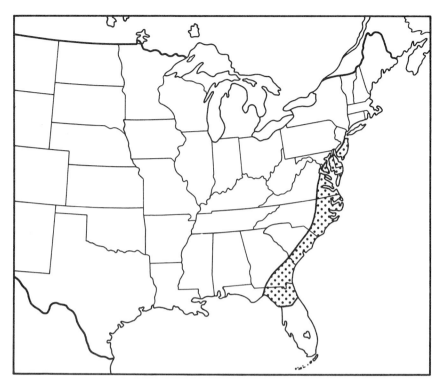

Map 21. Distribution of the Mud Sunfish (*Acantharchus pomotis*)

fewer spines in each fin (6 in anal, not more than 8 in dorsal). The Black Crappie differs from the White in having usually more spines in the dorsal fin (7 or 8) and in having *longitudinal rows of spots* on a silvery background. (Also see Key.) Total length up to 300 mm or more.

DISTRIBUTION: Manitoba and Quebec southward to eastern Texas and Florida (except in mountains and New England), ranging throughout the Florida mainland.

139. White Crappie, *Pomoxis annularis* Rafinesque

IDENTIFICATION: Similar in form to the Black Crappie, but differing in proportions (see Key), usually in having *fewer spines in the dorsal fin* (5 or 6), and with a pattern of *dark vertical bars* on a silvery background. Maximum length 250 to 300 mm.

DISTRIBUTION: Nebraska, southern Minnesota, and southern Ontario southward to Texas, North Carolina, and northwestern Florida, except absent from mountains and northeastern states. Florida records come from the upper Apalachicola River (Kilby et al., 1959) and the upper Choctawhatchee River (Yerger, personal communication).

140. Flier, *Centrarchus macropterus* (Lacépède)

IDENTIFICATION: Similar to the crappies (*Pomoxis*) in having *dorsal and anal fins light-spotted and of almost equal length*, but differs in having *at least 11 dorsal spines* and deeper body proportions. The ground color is also more greenish (less silvery) and the dark body spots smaller than those of the Black Crappie and not forming vertical bars like those of the White Crappie. The young are more brightly colored and show a *dark spot at the rear of the dorsal fin*. Total length up to 240 mm (FSU 7691) (Fig. 1).

DISTRIBUTION: Low elevations from southern Illinois, southern Indiana, and Virginia southward to eastern Texas and central Florida.

ORDER PERCIFORMES: XXXI. FAMILY PERCIDAE, Perches and Darters. The body form in this family is more elongate than in sunfishes, often nearly terete. Also there are *2 distinct dorsal fins*, spinous and soft. The *anal fin has only one or 2 spines*. Some species can be identified only with the aid of strong magnification, so that the fin rays and pored scales of the lateral line may be counted. Most species reach a maximum length of less than 150 mm.

141. Yellow Perch, *Perca flavescens* (Mitchill)

IDENTIFICATION: In having about 6 *broad, dark, often triangular, vertical bars* the Yellow Perch differs from most Florida fishes, as it does in having the *pelvic fins almost contiguous*. The caudal fin is forked. Body depth about one-third of standard length. Maximum length about 355 mm.

DISTRIBUTION: From much of eastern and central Canada southward to Kansas, Ohio, and northern Georgia. Introduced into southern Georgia and the upper Apalachicola River, Florida (Kilby et al., 1959).

142. Sauger, *Stizostedion canadense* (Smith)

IDENTIFICATION: A streamlined fish with 2 dorsal fins, a pointed head, and a *large mouth provided with canine teeth*. It differs from the Yellow Perch in its blotched color pattern and *greater distance between the pelvic fins*. Both have a forked caudal fin. Length up to 400 mm.

DISTRIBUTION: Much of central and eastern Canada southward to Arkansas, Louisiana, and northern Georgia; also introduced in eastern and western Georgia and spreading to northwestern Florida (Apalachicola River westward).

143. Logperch, *Percina caprodes* (Rafinesque)

IDENTIFICATION: Darters in this genus are somewhat larger than those in *Etheostoma*, and differ from most species in having a *complete lateral line*, the *pelvic fins farther apart*, and the *caudal fin notched* (see Key). The Logperch differs from other darters of the genus *Percina* in having a *longer snout and dark vertical bars* on a yellowish background. Maximum length about 150 mm.

DISTRIBUTION AND VARIATION: Alberta, Saskatchewan and Hudson Bay drainage southward to central California and the Rio Grande, northwestern Florida, and northwestern Georgia; Florida records range east to Washington County (*P. c. carbonaria*).

144. Black-banded Darter, *Percina nigrofasciata* (Agassiz)

IDENTIFICATION: Similar to the Logperch, but with a moderate snout and a *pattern of confluent dark blotches arranged longitudinally along lateral line* on an olive green background. Also 7 or 8 dark blotches straddling the back. Eyes normally placed. Maximum length about 125 mm.

DISTRIBUTION AND VARIATION: Coastal Plain and Piedmont from South Carolina to Louisiana, ranging southward to Orange County, Florida (*P. n. nigrofasciata*; Crawford, 1956).

145. Star-gazing Darter, *Percina uranidea* (Jordan and Gilbert)

IDENTIFICATION: Similar to the Black-banded Darter, but with the *lateral dark blotches separated* and the *eyes closer together* (see Key). Also about 4 dark blotches straddling the back. The anal fin is very long in adult males. Smaller, probably not exceeding 75 mm in length.

DISTRIBUTION: Southern parts of Missouri and Indiana southward to Louisiana, Georgia (?), and extreme western Florida (Escambia River).

146. Naked Sand Darter, *Ammocrypta beani* Jordan

IDENTIFICATION: Its extreme slenderness, virtual *lack of pigment and scales*, and darterlike form make this little fish almost unmistakable. Eyes dorsolateral, very close together. There is one anal spine, the lateral line is complete, and the premaxillary is protractile. Maximum length about 75 mm.

DISTRIBUTION: Sandy streams of Gulf Coastal Plain from Louisiana to the Choctawhatchee River drainage of Florida and Alabama (Map 22).

147. Speckled Darter, *Etheostoma stigmaeum* (Jordan)

IDENTIFICATION: Darters of this genus are separated from smaller specimens of *Percina* only with difficulty (see Key). The present species is similar to the recently described *E. davisoni*. Both species have a

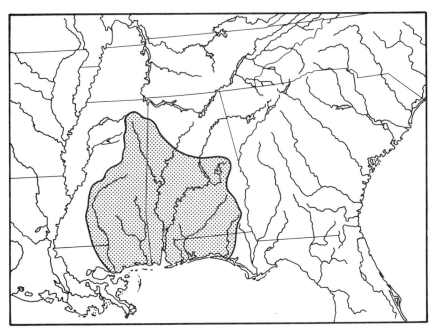

Map 22. Distribution of the Naked Sand Darter (*Ammocrypta beani*)

steeply sloped profile, a naked predorsal area in the midline, and a *protractile premaxillary*. Males of this species differ in having a *row of about 10 dark blotches along the side of the body*, each blotch tending to form a vertical bar; also the dorsal fin is more heavily pigmented, especially around the edge. Males are marked with red and blue in life. The color pattern in females is much like that in *E. davisoni*, but their long, free genital papilla should easily distinguish them. Maximum length about 65 mm.

DISTRIBUTION: Southern Missouri and Kentucky southward to Louisiana, northwestern Georgia, and northwestern Florida (Escambia River).

148. Choctawhatchee Darter, *Etheostoma davisoni* Hay

IDENTIFICATION: Color pattern similar to that of the Johnny Darter (dark dorsal blotches, W-shaped or M-shaped lateral markings), but differs in having 2 anal spines, 7 or 8 anal rays, and less heavily pigmented fins. The *protractile premaxillary* (separated from head by a groove) and *steeply down-sloped head* anterior to the eyes tend to separate it from other darters. No predorsal scales in midline. Formerly confused with *E. stigmaeum*. Maximum length about 65 mm.

DISTRIBUTION: Escambia and Choctawhatchee River systems of Alabama and Florida (Smith-Vaniz, 1968; Map 23).

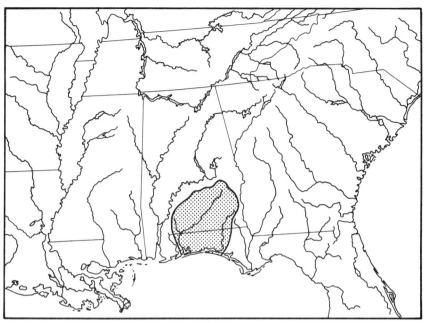

Map 23. Distribution of the Choctawhatchee Darter
(*Etheostoma davisoni*)

149. Johnny Darter, *Etheostoma nigrum* Rafinesque

IDENTIFICATION: The Johnny Darter differs from other Florida species in having *only one anal spine* and a *complete lateral line*. The *anal-ray count of 9* is also high. On its light brownish ground color a series of *W-shaped markings* can usually be found along the lateral line. Total length to 75 mm.

DISTRIBUTION AND VARIATION: Great Lakes and southeastern Canada southward to Oklahoma, Mississippi, Alabama, and Georgia, except in mountains. Also (*E. n. olmstedi*) in the Oklawaha River, Florida, probably by introduction.

150. Harlequin Darter, *Etheostoma histrio* Jordan and Gilbert

IDENTIFICATION: Another rather small darter with a *steeply down-sloped, blunt snout*. Predorsal midline scaly. *Lateral line virtually complete*. There are about 8 large lateral blotches *connected to dorsal blotches*, though tending to alternate with them in position. Frenum poorly developed. Maximum length about 50 mm.

DISTRIBUTION: Indiana and Kentucky southward to eastern Texas, southern Alabama, and western Florida (Escambia River; FSU 5911).

151. Gulf Darter, *Etheostoma swaini* (Jordan)

IDENTIFICATION: A heavy-bodied darter, formerly confused with *E. okaloosae*, but differing in its color pattern of dark spots on a light background. These are roughly arranged in longitudinal rows. There are *3 larger blotches arranged vertically at the caudal base*. The *lateral line is complete* or nearly so. Scales usually lacking along predorsal midline. Maximum length about 50 mm.

DISTRIBUTION: Eastern Louisiana to western and southern Georgia and western Florida (east to Apalachicola River system).

152. Gold-striped Darter, *Etheostoma parvipinne* Gilbert and Swain

IDENTIFICATION: Best distinguished by its color pattern of about *10 dark vertical bars, interrupting the broad, light lateral line*; often a *vertical row of large dark spots at base of caudal fin* and rows of smaller spots in the fins. The lateral line is *yellowish in life*. Breast scaly. Maximum length about 65 mm.

DISTRIBUTION: Southeastern Oklahoma and eastern Texas eastward to Tennessee, Alabama, western and southern Georgia, and northwestern Florida (east to Washington County; Yerger and Suttkus, 1962).

153. Brown Darter, *Etheostoma edwini* (Hubbs and Cannon)

IDENTIFICATION: Similar to the Okaloosa Darter in having no scales on the breast and the lateral line inconspicuous but almost complete. It differs in its lighter coloration and *normal profile of back of head*. In life it may be distinguished by the *red spots* on the body and median fins. Spot above base of pectoral fin usually faint or lacking. Up to 50 mm long.

DISTRIBUTION: Southern Alabama, southern Georgia, and northern Florida (west to Perdido River and south to Gilchrist and Alachua Counties; Map 24) (Collette and Yerger, 1962; Hellier, 1967).

154. Okaloosa Darter, *Etheostoma okaloosae* (Fowler)

IDENTIFICATION: An *unusually dark* darter with a *depression on the head behind the eyes*. The lateral line is inconspicuous but almost complete. The dorsal blotches are not distinct, but there are dark spots on the head and blotches, arranged in horizontal rows, on the sides. A good identification mark is a *dark spot above the pectoral base*. Breast not scaled. Length up to 50 mm.

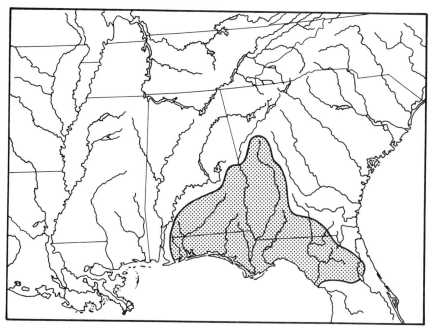

Map 24. Distribution of the Brown Darter (*Etheostoma edwini*)

DISTRIBUTION: Restricted to several streams that empty into Choctawhatchee Bay, Florida (Collette and Yerger, 1962).

155. Swamp Darter, *Etheostoma fusiforme* (Girard)

IDENTIFICATION: An unusually long, slender darter (depth less than 20 percent of standard length), with irregular dark blotches and a scaly breast. The *lateral line is* whitish (never yellow in life), conspicuous, and is *arched above the pectoral fin*. The scales are smaller than in most darters (47 to 60 along the lateral line). Maximum length about 50 mm.

DISTRIBUTION AND VARIATION: Swampy streams in the Coastal Plain from southern Maine to Florida, westward to Texas and Oklahoma; south in Florida to Dade and Collier Counties (*E. f. barratti*) (*Fla. Scientist* 37:122).

156. Cypress Darter, *Etheostoma proeliare* (Hay)

IDENTIFICATION: Distinguished from all other darters by its *rudimentary lateral line*, with no more than 7 pored scales. Also unusually small and slender, especially the peduncle, its total length probably never reaching 50 mm. Color pattern not distinctive (dark lateral blotches on olivaceous background).

DISTRIBUTION: Missouri, Illinois, and Kentucky southward to eastern Texas and northwestern Florida (east to Holmes County).

ORDER PERCIFORMES: XXXII. FAMILY ECHENEIDAE, Remoras. Moderately large, slender fishes, in which the *first dorsal fin is modified into a suckerlike organ* on the head, crossed by numerous laminae. The anal and second dorsal fins are long, about equally so, and the pelvic fins almost as far forward as the pectorals.

157. Sharksucker, *Echeneis naucrates* L.

IDENTIFICATION: (See description of family.) One of the most slender of Florida fishes, its maximum depth less than 10 percent of standard length. The disk contains about 20 to 28 laminae, the head is flattened, and the *lower jaw protrudes well beyond the upper* and has a flexible tip. The pectoral fins are unusually high on the body. Color grayish, with a broad, dark stripe down the side of the body, bordered by narrow light stripes. (Preserved specimens may be almost patternless.) Maximum length about 1 m.

DISTRIBUTION: Ranges widely through the warmer seas of the world, entering mouths of freshwater streams of Gulf drainage; northward on Atlantic Coast to South Carolina.

ORDER PERCIFORMES: XXXIII. FAMILY CARANGIDAE, Jacks and Pompanos.[1] A large group of marine fishes, often deep-bodied, and with the *soft dorsal and anal fins very long*. Caudal fin deeply forked, with slender lobes. *Posterior part of lateral line often keeled*. Peduncle slender.

158. Common Jack, *Caranx hippos* (L.)

IDENTIFICATION: (See family description.) This species has a protruding lower jaw, deep head and blunt snout, 2 free dorsal and anal spines (each), and a *largely naked breast*. Coloration silvery blue, with yellowish suffusion; a *dark spot on operculum*. Maximum length about 76 cm.

DISTRIBUTION: Atlantic Ocean as far north as Massachusetts; eastern Pacific Ocean; one record in Florida fresh water (Herald and Strickland, 1948).

ORDER PERCIFORMES: XXXIV. FAMILY LUTJANIDAE, Snappers. Heavy-bodied perciform fishes with the *dorsal fin single, but long* (10 spines, 14 rays); body depth at least one-third of standard length; and *protruding fangs* at front of upper jaw. Mostly dull colored, sometimes reddish, or with dark blotches.

---

1. *Oligoplites saurus* has been reported from the St. Lucie River (*Fla. Scientist* 37:123).

159. Gray Snapper, *Lutjanus griseus* (L.)

IDENTIFICATION: More slender than *L. apodus* (body *depth less than 40 percent of standard length*), with a *higher lateral-scale count* (more than 47), and usually a convex head. Each species has one or 2 pairs of enlarged fangs at the front of the upper jaw. Greenish or olive above, reddish or yellowish median fins and underparts, with rows of faint spots along sides. The young show light and dark vertical bars. Maximum length about 45 cm, rarely 90 cm.

DISTRIBUTION: Massachusetts to Bermuda and Brazil, invading fresh water in Florida.

160. Schoolmaster, *Lutjanus apodus* (Walbaum)

IDENTIFICATION: Similar to the Gray Snapper, but *body depth more than 40 percent of standard length*, with *fewer than 47 scales along sides*, and a concave or straight profile. One or 2 pairs of fangs at front of upper jaw. Greenish or olive above, reddish or yellowish below, with rows of faint spots along sides. Often a blue stripe on head below eye. Maximum length about 45 cm.

DISTRIBUTION: Massachusetts to Bermuda and Brazil, invading fresh water in Florida, where it ranges northwestward to Bay County.

ORDER PERCIFORMES: XXXV. FAMILY GERREIDAE, Mojarras.[1] Silvery perciform fishes with a single dorsal fin and the lateral line terminating short of the caudal fin. The first dorsal spine is shorter and the *premaxillary more protractile* than in other perciform families. The caudal fin is deeply forked, and the base of the dorsal fin is enveloped in a sheath.

161. Spotfin Mojarra, *Eucinostomus argenteus* Baird

IDENTIFICATION: A strongly compressed, *silvery fish with a slender peduncle* and gracefully forked caudal fin. It differs from its other Florida relatives in having the *second anal spine not especially enlarged*. The color pattern is silvery, with irregular blotches dorsally and a *dark smudge on the operculum*. Maximum length about 175 mm.

DISTRIBUTION: Atlantic and Gulf Coasts from North Carolina (rarely Massachusetts) to Brazil: enters Florida rivers and streams.

162. Yellowfin Mojarra, *Gerres cinereus* (Walbaum)

IDENTIFICATION: Similar to *Eucinostomus*, but larger and deeper. Second anal spine somewhat thickened, but less so than in *Diapterus*. *No serrations on preopercle*. Silvery, darker above, and with irregular,

1. *Eucinostomus gula* and *Diapterus plumieri* have also been reported from fresh water in south Florida (*Fla. Scientist* 37:123).

vertical bars, bluish in life. Paired and anal fins yellowish in life. Eyes large. Maximum length about 380 mm.

DISTRIBUTION: Atlantic Ocean from southern Florida, the Gulf of Mexico, the West Indies, and Bermuda to Brazil; also the Pacific Ocean, from Baja California to Peru. Euryhaline.

163. Irish Pompano, *Diapterus olisthostomus* (Goode and Bean)

IDENTIFICATION: A strongly compressed, silvery fish, with a slender peduncle. The *second anal spine is much thicker than the first*. As in the Spotfin Mojarra, the longest dorsal spines are much longer than the longest dorsal rays, but *the preopercle is serrate*. The young have a few *narrow, dark vertical bars*, but older specimens show no contrasting colors. Maximum length about 300 mm.

DISTRIBUTION: Coastal waters from Florida and the West Indies to Brazil, entering the St. Johns River, Florida.

ORDER PERCIFORMES: XXXVI. FAMILY POMADASYIDAE, Grunts. Deep-bodied, hump-backed perciform fishes with a long dorsal fin. The mouth is often large and the *lips thick*. From 10 to 12 spines in the dorsal fin, and 3 in the anal.

164. Burro Grunt, *Pomadasys crocro* (Cuvier)

IDENTIFICATION: (See description of family.) Front of head often depressed. Mouth not large. Preopercle serrate. Scales rather large. Eye relatively large. Dorsal fin notched. *Second anal spine long and greatly thickened*; anal rays shorter, numbering 6 or 7. About 54 scales in lateral line. Olivaceous above, silvery below, with 3 or 4 faint longitudinal dark stripes. Maximum length probably about 300 mm.

DISTRIBUTION: Gulf of Mexico and southern Florida to Brazil; euryhaline (Briggs, 1958).

165. Pigfish, *Orthopristis chrysopterus* (L.)

IDENTIFICATION: A silvery, deep-bodied fish with a strongly arched back. Differs from the Burro Grunt in its light color (bluish dorsally in life), its *more slender dorsal spines*, and its *larger number of anal rays* (11 to 13). Mouth small. About 60 scales in lateral line. Preserved specimens may show dark vertical bands. Maximum length about 380 mm.

DISTRIBUTION: Atlantic and Gulf Coasts from Long Island to the Rio Grande; recorded from fresh water in the St. Johns River, Florida (Tagatz, 1967).

ORDER PERCIFORMES: XXXVII. FAMILY SPARIDAE, Porgies. Deep-bodied perciform fishes with a long, continuous dorsal fin, *incisorlike*, *often protruding, front teeth*, and molariform teeth posteriad. Florida species show 12 dorsal and 3 anal spines. The lateral line is complete and the caudal fin forked.

166. Pinfish, *Lagodon rhomboides* (L.)

IDENTIFICATION: Distinguished from others in its family by having about *4 or 5 dark vertical bars* on the sides and a *dark spot above the back of the operculum*. The ground color is olive above, with yellowish stripes on the sides; fins mostly yellowish. The scales are small, numbering 65 to 70 along the lateral line. Body depth about 45 percent of standard length. Maximum length about 250 mm.

DISTRIBUTION: Atlantic and Gulf Coasts from Massachusetts to Texas and the West Indies (rarely Bermuda), entering Florida rivers.

167. Sheepshead, *Archosargus probatocephalus* (Walbaum)

IDENTIFICATION: Easily distinguished by the striking pattern of *about 10 alternating light and dark vertical bars on the side*, extending to the venter. The scales are unusually large, with no more than 50 along the lateral line. Body depth about 50 percent of standard length. Maximum length 300 to 380 mm.

DISTRIBUTION: Atlantic and Gulf Coasts from Massachusetts (rarely Maine) to Texas; enters Florida rivers.

168. Spot-tailed Pinfish, *Diplodus holbrooki* (Bean)

IDENTIFICATION: Differs from others in the family by its *black saddle on the peduncle*. It lacks the dark spot near the operculum of *Lagodon* and has somewhat larger scales (less than 60 along lateral line). The pattern of narrow, dark vertical bars becomes obscure with age. Maximum length about 250 mm.

DISTRIBUTION: Atlantic and Gulf Coasts from North Carolina to the Florida Panhandle, entering freshwater streams in Florida.

ORDER PERCIFORMES: XXXVIII. FAMILY SCIAENIDAE, Drums.[1] Large, marine fishes with a deeply notched dorsal fin (or 2 separate fins) and a *lateral line that extends onto the caudal fin*, which is neither forked nor notched in most species.

169. Silver Perch, *Bairdiella chrysura* (Lacépède)

IDENTIFICATION: A slender, silvery marine fish with little color pattern, but a prominent, arched lateral line. The combination of *elongate central rays in the caudal fin* and the *absence of black spots* on the body and fins distinguishes it from other members of the family. The snout is blunt, with the upper jaw protruding. The dorsal fins appear to be sep-

---

1. *Cynoscion arenarius* and *C. regalis* have also been reported from fresh water in south Florida (*Fla. Scientist* 37:123).

arate. Body depth about 30 percent of standard length. No barbels on chin. Maximum length about 230 to 250 mm.

DISTRIBUTION: Atlantic and Gulf Coasts from New York to Texas; occasional in mouths of Florida rivers.

170. Spotted Seatrout, *Cynoscion nebulosus* (Cuvier)

IDENTIFICATION: More slender than most other drums, with a *protruding lower jaw and rounded, dark spots* on body and fins on a yellowish to silvery background. Body depth about 20 percent of standard length. *Dorsal fins separate*. Maximum length 75 to 90 cm.

DISTRIBUTION: Atlantic and Gulf Coasts from New York to Texas, entering Florida rivers.

171. Red Drum, *Sciaenops ocellata* (L.)

IDENTIFICATION: The *reddish coloration* along with the *rounded black spot at the base of the caudal fin* will distinguish the "Redfish" in life. There are often black spots on the peduncle or body. The central rays of the caudal fin are the longest ones. Body proportions slender (depth less than 30 percent of standard length). Immatures have many irregular dark blotches on the body. Maximum length about 120 cm, rarely 150 cm.

DISTRIBUTION: Atlantic and Gulf Coasts from New York to Texas, entering mouths of Florida rivers.

172. Spot, *Leiostomus xanthurus* Lacépède

IDENTIFICATION: Most specimens may be distinguished in life by having about *15 oblique, yellowish bars alternating with dark ones*, and a *dark spot behind the operculum*. (Yellowish bars appear pale in preservative.) Deeper and more compressed than the Red Drum, and with a *forked caudal fin*. Maximum length about 300 mm.

DISTRIBUTION: Atlantic and Gulf Coasts from Massachusetts to Texas, entering Florida rivers.

173. Black Drum, *Pogonias cromis* (L.)

IDENTIFICATION: This species and the Atlantic Croaker are distinguished by the presence of *barbels on the chin*, but they are better developed in this species. The Black Drum has a *very high spinous dorsal fin*, and the young are marked with about 4 broad dark bars on each side. Body depth is more than 40 percent of standard length. About 22 rays in soft dorsal fin. Total length up to 120 cm.

DISTRIBUTION: Atlantic and Gulf Coasts from New York to the Mexican border, southward in Atlantic to Argentina; enters the St. Johns River in Florida.

174. Atlantic Croaker, *Micropogon undulatus* (L.)

IDENTIFICATION: Much smaller than the Black Drum and with *inconspicuous chin barbels*. It has more rays in the soft dorsal fin than that fish (at least 25), and its depth is about 35 percent of its standard length. There are *20 or more diagonal bars* on each side and a dark area at the pectoral base. Maximum length about 45 cm.

DISTRIBUTION: Massachusetts to Texas in coastal waters, entering Florida rivers.

ORDER PERCIFORMES: XXXIX. FAMILY CICHLIDAE, Cichlids. Similar to sunfishes (Centrarchidae) in size and form, but with *only one pair of nostrils* and usually an *interrupted lateral line*, the anterior part higher on the body than the posterior. Dorsal fin not divided. None are native to Florida. (Also see Addenda.)

175. Two-spotted Cichlid, *Cichlasoma bimaculatum* (L.)

IDENTIFICATION: Differs from the Oscar (*Astronotus*) in its smaller number of dorsal and anal rays and from *Tilapia* in its smaller number of lateral-line scales and greater number of pectoral rays. General coloration may be highly variable, but usually with *large, black blotches at midsides and near caudal base*; another may be under the eye, connecting with a dark longitudinal stripe on the side. Maximum length 175 to 200 mm (Sterba, 1962).

DISTRIBUTION: Native to most of northern South America; introduced into Florida and found (1972) in Palm Beach, Dade, Monroe, and probably Collier Counties (Walter Courtenay, personal communication).

176. Oscar, *Astronotus ocellatus* (Agassiz)

IDENTIFICATION: Differs from other Florida cichlids in the very *high numbers of dorsal and anal rays* (about 20 and 15, respectively). Also the body is less deep and the dorsal, anal, and caudal fins more rounded. Coloration brilliant, but variable, usually featuring a *rounded, black ocellus, surrounded with red, at the caudal base*. Maximum length 300 to 330 mm (Regan, 1905; Sterba, 1962).

DISTRIBUTION: Found over most of lowland South America, east of the Andes; introduced into Dade and Broward Counties, Florida (Vernon Ogilvie, personal communication).

177. Black-chinned Tilapia, *Tilapia melanotheron* (Rüppell)

IDENTIFICATION: Resembles *Cichlasoma* in form and size, differing in the number of lateral-line scales and rays in the pectoral fin. Upperparts brownish to greenish olive, with dark-edged scales, lighter ventrally, but coloration quite variable. Dark blotches often present. Operculum translucent in females. Young may have 6 dark vertical bars and a dark ocellus at base of soft dorsal. Maximum length about 300 mm (Regan, 1905; Sterba, 1962). See Key.

DISTRIBUTION: Brackish and fresh waters of West Africa (Senegal to Congo); introduced into Tampa Bay, Florida, and also collected at Eureka Springs (FSU 6197 and 6198) (Vernon Ogilvie, personal communication).

ORDER PERCIFORMES: XL. FAMILY MUGILIDAE, Mullets. Almost terete, blunt-nosed fishes with *high pectoral fins*, 2 well-separated dorsal fins, the first bearing *4 slender spines*, no lateral line, and a *mouth shaped like an inverted V*. The location of the pelvic fins is between thoracic and abdominal. The mouth is small and weak, the caudal fin strongly forked. Lower jaw not protruding beyond upper.

178. Striped Mullet, *Mugil cephalus* L.

IDENTIFICATION: All mullets in the genus *Mugil* have *adipose tissue around the edges of the eye*. The present species is gray blue above and silvery white below, the *sides of adults striped with dusky*, the stripes somewhat interrupted in some cases. The lateral-scale count is about 40, and there are few scales around the bases of the anal and soft dorsal fins. Total length up to about 75 cm.

DISTRIBUTION: Around the Atlantic Coast and Gulf of Mexico from Cape Cod (rarely Maine) to Brazil, southern Europe, and northern Africa; along the Pacific Coast from California to Chile; also on coasts of Haiti, Hawaii, and Japan. Ascends far up most Florida rivers and to Lake Okeechobee.

179. White Mullet, *Mugil curema* Valenciennes

IDENTIFICATION: This species and the Fan-tailed Mullet differ from the Striped Mullet in their *lack of stripes*, in having *less than 40 oblique scale rows along the side*, and the anal and soft dorsal fins are fully scaled. The White Mullet has a *greater number of oblique scale rows* (37–39) *and anal rays* (9) than the Fan-tailed Mullet. Body depth not more than 25 percent of standard length. General color silvery, but with a *dark spot at pectoral base*. Maximum length 60 to 90 cm, but probably not more than 45 cm in Florida.

DISTRIBUTION: Restricted to the Western Hemisphere, where its range is similar to that of the Striped Mullet. Ascends the larger Florida rivers.

180. Fan-tailed Mullet, *Mugil trichodon* Poey

IDENTIFICATION: Differs from the White Mullet chiefly in having *fewer oblique scale rows* along the side (32 to 34) and only 8 anal rays. It also shows "a dark blotch at the base of the pectoral fin" (Carr and Goin, 1955) and typically shows 11 longitudinal scale rows rather than 12. General color bluish above, lighter below; pelvic and anal fins pale, others dusky. Darker and less silvery than other mullets. Body depth more than 25 percent of standard length. Maximum length about 30 cm.

DISTRIBUTION: Salt and brackish water from *southern* Florida and Bermuda southward to Brazil, occasionally into fresh water; northward in Florida at least to Collier County.

181. Mountain Mullet, *Agonostomus monticola* (Bancroft)

IDENTIFICATION: The *adipose eyelid* of other mullets *is lacking in this species*, and the head is longer (more than 25 percent of standard length). Also a *lateral cleft extends from the corner of the mouth to the level of the orbit*. Dorsal coloration more brownish, with "yellow around the base of the caudal fin" (Carr and Goin, 1955). Immatures, at least, have a light stripe along the side and a dark spot at the caudal base. Total length up to 30 cm (probably not more than 20 cm in Florida).

DISTRIBUTION: Small, swift streams of the Caribbean area, westward in the Gulf of Mexico to Texas; entering Florida streams of Atlantic drainage and the Gulf drainage of Pinellas County and the Apalachicola River.

ORDER PERCIFORMES: XLI. FAMILY GOBIIDAE, Gobies and Sleepers.[1] Unlike most perciform fishes, the gobies and sleepers have *slender, flexible dorsal spines*, and the *gill membranes joined to the isthmus*. The pectoral fins are broad, the caudal fin rounded (in Florida species), and there are 2 dorsal fins. The lateral line is lacking. *In gobies*, but not sleepers, *the pelvic fins are united*.

182. Spiny-cheeked Sleeper, *Eleotris pisonis* (Gmelin)[2]

IDENTIFICATION: A long, slender fish with a broad head, large fins, and a large mouth. It bears some resemblance to the Big-mouth Sleeper (*Gobiomorus*), but differs in that the *gill openings do not extend as far forward*. Both species have smaller scales than the Fat Sleeper (*Dormitator*) with at least 50 along the lateral line. The anal fin has 8 rays, one less than in *Gobiomorus*. The general coloration is brown, sometimes with small darker spots on the sides. Maximum length about 250 mm.

DISTRIBUTION: Bermuda, South Carolina, and the Gulf of Mexico to Brazil and the West Indies; fresh and brackish water near the coast.

---

1. The sleepers are retained in the Family Eleotridae in the Amer. Fisheries Soc. bulletin (1970).
2. Includes *E. abacurus* of Carr and Goin, 1955.

183. Big-mouth Sleeper, *Gobiomorus dormitor* Lacépède

IDENTIFICATION: Similar to the Spiny-cheeked Sleeper in its slender build, large fins, and large mouth, but much larger (up to 600 mm in total length). The color pattern is also similar in both, but this species sometimes shows dark spots on the head. Specimens of all sizes can be distinguished by the fact that the *gill openings extend forward to the position of the eyes* and by the presence of 9 rays in the anal fin. (Each species has a slender anal spine!) The spinous dorsal fin is often dark-margined. A higher lateral scale count (55–57) will separate it from the Fat Sleeper (*Dormitator*).

DISTRIBUTION: Mexico and southern Florida to the West Indies and Central America; northward in Florida to Lake Okeechobee and St. Lucie County, entering canals.

184. Fat Sleeper, *Dormitator maculatus* (Bloch)

IDENTIFICATION: A much more heavy-bodied fish than the other 2 species of sleepers (depth more than 25 percent of standard length). Also the *depth of the head is greater than its width*, the *scales are larger* (less than 40 along lateral line), and the mouth is smaller. The color is a darker shade of brown than in the other sleepers. Total length up to more than 300 mm, but usually much smaller.

DISTRIBUTION: Coastwise from North Carolina through Mexico and the West Indies to Brazil, entering freshwater streams and canals in the Florida Peninsula.

185. Frillfin Goby, *Bathygobius soporator* (Valenciennes)

IDENTIFICATION: Distinguished from all other Florida fishes by its *fringe of free rays* on the upper part of the pectoral fin. (Preserved specimens in which these rays have been destroyed can be identified only with difficulty unless more than 100 mm long.) About 35 to 40 oblique scale rows. Color pattern variable, but usually large dark blotches on lighter ground color. The dark bases of the rays in the second dorsal fin often are in contrast to their lighter tips. Maximum length about 150 mm.

DISTRIBUTION: Tropical seas, northward to North Carolina. Enters freshwater streams and canals of southern Florida.

186. Darter Goby, *Gobionellus boleosoma* (Jordan and Gilbert)

IDENTIFICATION: The 2 species of *Gobionellus* are small gobies of moderate proportions, not differing markedly from one another. The present species has a *dark humeral spot*, but lacks the dark bar on the cheek of the Freshwater Goby (*G. shufeldti*). It has *fewer oblique scale* rows along the sides (30 to 33) and, as a rule, *one less ray in the anal and dorsal fins*. Typically there are about 3 V-shaped markings on each side. The orange tint of the fins in life helps to distinguish it. Maximum length about 50 mm (Ginsburg, 1932).

DISTRIBUTION: Atlantic and Gulf Coasts from North Carolina to Brazil. Enters freshwater streams in Florida Peninsula.

187. Freshwater Goby, *Gobionellus schufeldti* (Jordan and Eigenmann)

IDENTIFICATION: Like the other members of this genus, small and without unusual modifications for a goby. It is best distinguished from the Darter Goby by the *dark horizontal bar on the cheek* and a *lateral-scale count of 34 to 36*. Maximum length about 50 mm (Ginsburg, 1932).

DISTRIBUTION: Atlantic and Gulf Coasts from South Carolina to Texas, entering fresh water.

188. River Goby, *Awaous tajasica* (Lichtenstein)

IDENTIFICATION: Among the gobies, this species is distinguished by its *small scales* (at least 60 oblique rows along side of body) and *long snout* (almost 50 percent of head length). The color pattern of dark blotches on the body and dark streaks on the head helps to distinguish it. Maximum length about 300 mm.

DISTRIBUTION: Florida to Panama and the West Indies, entering Florida streams as far north as the St. Johns River and Choctawhatchee Bay.

189. Clown Goby, *Microgobius gulosus* (Girard)

IDENTIFICATION: Like the first 2 species of *Gobionellus*, the Clown Goby is small and grayish with dark mottling. It differs, however, in its *large mouth* and *long, soft dorsal fin* (at least 15 rays). The angle of the jaws is posterior to the level of the eyes. The dark edges of the soft dorsal (when present) will distinguish it. Eyes close together, dorso-lateral in position. Maximum length about 75 mm.

DISTRIBUTION: Duval County, Florida, to Texas, entering fresh-water streams and canals to Lake Okeechobee.

190. Crested Goby, *Lophogobius cyprinoides* (Pallas)

IDENTIFICATION: Large specimens are among the most distinctive of Florida fishes, with their *extremely enlarged fins* and *dark coloration*, a *fleshy crest on the head*, blunt snout, and thick-lipped oblique mouth. Smaller specimens, with inconspicuous head crests, often show *vertical bars on the peduncle* or *longitudinal stripes anteriad*. Total length up to 75 mm.

DISTRIBUTION: Bermuda and southern Florida through the West Indies to Panama; north in Florida to Charlotte and Martin Counties, entering freshwater streams and canals.

191. Robust Goby, *Gobiosoma robustum* Ginsburg

IDENTIFICATION: A small, heavy-set goby that resembles the Naked Goby in its *complete lack of scales*. It differs from that species in its *generally light color*, without clean-cut, light vertical bars; its *longer pelvic fin*; and smaller numbers of dorsal and anal rays (not more than 12 and 10, respectively). Preserved specimens may appear very light, with virtually no color pattern. Maximum length about 50 mm.

DISTRIBUTION: Atlantic and Gulf Coasts from Cocoa, Florida, to Bahia, Brazil. Euryhaline (Ginsburg, 1933).

192. Naked Goby, *Gobiosoma bosci* (Lacépède)

IDENTIFICATION: A small, heavy-bodied goby that differs from those of other genera in its *complete absence of scales*. The 12 or more dorsal spines separate it from most gobies, as does the color pattern of many *alternating light and dark vertical bars*. If these are not in evidence, there may be a *narrow dark bar at the base of the caudal fin*. Maximum length about 65 mm.

DISTRIBUTION: Atlantic and Gulf Coasts from Massachusetts to Mexico, entering fresh water in Florida (Ginsburg, 1933).

ORDER PLEURONECTIFORMES: XLII. FAMILY BOTHIDAE, Lefteye Flounders. All members of this order are *laterally compressed fishes that lie on one side*, the eye from the under side having rotated to the upper. *The under side is unpigmented*. They usually have no spines, and not more than 6 rays in the pelvic fins. The dorsal and anal fins, however, are extremely long and are confluent with the caudal fin. Flatfishes in the Family Bothidae *lie on the right side*. They have the rounded outline of most flatfishes (Norman, 1934).

193. Bay Whiff, *Citharichthys spilopterus* Günther

IDENTIFICATION: This little flatfish is distinguished from the other, larger members of its family by having *asymmetrical pelvic fins* (anteriad) and a *gently sloping lateral line*. It also has a much lower lateral-line scale count (less than 60), and seldom exceeds 125 mm in length. Color (above) grayish brown with darker blotches.

DISTRIBUTION: Coastwise from New Jersey to Brazil, reaching fresh water in Florida only in the St. Johns River.

194. Gulf Flounder, *Paralichthys albigutta* Jordan and Gilbert

IDENTIFICATION: Both Florida flounders in this genus have *symmetrical pelvic fins* and a *lateral line that arches strongly over the pectoral fin*. The lateral-line scale count is high (70 to 82 in this species). They tend to be a darker shade of brown and reach a larger size, in this species to 450 mm long. The *dark spots* on the body of the Gulf Flounder *form a triangle*, and the *eyes are closer together* than in the Southern Flounder. Maximum length about 450 mm.

DISTRIBUTION: South Atlantic and Gulf Coasts; recorded from fresh water in Florida only in the St. Johns River (Carr and Goin, 1955).

195. Southern Flounder, *Paralichthys lethostigma* Jordan and Gilbert

IDENTIFICATION: Differing from the Gulf Flounder in the *random arrangement of its dark spots*, the *greater distance between the eyes* (equal to diameter of eye), and its *high lateral-line scale count* (85 to 100). Maximum size greater than in *P. albigutta* (to about 900 mm long).

DISTRIBUTION: Atlantic Coast from New York to northern South America, penetrating fresh water in the St. Johns, Suwannee, and Apalachicola River systems.

ORDER PLEURONECTIFORMES: XLIII. FAMILY SOLEIDAE, Soles. General body form as in the Bothidae, but *wider* (standard length less than twice width of disk). *One or both of the pectoral fins may be lacking*. They differ from all other flatfishes known to enter fresh water in Florida in that they *lie on the left side*. The lateral line is straight. Both of the following species are small and typically marked with dark crossbars.

196. Lined Sole, *Achirus lineatus* (L.)

IDENTIFICATION: A small, rounded flatfish, with the right (upper) side and fins marked with numerous dark spots (often forming crossbars on the body) and a *pectoral fin on the right side only*. Maximum length 100 to 125 mm.

DISTRIBUTION: Florida and the Gulf of Mexico to Uruguay (Briggs, 1958).

197. Hogchoker, *Trinectes maculatus* (Bloch and Schneider)

IDENTIFICATION: Similar to the Lined Sole (*Achirus*), but *lacking both pectoral fins*. Upperparts (right side) heavily spotted and often lined with crossbars. Dorsal and anal fins often striated in larger specimens. Under side rarely spotted, but more frequently so north of Florida. Maximum length about 150 to 200 mm.

DISTRIBUTION: Atlantic and Gulf Coasts from Massachusetts to Panama, ranging far up Florida rivers and to Lake Okeechobee.

ORDER PLEURONECTIFORMES: XLIV. FAMILY CYNOGLOSSIDAE, Tonguefishes. These flatfishes differ from others chiefly in their *more elongate proportions* (standard length more than 3 times body depth) and from the Soleidae in that they *lie on the right side. Pectoral fins lacking.*

198. Black-cheeked Tonguefish, *Symphurus plagiusa* (L.)

IDENTIFICATION: More elongate than other flatfishes found in Florida fresh water (*width of disk*, exclusive of fins, *less than one-third of standard length*). In its complete *lack of pectoral fins* it resembles only the Hogchoker (*Trinectes*). The caudal fin is narrow and rather pointed. Color light grayish, sometimes with faint dark bars. The *lateral line is lacking.* Maximum length about 200 mm.

DISTRIBUTION AND VARIATION: Atlantic and Gulf Coasts from Cape Hatteras to Pensacola, Florida, southward to Argentina. Recorded from fresh water in Florida in the St. Johns River (*S. p. civitatum*) (Ginsburg, 1951).

198H. HYPOTHETICAL LIST

The following species, although included in the foregoing Key, do not appear in the species accounts because of doubt regarding their status in Florida fresh water.

The River Redhorse (*Moxostoma carinatum*; Catostomidae) has been collected in streams in southern Alabama very close to the Florida line (Yerger and Suttkus, 1962). Although it probably occurs in Florida too, definite records to substantiate the fact appear to be lacking.

The Violet Goby (*Gobioides broussonnetti*; Gobiidae), along with several fishes not in the Key, has been called euryhaline on one occasion (Gunter and Hall, 1963), but it seems unlikely that it occurs in truly freshwater situations.

CLASS AMPHIBIA: AMPHIBIANS

ORDER CAUDATA: XLV. FAMILY SIRENIDAE, Sirens.[1] Large to moderately small aquatic salamanders with *external gills and no hind limbs*; forelimbs reduced. Two of our species resemble *Amphiuma*, but differ in the absence of hind limbs, the wider head, and the presence of gills and of 3 or 4 toes per foot.

---

1. The sequence of salamander families follows that of Brame, 1967.

199. Greater Siren, *Siren lacertina* L.

IDENTIFICATION: A large, gray black, aquatic salamander with external gills and no hind limbs. Immatures are nearly identical to the Lesser Siren, but usually have *more costal grooves* (36 or more from forelimbs to vent) and *some light markings* (greenish or yellowish in life) *on the ventral surface*. Four toes per foot. Maximum length about 1 m. Larva: as in *S. intermedia*?

DISTRIBUTION: Atlantic and Gulf Coastal Plain from Maryland to southern and coastal Alabama, ranging throughout Florida mainland.

200. Lesser Siren, *Siren intermedia* Le Conte

IDENTIFICATION: Almost identical in appearance to the Greater Siren, but often distinguishable by the *lack of light markings* on the body. Four toes per foot. The number of costal grooves seldom exceeds 35, and the total length is seldom more than 380 mm in Florida. Larva: medium dark with darker streak through eye.

DISTRIBUTION AND VARIATION: Atlantic Coast of North Carolina through lower Piedmont of Georgia, most of Florida, and Coastal Plain of Alabama westward to southeastern Oklahoma, eastern Texas, and extreme northeastern Mexico; also ranging up Mississippi Rivey Valley to Illinois and Indiana; *Siren intermedia intermedia* ranges southward to Highlands, Polk, and Hillsborough Counties (Funderburg and Lee, 1967).

201. Dwarf Siren, *Pseudobranchus striatus* (Le Conte)

IDENTIFICATION: Like other sirens in absence of hind limbs and presence of external gills, but *much smaller and longitudinally striped light and dark*. Unlike the larger species, it has *only 3 toes*. The light stripes are usually yellowish or buffy in life, with 2 on each side. The back is uniformly dark, but the belly has light flecks. Costal grooves about 30 to 36. Maximum length up to 250 mm in north Florida, smaller southward (Freeman, 1959).

DISTRIBUTION AND VARIATION: Coastal Plain from southern South Carolina southwestward to Gulf County, Florida; to be expected in extreme southeastern Alabama. The following subspecies occur in Florida: *P. s. spheniscus*, Gulf County and Apalachicola River system eastward to Baker and Dixie Counties; *P. s. lustricolus*, Gulf Hammock (chiefly Levy County); *P. s. axanthus*, northeastern Florida southward to about Lake Okeechobee; *P. s. belli*, from about Lake Okeechobee southward to tip of mainland.

ORDER CAUDATA: XLVI. FAMILY PROTEIDAE, Mud Puppies and Waterdogs. Moderate-sized aquatic salamanders with 2 pairs of reduced limbs and external gills (3 pairs). Never more than 4 toes per limb. Eyes small and without lids. Tail strongly compressed. Costal grooves present.

202. Gulf Coast Waterdog, *Necturus beyeri* Viosca

IDENTIFICATION: Distinguished from other members of its family by the *numerous conspicuous spots*, some dark and some light. Dark brown above, lighter below. Maximum length about 200 to 230 mm.

DISTRIBUTION AND VARIATION: Eastern Texas, southern Louisiana, eastern Mississippi, most of Alabama, western Georgia, and northwestern Florida, eastward to the Apalachicola River valley (*N. b. alabamensis*).

203. Dwarf Waterdog, *Necturus punctatus* (Gibbes)

IDENTIFICATION: Despite the name *punctatus*, this waterdog is characterized by its *plain, unspotted coloration*, although some *pale* spots may be present. It is dark brown above, lighter below, with the throat much lighter. Maximum length 150 to 175 mm.

DISTRIBUTION AND VARIATION: Southeastern Virginia, Coastal Plain of the Carolinas, Georgia (except southern), and Alabama, and throughout northwestern Florida east to Apalachicola River (*N. p. lodingi*).

ORDER CAUDATA: XLVII. FAMILY SALAMANDRIDAE, Newts. Small, slender salamanders with no nasolabial grooves and no conspicuous costal grooves. Vomerine and palatine teeth are in 2 continuous, diverging, longitudinal rows. Typically there is a reddish *terrestrial* stage, with rough skin, interposed between greenish or olivaceous aquatic stages with smooth skin.

204. Spotted Newt, *Notophthalmus viridescens* Rafinesque

IDENTIFICATION: Terrestrial stage reddish, often with reddish, black-encircled spots; skin rough. Aquatic stages (one without gills) greenish or olivaceous above, yellowish below (darker when transforming), with *numerous black specks* and sometimes *reddish spots*. Caudal fin pronounced in breeding males. Total length about 75 to 100 mm.

DISTRIBUTION AND VARIATION: Southern Canada and eastern United States, westward to central Minnesota, eastern Iowa, all but northwestern Missouri, southeastern Kansas, eastern Oklahoma, and eastern Texas. In Florida *N. v. viridescens* occurs locally in the Panhandle (Neill, 1954); *N. v. louisianensis* occupies the remainder of northern Florida, both east and west of the Apalachicola River, and southward to Levy and St. Johns Counties; and *N. v. piaropicola* inhabits the remainder of the Peninsula.

205. Striped Newt, *Notophthalmus perstriatus* (Bishop)

IDENTIFICATION: Very similar to the corresponding stages of the Spotted Newt, but of more slender proportions and usually with a *red stripe down each side of the back*, bordered by blackish. Maximum length about 75 mm.

DISTRIBUTION: Southern Georgia and the northern Peninsula of Florida, westward at least to Leon County.

ORDER CAUDATA: XLVIII. FAMILY AMPHIUMIDAE, Amphiumas. Large aquatic salamanders with *no external gills and vestigial limbs*, never with more than 3 toes (probably not more than 2 in Florida). Coloration dark, sometimes much lighter below. Eyes small and without lids. Maximum length about one meter.

206. Two-toed Amphiuma, *Amphiuma means* Garden[1]

IDENTIFICATION: As described under the family, except only *2 toes per foot*, belly not conspicuously lighter than upperparts, and without a dark throat patch. The appressed forelimb extends *over the ear opening*. Maximum length about one meter. Larva: color pattern similar to adult's.

DISTRIBUTION: The Florida mainland northward and westward to southeastern Louisiana, south-central Alabama, central and eastern Georgia, South Carolina (below about 650 m in the mountains), the eastern half of North Carolina, and southeastern Virginia.

207. One-toed Amphiuma, *Amphiuma pholeter* Neill

IDENTIFICATION: A small, secretive Amphiuma with only *one toe per foot*. It is further distinguished by its almost identical dorsal and ventral coloration and the fact that the *appressed forelimb does not extend over the ear opening*. Maximum length about 250 mm.

DISTRIBUTION: Calhoun to Levy Counties, Florida (Neill, 1964a; Stevenson, 1967; Map 25).

ORDER CAUDATA: XLIX. FAMILY AMBYSTOMATIDAE, Mole Salamanders. Heavy-bodied, sluggish, broad-headed salamanders, terrestrial as adults, with prominent costal grooves, but *no nasolabial grooves*.

---

1. The possible occurrence of the Three-toed Amphiuma (*Amphiuma tridactylum*) in the northwestern corner of Florida should be considered.

Ground color usually blackish in Florida. Most Florida species scarcely exceed a length of 100 mm.

208. Flatwoods Salamander, *Ambystoma cingulatum* Cope

IDENTIFICATION: Somewhat more slender than other Florida members of its family, and with at least 13 costal grooves. It may be separated from others in its family by the *pinnately arranged grooves on the upper surface of the tongue*. The light dorsal markings on the blackish background tend to form a *reticulate pattern*. Ventrally there are scattered light spots on a dark background. The *larvae are longitudinally striped*, somewhat resembling *Pseudobranchus*. Maximum length about 100 mm.

DISTRIBUTION: Southeastern North Carolina through the Coastal Plain to extreme southeastern Mississippi, southward in Florida to Marion County (Neill, 1954). There appears to be a hiatus in the Tallahassee region.

209. Mole Salamander, *Ambystoma talpoideum* (Holbrook)

IDENTIFICATION: An unusually heavy-bodied and big-headed salamander with a *dull grayish color pattern*. The 10 costal grooves are the

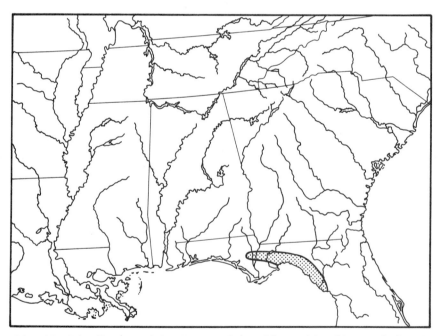

Map 25. Distribution of the One-toed Amphiuma (*Amphiuma pholeter*)

fewest of any member of this family. Legs and feet well developed. Brownish gray with paler spots. Larvae and post-larvae have a *dark midventral stripe*. Maximum length about 100 mm.

DISTRIBUTION: Upper Coastal Plain from South Carolina to northern Mississippi and Louisiana, southward to north-central Florida; also in a few widely separated areas farther north.

210. Marbled Salamander, *Ambystoma opacum* (Gravenhorst)

IDENTIFICATION: One of the most strikingly colored of Florida salamanders. Ground color typically *jet black, crossed dorsally with saddles of white or light gray* in strong contrast. It is also one of the most heavy-bodied and sluggish of salamanders. Costal grooves usually 11, sometimes 12. The larvae and subadults are blackish above and lighter below, with lighter spots on the sides often in rows as in adults of *Desmognathus*. Maximum length about 125 mm.

DISTRIBUTION: Southern Massachusetts and Lake Erie southwestward to southeastern Oklahoma, eastern Texas, and Florida (southward to near Tampa, Gainesville, and Fernandina).

211. Tiger Salamander, *Ambystoma tigrinum* (Green)

IDENTIFICATION: Ground color above blackish, but heavily overlaid by large, light-colored spots of irregular size and distribution (olive to yellowish in life). Some of these are on the *lower sides*; lighter below. Newly transformed specimens are medium gray with *small*, light spots on the sides, often arranged in one row per side—a color pattern resembling that in *A. talpoideum*. Few other Florida members of the family have as many as 12 costal grooves, and none reaches so great a size (maximum length up to 250 mm). Larva: Blackish above and whitish below, with a white lateral stripe; tail fin light with dark spots.

DISTRIBUTION AND VARIATION: Almost throughout the United States and much of Mexico, but *absent from* the more arid parts of the West and Southwest, much of the Great Lakes region, and from New England southward through the mountains to northern Georgia; also not recorded south of Hernando County, Florida (*A. t. tigrinum*).

ORDER CAUDATA: L. FAMILY PLETHODONTIDAE, Lungless Salamanders. As in the Ambystomatidae, there are prominent *costal grooves*, but the lungless salamanders also have nasolabial grooves (nostril to lip). (Magnification may be necessary to see the latter grooves.) Their proportions are more similar to those of the Salamandridae than to the heavy-set ambystomatids. Both vomerine and parasphenoid teeth are present.

212. Slimy Salamander, *Plethodon glutinosus* (Green)

IDENTIFICATION: The *large, white spots on a blue black background* will distinguish this salamander from others of its family in Florida. Rarely the spots are lacking, and such specimens are confusable with *Desmognathus auriculatus*, but that species is not so jet black. Although *Desmognathus* also shows white spots, these are much smaller and usually in a longitudinal row on the lower side of the body. Finally, the white spots in *Plethodon* are smaller than those in the more heavy-set, white-spotted species of *Ambystoma*. The underparts are slate gray, lighter on the chin. Costal grooves usually 15. Maximum length about 125 mm. No free larval stage.

DISTRIBUTION AND VARIATION: New York, eastern and southern Ohio, and the southern parts of Indiana, Illinois, and Missouri southward to northeastern (locally to central) Texas, Louisiana, and (east of the Mississippi River) to the Gulf Coast and central Florida. *P. g. glutinosus* south to northeastern Manatee County (USF), Lakeland, and Orlando (Highton, 1962; R. Sanderson, personal communication).

213. Four-toed Salamander, *Hemidactylium scutatum* (Schlegel)

IDENTIFICATION: The *white belly with large black spots* and the *constriction at the base of the tail* distinguish this salamander from all others in Florida. Only one other terrestrial species, the Dwarf Salamander, has *just 4 toes on the hind foot*. Upperparts brownish, with fewer spots. Maximum length about 75 mm. The tail is proportionately short and thick.

DISTRIBUTION: New Brunswick and Nova Scotia, southwestern Maine, southern Ontario, and the Great Lakes region southward to the northern parts of Alabama and Georgia; apparently disjunct populations occur in southeastern Missouri, eastern Oklahoma, Arkansas, Louisiana, Mississippi, Georgia, and north Florida (upper Ochlockonee River to southern Okaloosa County; (Stevenson, 1958a; Fugler and Folkerts, 1967; Means, personal communication).

214. Georgia Blind Salamander, *Haideotriton wallacei* Carr

IDENTIFICATION: An aquatic, pigmentless salamander, with long external gills and *slender, frail-looking legs* that overlap (fore and hind) when appressed. Eyes vestigial. Found only in caves, wells, and other underground water. Maximum length about 75 mm.

DISTRIBUTION: Known only from an artesian well near Albany, Georgia, and caves near Climax, Georgia, and Marianna, Florida (Pylka and Warren, 1958).

215. Red Salamander, *Pseudotriton ruber* (Latreille)

IDENTIFICATION: A large terrestrial salamander with a *reddish or salmon ground color*. Upperparts with discrete or coalescent darker spots. Underparts lighter, but with smaller dark spots. It is most like the Mud Salamander, but differs in its relatively small eye and in having a *light streak from eye to nostril*. Also the costal-groove count tends to be lower in this species (15–16). Maximum length 150 to 175 mm (only 40 to 45 percent tail). Larva: brownish above, lighter below, usually without white spots on sides.

DISTRIBUTION AND VARIATION: Southeastern New York, Pennsylvania, and eastern Ohio southwestward to southeastern Louisiana and the Gulf Coast. In Florida *P. r. vioscai* ranges eastward to the Apalachicola River.

216. Mud Salamander, *Pseudotriton montanus* Baird

IDENTIFICATION: In its *reddish background with darker markings* this animal resembles the larger Red Salamander, but differs in having *no light streak from eye to nostril* and a relatively large eye (see Key). The number of costal grooves is 16 to 18. Maximum length 125 to 150 mm (only 40 to 45 percent tail). Larva: Dark above and lighter below, with small light spots dorsally.

DISTRIBUTION AND VARIATION: Southern New Jersey southward through the Coastal Plain and Piedmont to central Florida and westward to southeastern Louisiana. *P. m. floridanus* occurs from Jackson to Highlands Counties, Florida (Means, personal communication); *P. m. flavissimus* occupies that part of the Panhandle west of Jackson County.

217. Long-tailed Salamander, *Eurycea longicauda* (Green)

IDENTIFICATION: A large, *slender, black-and-yellow-striped salamander with a long tail* (usually more than 60 percent of total length). The yellowish back is set off by a broad black stripe on each side and divided middorsally by a more narrow black stripe. Ventrally the yellowish ground color is heavily mottled with blackish. Maximum length up to 175 mm. Larva: dark above and whitish below, with lateral row of light spots as in *Desmognathus auriculatus* adults.

DISTRIBUTION AND VARIATION: Southern parts of New York, Ohio, Indiana, and Illinois southward to the Gulf Coast and southwestern Georgia, occurring in Florida from Leon County westward (*E. l. guttolineata*); also southern Missouri, northern Arkansas, and eastern Oklahoma; absent from southern New Jersey and the Delmarva Peninsula.

218. Two-lined Salamander, *Eurycea bislineata* (Green)

IDENTIFICATION: The generally yellowish ground color suggests the larger Long-tailed Salamander, but may be tinged with reddish or orange in this species. The broad, *yellowish dorsal stripe is not broken in the midline by a definite dark stripe* as in *E. longicauda*. Its color pattern is somewhat more suggestive of the darker, smaller, more slender Dwarf Salamander (*Manculus*), but typical specimens are separated on the basis of their *5 hind toes*. Maximum length about 100 mm (55–60 percent tail). Larva: Small specimens undistinguished; larger ones suggestive of form and color of adults.

DISTRIBUTION AND VARIATION: Southeastern Canada, New England, most of Ohio and Indiana, and southeastern Illinois southward to southeastern Louisiana and the Gulf Coast, ranging eastward in Florida to Dixie and Hamilton Counties (*E. b. cirrigera*).

219. Dwarf Salamander, *Manculus quadridigitatus* (Holbrook)

IDENTIFICATION: A small, slender salamander of yellowish brown ground color and a *broad black stripe down each side* (and sometimes a narrow one middorsally). The color pattern, though darker, resembles that of the Two-lined Salamander, and males of both these species have cirri in the nostrils. *Manculus* normally differs from most terrestrial salamanders in having *only 4 toes on the hind foot*. Probably the most slender of terrestrial Florida salamanders. Maximum length about 75 mm (55–60 percent tail). Larva: small specimens undistinguished; larger ones similar in form and color to adult.

DISTRIBUTION: Atlantic and Gulf Coastal Plains, from North Carolina to eastern Texas, northward through western Arkansas to southwestern Missouri; ranges southward in Florida to Charlotte County and Lake Okeechobee (Amer. Soc. Ichth. and Herp., 1963ff.).

220. Northern Dusky Salamander, *Desmognathus fuscus* (Rafinesque)[1]

IDENTIFICATION: Head relatively small, depressed, and often carried in a down-sloped position; *swellings on each side of throat*. Ground color mostly dusky gray, but rarely black dorsally. Young specimens have large light (sometimes reddish) areas dorsally, bounded laterally by a wavy dark line, producing a *scalloped effect*. This pattern largely disappears in old specimens, except for the dark line; *numerous dark spots* are often present at this age. Like others in this genus, it has a *light line from the eye to the angle of the jaws*. Tail more slender near tip than in *D. auriculatus* (see Key; Means, personal communication).

---

1. The Florida population is taxonomically closer to *D. ochrophaeus* than to the northern population of *D. fuscus* (Means, personal communication).

Costal grooves usually 14. Maximum length about 115 mm (Valentine, 1963). Larva: dorsal pattern similar to that of young adults.

DISTRIBUTION AND VARIATION: Quebec and New England southward and southwestward to southern Illinois, western Tennessee, northeastern and southern Mississippi, coastal Alabama, southwestern Georgia, northwestern South Carolina, and central North Carolina; Florida records are from ravines near the Ochlockonee River and westward (*D.f.fuscus*; Means, personal communication).

221. Southern Dusky Salamander, *Desmognathus auriculatus* (Holbrook)

IDENTIFICATION: Similar to *D. fuscus* in shape of body and head, and in having a light line from eye to angle of jaws, but *base of tail more slender and tip more bladelike*. The *general coloration is darker*, approaching that of the Slimy Salamander (*Plethodon*). Some individuals have a dorsal reddish stripe near the base of the tail, and most show a *series of small, light spots along each side, often arranged in a row*. They are generally smaller than the spots in *Plethodon*. Costal grooves usually 14. Maximum length about 150 mm (Funk, 1964). Larva: dark above, lighter below, with one or more lateral rows of light spots; sometimes resembles adults.

DISTRIBUTION: The Coastal Plain from the Mississippi River (possibly eastern Texas) to southeastern Virginia, ranging northward to southwestern Georgia and southward to central Florida (Map 26) (Rossman, 1959; Valentine, 1963; Means, personal communication).

222. Seal Salamander, *Desmognathus monticola* Dunn

IDENTIFICATION: A rather large, blackish, heavy-bodied salamander with a *greatly compressed tail*; underparts whitish to medium dark. There may be numerous black spots on a dark gray dorsum, or in younger specimens a double row of larger, rounded, brownish spots. As in *D. auriculatus*, there may be a single row of light spots along each side of the body. Maximum length over 125 mm.

DISTRIBUTION AND VARIATION: Mountains from western Pennsylvania to northeastern Alabama and northern Georgia; also (disjunct?) in southern Alabama and extreme northwestern Florida (*D. m. monticola*; Map 27; Means and Longden, 1970).

ORDER SALIENTIA: LI. FAMILY PELOBATIDAE, Spadefoot Toads. Similar to the true toads (Bufonidae) in their squat body form, but *skin not warty and pupils vertical*. The only Florida species differs also in having *no cranial crests*, and the *parotoid glands are rounded*. Tympanum present.

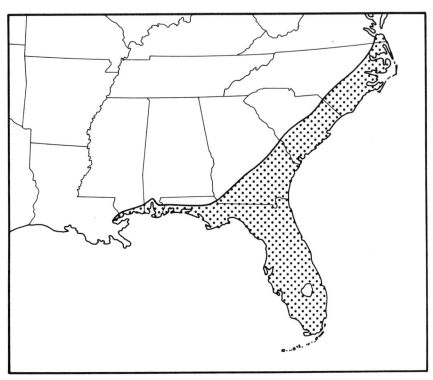

Map 26. Distribution of the Southern Dusky Salamander
(*Desmognathus auriculatus*)

223. Eastern Spadefoot, *Scaphiopus holbrooki* (Harlan)

IDENTIFICATION: (See description of family.) Color light to medium brown or grayish, with a broad light stripe middorsally. *The metatarsal tubercle* is a dark horny ridge on the medial side of each foot; a corresponding structure in the Bufonidae is lacking or less developed. Maximum length about 65 mm.

DISTRIBUTION: Massachusetts and the southeastern parts of New York and Pennsylvania, the southern parts of Ohio, Indiana, and Illinois, and southeastern Missouri southward to the Gulf Coast from eastern Louisiana through Florida (except Everglades; see Duellman, 1955); absent from mountainous areas.

ORDER SALIENTIA: LII. FAMILY BUFONIDAE, Toads. Anurans with a squat body form, prominent warts and cranial crests, and elliptical parotoid glands. The *pupil is rounded* and the horny ridge on the hind foot lacking or poorly developed. Tympanum present.

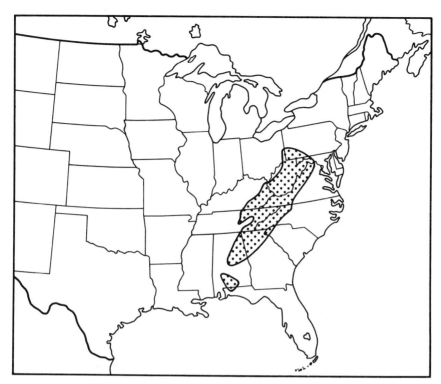

Map 27. Distribution of the Seal Salamander
(*Desmognathus monticola*)

224. Southern Toad, *Bufo terrestris* (Bonnaterre)

IDENTIFICATION: A medium-sized toad of *variable coloration*. Because of the similarity in size and coloration, most likely to be confused with *B. woodhousei* of west Florida. The most reliable distinction is that the Southern Toad's *postorbital ridge does not make direct contact with the parotoid gland* (although its backward-projecting spur may do so). Toward their posterior ends, the cranial crests are enlarged, projecting upward and backward. As a rule the warts in a given pigment spot are of irregular size in the Southern Toad. Color typically grayish with darker brown pigment spots scattered over the dorsal surface, sometimes with a faint lighter middorsal stripe. Underparts whitish. Maximum size about 100 mm (Riemer, 1958).

DISTRIBUTION: Extreme southeastern Virginia southwestward through upper Coastal Plain to the lower Mississippi River, ranging throughout the Florida mainland and on the lower Keys. Possibly only a subspecies of the American Toad (*B. americanus*).

225. Woodhouse's Toad, *Bufo woodhousei* Girard

IDENTIFICATION: Very similar in size and coloration to the Southern Toad, but differs in that the *postorbital ridge is in direct contact with the parotoid gland*. Also the posterior ends of the cranial crests are less modified. Typically the warts in a single pigment spot are of nearly uniform size. The light middorsal stripe is faint, as in *B. terrestris*. Maximum length about 75 mm.

DISTRIBUTION AND VARIATION: Throughout most of the United States and northern Mexico, but generally *absent from* the Pacific States and Nevada, areas near the Canadian border (except Michigan and Ontario), and the extreme southeastern United States. *B. w. fowleri* (sometimes called Fowler's Toad) occurs in Florida west of the Apalachicola River (considered a distinct species by some authorities).

226. Oak Toad, *Bufo quercicus* Holbrook

IDENTIFICATION: A miniature toad scarcely an inch long with a *prominent middorsal white line*. Although a less conspicuous line may be seen in the young of *B. terrestris*, the Oak Toad may be distinguished with certainty by its *diverging parotoid glands*. The cranial crests are less conspicuous than in other toads. The ground color is light to dark gray. Maximum length about 30 mm.

DISTRIBUTION: Upper Coastal Plain from North Carolina to the Mississippi River in southern Mississippi and southeastern Louisiana, and to the tip of the Florida mainland; an isolated population inhabits the lower Keys.

227. Giant Toad, *Bufo marinus* (L.)

IDENTIFICATION: A toad the size of a Bullfrog or larger. The dorsal coloration is often browner than in the Southern Toad, but is not critical. However, the *warts are larger and more flattened* than in that species, and the *tips of the digits are dark*. Also the *greatest width of the parotoid glands* (anteriad) *is much greater than the distance between the orbits*. Maximum length about 200 mm.

DISTRIBUTION: Southern tip of Texas southward to southern South America; widely introduced in warmer parts of the world, including southern Florida, where it ranges from Homestead to southern Broward County, westward to the Everglades; also near West Palm Beach (Krakauer, 1968).

ORDER SALIENTIA: LIII. FAMILY LEPTODACTYLIDAE, Tropical Frogs. External characters variable, but most species are small.

228. Greenhouse Frog, *Eleutherodactylus planirostris* Cope[1]

IDENTIFICATION: A small frog with slightly enlarged digital disks. The ground color is dark with *longitudinal light stripes* in one color phase; in the other the pattern is mottled, and there may be a dark triangle between the eyes as in the cricket frogs. An absolute identification may be made by examining the *subarticular pads* on the toes, which in this species *are elongate and point toward the tips of the toes*. Maximum length about 25 mm.

DISTRIBUTION AND VARIATION: Greater Antilles, Bahama Islands, and Peninsular Florida, *E. p. planirostris* ranging northward to Tallahassee (Reichard and Stevenson, 1964).

ORDER SALIENTIA: LIV. FAMILY HYLIDAE, Tree Frogs. Rather small frogs without conspicuous warts, cranial crests, or (usually) tympani. Larger species, especially, are distinguished by their prominent *digital disks*, and most also have *webbing between the toes*. Small warts, at least, are on the ventral side. Live immatures of several hylids are grassy green.

229. Spring Peeper, *Hyla crucifer* Wied

IDENTIFICATION: Members of this genus have the *toes well webbed* and the *digital disks strongly developed*. The present species is best distinguished by its color pattern—an *irregular dark X on the back* on a lighter background. Usually there is a dark stripe through the eye continuing along the side of the body, as in some other hylids. Maximum length hardly more than 25 mm.

DISTRIBUTION AND VARIATION: East of the Great Plains from southern Canada to northern Florida and the Gulf Coast. *H. c. crucifer* occupies the Panhandle west of the Apalachicola River, and *H. c. bartramiana* the remainder of north Florida, ranging southward to Lake County.

230. Pine-woods Tree Frog, *Hyla femoralis* Latreille

IDENTIFICATION: A small frog of variable color, but always with *rounded white or yellow spots on the back of the thigh*. The ground color is often dark gray with darker blotches above. Maximum length about 35 to 40 mm. (Also see Key.)

DISTRIBUTION: Coastal Plain from southern Maryland (west side of Chesapeake Bay) through central Alabama to the lower Mississippi River, southward to the northern edge of the Everglades and to Broward and Collier Counties.

1. See Schwartz, 1965.

231. Barking Tree Frog, *Hyla gratiosa* Le Conte

IDENTIFICATION: The largest tree frog that occurs over most of Florida, sometimes exceeding 65 mm in length. The ground color may change from green to gray or brown, but *rounded darker spots* and *tiny golden flecks* are usually in evidence dorsally. The underside is whitish. Resembles the Cuban Tree Frog, but the *skin is not fused to the skull*.

DISTRIBUTION: Upper Coastal Plain from North Carolina to the lower Mississippi River, southward to Broward and Collier Counties (Duellman and Schwartz, 1958); isolated colonies in northern parts of Alabama and Georgia, and in Tennessee and Kentucky.

232. Squirrel Tree Frog, *Hyla squirella* Latreille

IDENTIFICATION: A small frog difficult to identify with certainty. The ground color may change from green to brown, and there are often small dark spots on the back, sometimes arranged in longitudinal rows. Usually there is a *dark blotch or bar between the eyes* and a *light line with an irregular border along the side of the body*. Preserved specimens may strongly resemble *Pseudacris nigrita* in color pattern. Maximum length about 35 mm.

DISTRIBUTION: Upper Coastal Plain from extreme southeastern Virginia to eastern Texas, southward throughout Florida; introduced on Grand Bahama.

233. Southern Gray Tree Frog, *Hyla chrysoscelis* Le Conte[1]

IDENTIFICATION: Another tree frog with a highly changeable color pattern, but always with *yellow or orange on the back of the thigh* in life. The most frequent ground color is a light gray, with darker mottling above, and a *light spot can usually be seen under the eye*. The *upper surface is much more warty* than in most tree frogs. Maximum length about 50 mm.

DISTRIBUTION AND VARIATION: East of the Great Plains from southeastern Manitoba, Ontario, and southern Maine southward to the Gulf Coast and Marion County, Florida (*H. v. versicolor*).

234. Bird-voiced Tree Frog, *Hyla avivoca* Viosca

IDENTIFICATION: Similar to *H. chrysoscelis*, but the *back of the thigh is green* in life rather than yellowish. Also the ground color is a darker shade of gray. Maximum length about 50 mm.

DISTRIBUTION AND VARIATION: Southeastern Louisiana, all of Mississippi, southern and central Alabama, western Florida, and south-

---

1. See *Science*, 167:385–386 and *Southwestern Naturalist*, 13:283–300.

eastern Georgia, up the Mississippi River valley to western Kentucky and southern Illinois (also one record in southeastern Oklahoma; *H. a. avivoca*, ranging east in Florida to the upper Ochlockonee and lower Apalachicola Rivers); also a disjunct population in inner Coastal Plain of southern South Carolina and eastern Georgia (*H. a. ogechiensis*).

235. Green Tree Frog, *Hyla cinerea* (Schneider)

IDENTIFICATION: A small tree frog with a bold, rather constant color pattern. The upperparts vary only from *yellowish to grayish green* interrupted by a *distinct light line along each side of the head and body*, rarely lacking. Underparts whitish. Small specimens of *H. squirella* may be similar in life. Maximum length 50 mm or a little more.

DISTRIBUTION AND VARIATION: Coastal Plain from Delaware and Maryland southward to the Florida Keys, westward to eastern Texas, and northward to southeastern Missouri and southern Illinois.

236. Cuban Tree Frog, *Hyla septentrionalis* Boulenger[1]

IDENTIFICATION: By far the largest of Florida tree frogs, both sexes reaching a length of more than 100 mm (Mittelman, 1950). Aside from its larger size, it may be distinguished by its *enlarged digital disks* (almost equal in diameter to tympanum). Ground color green or bronze dorsally, without a lateral stripe of any kind, but with small dark spots and larger areas of medium dark color dorsally. More warty than most other Florida tree frogs. Underparts light colored. Unlike all other Florida hylids, it has the *skin firmly fused to the top of the skull*.

DISTRIBUTION: Southern Florida, Cuba, the Bahamas, and locally on other islands of the Caribbean; Florida records range north near the east coast to Broward County (but this species has been heard calling in Highlands, Charlotte, and De Soto Counties).

237. Little Grass Frog, *Limnaoedus ocularis* (Daudin)[2]

IDENTIFICATION: This frog and those in the genus *Pseudacris* have *less webbing between the toes and smaller digital disks* than other tree frogs. Their proportions are more slender than in frogs of the genus *Hyla*. As the name suggests, the Little Grass Frog is the smallest Florida amphibian, adults averaging *about 13 mm in length*. Ground color light brownish, with a *dark stripe running through the eye*. The muzzle is more truncate in profile than in most frogs, and the hind legs are disproportionately long.

---

1. Trueb and Tyler (1974), Occ. Papers Mus. Nat. History, Univ. Kansas, 24:1–60, have proposed the erection of the genus *Osteopilus* for this species.

2. See Mittelman and List, 1953; also Lynch, 1963.

DISTRIBUTION: Coastal Plain from southeastern Virginia to southeastern Alabama and western Florida (to Holmes and Bay Counties), southward throughout the Peninsula.

238. Upland Chorus Frog, *Pseudacris triseriata* Wied[1]

IDENTIFICATION: A slender hylid with reduced webbing and virtually no digital disks. Most similar to the Southern Chorus Frog, but differing in its lighter ground color and in various proportions. It appears thicker than that species and has a *shorter tibia* (appressed knee not touching the appressed elbow). The width of the head at the angle of the jaws is more than 33 percent of the snout-vent length. Maximum snout-vent length often more than 30 mm (Schwartz, 1957; Crenshaw and Blair, 1959; Mecham, 1959).

DISTRIBUTION AND VARIATION: Southern and central Canada (except toward Atlantic and Pacific sides) southward to southern Idaho, northern and central Arizona and New Mexico, northern and southeastern Kansas, the Gulf Coast of Texas and Louisiana, southern Mississippi and Alabama, northern Florida, southwestern and northeastern Georgia, central and coastal South Carolina, and southeastern Virginia; Florida records in Jackson, Liberty, and Gadsden Counties.

239. Southern Chorus Frog, *Pseudacris nigrita* (Le Conte)

IDENTIFICATION: Similar to the Upland Chorus Frog but somewhat darker, usually without a triangle between the eyes, more slender, and with a longer tibia. The width of the head at the angle of the jaws is usually less than 33 percent of snout-vent length. Also similar to *Acris* in form, but differs in having a *light line along the edge of the mouth* and *3 or 5 longitudinal dark lines or rows of spots dorsally*. In both of these respects it may resemble the Squirrel Tree Frog, but it has a more slender head, less webbing, and smaller disks than that species. The skin is somewhat rougher than in other small hylids. Maximum length usually less than 30 mm.

DISTRIBUTION AND VARIATION: Coastal Plain from North Carolina to Mississippi, southward throughout Florida (except in the Everglades and Keys); *P. n. nigrita* occupies northern Florida and is replaced by *P. n. verrucosa* from the lower Suwannee River and Jacksonville southward.

240. Ornate Chorus Frog, *Pseudacris ornata* (Holbrook)

IDENTIFICATION: Coloration variable, but often reddish brown dorsally. Best distinguished by the *black line running through the eye*

---

1. Neill (1954) referred 7 specimens from Okaloosa County to "*cf. Pseudacris brachyphona*" (a species not known to Florida), but his 2 specimens now in the Florida State Museum (Gainesville) appear instead to be *P. triseriata*.

and the *3 large, rounded, dark, light-bordered spots near the groin*. The spots on the back of the thigh are more greenish than those in *Hyla femoralis*, and there is also some yellowish green in the groin. Maximum length a little more than 25 mm.

DISTRIBUTION: Upper Coastal Plain from North Carolina to the lower Mississippi River and central Florida (Lake County).

241. Southern Cricket Frog, *Acris gryllus* (Le Conte)

IDENTIFICATION: The 2 species of cricket frogs resemble chorus frogs (*Pseudacris*) in their small size, body form, wartiness, and somewhat in color pattern. They differ in having virtually no tympanum and more webbing between the toes. Also both species of cricket frogs show more or less *striping on the hind surface of the thigh*, but that of *gryllus* is *more clean-cut*. General coloration variable, but with a dark triangle between the eyes. The Southern Cricket Frog also has a more pointed snout than the northern species, whereas anal warts are small or lacking. Maximum length about 25 mm.

DISTRIBUTION AND VARIATION: Southwestern Tennessee, northern parts of Mississippi, Alabama, and Georgia, central South Carolina, and southeastern Virginia southward to southeastern Louisiana (east of Mississippi River), the Gulf Coast, and southern Florida. *A. g. gryllus* occurs from Leon and Wakulla Counties westward, and *A. g. dorsalis* from those counties southward throughout the Peninsula, as well as on Dog and St. Vincent Islands (Franklin County; Blaney, 1971).

242. Northern Cricket Frog, *Acris crepitans* Baird[1]

IDENTIFICATION: Very similar to the Southern Cricket Frog, but with a more rounded snout, shorter legs, and more webbing between the toes. Perhaps the best character is the *more irregular* light and dark stripes on the back of the thighs. There are also 2 prominent anal warts and many smaller warts on the back. The ground color is usually a pale gray, and there is often a dark triangle between the eyes. Formerly considered conspecific with *A. gryllus*, but sympatric with it in parts of north Florida. Maximum length slightly more than 25 mm.

DISTRIBUTION AND VARIATION: Eastern Colorado, southeastern South Dakota, southern Wisconsin, lower Michigan, western Ohio, southeastern Pennsylvania, and New Jersey southward into eastern Mexico and to the Gulf Coast as far east as northern Leon County, Florida (*A. c. crepitans*); absent from coastal parts of the Carolinas and Georgia, and from the Florida Peninsula and much of Mississippi (Boyd, 1964).

---

1. See Mecham, 1964.

ORDER SALIENTIA: LV. FAMILY RANIDAE, True Frogs. Anurans of larger size, lacking digital disks, and with the hind toes fully webbed. Skin usually smooth and tympanum conspicuous.

243. Bullfrog, *Rana catesbeiana* Shaw

IDENTIFICATION: The largest of Florida frogs, reaching 175 mm or more in length. There is usually some green in the otherwise dark dorsal color, and grayish mottling on the white underparts. The latter markings are not so dark as in the River Frog, nor is the skin so rough dorsally. Some individuals may resemble Bronze Frogs, but the *dorsolateral fold is lacking*. The more slender Pig Frog differs in having the *back of the thigh longitudinally striped* (usually), and the *hind toes webbed to their tips*. When the hind surface of the Bullfrog's thigh is striped, the stripes are ragged and interrupted.

DISTRIBUTION: Southeastern Canada, southern Maine, Wisconsin, southern Iowa, and southern South Dakota southward through eastern Colorado and New Mexico into eastern Mexico, to the Gulf Coast, and southward in Florida to Highlands and northern Charlotte Counties; introduced into Italy, Japan, and elsewhere.

244. River Frog, *Rana heckscheri* Wright

IDENTIFICATION: Resembles a Bullfrog in its lack of dorsolateral ridges, but darker above and below. The *extensive blackish mottling of the underside* helps to distinguish it, as do the *light spots along the jaws*. The skin is rougher dorsally than in most other frogs. Maximum length about 125 mm.

DISTRIBUTION: Coastal Plain from southeastern North Carolina (Simmons and Hardy, 1959), South Carolina, and Augusta, Georgia, westward to southern Mississippi and southward to Marion County, Florida (Map 28).

245. Pig Frog, *Rana grylio* Stejneger

IDENTIFICATION: Most likely to be confused with the Bullfrog because of similarity in size, coloration, smooth skin, and a lack of dorsolateral ridges. It differs from that species in its more slender head and the fact that the *back of the thigh is usually longitudinally striped with black and white*. Also *2 faint, broad, light stripes can be discerned on the back*. Other differences in the Pig Frog are that the first finger is slightly shorter than the second, and the fourth toe is longer than the fifth. Hind toes webbed to their tips. Maximum length about 150 mm.

DISTRIBUTION: Coastal Plain from southern South Carolina and

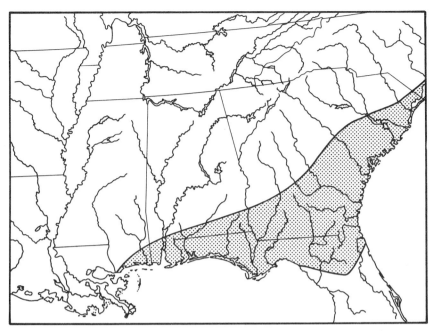

Map 28. Distribution of the River Frog (*Rana heckscheri*)

southeastern Georgia westward to extreme southeastern Texas, including all of the Florida mainland. Introduced in Bahama Islands.

246. Carpenter Frog, *Rana virgatipes* Cope

IDENTIFICATION: Like the preceding, much larger species, this frog has no dorsolateral ridges. *Two light stripes on the back* (yellowish brown in life) and *2 more on the sides* distinguish it from any other *Rana* in Florida. Ground color brownish with rounded dark spots dorsally; whitish below. Maximum length about 65 mm.

DISTRIBUTION: Coastal Plain from southern New Jersey to the Okeefenokee Swamp, including northwestern Baker County, Florida (Stevenson, 1969b).

247. Bronze Frog, *Rana clamitans* Latreille

IDENTIFICATION: The presence of *incomplete dorsolateral ridges* sets this species apart from all other frogs except the Gopher Frog, which is quite different in color pattern (except in extreme west Florida). The Bronze Frog is *uniformly brownish bronze dorsally* and largely white underneath (northern race greenish above). Maximum length about 75 mm.

DISTRIBUTION AND VARIATION: East of the Great Plains, from southern Canada southward to the Gulf Coast of eastern Texas and northern Florida, ranging southward in the Peninsula to Lake County (*R. c. clamitans*).

248. Leopard Frog, *Rana pipiens* Schreber[1]

IDENTIFICATION: A rather large, slender frog with *conspicuously light, complete dorsolateral ridges*. The *green dorsal coloration overlaid with rounded dark spots* also tends to separate it from other Florida frogs, as does a light line on the upper jaw. Underparts uniformly whitish. Specimens from the Keys are darker (Duellman and Schwartz, 1958). Maximum length about 115 mm.

DISTRIBUTION AND VARIATION: Southern and northwestern Canada southward through Mexico to Panama, to the northern Gulf Coast, and the Florida Keys; the race in Florida is *R. p. sphenocephala*.

249. Gopher Frog, *Rana areolata* Baird and Girard

IDENTIFICATION: The 2 subspecies (sometimes treated as distinct species) vary so much in color, body proportions, and shape of the dorsolateral ridges that a common description for both is impossible. One feature they have in common, though, is their almost *toadlike body form*. The dorsolateral ridge is either *complete or almost so*, but is wide and flattened in *R. a. aesopus*, narrow and high in *R. a. sevosa*. Also the former race usually has a *pale, almost white, ground color*, whereas *R. a. sevosa is quite dark*. Except when the ground color is too dark, contrasting *rounded dark spots show dorsally*. There are also dark spots on the chin and throat of both, continuing onto the belly in *sevosa*. Maximum length about 110 mm.

DISTRIBUTION: Southern Iowa, southern Illinois, and southwestern Indiana discontinuously southward to eastern Texas, southwestern Louisiana, and north-central Alabama; also from eastern Louisiana (east of the Mississippi River) eastward through the lower Coastal Plain to southeastern North Carolina and southern Florida; *R. a. sevosa* occupies extreme west Florida, and *R. a. aesopus* occurs in the eastern Panhandle and the Peninsula except for the Everglades (Neill, 1957).

ORDER SALIENTIA: LVI. FAMILY MICROHYLIDAE, Narrow-mouthed Toads. Small frogs with pointed heads and a *fold of skin across the back of the head*. The toes are not webbed, and the tympanum is absent.

---

1. *R. utricularia* in Conant's new (1975) field guide.

250. Eastern Narrow-mouthed Toad, *Gastrophryne carolinensis* (Holbrook)[1]

IDENTIFICATION: This small anuran may be distinguished at a glance by its uniformly brownish gray dorsal color and the *triangular outline of the head and body* from a dorsal aspect. Underparts lighter with dark speckling. Maximum length about 30 mm.

DISTRIBUTION: Eastern Oklahoma, southern Missouri, southern Kentucky, Virginia, and North Carolina (except in mountains) southward to the Gulf Coast from Texas to the Florida Keys.

250H. HYPOTHETICAL LIST

Three species of salamanders and one frog have been collected in southern Alabama within about 80 km (50 miles) of the Florida line. As our knowledge of the herpetofauna of western Florida is quite incomplete, it appears likely that any or all of these species may ultimately be found in Florida.

In the Family Ambystomatidae the Spotted Salamander (*Ambystoma maculatum*) has been collected near Eufaula (Neill, 1954). Two species in the Family Plethodontidae occur in the Coastal Plain of Alabama —the Zigzag Salamander, *Plethodon dorsalis* (Blaney and Relyea, 1967), and the newly described Red Hills Salamander, *Phaeognathus hubrichti* (Highton, 1961).

The Pickerel Frog, *Rana palustris* (Ranidae), has been collected a short distance west of Andalusia, Alabama (Brown and Boschung, 1954), about 30 km (20 miles) from Florida.

CLASS REPTILIA: REPTILES

ORDER CHELONIA:[2] LVII. FAMILY DERMOCHELYIDAE, Leatherbacks. Highly modified (externally), gigantic sea turtles distinguishable from all other turtles by the absence of toes and claws and several other features (see description of species). Monotypic.

251. Leatherback, *Dermochelys coriacea* (L.)

IDENTIFICATION: Reaching a shell length of about 2 m and a weight of at least 680 kg (1,500 lbs), this is the *largest living turtle*. In several other respects it is distinctive or unique. The feet are modifid to form flippers that *lack toes and claws*. The *shell is not covered with scales* (except in young), but with a thick, leathery skin. There are *7 prominent*

1. This genus possibly should be merged with *Microhyla*.
2. Testudines in A.S.I.H. *Catalogue*.

longitudinal ridges on the carapace and 5 on the plastron. Adults are chiefly dark, but with irregular light patches.

DISTRIBUTION AND VARIATION: Almost cosmomarine, but absent from Arctic and subarctic waters. The Atlantic subspecies (*D. c. coriacea*) has been found occasionally on various parts of the Florida coast, both Gulf and Atlantic; several nestings on southeast coast.

ORDER CHELONIA: LVIII. FAMILY CHELYDRIDAE, Snapping Turtles. Ugly, large-headed, rough-shelled turtles of large size. The carapace is rough, dark, and often 3-ridged; the plastron smooth, greatly reduced, and cross-shaped. There are 10 plastral plates.

252. Common Snapping Turtle, *Chelydra serpentina* L.

IDENTIFICATION: Differs from the Alligator Snapper in its smaller size, *single row of marginals*, and *more dorsal position of the eyes*. There are *2 rows of enlarged, flat scales under the tail*. Maximum shell length about 30 to 38 cm.

DISTRIBUTION AND VARIATION: Southern Canada (southern Saskatchewan to Nova Scotia) southward to about Monterrey, Mexico, southern Texas, the Gulf Coast, and the southern tip of the Florida mainland. The nominate race (*C. s. serpentina*) occupies north Florida, and *C. s. osceola* ranges from about Alachua County southward.

253. Alligator Snapping Turtle, *Macroclemys temmincki* (Troost)

IDENTIFICATION: Similar to the Common Snapping Turtle, but with an incomplete *second, inner row of marginals* on the carapace (hard to see in young), the *eyes lateral in position*, and *many small scales underneath the tail, the basal ones hemispherical*. The skin of the head and neck is much more spinose that that of the Common Snapping Turtle. Maximum shell length about 61 cm.

DISTRIBUTION: Gulf Coastal Plain from southern Georgia and northern Florida to Texas, northward to southeastern Indiana, northwestern Illinois, northeastern and southern Missouri, and southeastern Kansas.

ORDER CHELONIA: LIX. FAMILY KINOSTERNIDAE, Mud and Musk Turtles. Smaller than the snapping turtles (shell length less than 150 mm) and with the carapace smoother, more arched, and its plates more obvious. Only 10 or 11 plastral plates. Tail relatively shorter than in Chelydridae. Width of carapace about two-thirds its length in adults.

254. Stinkpot, *Sternotherus odoratus* (Latreille)

IDENTIFICATION: Most individuals clearly show *2 light stripes on the side of the otherwise-dark head*. Older ones (in which the stripes may be obscure) *do not have the ridged carapace* of the other musk turtle, nor the highly flexible plastron of a mud turtle. Shell brownish, lighter below; plastron yellow and black in young. *Barbels are present both on the chin and throat*. The pectoral scute is quadrangular. Maximum shell length about 125 mm, but much smaller in Florida.

DISTRIBUTION: Southern Wisconsin, Ontario, southern New York, and southern Maine southward through Missouri to the eastern parts of Kansas, Oklahoma, Texas, and Chihuahua, Mexico, the Gulf Coast, and the southern tip of the Florida mainland.

255. Loggerhead Musk Turtle, *Sternotherus minor* (Agassiz)

IDENTIFICATION: Most easily distinguished from the Stinkpot by its *head pattern—dark spots or bars on a lighter background*. Also *barbels are* on the chin but *not on the throat*. With the other musk turtle (and unlike the mud turtles) it shares the *quadrangular pectoral plates* and scarcely movable plastron. The carapace is lighter than in *S. odoratus*, and its *middorsal ridge persists in adults*. The plastron is red in the young. Maximum shell length about 140 mm.

DISTRIBUTION AND VARIATION: Southeastern Kentucky and southwestern Virginia southward and westward to central Florida (*S. m. minor*, to Lake County—Carr and Goin) and central and coastal Mississippi (*S. m. peltifer*, which also occurs in extreme western Florida).

256. Striped Mud Turtle, *Kinosternon bauri* Garman

IDENTIFICATION: Mud turtles have a larger, more movable plastron than musk turtles and virtually *triangular pectoral plates*. This species is well named, with *3 light stripes* (sometimes obscure) *on the carapace and 2 on each side of the head*. Shell otherwise brownish. Young mud turtles have a keel on the carapace similar to that in the Loggerhead Musk Turtle, but it is lacking in adults. Maximum shell length about 110 mm.

DISTRIBUTION AND VARIATION: Southeastern Georgia, peninsular Florida, and the Keys, westward to Wakulla County (TT 693). *K. b. palmarum* occupies the mainland and upper Keys, but the race on the lower Keys is *K. b. bauri* (Duellman and Schwartz, 1958).

257. Common Mud Turtle, *Kinosternon subrubrum* (Lacépède)

IDENTIFICATION: A common, widely distributed aquatic turtle with a *uniformly dark, smooth carapace* so arched that the sides are almost vertical; no ridge on carapace of adults. The *plastron is* well developed, *fully movable at the bases of the forelimbs, and typically yellowish or reddish brown* (red and black in young). The color pattern of the head usually bears some resemblance to that of the Loggerhead Musk Turtle, but in the Tallahassee area it may be striped. Maximum shell length about 125 mm.

DISTRIBUTION AND VARIATION: New York City area, Maryland, the eastern parts of Virginia and North Carolina, northern South Carolina and Georgia, Tennessee (except mountains), southwestern Indiana, southern Illinois, northern Arkansas, and eastern Oklahoma southward to southern Texas, the Gulf Coast, and the southern tip of the Florida mainland, except most of the Everglades (Duellman and Schwartz, 1958). *K. s. steindachneri* occupies the Peninsula from central Florida southward, and *K. s. subrubrum* the remainder of northern and western Florida, intergrading with *hippocrepis* in west Florida.

ORDER CHELONIA: LX. FAMILY EMYDIDAE, Terrapins. Carapace and plastron well developed, very hard, and covered with obvious epidermal plates. The number of plastral plates (12) is greater than in the Kinosternidae, and the species average somewhat larger (shell length of largest up to 400 mm). Width of carapace usually more than 70 percent of its length.

258. Spotted Turtle, *Clemmys guttata* (Schneider)

IDENTIFICATION: A small turtle with the proportions of a mud turtle, but with rounded, *light spots* (often yellow or orange in life) *on the carapace and head*. (Rarely those on the carapace may be lacking.) There is also some yellow on the plastron, and the ground color of the entire shell is blackish. Maximum length about 110 mm.

DISTRIBUTION: Southern Michigan and Ontario and northern Indiana eastward to New England (except northern parts) and thence southward, east of the mountains, to southeastern Georgia (Folkerts, 1967). There are records, some perhaps based on escaped captives, from northeastern Florida southward to Polk County and westward to Wakulla County (TT 686).

259. Box Turtle, *Terrapene carolina* (L.)

IDENTIFICATION: A terrestrial turtle with a plastron so hinged that the animal may withdraw itself into the shell more completely than even a mud turtle. The *hinged plastron and unwebbed front toes* will separate it from all other Florida species in the family, as will the relative depth of the shell (about 65 percent of maximum width). Although the color of the shell is variable, it usually contains some yellow or orange. Maximum shell length about 175 mm.

DISTRIBUTION AND VARIATION: Southern Michigan, most of Pennsylvania, southeastern New York, and southern New England southward and westward to the Florida Keys, the Gulf Coast, and the eastern parts of Kansas, Oklahoma, and Texas. *T. c. carolina* reaches Florida only in the Tallahassee area; *T. c. major* ranges from there throughout the Panhandle; *T. c. bauri* occupies the remainder of the state except for the Everglades.

260. Diamondback Terrapin, *Malaclemys terrapin* (Schoepff)

IDENTIFICATION: Confined to salt or brackish marshes, therefore *not of normal occurrence away from the coast*. There are usually *concentric rings, a ridge, or a series of knobs on the carapace*. The large head is usually light gray with darker spots (or stripes on the Keys), but in west Florida the ground color is darker. The shell color is variable, but the knobs are often lighter than the rest of the carapace. Maximum shell length about 200 mm.

DISTRIBUTION AND VARIATION: Salt marshes of Atlantic and Gulf Coasts from Cape Cod to southern Texas. The following races occur in Florida: *M. t. centrata*, Nassau and Duval Counties; *M. t. tequesta*, remainder of Atlantic Coast, exclusive of Keys; *M. t. rhizophorarum*, the Keys (in mangroves); *M. t. macrospilota*, Florida Bay around Gulf Coast to Panhandle; *M. t. pileata*, extreme west Florida.

261. Barbour's Map Turtle, *Malaclemys barbouri* Carr and Marchand

IDENTIFICATION: A freshwater turtle with the proportions of *Chrysemys*, but with a *row of prominent knobs on the midline of the carapace*. Although this feature is shared with the Alabama Map Turtle, it may be distinguished from that species by its *curved, light-colored, transverse bar on the horny part of the chin* and its C-shaped light markings on the carapace. Plastron mostly yellowish except in largest specimens. Both species of map turtles have a large yellowish area behind the eye. Maximum shell length about 200 mm.

DISTRIBUTION: Southern Alabama, southwestern Georgia, and the Apalachicola River system in Florida.

262. Alabama Map Turtle, *Malaclemys pulchra* Baur

IDENTIFICATION: With Barbour's Map Turtle this species shares the *Chrysemys*-like shape, the large yellowish area behind the eye, and the *saw-back knobs on the carapace*. It differs in the *longitudinal orientation of the straight, light bar on the chin*, and in the *reticulate pattern of light markings on the carapace*, but the latter may be indistinct. The plastron is usually yellowish, with darker pigment along the seams. Maximum shell length about 200 mm.

DISTRIBUTION: East of the Mississippi River in southeastern Louisiana, southern Mississippi, southwestern Alabama, and the Escambia and Yellow River systems of northwestern Florida (Dobie, 1972).

263. Painted Turtle, *Chrysemys picta* (Schneider)

IDENTIFICATION: Turtles in this genus are somewhat flattened, of rounded outline, and have the carapace wrinkled and with a low median keel. The Painted Turtle, however, has a smoother shell and *lacks the conspicuous serrations* of the other sliders *on the hind edge of the carapace*. Rather than a median keel, the *carapace often has a light line*. Red pigment is frequent on the shell, head, and neck. Maximum shell length about 175 mm in the East (Fig. 4).

DISTRIBUTION AND VARIATION: Southern Canada southward to northern Oregon, central Idaho, northern and eastern Wyoming, northeastern Colorado, Kansas, southern Missouri, northern Mississippi, the central parts of Alabama, Georgia, and South Carolina, and most of North Carolina; also down the Mississippi River Valley to the Gulf Coast. Also isolated populations in the Southwest. There is a single specimen of *C. p. picta* from Jackson County, Florida, in the Florida State Museum (Gainesville).

264. Yellow-bellied Turtle, *Chrysemys scripta* (Schoepff)

IDENTIFICATION: The Yellow-bellied Turtle differs from others in its genus in the presence of a *large yellow (or red) blotch behind the eye*; also the *lower jaw is rounded ventrally*. The *vertical yellow bars on the costals are wider than in most other turtles*. Plastron yellow with dark smudges. With the Chicken Turtle (*Deirochelys*) it shares vertical yellow bars on the back of the thigh. Old, melanistic specimens have little or no color pattern. Maximum shell length about 265 mm.

DISTRIBUTION AND VARIATION: Central and northeastern Kansas, northern Missouri, southeastern Iowa, western Indiana, most of Kentucky, the southern parts of Ohio and West Virginia, and from South Carolina and eastern North Carolina southward into northern South America, to the northern Gulf Coast, and to Alachua and Levy Counties, Florida (*C. s. scripta*, intergrading with *C. s. elegans* in the Panhandle); according to King and Krakauer (1966), the latter race has been introduced in the Miami area.

265. River Cooter, *Chrysemys concinna* Le Conte

IDENTIFICATION: A slider with stripes on the head, neck, and legs and a *light C-shaped marking on the second costal scute*, as in

Malaclemys barbouri. Other *yellow markings on the carapace are often arranged concentrically*. The carapace markings and head stripes are usually yellowish, but may be orange or reddish. Unlike *C. floridana*, it has dark markings on the plastron. Maximum shell length about 400 mm.

DISTRIBUTION AND VARIATION: Southeastern Kansas, southern Missouri and Illinois, western Kentucky and Tennessee, extreme northern Alabama and Georgia, North Carolina (except mountains), and southeastern Virginia southward to northeastern Mexico, the Gulf Coast, and central Florida. *C. c. suwanniensis* occurs from the lower Apalachicola River eastward to Marion County and southward to Sumter and Pinellas Counties; *C. c. mobilensis* occupies that part of Florida west of the Apalachicola River. A specimen collected north of Tallahassee (TT 697) is close to *C. c. concinna*.

266. Florida Cooter, *Chrysemys floridana* (Le Conte)

IDENTIFICATION: This species resembles the River Cooter rather than the Yellow-bellied Turtle in having the *head yellow-striped* and the *chin flattened ventrally*, but differs in lacking the C on the second costal scute. Instead it has *dark O-shaped markings on the ventral surface of the marginals*, but in one subspecies (*peninsularis*) these circles are solid. The *plastron is without dark markings*. Maximum shell length about 380 mm.

DISTRIBUTION AND VARIATION: Southeastern Kansas, southern Missouri and Illinois, western Tennessee, and the central parts of Alabama, Georgia, South Carolina, and North Carolina, and coastal Virginia southward to Aransas Bay, Texas, the Gulf Coast, and southern Florida; also West Virginia (introduced?). Northern Florida is inhabited by *C. f. floridana*, but south of Alachua County it is replaced by *C. f. peninsularis*, which ranges to the tip of the Peninsula.

267. Florida Red-bellied Turtle, *Chrysemys nelsoni* Carr

IDENTIFICATION: This species and the Alabama Red-bellied Turtle resemble one another and differ from other Florida cooters in the following respects: the notch at the tip of the upper jaw is deep and bordered on each side by a cusp; the lower jaw is strongly serrate; the plastron is usually reddish or orange and usually without dark markings; there is a *light arrow-shaped marking on the head, pointing anteriad*. The chief distinction between the 2 species is that the paramedian stripe on the neck stops short of the eye in *C. nelsoni* except in small specimens. (See also Distribution.) Maximum shell length about 300 mm.

DISTRIBUTION: Restricted to the Florida Peninsula, ranging from Alachua and Levy Counties (possibly Wakulla) to the southern tip (Carr and Crenshaw, 1957).

268. Alabama Red-bellied Turtle, *Chrysemys alabamensis* Baur

IDENTIFICATION: Very similar to the Florida Red-bellied Turtle
and perhaps better considered only a subspecies of it. Large specimens
can be distinguished, however, by the fact that the *paramedian light
stripe extends anterior to the eye*. Maximum shell length probably about
300 mm.

DISTRIBUTION: Near Gulf Coast from eastern Texas to Wakulla
and Franklin Counties, Florida (Carr and Crenshaw, 1957).

269. Chicken Turtle, *Deirochelys reticularia* (Latreille)

IDENTIFICATION: At first glance this turtle may appear to be
another *Chrysemys*, but several points of difference can be seen on
closer examination. The *black and yellow vertical bars on the back of
the thigh* are found elsewhere only in *C. scripta*. The carapace has a
reticulate pattern of light lines and is smoother than in *Chrysemys*. Most
noticeable is the *length of the neck*, if it can be extended, as it
approximates the length of the plastron. Maximum shell length about
175 mm, rarely more.

DISTRIBUTION AND VARIATION: Southeastern Oklahoma, southeastern Missouri, central and northeastern Mississippi, north-central
Alabama, the central parts of Georgia and South Carolina, and southeastern North Carolina southward to the Gulf Coast from eastern Texas
to Florida. *D. r. reticularia* ranges across northern Florida, and *D. r.
chrysea* throughout the rest of the Peninsula (Schwartz, 1956).

ORDER CHELONIA: LXI. FAMILY TESTUDINIDAE, Land Tortoises.
These are large, terrestrial turtles with well-developed shells, broad feet,
flattened claws, and no trace of webbing between the toes.

270. Gopher Tortoise, *Gopherus polyphemus* (Daudin)

IDENTIFICATION: A large, heavy-bodied, thick-limbed, terrestrial
turtle that spends much of its time in burrows. The carapace is distinctively inscribed with *concentric, nearly square growth rings on each
scale*. The plastron is smooth, yellowish, not hinged, and the gular
plates project far forward. As the animal withdraws under the carapace,
the heavily scaled limbs afford protection for the softer skin at the axils.
The *width of the forelimbs*, especially, and their *flattened claws* help to
confirm the identification. Maximum shell length about 35 cm.

DISTRIBUTION: Coastal   Plain   from   extreme   southern   South
Carolina, across southern Georgia and Alabama, to the lower Missis-
sippi River, and southward in Florida to northern Dade and Collier
Counties (Duellman and Schwartz, 1958). Terrestrial.

ORDER CHELONIA: LXII. FAMILY CHELONIIDAE, Sea Turtles. These gigantic marine animals somewhat resemble the Leatherback in form, but not in certain details of structure. Although the limbs are flipperlike, the *toes and claws are discernible* to some extent. The most useful distinction is the *division of the shell into horny scutes*. Also note the *general absence of ridges on the carapace* of adults (immatures never have more than 3 ridges).

271. Green Turtle, *Chelonia mydas* (L.)

IDENTIFICATION: The plain brownish shell color of most adult sea turtles offers little help to their identification. Perhaps the best clue to identification in the Green Turtle is the presence of *only 1 pair of pre-frontal plates* on the head. Also of value are the *4 pairs of costal plates, the first of which does not touch the nuchal plate*. The lower jaw is strongly serrate in adults. The young may be distinguished by the color of the flippers, which are black-tipped with white edges. Maximum shell length about 150 cm.

DISTRIBUTION AND VARIATION: The warmer oceans and large seas of the world. The Atlantic race (*C. m. mydas*) also inhabits the Gulf of Mexico, therefore may be found on any part of the Florida coast. It has nested near Vero Beach (Carr and Ingle, 1959).

272. Hawksbill, *Eretmochelys imbricata* (L.)

IDENTIFICATION: Resembles the Green Turtle in having only 4 pairs of costal plates, the first not touching the nuchal. Unlike that species, it has *2 pairs of prefrontal scales*, and the *lower jaw is never more than weakly serrate*. There is a better developed keel than on the carapace of most adult sea turtles. Typical shell length about 50 to 63 cm in adults.

DISTRIBUTION AND VARIATION: The warmer oceans and large seas of the world. The Atlantic race (*E. i. imbricata*) also frequents the Gulf of Mexico, thus has occurred on both sides of the Florida Peninsula and on the Keys.

273. Loggerhead, *Caretta caretta* (L.)

IDENTIFICATION: The *color of the upperparts in this turtle is more reddish brown* than in other sea turtles, and this often contrasts with the light-colored margins of the plates on the carapace. The presence of *5 pairs of costal plates, the first touching the nuchal*, eliminates the Green Turtle and the Hawksbill. Unlike the Ridley, it has *only 3 enlarged inframarginals*. Young may be distinguished by their *3 dorsal and 2 plastral keels*. The middorsal keel tends to persist in adults. Typical shell length up to 125 cm or more.

DISTRIBUTION AND VARIATION: The warmer oceans and large seas of the world. The Atlantic subspecies (*C. c. caretta*) also inhabits the Gulf of Mexico and thus has been found on all parts of the Florida coast.

274. Atlantic Ridley, *Lepidochelys kempi* (Garman)

IDENTIFICATION: The *general coloration* of the shell and body *is grayer* than that of the other sea turtles. In the number and arrangement of the costal scutes it resembles the Loggerhead, but differs in having *4 enlarged inframarginals* on the bridge; the middorsal keel is prominent even in adults. Maximum shell length about 65 cm.

DISTRIBUTION: North Atlantic Ocean, ranging along North American coast from Nova Scotia to the Gulf of Mexico, including all parts of the Florida coast.

ORDER CHELONIA: LXIII. FAMILY TRIONYCHIDAE, Softshell Turtles. With their *flexible shell*, *covered with skin*, and their *tubular snout*, these are among the most distinctive of American turtles. They are flatter than any others and more nearly circular in shell outline than most. With the exception of snapping turtles, they are about as large as any freshwater species in Florida, females often attaining a shell length of more than 40 cm.

275. Smooth Softshell, *Trionyx muticus* (Lesueur)

IDENTIFICATION: Any softshell turtle is so distinct from other turtles (see preceding paragraph) that the only problems of identification are between the various species and subspecies in the family. The Smooth Softshell, absent from most of Florida, can be distinguished by the *absence of a ridge in the nostril and of tubercles on the carapace*. The carapace features a shallow trough middorsally. Older specimens are plain olive brown on carapace and plastron. Maximum shell length over 30 cm.

DISTRIBUTION AND VARIATION: Gulf Coast from eastern Texas to the Escambia River, Florida, northward to southern Nebraska, Iowa, Illinois, Indiana, and Ohio (and farther along major streams), but not eastward to Georgia or the higher mountains. The race in Florida is *T. m. calvatus* (Webb, 1959).

276. Spiny Softshell, *Trionyx spiniferus* Lesueur[1]

IDENTIFICATION: This species and the Florida Softshell differ from the Smooth Softshell in having a *fleshy ridge on the median wall of each*

---

1. See Schwartz, 1956a.

nostril and *tubercles toward the front of the carapace* (minute in young). These tubercles are spinose to conical in this species, unlike those in *T. ferox*. Also the carapace is *light gray brown*, the young showing a *pattern of small, dark spots and concentric dark lines* around the edge. Maximum shell length over 38 cm.

DISTRIBUTION AND VARIATION: Central Minnesota, Wisconsin, Central Michigan, southern Ontario, and western New York, to western Texas (locally farther west), northeastern Mexico, the Gulf Coast, the Florida Panhandle (east to Ochlockonee River), and southern Georgia, thence northeastward to southern North Carolina. The Florida race is *T. s. asper*.

277. Florida Softshell, *Trionyx ferox* Schneider

IDENTIFICATION: Like the Spiny Softshell in having a *septum in the nostril and tubercles at the front of the carapace*, but the tubercles are shaped like *flattened hemispheres* in this species. The *carapace is always dark*, but relieved by a *reticulate pattern of light lines in the young*. Maximum shell length about 60 cm.

DISTRIBUTION: The Florida Peninsula, westward to Covington County, Alabama (Mount and Folkerts, 1968) and southern Okaloosa County, Florida (FSU 354); also northward to southern and southeastern Georgia and southern South Carolina, rarely entering salt or brackish water (Map 29).

ORDER SQUAMATA: SUBORDER LACERTILIA.[1] LXIV. FAMILY IGUANIDAE, Iguanas and allies. The Order Squamata includes lizards and snakes, the Suborder Lacertilia lizards only. Most of the latter differ from snakes in the presence of limbs, and the few that are limbless differ in having external ear openings, normal eyelids, and more than one row of ventral plates. Iguanid lizards differ from most others in Florida in having the *scales keeled or granular*. Femoral pores are also present in some species.

278. Bahaman Bark Anole, *Anolis distichus* Cope

IDENTIFICATION: Anoles have *smaller scales* than other lizards, and this species differs from other anoles in its *granular, unkeeled scales on the back*, blunter head, and protruding eyes. It is further set apart by a *dark crossbar between the eyes* and *4 dark V's on the back*. Dewlap yellowish. Snout-vent length about 40 mm, the tail somewhat longer.

DISTRIBUTION AND VARIATION: Originally Hispaniola and the Bahamas; *A. d. floridanus* and *A. d. dominicensis* are established at Miami (King and Krakauer, 1966; Schwartz, 1968).

---

1. Sauria of A.S.I.H. *Catalogue*.

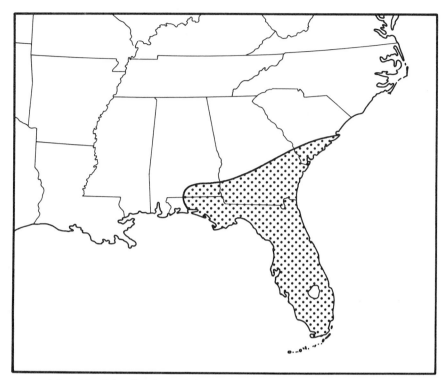

Map 29. Distribution of the Florida Softshell (*Trionyx ferox*)

279. Green Anole, *Anolis carolinensis* Voigt

IDENTIFICATION: This anole differs from many others in its *green coloration*, but it can quickly change to gray or brown. The *keeled scales* in conjunction with the *rounded tail* will separate it from many other species. Males have a *gular fold* (dewlap), *pink* when extended. Snout-vent length up to almost 75 mm, but tail longer.

DISTRIBUTION: Southeastern Oklahoma, central Arkansas, northern Mississippi, southern Tennessee, and southern North Carolina (except high elevations) southward to the Gulf Coast from eastern Texas to the Florida Keys (*A. c. carolinensis*); also other subspecies in Cuba and the Bahamas.

280. Knight Anole, *Anolis equestris* Merrem

IDENTIFICATION: A very large, long-tailed lizard with a *white slash over each shoulder* (yellowish in life). Like the Green Anole, it is capable of color changes ranging from green to gray or brown. The body scales are small (though larger than in other anoles), rectangular, and not keeled. The top of the head is concave and bears larger scales. Immatures have white bands across the back. Dewlap pink. Maximum length about 45 cm, two-thirds of which may be tail (Barbour and Ramsden, 1919).

DISTRIBUTION AND VARIATION: Cuba and the Isle of Pines; *A. e. equestris* is established and spreading in the Miami area (King and Krakauer, 1966) and recently was found on Elliott Key (Brown, 1972).

281. Brown Anole, *Anolis sagrei* Cocteau

IDENTIFICATION: Similar in some respects to the Green Anole and not always distinguished by its color, as *A. carolinensis* may change to brown. The *laterally compressed, keeled tail* should be an infallible distinction. The brown color varies somewhat according to sex and subspecies. Dewlap orange. Snout-vent length about 50 mm, but tail longer.

DISTRIBUTION AND VARIATION: Original range the western Antilles, from the Bahamas to Jamaica and the Swan Islands; also the Gulf Coast from Mexico to Honduras. Now established in Florida as follows: *A. s. sagrei*, St. Petersburg, Tampa, West Palm Beach, Miami, Pigeon and Cudjoe Keys, and Key West; *A. s. ordinatus*, Miami and Palm Beach County (Bell, 1953; King and Krakauer, 1966); Everglade City (subsp.?).

282. Curly-tailed Lizard, *Leiocephalus carinatus* Gray

IDENTIFICATION: A rough-scaled lizard with a *middorsal row of enlarged spinose scales on the tail*. There are about 13 dorsal rows of smaller spinose scales on the body. The ground color is yellowish brown, with dark crossbars dorsally and longitudinal dark streaks on the throat. The *tail is often tightly curled over the back* in life. Maximum snout-vent length in Florida about 90 mm.

DISTRIBUTION AND VARIATION: Western Caribbean islands, from the Bahamas to the Swan Islands; also introduced around Palm Beach and Miami, Florida (*L. c. armouri*; King and Krakauer, 1966).

283. Eastern Fence Lizard, *Sceloporus undulatus* (Bosc)[1]

IDENTIFICATION: Members of this genus have *rougher scales* than any others native to Florida. They also share *grayish upperparts*, with some *blue underneath*, especially in males. In the limited area of the Florida Peninsula where 2 species occur, this one may be distinguished by the *absence of a uniformly brown band along each side*, though it often does have an indistinct *blackish* band. It also has more black

---

1. See Banta, 1961.

pigment on the ventral side than the Florida Scrub Lizard. Maximum snout-vent length about 75 mm, the tail slightly longer.

DISTRIBUTION AND VARIATION: Western United States (except in high mountains and border states), central Missouri, southern Illinois, Indiana, Ohio, Pennsylvania, and New Jersey southward into northern Mexico, to the northern Gulf Coast, and central Florida (*S. u. undulatus*, to Hardee County; Neill, 1954).

284. Florida Scrub Lizard, *Sceloporus woodi* Stejneger

IDENTIFICATION: A rough-scaled lizard of more brownish aspect than *S. undulatus*. Its chief mark of distinction is the *clear-cut, uniformly brown stripe down each side*, but the underparts are more extensively white than in the Fence Lizard. The middle of the back may or may not have the wavy crossbars of *S. undulatus*. Maximum snout-vent length about 55 mm, with the tail somewhat longer.

DISTRIBUTION: Florida Peninsula: near Gulf Coast in Lee and Collier Counties, on central ridge from southern Putnam County to northern Glades County, and near Atlantic Coast from Brevard County to northern Dade County, in sand pine–rosemary scrub (Jackson, 1973).

285. Texas Horned Lizard, *Phrynosoma cornutum* (Harlan)

IDENTIFICATION: With its *squat body, short tail, and spines on the head and sides of the body*, the "horned toad" can hardly be confused with any other lizard known to Florida. The ground color is a desert gray, and darker blotches are scattered over the back. Maximum snout-vent length about 100 mm, more than twice length of tail.

DISTRIBUTION: Original range from southeastern Colorado and most of Kansas southward and southwestward to southeastern Arizona, northern and northeastern Mexico, extreme northwestern Arkansas, and the Texas coast. Widely released in Florida (and some other states), it has been collected in Escambia, Lake, Orange, Putnam, Duval, and Dade Counties (Carr and Goin, 1955), and is considered established near Pensacola (Allen and Neill, 1955). It appears likely that some escaped Florida specimens may represent other species of horned lizards.

ORDER SQUAMATA: SUBORDER LACERTILIA: LXV. FAMILY GEKKONIDAE, Geckos. Most geckos differ from other Florida lizards in their large eyes, with *vertical pupils and immovable, transparent lids* (like the eyes of snakes). Most of them also have expanded pads on their toes. The scales are uniformly small.

286. Yellow-headed Gecko, *Gonatodes albogularis* (Duméril and Bibron)

IDENTIFICATION: Differs from other geckos in having *round pupils* and *no expanded digital pads*. The original *tail is light-tipped*. The color pattern of the male is distinctive, *very dark with a contrastingly yellow head*; females usually show a *light partial collar* near the shoulder and light lateral spots on a brownish background. Maximum snout-vent length about 38 mm, with tail about same length.

DISTRIBUTION AND VARIATION: Original range Cuba, Jamaica, southern Mexico, most of Central America, and part of Colombia; *G. a. fuscus* is introduced and established at Key West and, perhaps, at Miami (King and Krakauer, 1966).

287. Tokay Gecko, *Gekko gecko* L.

IDENTIFICATION: A large, stocky lizard with expanded disks under the digits and no eyelids; head relatively large and triangular. The *transverse lamellae* under the digits *are not divided medially*. *Upperparts ultramarine with numerous spots of rusty red or orange*; tail banded with dark pigment. Total length up to 350 mm (Taylor, 1963).

DISTRIBUTION: Southeastern Asia; introduced at Gainesville and Miami (King and Krakauer, 1966).

288. Mediterranean Gecko, *Hemidactylus turcicus* (L.)

IDENTIFICATION: This gecko has the largest eyes and toe pads of any in Florida except the restricted Tokay Gecko. This species and the Indo-Pacific Gecko differ from all others in having the *transverse lamellae divided medially*. Also distinctive are its *color pattern of dark markings on a whitish background*, *the warts on its back*, and its comparatively large size (maximum snout-vent length about 60 mm, the tail about the same).

DISTRIBUTION AND VARIATION: Original range the Mediterranean area, eastern Africa (except southward), and southwestern Asia. Introduced and established in eastern and north-central Mexico, southern and central Arizona, south and south-central Texas, Louisiana (New Orleans), and the Florida Peninsula and Keys and Cuba. *H. t. turcicus* ranges as far north as Gainesville (King, 1968) and Tampa (Brown and Hickman, 1970) (Map 30).

289. Indo-Pacific Gecko, *Hemidactylus garnoti* Duméril and Bibron

IDENTIFICATION: Similar in size, form, and condition of the subdigital pads and lamellae to the Mediterranean Gecko, but differing in its *lack of warts* (low spines) *on the back*. The color also is more of a uniform grayish fawn, with small, scattered spots, white in preservative. Unisexual (Taylor, 1963).

DISTRIBUTION: Southeastern Asia; introduced locally in Miami (King and Krakauer, 1966).

290. Reef Gecko, *Sphaerodactylus notatus* Baird

IDENTIFICATION: Geckos in this genus are smaller and have more pointed snouts than others in Florida. Like the Mediterranean Gecko, they have expanded disks on the toes, but *most of the transverse lamellae are undivided medially*. This species may be confused with the Ashy Gecko, but the *scales are larger and keeled* (about 2 per mm in adults). All individuals have dark spots on a lighter, somewhat yellowish background, but females may have *3 prominent dark stripes on the head and neck and 2 light spots on the nape*. Maximum snout-vent length about 30 mm, the tail almost as long.

DISTRIBUTION AND VARIATION: Cuba, the Isle of Pines, and the Bahamas; also Florida Keys and extreme southern mainland (Dade, Broward, and Monroe Counties; *S. n. notatus*; see Schwartz, 1966).

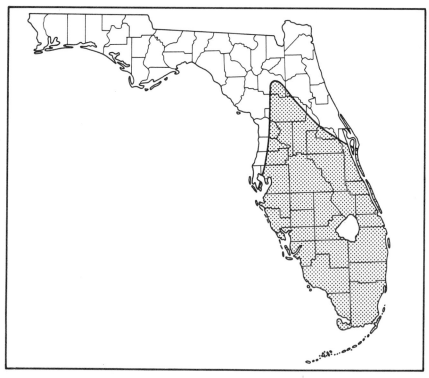

Map 30. Distribution of the Mediterranean Gecko
(*Hemidactylus turcicus*)

291. Ocellated Gecko, *Sphaerodactylus argus* Gosse

IDENTIFICATION: A tiny, mottled, grayish-to-blackish lizard with minute, *flattened, sometimes keeled scales* (granular toward back of head). It has the expanded digital pads and wide-open eyes of other small geckos. Total length slightly more than 50 mm (snout-vent length about 30 mm). (Also see Key.)

DISTRIBUTION AND VARIATION: Cuba and Jamaica; *S. a. argus* has been introduced at Key West, Florida (Savage, 1954; King and Krakauer, 1966) (Map 31).

Map 31. Distribution of the Ocellated Gecko (*Sphaerodactylus argus*)

292. Ashy Gecko, *Sphaerodactylus cinereus* Wagler

IDENTIFICATION: The Ashy Gecko has *unusually small, granular scales* (at least 4 per mm in adults) and *several, irregular, dark stripes on the head and neck*. The back is spotted with dark and yellowish tiny spots. The young are gaudily marked with black crossbars on a lighter body and are red on the tail and limbs. Maximum snout-vent length about 35 mm, about same as tail length.

DISTRIBUTION: Originally Cuba, Hispaniola, the Isle of Pines, and nearby islands; now established at Key West and Boca Chica (Duellman and Schwartz, 1958).

ORDER SQUAMATA: SUBORDER LACERTILIA: XVI. FAMILY ANGUIDAE, Glass Lizards and allies. Best characterized by a *deep groove running along each side of the body*, although Florida species are *limbless*.

293. Eastern Glass Lizard, *Ophisaurus ventralis* (L.)

IDENTIFICATION: Glass lizards are stiff, slick, snakelike lizards *without limbs*, but with ear openings and movable eyelids. Adults of the Eastern Glass Lizard are distinctively colored with *white flecks on a dark green background*, these flecks often being at the *posterior* edges of the scales. Young must be carefully distinguished from other species. At any age, *O. ventralis* has *no dark pigment underneath, only one frontonasal plate*, and the width of the body is more than 80 percent of its depth. Also the distance between the eyes is about equal to the snout length. When there is a single dorsolateral stripe it involves the fourth through sixth scale rows (counted from the middorsal line). Maximum snout-vent length about 280 mm; the tail, unless regenerated, may be more than twice this length.

DISTRIBUTION: The southeastern states, ranging northward and westward to northeastern and central North Carolina, Georgia and South Carolina (except in mountains), north-central Alabama, and the southeastern parts of Mississippi and Louisiana; outlying records in southeastern Oklahoma and St. Louis, Missouri.

294. Island Glass Lizard, *Ophisaurus compressus* Cope

IDENTIFICATION: Smaller than other glass lizards, with a *very light ground color* and a *single, dark dorsolateral stripe on each side*. A middorsal stripe is not quite so dark, or may be broken into a row of spots. Also it usually differs in having *2 frontonasal plates*. Distance between eyes at their center about 75 percent of snout length. Maximum snout-vent length about 150 mm, but tail up to 2 or 3 times that length.

DISTRIBUTION: Coast and offshore islands of South Carolina, Georgia, and the Florida Peninsula (partly inland), north on the west coast at least to Wakulla County (TT 695); also on St. George Island, Franklin County; Blaney, 1971. (Map 32).

295. Slender Glass Lizard, *Ophisaurus attenuatus* Baird

IDENTIFICATION: Similar to the Eastern Glass Lizard in general form, in color of the young, and in having *only 1 frontonasal scale*. However, it is the only species of glass lizard with *ventrolateral dark stripes* (that is, *below* the lateral fold; more prominent on tail than on body). Adults are browner than those of *O. ventralis* and have alternating light and dark crossbars on the back. Distance between eyes at their

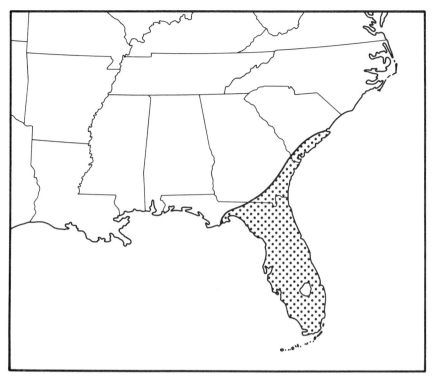

Map 32. Distribution of the Island Glass Lizard
(*Ophisaurus compressus*)

centers about equal to length of snout. Maximum snout-vent length
about 280 mm, but tail at least twice that long. The regenerated tails
of these lizards, however, are much shorter and of a different color.

DISTRIBUTION AND VARIATION: The most widespread of glass
lizards, ranging from southeastern Nebraska, southern Iowa and Wisconsin, western Indiana, southern Kentucky, southwestern and southeastern Virginia (except at high elevations) southward to the mouth of
the Rio Grande, the Gulf Coast, and southern Florida, except the
Everglades (Duellman and Schwartz, 1958). The eastern race, *O. a.
longicaudus*, is the only one in Florida. (McConkey, 1954).

ORDER SQUAMATA: SUBORDER LACERTILIA: LXVII. FAMILY
TEIIDAE, Whiptail Lizards. Distinguished from most other North
American lizards by the *8 to 10 rows of enlarged rectangular scales on
the ventral side*, each scale *arranged transversely*. The tail, if undam-
aged, is proportionately longer than for any other Florida lizards with
limbs (at least twice snout-vent length).

296. Colombian Ground Lizard, *Ameiva ameiva* L.

IDENTIFICATION: *Larger than most Florida lizards* (total length 380 to 500 mm, the tail making up at least two-thirds). Upperparts dark, with light spots, except for a light head; underparts dark, except that legs and tail are light with some dark spotting. There are *few or no stripes present. Ten rows of ventral scales.*

DISTRIBUTION AND VARIATION: The southern Lesser Antilles and much of South America; *A. a. petersi* and *A. a. ameiva* are established in the Miami area (King and Krakauer, 1966).

297. Six-lined Racerunner, *Cnemidophorus sexlineatus* (L.)

IDENTIFICATION: With its 6 light stripes above and the whitish belly of the female, this lizard may be confused with skinks in the field, although in hand the scalation of the under side will readily distinguish it (see description of family). As they are difficult to catch, however, it may be more practical to note their *duller appearance* (skinks being smooth and shiny) and the *absence of blue on the tail* of adults. Males are further recognizable by their *light blue underparts*. The dry, open, often sandy places they usually inhabit are not normally frequented by skinks. Maximum snout-vent length about 75 mm, the tail often twice that length.

DISTRIBUTION AND VARIATION: Southeastern Colorado, southern South Dakota, southwestern Iowa, northern Missouri, southern Illinois (much farther north in river valleys), southwestern Indiana, western Kentucky, Tennessee, and (east of the mountains) southern Maryland southward to the lower Rio Grande, the Gulf Coast, and the Florida Keys, but absent from the Everglades (*C. s. sexlineatus*).

ORDER SQUAMATA: SUBORDER LACERTILIA: LXVIII. FAMILY SCINCIDAE, Skinks. Easily distinguished from other Florida lizards by their *smooth, shiny scales, uniformly small and rounded* in outline. Florida species are also *striped*, at least when young.

298. Ground Skink, *Scincella lateralis* (Say)[1]

IDENTIFICATION: One of the easiest of Florida skinks to identify. First, it is the smallest, its *total* length scarcely exceeding 100 mm. Also the color pattern of a *broad, brown stripe down the back, bordered by a narrow black line on each side* differs from that of most other small skinks. Most other species also have white dorsal stripes and blue or red in the tail. Finally, under magnification *a transparent section may be*

---

1. Sometimes placed in *Lygosoma* or *Leiolopisma*.

seen in the lower eyelid, almost unique among Florida lizards, Maximum snout-vent length almost 50 mm, the tail sometimes much longer.

DISTRIBUTION: Eastern Kansas, central Missouri, southern Illinois, Indiana, Ohio, and West Virginia, all of Virginia, and the southern parts of Maryland and Delaware southward to central Texas, the lower Rio Grande, the Gulf Coast, and the Florida Keys (but not Everglades).

299. Northern Five-lined Skink, *Eumeces fasciatus* (L.)

IDENTIFICATION: Three of the 5 skinks in Florida are virtually impossible to identify unless in the hand. All are *conspicuously striped and have blue tails when young*. They become more obscurely colored, losing the blue in the tail, in old age. Often the old males develop *red on the head*. The present species and *E. laticeps* are the most difficult to separate except for old males of *laticeps*, which have *very wide heads*. Both, however, differ from *E. inexpectatus* in having the midventral scale row of the tail much wider than the adjacent rows. *E. fasciatus* may be distinguished from the young or females of the Broad-headed Skink (*E. laticeps*) by its *2 enlarged postlabial plates*, as *laticeps* has no more than one enlarged scale in this position. The white stripe across the ear in this species does not involve the scales *above* the ear, and the number of scale rows at mid-body is not more than 30; the light lateral stripe (when present) is on the *third and fourth scale rows*, counted from the middorsal line; 5 such lines may be present. A few pointed auricular scales project over the anterior part of the ear opening. Maximum snout-vent length about 75 mm, the tail slightly longer. (Also see "Distribution.")

DISTRIBUTION: Chiefly from southeastern Nebraska, northern Missouri, central Wisconsin, lower Michigan, southern Ontario, northern Pennsylvania, southeastern New York, and Massachusetts (except northeastern) southward to Aransas Bay, Texas, the Gulf Coast of Louisiana and Mississippi, the southern parts of Alabama and Georgia, and northwest Florida (Liberty County westward; FSU 153 is from Okaloosa County); also an unconfirmed record from Jefferson County (Blaney, personal communication).

300. Broad-headed Skink, *Eumeces laticeps* Schneider

IDENTIFICATION: Except for old males, with their *greatly widened, coppery-colored heads*, this lizard is identical in its general appearance to *E. fasciatus*, and both have a widened median row of subcaudal scales. On closer inspection, however, it shows *one or no enlarged postlabial*, some white on the scales immediately above the ear, and 30 or more scale rows at midbody. The light dorsolateral stripe (if present) involves the fourth and fifth scale rows; *5 or 7 such stripes may be present*. Auriculars unmodified and not projecting over anterior part of ear opening. This is Florida's largest widespread lizard (other than the burrowing glass lizards), the snout-vent length up to at least 125 mm, and the tail somewhat longer (unless regenerated). Thus a large specimen may reach a length of about 300 mm. (Fig. 5).

DISTRIBUTION: Eastern Kansas, central Missouri, western and southern Illinois, central Indiana, the southern parts of Ohio and West Virginia, and Maryland (except western) southward to central and southeastern Texas, southern Mississippi, and Florida (southward to Lake County).

301. Southeastern Five-lined Skink, *Eumeces inexpectatus* Taylor

IDENTIFICATION: Grossly identical to *E. fasciatus* and *E. laticeps*, but easily distinguished by examining the scales under the tail. If it is *difficult to detect any difference in the width of the median scale row* and those on each side of it, this is the species. In the other 2 species the median row is at least 50 percent wider than adjacent rows and gives the impression of being twice as wide. A second difference from *laticeps* is that the light lateral stripe involves the *fifth scale row* (sometimes fourth also) down from the middorsal row. Maximum length about 200 mm.

DISTRIBUTION: Southeastern Kentucky and the southern half of Virginia southward to the Gulf Coast and Florida Keys and westward through central Tennessee and northern Mississippi to southeastern Arkansas and eastern Louisiana.

302. Coal Skink, *Eumeces anthracinus* (Baird)

IDENTIFICATION: Slightly suggestive of the Ground Skink (*Scincella*), but easily distinguished from all other Florida skinks. The black lateral stripe is wide and *bordered above and below by a light streak*, but there are *no light stripes on the top of the head*. The *absence of blue or red in the tail* will eliminate most skinks of comparable size. From the Ground Skink it differs in having *light stripes but no transparent area in the lower lid*. Maximum snout-vent length almost 75 mm, the tail somewhat longer.

DISTRIBUTION AND VARIATION: Discontinuously distributed from Lake Ontario southwestward, chiefly in mountainous and hilly areas, to the coast of Mississippi, adjacent parts of Alabama and Louisiana, and west-central Georgia (Mecham, 1960); also from southeastern Kansas and southern Missouri southward to northeastern Texas and northwestern Louisiana. A small population of *E. a. pluvialis* occurs in Florida from Franklin County westward (Seibert, 1964; Stevenson, 1968a).

303. Red-tailed Skink, *Eumeces egregius* (Baird)

IDENTIFICATION: The *entirely orange or reddish color of the tail* usually sets this skink apart from all other Florida lizards, but in *E. e. lividus* the tail of immatures is bright blue. Like most other skinks, it has alternating light and dark stripes on the body, but is smaller and more slender than other striped species, having *only 22 longitudinal scale rows*. Maximum snout-vent length a little over 50 mm, the tail slightly longer.

DISTRIBUTION AND VARIATION: Lower Savannah River, Georgia, across Coastal Plain and lower Piedmont to Alabama coast, southward to Florida Keys. *E. e. similis* ranges across north Florida southward to the Suwannee and Santa Fe Rivers; *E. e. lividus*, Lake Wales ridge of Polk and Highlands Counties; *E. e. insularis*, Cedar and Seahorse Keys; *E. e. onocrepis*, most of remaining Peninsula, except for Everglades; *E. e. egregius*, only on the Keys, including the Dry Tortugas (Mount, 1965).

304. Sand Skink, *Neoseps reynoldsi* Stejneger

IDENTIFICATION: Easily distinguished from all other Florida lizards by its *vestigial limbs*, none of which bears more than 2 toes. With *Scincella*, it shares a *partly transparent lower eyelid*. Although the *color is pale*, it is not pigmentless like the Worm Lizard; some specimens have faint lateral stripes. Maximum snout-vent length slightly over 50 mm, the tail about equally as long.

DISTRIBUTION: Sand dunes in the central Florida Peninsula from Alachua County to Polk and Highlands Counties.

ORDER SQUAMATA: SUBORDER AMPHISBAENIA: LXIX. FAMILY AMPHISBAENIDAE, Ringed Lizards. Peculiar, burrowing lizards that look superficially as much like earthworms as lizards or snakes. They appear to be naked and externally segmented, but closer examination shows that they are covered with scales like those of lizards. Their extreme adaptations to burrowing commonly include *loss of limbs, loss of color, loss of external eyes and ear openings, and a wedge-shaped head.*

305. Worm Lizard, *Rhineura floridana* Baird

IDENTIFICATION: An annulated reptile with the appearance of a white or pinkish earthworm, *having no limbs or ear openings*. The tail is short and scarcely distinguishable from the body, but is dorsoventrally flattened. Scales of body and tail square or rectangular. Mouth ventral.

(Also see Family description.) The maximum length is about 300 to 380 mm (Telford, 1955).

DISTRIBUTION: Northern and central portions of the Florida Peninsula, from Columbia and Alachua Counties southward to Highlands County.

ORDER SQUAMATA: SUBORDER SERPENTES: LXX. FAMILY COLUBRIDAE, Colubrid Snakes. Snakes are long, slender, limbless vertebrates that differ from limbless lizards in having *transparent, immovable eyelids (apparently no* eyelids) and *no ear openings.* Moreover, Florida species have a single row of widened scales on the under side of the body. All of our nonpoisonous species are in this family. They can be separated from the pit vipers by the *triangular shape of the head* and the *pit between each eye and nostril of the latter group.* From the coral snakes they differ in *not having red, yellow, and black rings unless the snout is also red and the red rings contact the yellow ones.*

306. Glossy Water Snake, *Regina rigida* (Say)[1]

IDENTIFICATION: The *glossy appearance* aids identification in life. The color pattern is a greenish brown above, with faint darker stripes laterally, and a whitish ground color underneath. The *dark spot on each side of each ventral scale* gives the effect of *2 dark rows*. This species and the Queen Snake differ from other water snakes in having only 19 longitudinal scale rows. Maximum length 63 cm or more.

DISTRIBUTION AND VARIATION: Southeastern Oklahoma, central and southeastern Arkansas, and the central parts of Mississippi, Alabama, Georgia, and South Carolina southward to Galveston Bay, the Gulf Coast, and Marion County, Florida. *R. r. rigida* inhabits the Peninsula, and *R. r. sinicola* the Panhandle eastward to Leon and Wakulla Counties (Huheey, 1959).

307. Queen Snake, *Regina septemvittata* (Say)

IDENTIFICATION: Similar to the Glossy Water Snake in general appearance and number of scale rows, but not glossy in life and with a *light stripe* (yellow in life) *low on each side*, though not touching ventral plates. In ventral aspect it shows *2 narrow and 2 broad dark stripes on a yellowish background*. Dorsal scale rows 19. Maximum length 75 to 90 mm.

DISTRIBUTION AND VARIATION: The southern Great Lakes area southward through northeastern Illinois and western Indiana, western

---

1. See Rossman (1963b) for use of *Regina*.

Kentucky and Tennessee to Mississippi, southern Alabama, and the Florida Panhandle, and eastward to the edge of the upper Coastal Plain in Virginia, North and South Carolina, and Georgia. Specimens have been found in Florida from the Ochlockonee River westward to Escambia County (Spangler and Mount, 1969) (Map 33).

308. Striped Swamp Snake, *Regina alleni* (Garman)

IDENTIFICATION: Ground color brownish, with a broad, but inconspicuous, middorsal stripe and a *conspicuous yellow stripe on each side*. The underparts are yellowish, sometimes lightly marked with dark pigment laterally. It differs from other Florida snakes with a similar color pattern in its combination of a small head, divided anal plate, and *smooth scales*. The young, however, may lack yellow stripes and may have slightly keeled scales on the middorsal row. Maximum length about 60 cm.

DISTRIBUTION: Southeastern Georgia and Peninsular Florida, from Leon County to the tip of the mainland.

309. Green Water Snake, *Natrix cyclopion* (Duméril, Bibron, and Duméril)

IDENTIFICATION: It requires a stretch of the imagination to detect green in the coloration of this snake, but the *dorsal color is often close to olive*. On this background there are usually faint blackish markings. The underparts are whitish, marked with light gray in north Florida. There is also a reddish color phase in south Florida. Whatever its color, though, this species can readily be separated in the hand from all other water snakes by its *row of subocular scales*. The scales on the sides of the body are more weakly keeled than in other water snakes, and there are more scale rows (27 to 31) than in most other species. Maximum length about 180 cm.

DISTRIBUTION AND VARIATION: Coastal Plain from Aransas Bay, Texas, to southern South Carolina, northward through Louisiana, eastern Arkansas, and western Mississippi to extreme southern Illinois. That portion of the Panhandle from Leon County and Apalachicola westward is populated by *N. c. cyclopion*, and the remainder of the state (except the Keys) by *N. c. floridana*.

310. Red-bellied Water Snake, *Natrix erythrogaster* (Forster)

IDENTIFICATION: Though the ground color may vary among individuals, this large water snake displays a *plain color pattern* in this part of its range. The *underparts* are most distinctive, ranging from *red to yellow*. The upperparts may be brownish or grayish green. The *young*,

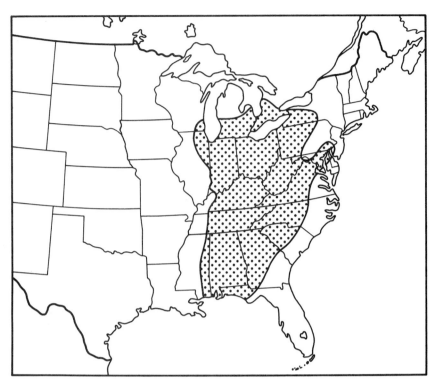

Map 33. Distribution of the Queen Snake (*Regina septemvittata*)

however, *show distinct, but small, dark blotches above*. The number of ventral plates (more than 140) is greater than in most water snakes. Maximum length about 150 cm.

DISTRIBUTION AND VARIATION: Southeastern half of Kansas, western and southeastern Missouri, northern Arkansas, southern Illinois and Indiana, western Kentucky and Tennessee, northern Alabama and Georgia, central parts of the Carolinas, southeastern Virginia and Maryland, and southern Delaware southward to southeastern New Mexico, northeastern Mexico, the Gulf Coast, and northern Florida. *N. e. flavigaster* occurs in extreme western Florida (Conant, 1958), and *N. e. erythrogaster* across the remainder of north Florida as far south as Levy and Alachua Counties.

311. Brown Water Snake, *Natrix taxispilota* (Holbrook)

IDENTIFICATION: When the *dark dorsal blotches* (2 rows alternating in position, usually not contacting one another) are distinct, they will separate this species from most other Florida water snakes. Each blotch is surrounded by a *light border*. However, the blotches become indistinct in large specimens. The *large number of scale rows* (27 to 33) should help to confirm the identification. The underparts are yellowish with irregularly distributed dark blotches. The head is relatively small, but widened posteriad. Maximum length about 150 to 170 cm.

DISTRIBUTION: Coastal Plain (into lower Piedmont of Georgia and South Carolina) from Virginia to Alabama, southward to the tip of the Florida mainland.

312. Banded Water Snake, *Natrix fasciata* L.

IDENTIFICATION: Differs from most other water snakes in having 21 to 25 rows of scales, strongly keeled lateral scales, less than 140 ventral plates, and some dark markings ventrally. Although the subspecies in Florida vary in color among themselves, the *top of the head is usually dark brown or blackish* and there is a *black band from the eye to the corner of the mouth*. Otherwise they may be blotched, striped, or plain dorsally. Maximum length about 150 cm.

DISTRIBUTION AND VARIATION: Coastal Plain from North Carolina to eastern Texas, northward to southeastern Oklahoma, southeastern Missouri, and western Tennessee and Mississippi, and southward to the Florida Keys. *N. f. fasciata* ranges across the Panhandle eastward to about Lake County; *N. f. pictiventris* occupies the remainder of the Peninsula (both of these being freshwater races); *N. f. clarki* inhabits salt marshes from Cedar Keys to southern Texas; *N. f. compressicauda* lives south of Cedar Keys along the coast to Dade County and the Florida Keys; and *N. f. taeniata* is limited to a small part of the Atlantic Coast from Volusia to Indian River Counties.

313. Common Water Snake, *Natrix sipedon* (L.)

IDENTIFICATION: Very similar to the Banded Water Snake in scutellation and until recently considered conspecific with it (see Conant, 1963). However, it *lacks the black band* from the eye to the corner of the mouth *and the dark dorsal coloration of the head* typically found in all races of *N. fasciata*. When directly compared with that species it shows a broader head. (Note also Key and Distribution.) Maximum length about 125 cm.

DISTRIBUTION AND VARIATION: Eastern Colorado, most of Nebraska and Iowa, southeastern Minnesota, Wisconsin, lower Michigan, southern Ontario, northern Vermont and New Hampshire, and southwestern Maine southward to the Gulf Coast of Mississippi, Alabama, and western Florida (*N. s. pleuralis*, east to Ochlockonee River); also westward to eastern Oklahoma, Arkansas, and eastern Louisiana.

314. Black Swamp Snake, *Seminatrix pygaea* Cope

IDENTIFICATION: *A small, smooth snake, blackish above and red below.* Other snakes with a roughly similar color pattern (Red-bellied Snake and Red-bellied Water Snake) have keeled scales. Faint, narrow light stripes may be seen on the lower sides. Maximum length 38 to 45 cm.

DISTRIBUTION AND VARIATION: Southeastern North Carolina, central South Carolina, eastern and central Georgia, and extreme southern Alabama southward to the Gulf Coast and southern tip of the Florida mainland. *S. p. pygaea* is restricted to north Florida, ranging as far west as Gulf and Jackson Counties, southward to near Orlando; *S. p. cyclas* inhabits most of the southern Peninsula (Map 34).

315. Brown Snake, *Storeria dekayi* (Holbrook)

IDENTIFICATION: A small snake with a combination of *keeled scales, brownish upperparts* (with black markings), and lighter underparts. In preservative it may be confused with the Red-bellied Snake, but differs in the shape of the scales in the lowermost row of dorsals (see Key). There is often a row of dark spots toward each side of the ventral plates, especially near the head, and a dark blotch at the angle of the jaws. Faint, dark crossbars can usually be discerned. Many individuals (especially in the Peninsula) show one or 2 *light spots at the back of the head, bordered behind with blackish.* Maximum length about 50 cm.

DISTRIBUTION AND VARIATION: Southern Minnesota and Wisconsin, lower Michigan, southern Ontario, and southeastern Maine southward through central parts of Kansas, Oklahoma, and Texas to Honduras, the northern Gulf Coast, and the lower Florida Keys.[1] The race in the Panhandle (Gadsden and Liberty Counties westward) is *S. d. wrightorum*; throughout the Peninsula and on the lower Keys *S. d. victa* is found. It has only 15 scale rows, as against 17 in *wrightorum*.

316. Red-bellied Snake, *Storeria occipitomaculata* (Storer)

IDENTIFICATION: Because of its *small size, red belly, and frequently blackish upperparts*, this snake can be confused with the Black Swamp Snake (*Seminatrix*), but differs in having *keeled scales*. Like *Storeria dekayi*, it has a light area at the back of the head and in preservative might be mistaken for that species; the square shape of the scales in the lowermost row of dorsals, however, will separate it. The pattern of the head and neck also may suggest *Tantilla* and *Diadophis* (preserved), but these snakes also differ in having smooth scales. Although the dorsal coloration (in life) has been described as gray, brown,

---

1. *Storeria tropica* is included; see Sabath and Sabath, 1969.

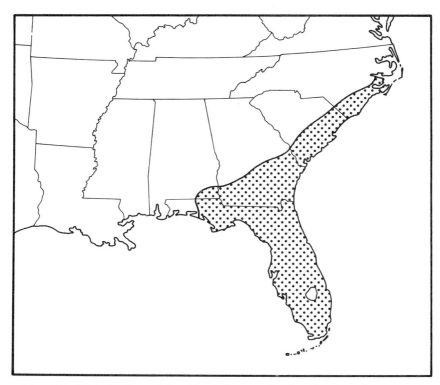

Map 34. Distribution of the Black Swamp Snake (*Seminatrix pygaea*)

or black, it is often *reddish* in the Tallahassee area. Maximum length about 38 cm.

DISTRIBUTION AND VARIATION: From southern Manitoba across southern Canada to Nova Scotia and southward to the Gulf Coast and central Florida; ranges westward to eastern North Dakota, northeastern Iowa, western Illinois, eastern Kansas and Oklahoma, and western Louisiana. *S. o. occipitomaculata* has been ascribed to that part of the Florida Panhandle west of the Apalachicola River (Conant, 1958); *S. o. obscura* occurs southeastward as far as Marion County, possibly farther.

317. Common Garter Snake, *Thamnophis sirtalis* (L.)

IDENTIFICATION: Throughout most of its range this snake has alternating light and dark longitudinal stripes and can be distinguished by the fact that the light *lateral stripes are confined to the second and third scale rows* (counting upward from the ventrals). In some individuals, however, no lateral stripes are discernible, but numerous, fine, black specks or larger spots are present on a lighter background. In such cases the *relative tail length* (less than 30 percent of total length), the *number of subcaudals* (less than 85 pairs), and the *light underparts* should separate it from the similar Ribbon Snake, and the *single anal plate* from *Natrix fasciata clarki*. Maximum length probably not over 1 m.

DISTRIBUTION AND VARIATION: One of the most wide-ranging of North American snakes; found from eastern British Columbia, southern Northwest Territory, and across southern Canada southward to southern Utah and New Mexico, Texas (northern Panhandle and eastern portion), the Gulf Coast, and the southern tip of the Florida mainland. *T. s. sirtalis* inhabits most of the state, but the form near the coast from eastern Wakulla County to Levy County is *T. s. similis* (Rossman, 1964).

318. Eastern Ribbon Snake, *Thamnophis sauritus* (L.)

IDENTIFICATION: This very slender, striped snake differs in *proportionate tail length* (more than 30 percent of total length) from all other striped snakes in Florida. Unlike most, it has a *single anal plate* and differs from *T. sirtalis* in having the *light lateral stripe usually located on the third and fourth scale rows*. This stripe may be yellow, brown, or blue in life and the underparts dark. Maximum length 60 to 75 cm.

DISTRIBUTION AND VARIATION: Chiefly east of the Mississippi River, from southern Wisconsin and Michigan, southern Ontario, central New York, southern New Hampshire and Vermont, and southwestern Maine southward to the Gulf Coast and Florida Keys. *T. s. sauritus* occupies the western Panhandle eastward to the Apalachicola River; *T. s. nitae*, near the coast from eastern Wakulla and Franklin Counties to Levy County; and *T. s. sackeni* the remainder of the state, including the Keys and "apparently" St. Vincent Island (Rossman, 1963a; Blaney, 1971).

319. Rough Earth Snake, *Virginia striatula* (L.)

IDENTIFICATION: There are 2 *small, grayish brown Florida snakes without distinctive markings*. This species differs from the other (Smooth Earth Snake) in having *keeled scales*. It is usually more brown than gray, with the underparts whitish. The head is more pointed than in *V. valeriae*, and there are only 5 upper labials; internasal single. Other small snakes with keeled scales have distinctive markings (for example, *Storeria*). Maximum length about 30 cm.

DISTRIBUTION: Eastern Oklahoma, southeastern Kansas, southern Missouri, southwestern Tennessee, northern Alabama and Georgia, South Carolina (except mountains), eastern North Carolina and Virginia southward to Aransas Bay, Texas, the Gulf Coast, and northern Florida (Alachua County).

320. Smooth Earth Snake, *Virginia valeriae* Baird and Girard

IDENTIFICATION: A *small, grayish brown* snake with whitish underparts and *no distinctive markings*. Most easily confused with the Rough Earth Snake, but it is grayer, the head is less pointed, there are usually 2 internasals, 6 upper labials, and the *scales are smooth*. Most other small Florida snakes, including *V. striatula*, have either keeled scales or distinctive markings. Maximum length about 30 cm.

DISTRIBUTION AND VARIATION: Eastern Kansas, central Missouri, southern Illinois, southwestern Indiana, southern Ohio, western Virginia, eastern Maryland, southeastern Pennsylvania, and New Jersey (except northern) southward to central Texas, the Gulf Coast, and central Florida (Highlands County, Campbell, 1962; *H. v. valeriae*); absent from the mountains except an isolated population in western Pennsylvania, northeastern West Virginia, and extreme western Maryland (*H. v. pulchra*).

321. Yellow-lipped Snake, *Rhadinaea flavilata* (Cope)

IDENTIFICATION: A small brownish (typically *golden* brown) snake with a dark line through the eye and, sometimes, 2 or 3 faint dark lines down the body. It may be differentiated from other small snakes of similar color pattern by its *divided anal plate* and *smooth scales*. Although there is *only one preocular*, there is a small scale (resembling a second preocular) below it in a small percent of specimens. The under side of the body and the upper labials are yellowish in life. There are *17 scale rows* on all parts of the body. Maximum length about 38 cm (Myers, 1967).

DISTRIBUTION: Coastal Plain from southeastern North Carolina to southeastern Louisiana southward to Glades, Okeechobee, and Palm Beach Counties, Florida (Myers, 1967).

322. Ring-necked Snake, *Diadophis punctatus* (L.)

IDENTIFICATION: A small snake with blackish dorsal coloration, *yellow to reddish ventrally, the latter color extending dorsally to form a ring around the neck*, sometimes interrupted middorsally. (Upperparts may be lighter in the West and ring lacking or faint in Southwest.) In preservative the Red-bellied Snake is similar, but differs in having keeled scales. Maximum length about 38 cm.

DISTRIBUTION AND VARIATION: Southeastern Canada, Wisconsin, northern Iowa, eastern and southern Nebraska, eastern Colorado, southwestern Idaho, and southeastern Washington southward into northern Mexico (Sonora and Vera Cruz), the northern Gulf Coast, and the Florida Keys; absent from large areas of the West and Midwest.[1] *D. p. punctatus* occurs over most of Florida, but the race on Big Pine Key is *D. p. acricus* (Paulson, 1966) (Map 35).

323. Rainbow Snake, *Farancia erytrogramma* (de Beauvois)

IDENTIFICATION: Unique among Florida snakes in its *black and red longitudinal stripes*; underparts red with black spots arranged in 2 rows. Maximum length about 140 cm.

DISTRIBUTION AND VARIATION: *F. e. erytrogramma* ranges from southeastern Louisiana (east of Mississippi River) southward to northeastern Lake County, Florida; *F. e. seminola* is known only from Fish-eating Creek (Glades County), Florida (Neill, 1964b).

324. Mud Snake, *Farancia abacura* (Holbrook)

IDENTIFICATION: A large smooth snake, *black above and reddish below*, with the latter color extending up onto the sides at regular intervals. Conversely, the black dorsal color may extend onto the edges of the underside, especially anteriad, or even meet black bands from the opposite side midventrally. A few *much smaller* snakes in Florida have a roughly similar color pattern, though hardly similar enough to cause confusion. Maximum length about 180 cm.

DISTRIBUTION AND VARIATION: Southeastern Oklahoma, southern and eastern Arkansas, southeastern Missouri, southern Illinois, western parts of Kentucky and Tennessee, central Alabama, Georgia, and South Carolina, and the Coastal Plain of North Carolina and Virginia southward to Aransas Bay, Texas, the Gulf of Mexico, and the tip of the Florida mainland. *F. a. reinwardti* occupies extreme western Florida and *F. a. abacura* the remainder of the state.

325. Racer, *Coluber constrictor* L.

IDENTIFICATION: In most parts of Florida this snake is *uniformly black above*, somewhat lighter below, and whitish on the chin (except immatures). Near the Apalachicola River the chin may be brownish, but in the Everglades and on Merritt Island the black upperparts appear bleached to a *bluish, grayish, or brownish color*. Immatures in any part of the state differ strikingly from adults in color pattern, having dark brown dorsal and smaller ventral blotches on a light gray background. These may have to be keyed out to the correct species, although all

---

1. Four nominal species of *Diadophis* are included here; see Croulet, 1965.

racers have *large eyes and very slender proportions*. The combination of *less than 19 rows of smooth scales (15 just anterior to the vent) and 2 preoculars* should separate this species from all other Florida snakes. Maximum length about 170 cm.

DISTRIBUTION AND VARIATION: Almost country-wide, but absent from extreme northern United States and unsuitable habitat in the West. *C. c. priapus* occurs in all parts of Florida except where replaced by the

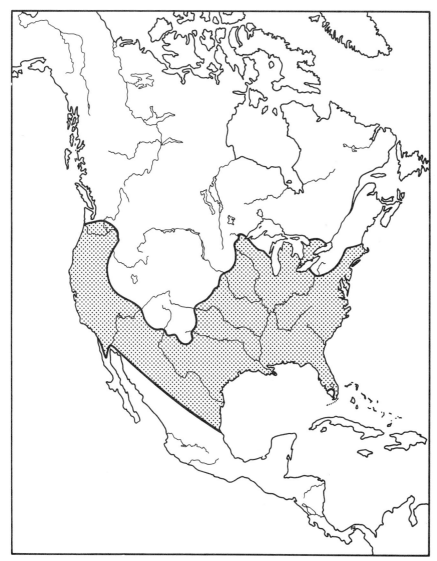

Map 35. Distribution of the Ring-necked Snake (*Diadophis punctatus*)

2 following subspecies: *C. c. helvigularis*, near the Apalachicola and Chipola Rivers; *C. c. paludicola*, coast of Brevard County and from the Everglades to Key Largo (Auffenberg, 1954; Duellman and Schwartz, 1958).

326. Coachwhip, *Masticophis flagellum* (Shaw)

IDENTIFICATION: Very similar to the Racer in form and proportions. Adults differ from all other Florida snakes in their *gradual transition from a uniformly dark to a uniformly light (grayish tan) coloration from front to back*. Immatures have irregular dark crossbars on a yellowish background above, and 2 longitudinal dark streaks below, but can also be identified by their combination of *less than 19 rows of smooth scales (not more than 13 just anterior to the vent) and the 2 preoculars*. Maximum length about 250 cm.

DISTRIBUTION AND VARIATION: Widespread across southern and central United States; *M. f. flagellum* ranges from eastern Kansas, central Missouri, southwestern Tennessee, northern Alabama and Georgia, and southern and coastal North Carolina southward to Aransas Bay, Texas, the Gulf Coast (except southern Louisiana), and the southern tip of the Florida mainland.

327. Rough Green Snake, *Opheodrys aestivus* (L.)

IDENTIFICATION: The only *bright green* snake in Florida; underparts whitish. Immatures, however, are *grayish* green. Unlike a related species north of Florida, it has *keeled scales*. Both are exceptionally slender. Maximum length about 90 to 125 cm.

DISTRIBUTION: From about 39 or 40 degrees North latitude southward through eastern Kansas, central Oklahoma, and west-central Texas into northeastern Mexico, to the Gulf Coast, and the Florida Keys, but not in the Everglades (Duellman and Schwartz, 1958).

328. Indigo Snake, *Drymarchon corais* (Daudin)

IDENTIFICATION: One of the largest and bulkiest of Florida snakes. Its *glossy, blue black coloration*, lighter blue underneath, distinguishes it from most other Florida snakes. From the other black snake, the Racer, it differs in having a *single anal plate* and, usually, in having some brown under the head. Maximum length about 215 cm.

DISTRIBUTION AND VARIATION: Peninsular Florida and the Keys, with separated populations in southeastern Georgia, western Florida, and southern Alabama (*D. c. couperi*); another race (*D. c. erebennus*) occurs in southern Texas and northeastern Mexico.

329. Corn Snake, *Elaphe guttata* (L.)

IDENTIFICATION: Live or freshly killed specimens are easily recognized by their color pattern of *reddish brown dorsal patches* on a gray to yellow (orange on Keys) background. In immatures, however, or in preservative, these patches are so dull as to cause confusion with the Rat Snake. In any case, it can be distinguished from that snake by the arrangement of the V-shaped marking on the head (see Key). The mainland form has conspicuous black blotches, alternating in position, on a whitish belly. Maximum length about 150 to 175 cm.

DISTRIBUTION AND VARIATION: Eastern Utah, southern Colorado, most of Nebraska, southern Missouri, southern Tennessee, most of Virginia, and the Coastal Plain of southern New Jersey southward into northern Mexico, to the Gulf Coast, and the Florida Keys, except absent from much of the Mississippi River Valley. *E. g. guttata* populates the mainland of Florida, and *E. g. rosacea* the Keys (Paulson, 1966).

330. Rat Snake, *Elaphe obsoleta* (Say)

IDENTIFICATION: Although the color patterns of the various subspecies vary widely in adults, immatures are all similar to the adult of *E. o. spiloides—light gray in ground color, with darker blotches outlined in white* dorsally; the underparts are whitish, with smaller and fainter dark blotches. The color patterns of other adults are as follows: *E. o. williamsi*, similar, but with 4 dark longitudinal stripes; *E. o. quadrivittata*, brown and yellowish longitudinal stripes; *E. o. rossalleni*, longitudinal stripes gray and indistinct on orange background; *E. o. deckerti*, similar to *E. o. quadrivittata*, but having dark spots as well as stripes. Maximum length more than 200 cm.

DISTRIBUTION AND VARIATION: Southeastern Nebraska, southern and eastern Iowa, southwestern Wisconsin, western and southern Illinois, southern Michigan and Ontario, central New York, and southwestern New England southward to central, western, and southern Texas, the Gulf Coast, and the Florida Keys. Five races inhabit Florida: *E. o. spiloides*, Panhandle eastward to about Madison County; *E. o. williamsi*, mostly Levy and Dixie Counties; *E. o. quadrivittata*, northeastern Florida southward to northern Collier County; *E. o. rossalleni*, the Everglades; and *E. o. deckerti*, the Keys (Paulson, 1966).

331. Pine Snake, *Pituophis melanoleucus* (Daudin)

IDENTIFICATION: A large, light-colored snake with a pattern of indistinct, dark, dorsal blotches. (In some specimens the more anterior blotches are quite distinct, as they are also in the immature.) It is best recognized by the shape of the *rostral scale*, which is *almost twice as deep as wide*, but it is not upturned like that of the hog-nosed snakes. The scales are strongly keeled at all ages. Maximum length 180 to 215 cm.

DISTRIBUTION AND VARIATION: Widespread in the West; farther east one race (*P. m. sayi*, the Bull Snake) inhabits the Great Plains, much of Missouri, Iowa, Minnesota, and Wisconsin, northwestern Illinois, Oklahoma (except southeastern), Texas (except eastern), and northeastern Mexico; an isolated population (*P. m. ruthveni*) lives in western Louisiana and eastern Texas; and 3 subspecies inhabit the Southeast—southern Kentucky and western Virginia, through Georgia to the Gulf Coast of Alabama and Mississippi and to southern Florida (except Everglades and Keys); only *P. m. mugitus* occurs in Florida.

332. Prairie Kingsnake, *Lampropeltis calligaster* (Harlan)

IDENTIFICATION: A moderately large snake with *reddish brown, black-bordered dorsal blotches on a light brown to yellowish background*. There is a row of smaller dark blotches on each side of the body, and the underparts are whitish with dark blotches, sometimes faint. Old specimens may become so dark as to obscure this color pattern, or may develop faint longitudinal stripes. Although several other Florida snakes have a color pattern like one of these, this snake's combination of *smooth scales, 21 to 23 scale rows, undivided anal plate, loreal and single preocular scales* should eliminate the others. Maximum length about 90 to 110 cm. (Fig. 6.)

DISTRIBUTION AND VARIATION: Western Oklahoma, central Kansas, southeastern Nebraska, southern Iowa, northern Illinois, western Indiana and Kentucky, most of Virginia, and Maryland (west of Chesapeake Bay) southward to southern Texas, the Gulf Coast, and northwestern Florida (east to Liberty County; TT 288; *L. c. rhombomaculata*) (Map 36).

333. Common Kingsnake, *Lampropeltis getulus* L.

IDENTIFICATION: The color pattern of this snake over most of northern Florida is of *narrow, white chainlike dorsal markings on a black background*. Some indication of the same pattern may be seen over most of the state although the white markings are often wider. However, in extreme southern Florida and the Apalachicola River Valley the pattern is virtually lost, but the impression of *whitish scales outlined in black is distinctive*. The young may have tan replacing white, or may show reddish on the sides. The combination of scale characters listed for the Prairie Kingsnake also applies here. Maximum length about 150 to 180 cm.

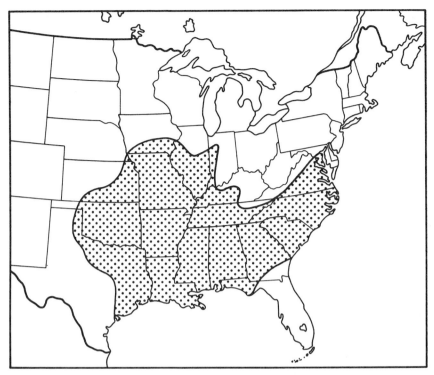

Map 36. Distribution of the Prairie Kingsnake
(*Lampropeltis calligaster*)

DISTRIBUTION AND VARIATION: Southwestern United States and northern Mexico; also from the southern parts of Nebraska, Iowa, Illinois, Indiana, and Ohio, western West Virginia, most of Virginia, Maryland, and Delaware, and southern New Jersey southward to the Gulf Coast and the southern tip of the Florida mainland. *L. g. goini* occupies the Apalachicola River Valley eastward to the Ochlockonee River (Means, personal communication); and *L. g. getulus* the remainder of northern Florida, south to Orange and Lake Counties; from there southward *L. g. floridana* is found (Duellman and Schwartz, 1958).

334. Scarlet Kingsnake, *Lampropeltis triangulum* (L.)[1]

IDENTIFICATION: The combination of *complete rings of red, black, and yellow—wide red bands being separated from yellow—along with the red snout*, should distinguish this snake from any other in Florida. Extreme care should be taken in the identification, as one of the similar

1. See Smith, Lynch, and Browne, 1965.

species is the highly venomous Coral Snake. Should the rings be in-complete ventrally, the presence of black underneath will distinguish it from the Scarlet Snake (*Cemophora*). Maximum length about 60 cm.

DISTRIBUTION AND VARIATION: Various races of "milk snakes" range widely over the United States, eastern races as far north as south-eastern Minnesota, central Wisconsin, lower Michigan, Ontario, and southeastern Maine southward into Mexico and to the Gulf Coast and southern tip of the Florida mainland (*L. t. triangulum*). (See Duellman and Schwartz, 1958.)

335. Short-tailed Snake, *Stilosoma extenuatum* Brown

IDENTIFICATION: Both in color and form this locally distributed species is distinct from all other Florida snakes. Although the body is long and slender, its *tail is proportionately the shortest of any Florida snake* (less than 10 percent of total length). The color pattern of dark blotches on a light gray background would not be unique except that the *spaces between the blotches* (middorsally) *are yellow to red in life*. Max-imum length about 60 cm.

DISTRIBUTION AND VARIATION: North-central Florida, the 3 sub-species as follows: *S. e. extenuatum*, from southeastern Putnam and northeastern Marion Counties southward to Lake, Seminole, Orange, and Polk Counties (throughout?); *S. e. arenicola*, Citrus to northern Pinellas and Sumter Counties; *S. e. multistictum*, near Gainesville (Alachua County) and Chiefland (Levy County) (Highton, 1956). How-ever, 2 specimens from Hillsborough County (Woolfenden, 1962) cast doubt on the validity of *S. e. arenicola*.

336. Scarlet Snake, *Cemophora coccinea* (Blumenbach)

IDENTIFICATION: Very similar to the Scarlet Kingsnake, but the *red, yellow, and black rings are incomplete*, the underparts being a plain whitish. From the Coral Snake it differs further in having a *red snout*. Maximum length about 75 cm.

DISTRIBUTION AND VARIATION: Eastern Oklahoma, southern Mis-souri, northern Tennessee, Indiana, North Carolina and Virginia (ex-cept in mountains), and southern Maryland, Delaware, and New Jersey southward to the Gulf Coast from eastern Louisiana (and 3 isolated localities in Texas) to south Florida (except Everglades; Duellman and Schwartz, 1958). *C. c. copei* ranges across north Florida, and *C. c. coccinea* occurs from Tampa Bay, Marion County, and Merritt Island southward (Williams and Wilson, 1967) (Map 37).

337. Crowned Snake, *Tantilla coronata* Baird and Girard[1]

IDENTIFICATION: A very small, grayish snake with a *black crown and smooth scales*. Other small snakes with a similar color pattern (for

---

1. See footnote in Key (p. 75).

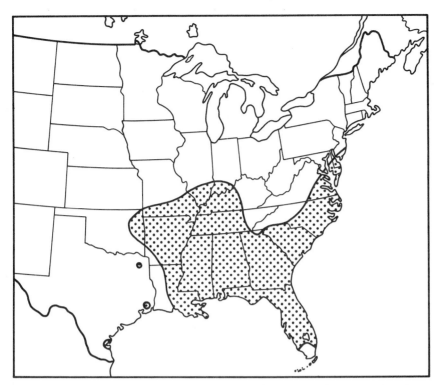

Map 37. Distribution of the Scarlet Snake (*Cemophora coccinea*)

example, *Storeria*) have keeled scales. In north Florida a light transverse band interrupts the black crown. Maximum length about 25 to 30 cm.

DISTRIBUTION AND VARIATION: East of the Mississippi River from southern Indiana, Kentucky (except northeastern), Tennessee and North Carolina (except above 600 m or north of Wilmington), southward to southeastern Louisiana, the Gulf Coast, and the upper Florida Keys. Most of this range, including all of northern Florida, is occupied by *T. c. coronata*; throughout the Peninsula southward to Lake Okeechobee and Key Largo *T. c. wagneri* is found (Duellman and Schwartz, 1958).

338. Eastern Hog-nosed Snake, *Heterodon platyrhinos* Latreille

IDENTIFICATION: The most stockily built of nonpoisonous snakes and the only Florida snakes with an *upturned snout* belong to this genus. Their actions betray their identity as well their structure does, as they expand the neck, feign striking, but may eventually feign death if handled roughly. The usual color pattern in this species is a *yellowish background with black blotches*, but melanism may be so intense as to produce a *completely black dorsal color*. The snout is less pronounced than in the Southern Hog-nosed Snake, and the *prefrontal scales are partly in contact* with each other. The under side of the tail is often lighter than the belly. Maximum length in Florida about 90 cm.

DISTRIBUTION: Southeastern South Dakota, eastern Minnesota, Great Lakes region, northeastern Ohio, northern Pennsylvania, southeastern New York, and Massachusetts southward to southern Texas, the Gulf Coast, the tip of the Florida mainland, and possibly Key Largo (Duellman and Schwartz, 1958).

339. Southern Hog-nosed Snake, *Heterodon simus* (L.)

IDENTIFICATION: Very similar to the Eastern Hog-nosed Snake (and different from all other snakes) in form and behavior—with an *especially well-developed upturned snout*. It differs from the other *Heterodon* in several respects: the *prefrontal scales are separated by several small scales*; the *under side of the tail is not lighter than the belly*; and the color pattern consists of *smaller* dark blotches on a *light gray* background. Usually smaller than *H. platyrhinos*, sometimes attaining a length of 50 cm.

DISTRIBUTION: Southeastern North Carolina, northern Alabama, and central Mississippi southward to about Lake Okeechobee, Florida (Map 38).

ORDER SQUAMATA: SUBORDER SERPENTES: LXXI. FAMILY ELAPIDAE, Coral Snakes and allies. Highly variable, but representatives in the Western Hemisphere have a pattern of red rings interspersed by rings of yellow and/or black.

340. Eastern Coral Snake, *Micrurus fulvius* (L.)

IDENTIFICATION: This dangerously poisonous, but usually inoffensive, snake can be invariably recognized by its *black snout and pattern of complete, red, yellow, and black rings, the red and yellow in contact with each other*. In 2 other snakes with similar patterns (Scarlet Snake and Scarlet Kingsnake) red and yellow are not in contact and the snout is red. Unlike the case with Florida's other poisonous snakes, the head is scarcely wider than the neck in this species. Maximum length up to 110 cm or more, but usually less than 60 cm.

DISTRIBUTION AND VARIATION: Central Texas, southern Arkansas, northern Mississippi, central Alabama, and the Coastal Plain of Georgia and the Carolinas southward into northeastern Mexico, to the Gulf Coast, and southern Florida. East of the Mississippi River, including the entire Florida Peninsula and Key Largo, the subspecies is *M. f. fulvius* (Duellman and Schwartz, 1958).

ORDER SQUAMATA: SUBORDER SERPENTES: LXXII. FAMILY CROTALIDAE, Pit Vipers. Heavy-bodied snakes with a relatively short tail, *wide, triangular head, and a pit between the eye and nostril.* The fangs are long, attached at the front of the mouth, and can be laid back against the roof of the mouth. The *subcaudals are in a single row,* at least near the base of the tail.

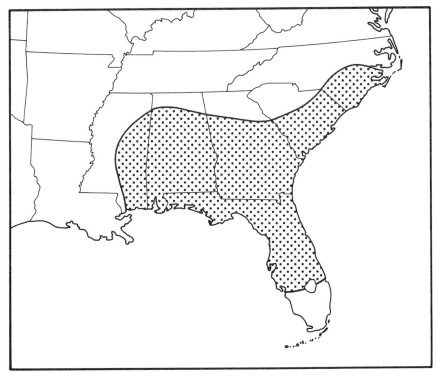

Map 38. Distribution of the Southern Hog-nosed Snake

341. Copperhead, *Agkistrodon contortrix* (L.)

IDENTIFICATION: Likely to be confused only with the Cotton-mouth, this snake differs in its *pattern of dark, hourglass markings arranged across the back* on a light grayish brown background. This pattern may be approached in young Cottonmouths, which differ in having a *broad, dark mask over the eye.* In both species, immatures have the tail tip yellow. At any age the Copperhead, unlike the Cotton-mouth, has *loreal and subocular scales.* Maximum length about 125 cm.

DISTRIBUTION AND VARIATION: Western Texas, central Okla-homa, eastern Kansas, southeastern Nebraska, northern Missouri, western and southern Illinois, southern Indiana and Ohio, most of Pennsylvania, southeastern New York, and Massachusetts southward to the Big Bend area of Texas, the Gulf Coast, and northwestern Florida. The subspecies in Florida is *A. c. contortrix,* which occurs in Liberty and Gadsden Counties (TT 683).

342. Cottonmouth, *Agkistrodon piscivorus* (Lacépède)

IDENTIFICATION: Like the Copperhead, a large, heavy-bodied pit viper without rattles and with *brown blotches on the back*, but these become obscure in old age. The *broad dark mask across the eye* usually persists, however, and the snake may be distinguished at any age by the *absence of loreal and subocular scales*. Maximum length about 170 to 185 cm.

DISTRIBUTION AND VARIATION: Eastern Oklahoma, southern Missouri and Illinois, western Kentucky and Tennessee, below 2000 feet in Alabama, Georgia, and South Carolina, coastal North Carolina, and southeastern Virginia southward to central Texas, the Gulf Coast, and the Florida Keys; the race in Florida and southern Georgia is *A. p. conanti* (Gloyd, 1969).

343. Pigmy Rattlesnake, *Sistrurus miliarius* (L.)

IDENTIFICATION: A small rattlesnake with *many small and rounded dark blotches*, dorsally and ventrally, *on a yellowish gray background*. Often there is a middorsal brown stripe. If it cannot be entirely distinguished from other rattlers by color pattern alone, the *enlarged plates in the frontal region of the head* will confirm the identification. Maximum length about 60 to 75 cm.

DISTRIBUTION AND VARIATION: Eastern Oklahoma, southern Missouri, southwestern Tennessee, north-central Alabama, northern Georgia, most of South Carolina, and southeastern North Carolina southward to Aransas Bay, Texas, the Gulf Coast, and the tip of the Florida mainland (*S. m. barbouri*).

344. Timber Rattlesnake, *Crotalus horridus* L.

IDENTIFICATION: The typical color pattern of this large rattler in the deep South is of *narrow, black, often V-shaped bands on a yellowish gray background, the bands separated on the middorsal line by a narrow reddish stripe*. The color becomes progressively darker toward the tail, losing all pattern in older specimens. In this species and the next the *scales on the frontal region of the head are not enlarged*. Maximum length about 170 to 185 cm.

DISTRIBUTION AND VARIATION: Southeastern Nebraska, southern Iowa (and up Mississippi River valley to Minnesota and Wisconsin), western and southern Illinois, southern Indiana and Ohio, most of Pennsylvania, New York (except northern), and southern Vermont and New Hampshire southward to central Texas and Aransas Bay, the Gulf Coast, and northern Florida (*C. h. atricaudatus*, the Canebrake Rattlesnake, to Alachua County); absent from Delaware, Long Island, and parts of New Jersey, Maryland, and Virginia.

345. Eastern Diamondback Rattlesnake, *Crotalus adamanteus* Beauvois

IDENTIFICATION: Although resembling the Timber Rattlesnake in its numerous small scales on the frontal region of the head, it differs in its color pattern of *dark, diamond-shaped markings, outlined by yellowish*, against a background of yellowish gray upperparts. The *2 diagonal yellowish lines on each side of the head* are also diagnostic. A very large and dangerous snake, occasionally attaining a length of 215 cm or more. (Color pattern rarely lacking.)

DISTRIBUTION: Lower Coastal Plain from North Carolina to southeastern Louisiana (east of the Mississippi River) and southward to the Florida Keys.

ORDER CROCODILIA: LXXIII. FAMILY CROCODYLIDAE, Crocodilians. Alligators and their allies are too well known to most Floridians to require description. To the uninitiated, if any, they may be separated from other reptiles by the presence of powerful, though short, limbs; the absence of a hard shell; their well-developed, differentiated teeth; a *longitudinal anal slit*; *webbed toes*; and a long, compressed, but powerful tail. Their scales are generally enlarged, quadrangular, and thickened compared with those of most reptiles.

346. American Crocodile, *Crocodylus acutus* Cuvier

IDENTIFICATION: A large crocodilian with a *narrow snout* (see Key) should be this species, but it is now rare and confined to the extreme southern part of the state. At close range the *fourth tooth of the lower jaw can be seen when the mouth is closed*. The color is a dusky gray, banded or spotted in the young, but almost patternless in adults. Maximum length in Florida about 4.5 m.

DISTRIBUTION AND VARIATION: Collier, Monroe, and Dade Counties, Florida (including the Keys), rarely northward near Gulf to Sarasota County (Le Buff, 1956); also through the Greater Antilles and from southern Mexico to northwestern South America. Original range in the United States probably more extensive, but exact limits probably unknown. The Florida race, *C. a. acutus*, inhabits fresh, brackish, and salt water.

347. American Alligator, *Alligator mississipiensis* (Daudin)

IDENTIFICATION: Far more common and widespread than any other Florida crocodilian. It may be readily distinguished from the American Crocodile by its *broader snout and hidden fourth tooth of the lower jaw when the mouth is closed*. From the smaller Spectacled Caiman it can be distinguished by the *absence of a curved ridge in front of the eyes* and by the wider dark bands in the tail. The young show much black dorsally, interspersed with yellowish, the latter color more prominent on the tail. This pattern may persist for several years, but large specimens are usually uniformly dark above. Maximum length about 4.5 to 5 m.

DISTRIBUTION: Originally from central Texas, southern Arkansas, northwestern and central Mississippi, central Alabama and Georgia, and the Coastal Plain of the Carolinas southward to the mouth of the Rio Grande and the Florida Keys; now extirpated from extensive areas northward.

348. Spectacled Caiman, *Caiman sclerops* (Schneider)

IDENTIFICATION: This is the baby "Alligator" on sale in pet shops and zoos. Although bearing a strong resemblance to the Alligator, it differs from other Florida crocodilians in its *curved, bony ridge in front of the eyes*. The *black markings on the tail are more narrow* than those of the American Alligator and are *separated by wider gray areas* rather than narrow *yellow* ones. Maximum length over 200 cm.

DISTRIBUTION: Native range southern Mexico to Brazil, but said to be established in Dade and Monroe Countries, Florida, and around the Okefenokee Swamp in Georgia.

348H. HYPOTHETICAL LIST

Two species of turtles and 2 of snakes have been found close enough to northwestern Florida to suggest their possible occurrence in that part of the state. The Black-knobbed Map Turtle (*Malaclemys nigrinoda*) occurs in rivers in south Alabama within 47 km (30 miles) of Escambia County, Florida, and the Mississippi Map Turtle (*M. kohni*) has been found at Mobile, Alabama (Conant, 1958). Also found in the same part of Alabama is the Diamond-backed Water Snake (*Natrix rhombifera*), and the Worm Snake (*Carphophis amoena*) has recently been collected within 10 km (7 miles) of the Florida line (*Herpetologica* 28:263–266).

CLASS AVES: BIRDS

ORDER GAVIIFORMES: LXXIV. FAMILY GAVIIDAE, Loons. Large swimming and diving birds with *webbed front toes* and a *spear-shaped bill*. The color pattern is largely gray to blackish above and white below. The tarsus is markedly compressed, and the outer toe is the longest one.

349. Common Loon, *Gavia immer* (Brünnich)

IDENTIFICATION: Total length 70 to 90 cm. *Culmen convex*. Breeding plumage (occasionally seen in Florida in late spring): iridescent black dorsally, with squarish white patches on wings and upper back, rounded spots on sides, and streaks on neck and sides of breast; mostly white ventrally, except for the black throat. Winter plumage: almost uniformly medium gray above (sometimes blotched) and white below; summering birds in this plumage have the head and neck whitish. As in other loons, iris reddish, bill grayish, and feet partly grayish.

DISTRIBUTION: Breeds from the Arctic Circle of North America southward to northeastern California, northwestern Montana, North Dakota, the northern parts of Indiana, Ohio, Pennsylvania, and New York, and to Connecticut, New Hampshire, Maine, and southeastern Canada (formerly to the northern parts of Iowa and Illinois); also in Iceland and possibly other parts of the Old World Arctic. Winters along Atlantic and Pacific Coasts from southern Alaska and Newfoundland southward to northwestern Mexico, Texas, the Gulf Coast, and the Florida Keys (occasionally to Cuba and in the interior northward to the Great Lakes); also winters in western Europe. Chiefly on salt water in Florida, October to late April (occasionally in summer).

350. Arctic Loon, *Gavia arctica* (L.)

IDENTIFICATION: Total length 58 to 74 cm. Culmen straight near base, convex distally. Similar to the Common Loon, but in breeding plumage differs in having a *gray nape, black throat, and white stripes on sides of neck*. Winter plumage almost identical to that of Common Loon, but *bill more slender*, and head and neck lighter than back.

DISTRIBUTION AND VARIATION: Breeds from the Arctic Circle southward to northern Europe, northern Asia, southern Alaska, and central Canada. Winters from southeastern Alaska to western Mexico; and in the Old World to the Mediterranean, Black, and Caspian Seas, and to India and Japan; accidental in southern Canada, the interior of the United States, and in New York, New Hampshire, and Florida (2 records of *G. a. pacifica*; Langridge, 1960; TT 2783).

351. Red-throated Loon, *Gavia stellata* (Pontoppidan)

IDENTIFICATION: Total length 60 to 70 cm. Bill slender and *culmen concave*. Breeding plumage (unknown in Florida?): gray above with white edgings to feathers, except for the neck, which is streaked with black and white; sides of head and neck gray; *throat reddish brown*; remainder of underparts mostly white. Winter plumage; similar to breeding plumage, but neck medium gray and throat light gray. *Upperparts lighter than in Common Loon*.

DISTRIBUTION: Breeds from the Arctic Circle southward to British Columbia, southern Mackenzie, northern Manitoba, Newfoundland, Ireland, Scotland, southern Sweden, and Russia. Winters from the Aleutian Islands and British Columbia southward along Pacific Coast to northwestern Mexico (Sonora), and on the Atlantic and Gulf Coasts from Maine to Florida; also in the Old World from Iceland, the British Isles, and the Baltic Sea to the Mediterranean, Black, and Caspian Seas, and to Turkestan, China, and Japan; casually in the Great Lakes. In migration casual in the interior of the United States. Rather rare on Florida coasts (less rare north of Daytona Beach) from November to April.

ORDER PODICIPEDIFORMES: LXXV. FAMILY PODICIPEDIDAE, Grebes. Small to medium-sized swimming and diving birds with lobed toes and usually pointed bills. Head and neck relatively slender. *Tarsus compressed*. Outer toe longest; *claws flattened*.

352. Red-necked Grebe, *Podiceps grisegena* (Boddaert)

IDENTIFICATION: Total length 46 to 56 cm. Breeding plumage (unknown in Florida?): upperparts brownish gray, except darker on crown and nape; *lower throat rufous*; upper throat and sides of head light gray; remainder of underparts white, shading to rufous on sides of breast. Winter plumage: adults as in summer, except paler above and with rufous of throat and breast replaced by gray. Immature: dark gray above, whitish below, including patch on side of head. In all plumages the *bill is yellowish toward the base and dark toward the tip*, and the iris is red. Only the Western Grebe, also very rare in Florida, is of similar size.

DISTRIBUTION AND VARIATION: Breeds in Alaska, Canada, and the northern parts of the United States, Europe, and Asia. Winters from southern Alaska southward to central California, and from Newfoundland to the Florida Keys (*P. g. holboelli*); also southward in the Old World to northern Africa, northern Iran, China, and Japan (including *P. g. grisegena*). Florida sight records range from November to March, but *there are no specimens*.

353. Horned Grebe, *Podiceps auritus* (L.)

IDENTIFICATION: Small (total length about 35 cm). Breeding plumage (frequent in spring): crest blackish, except for a broad stripe of creamy buff running through eye; throat and sides of body rufous; back dark brown; remainder of underparts white. Winter plumage: *dorsum gray*; *underparts, including patch on side of head, white*, providing more contrast than the Eared Grebe in corresponding plumage, but throat becomes gray in late winter. Iris red.

DISTRIBUTION AND VARIATION: Breeds from Alaska, northwestern and central Canada, and Nova Scotia southward to southern British Columbia, northern South Dakota, northeastern Iowa, and Wisconsin; also in northern Europe and Asia. The North American race (*P. a.*

cornutus) winters from the southern edge of its breeding range southward to southern California, New Mexico, southern Texas, and the Florida Keys. Most common on salt water, October to May (rarely earlier or later), but sometimes common on larger lakes.

354. Eared Grebe, *Podiceps nigricollis* (Hablizl)

IDENTIFICATION: Small (total length about 33 cm). Similar to the Horned Grebe, but differs in breeding plumage in its *pointed crest and black neck*. Winter birds are distinguished by their *dark throat and bill*, showing little contrast of dark and white. The bill is also more slender basally. Iris red.

DISTRIBUTION AND VARIATION: Breeds from central British Columbia, southern Manitoba, and western Minnesota southward to northern Baja California, New Mexico, and southern Texas; also from northern Europe and central Asia southward to southern Africa, Palestine, and Pakistan. The North American race (*P. n. californicus*) winters from southern British Columbia to northwestern South America (Colombia), and eastward to Iowa and Texas; casually to the Atlantic Coast, including Florida (Weston, 1960; Olson and Stimson, 1966); November to April.

355. Western Grebe, *Aechmophorus occidentalis* (Lawrence)

IDENTIFICATION: Large (total length 56 to 74 cm). Dark gray above; underparts and cheek patch white (duller in winter); *bill slender and entirely yellowish to olive buff*; feet greenish. It also differs from the Red-necked Grebe in its *longer neck and greater contrast of dark and white*. Iris red.

DISTRIBUTION: Breeds from south-central British Columbia, northwestern Alberta, south-central Saskatchewan, and south-central Manitoba southward to central California, western Nevada, northern Utah, west-central Wyoming, and northern North Dakota, locally southward and eastward. Winters coastwise from southeastern Alaska and British Columbia to Baja California and Jalisco, occasionally inland to western Nevada and Mexico (Puebla). Stray individuals have appeared in most eastern states, Florida records occurring from Pensacola to Tampa Bay and Cocoa (November to April); no specimens, but recognizable photographs examined.

356. Pied-billed Grebe, *Podilymbus podiceps* (L.)

IDENTIFICATION: Small (total length about 35 cm). Dark brown above, lighter underneath; *bill whitish and more blunt than in other grebes*. Adults in breeding season have a black throat patch and a *black band on the bill*. Iris brown. Downy young: longitudinally striped with black and white, as are juvenals[1] on the head.

DISTRIBUTION AND VARIATION: Resident locally from southern Canada to Argentina and the West Indies, except withdrawing in winter from most of Canada and the northern and central United States. The North American race (*P. p. podiceps*) occurs throughout Florida, chiefly on fresh water, commonly in winter, more irregularly in summer; not unusual on salt water in migration, also in winter in south Florida. Nests on smaller ponds and extensive marshes, especially in wet years.

ORDER PROCELLARIIFORMES: LXXVI. FAMILY DIOMEDEIDAE, Albatrosses. Members of this order are distinguished by their tubular nostrils, partitioned bill covering, and long, narrow wings. Albatrosses are among the largest of flying birds, with the wingspread sometimes exceeding 380 cm (11 feet); total lengths 70 to 130 cm; they differ from the other families in the order in having the *nostril tubes separated by the culmen*. Two species are on Florida's Hypothetical List.

ORDER PROCELLARIIFORMES: LXXVII. FAMILY PROCELLARIIDAE, Shearwaters and Petrels. Similar to albatrosses, but smaller (total length mostly under 90 cm) and with the *2 nostril tubes contiguous*. Their distribution is chiefly oceanic, thus they are seldom seen from land except during or immediately after tropical storms.

357. Cory's Shearwater, *Puffinus diomedea* (Scopoli)

IDENTIFICATION: About size of Ring-billed Gull (total length 46 to 56 cm). Plain grayish brown above, white below, gradually merging on side of head and *showing no contrast. Bill and feet yellowish*, with dusky spot near tip of bill.

DISTRIBUTION AND VARIATION: Atlantic Ocean, Mediterranean and Adriatic Seas, and the western Indian Ocean. *P. d. borealis* is the race usually listed for North America, but specimens from the Florida Keys are referred to *P. d. diomedea* (Stevenson and Baker, 1970). Formerly considered rare in Florida, May to November, but now frequent in Atlantic (Langridge, 1959; Mason, 1964a).

358. Greater Shearwater, *Puffinus gravis* (O'Reilly)

IDENTIFICATION: About size of Ring-billed Gull (total length 45 to 50 cm). Chiefly fuscous above and white below, somewhat darker on

---

1. Throughout this work the term *juvenal* is used, as either a noun or adjective, in reference to birds, and refers to a specific plumage stage—as recommended by E. Eisenmann (*Auk* 82:105).

head, where the *dark cap contrasts abruptly with the white chin*. Other differences from Cory's Shearwater are its *dark bill* and more *whitish upper tail coverts*. Feet partly whitish or pinkish.

DISTRIBUTION: Breeds in Tristan da Cunha Islands (South Atlantic Ocean), wandering northward as far as Greenland and Norway. Florida records cover all major coastal areas and all seasons of the year.

359. Sooty Shearwater, *Puffinus griseus* (Gmelin)

IDENTIFICATION: Medium-large (total length 38 to 41 cm). Entirely dark gray brown, except underside of wings light colored. The *dark underparts* distinguish it from all other shearwaters known to Florida. Bill dark gray, feet bluish gray.

DISTRIBUTION: Breeds on scattered islands in the southern Pacific Ocean (eastward to Falkland Islands, south Atlantic). Wanders over most of Atlantic and Pacific Oceans, rarely appearing in Florida waters at any season of the year.

360. Manx Shearwater, *Puffinus puffinus* (Brünnich)

IDENTIFICATION: Uniformly blackish above and whitish below, as is the very similar Audubon's Shearwater. It differs from that species, however, in its *shorter, less wedge-shaped tail* and its *larger size* (total length about 36 to 38 cm). Webs and part of foot pinkish in life.

DISTRIBUTION AND VARIATION: Breeds on scattered islands of the northern Atlantic and Pacific Oceans (near Baja California), including coast of Massachusetts. Strays to other parts of western Europe, North America, and northern South America. There is one Florida specimen of the Atlantic race, *P. p. puffinus*, at West Palm Beach (Green, 1961) and a sight record of the species off Grant (Robertson and Ogden, 1968).

361. Audubon's Shearwater, *Puffinus lherminieri* Lesson

IDENTIFICATION: Identical in general coloration to the Manx Shearwater, but with a *long wedge-shaped tail* and a *smaller size* (total length about 30 cm). Tarsus, toes, and webs partly flesh-colored.

DISTRIBUTION AND VARIATION: Breeds on scattered islands in the eastern Atlantic, the Pacific, and the Indian Oceans. *P. l. lherminieri*, of the Atlantic, regularly wanders to the coast of the eastern United States, records from the east coast of Florida ranging from March to December.

362. Black-capped Petrel, *Pterodroma hasitata* (Kuhl)

IDENTIFICATION: Near size of Sooty Shearwater (total length about 40 cm). The general color pattern—dark above, light below—suggests a small Greater Shearwater, but with *whitish on the forehead and nape*. Also dusky on sides of breast, and *feet largely flesh-colored*.

DISTRIBUTION AND VARIATION: Breeds in the West Indies (Dominica only?) and in Bermuda (*P. cahow* of the A.O.U. *Check-list*), wandering to the eastern United States, Ontario, and England. Two of the 3 Florida records refer to *P. h. hasitata* (Stoddard and Norris, 1967).

ORDER PROCELLARIIFORMES: LXXVIII. FAMILY HYDROBATIDAE, Storm-Petrels. Smaller than shearwaters, not exceeding 23 cm in total length, and with a *single, medially divided nostril tube*. The 3 Florida species have almost entirely dark plumage.

363. Leach's Storm-Petrel, *Oceanodroma leucorhoa* (Vieillot)

IDENTIFICATION: Small (total length about 20 cm). Mostly sooty brown, but with a white rump patch; bill and *feet black*; tarsus reticulate; *tail slightly forked*.

DISTRIBUTION AND VARIATION: Breeds on northern coasts of the Atlantic and Pacific Oceans, south in the United States to Massachusetts. Winters southward as far as the Equator or farther. The few Florida records fall in May, June, and September (*P. l. leucorhoa*; TT 2808).

364. Harcourt's Storm-Petrel, *Oceanodroma castro* (Harcourt)

IDENTIFICATION: The dark coloration, broken by a white rump patch, makes this petrel very similar to Leach's and Wilson's Storm-Petrels, but in Harcourt's the *white upper tail coverts are contrastingly tipped with blackish*. Unlike Leach's Storm-Petrel, it has a *square-tipped tail* and also lacks the yellowish webs of Wilson's Storm-Petrel. The tarsus is reticulate. Total length 18 or 20 cm.

DISTRIBUTION AND VARIATION: Breeds on islands in the eastern Atlantic Ocean and the eastern and central Pacific Ocean. Strays to North America, Cuba, western Africa, and England. *O. c. castro* has occurred in Florida at Key West, off Panama City, and at Pensacola, the records ranging from August to December (Hames, 1959; Hallman, 1966b; Baxter, 1970).

365. Wilson's Storm-Petrel, *Oceanites oceanicus* (Kuhl)

IDENTIFICATION: Another small, dark petrel with a white rump patch (total length about 18 cm). It differs from Leach's and Harcourt's Storm-Petrels in having *yellowish webbing* and a *booted tarsus*. Unlike Leach's Storm-Petrel, it has a *square-tipped tail*, and the white upper tail coverts are not tipped with black as in Harcourt's Storm-Petrel.

Like those 2 species, it shows some white on the flanks and under tail coverts.

DISTRIBUTION AND VARIATION: Breeds on islands in the southern Atlantic Ocean. Strays northward regularly to the Atlantic Coast of North America (*O. o. oceanicus*) and casually to California (*O. o. chilensis*). Regular offshore in Florida waters, late April to September.

ORDER PELECANIFORMES: LXXIX. FAMILY PHAETHONTIDAE, Tropicbirds. Marine birds of medium to large size with *webbing attached to all 4 toes* and usually some evidence of a *gular pouch*. The latter feature, however, is lacking in the tropicbirds, the adults of which can easily be distinguished by their *extremely long central rectrices* and chiefly white plumage.

366. White-tailed Tropicbird, *Phaethon lepturus* Daudin

IDENTIFICATION: A medium-sized white sea bird with some black dorsally and *extremely long central rectrices* in adults. Unlike the Red-billed Tropicbird (Hypothetical List), it has a *yellowish to orange bill*, and the *shafts of the middle tail feathers are dark above*. Tarsus gray. Total length 64 to 81 cm.

DISTRIBUTION AND VARIATION: Breeds on scattered islands in the Atlantic, Pacific, and Indian Oceans within 30 degrees of the Equator. Strays northward and southward. *P. l. catesbyi* rarely wanders to the Carolinas, Georgia, and Florida from March to September.

ORDER PELECANIFORMES: LXXX. FAMILY PELECANIDAE, Pelicans. Very large swimming birds with *totipalmate feet*, a long, depressed bill, and a *huge gular pouch*.

367. White Pelican, *Pelecanus erythrorhynchos* Gmelin

IDENTIFICATION: One of North America's largest birds (total length 100 to 165 cm). Adults: *white with black primaries*; bill and pouch yellow to orange; feet yellowish green. Young similar to adults, but brownish gray on top of head. Iris blue gray.

DISTRIBUTION: Breeds from central British Columbia, southern Manitoba, and southwestern Ontario southward to eastern Oregon and California, west-central Nevada, Great Salt Lake, Yellowstone National Park, southern Montana, east-central North Dakota, and central South Dakota; also on Texas coast. Winters from north-central California, central Arizona, the northern Gulf of Mexico, and Florida southward to southern Mexico and Guatemala. Locally common in Florida, especially near coast, from October to April, occasionally throughout summer.

368. Brown Pelican, *Pelecanus occidentalis* L.

IDENTIFICATION: Similar in size and form to the White Pelican (total length 110 to 140 cm), but *plumage dark except on head and neck*. Adults in breeding plumage: chiefly gray, with silver above, but head mostly white and neck seal brown, interrupted by a white streak. Bill yellowish to gray; iris yellow. Winter adults have the neck mostly white; immatures have a gray head and neck and a white belly. Newly hatched young are naked.

DISTRIBUTION AND VARIATION: Breeds on Pacific Coast from central California to southern Chile, and on the Atlantic Coast from North Carolina to central America and northern South America; on the Gulf Coast from Gulf County, Florida, eastward and southward: formerly in Texas and Louisiana; nonbreeding birds still summer as far west as Texas occasionally. Strays are known from the Bahama Islands northward to British Columbia and Long Island, and southward to Brazil. The population breeding in Florida is *P. o. carolinensis*, but *P. o. occidentalis* (from the West Indies) has appeared once at Pensacola. The species rarely strays inland.

ORDER PELECANIFORMES: LXXXI. FAMILY SULIDAE, Boobies and Gannets. Large diving birds with totipalmate feet and a reduced gular pouch. Unlike most members of this order, they have a *bill that is serrate and not hooked*.

369. Blue-faced Booby, *Sula dactylatra* Lesson

IDENTIFICATION: A large, diving bird with a totipalmate foot, an inconspicuous gular pouch, and a moderately large, unhooked bill (total length 63 to 73 cm). Adults are *white except for the black tips of the wings and tail* (much more black than on wing of Gannet). Bill greenish yellow; skin around its base dark blue; feet and iris yellowish. Immature: dark gray brown above with a *light patch on back*, white underneath (see adult Brown Booby); feet dark. Intermediates have white heads and a *white lining under the wings*.

DISTRIBUTION AND VARIATION: Breeds on islands in tropical and subtropical parts of Atlantic, Pacific, and Indian Oceans. Strays occasionally to northern Gulf of Mexico and irregularly to the Dry Tortugas (*S. d. dactylatra*).

370. Brown Booby, *Sula leucogaster* (Boddaert)

IDENTIFICATION: Size and form similar to that of the Blue-faced Booby (total length 65 to 75 cm). Adult: head, neck, and upperparts dark brown, contrasting with white of underparts; differs from *S. dactylatra* (immatures) in the *lack of a light patch on the back*; feet, bill, and skin around its base yellowish. Immature: *plumage entirely brownish gray*, lighter on abdomen; bill bluish and feet yellowish. Iris white at all ages.

DISTRIBUTION AND VARIATION: Breeds on islands in tropical and subtropical parts of Atlantic, Pacific, and Indian Oceans. Strays occasionally to northern Gulf of Mexico, the Atlantic Coast of the United States (north to Cape Cod), and regularly to the Dry Tortugas (*S. l. leucogaster*).

371. Red-footed Booby, *Sula sula* L.

IDENTIFICATION: Similar in size and form to other Florida boobies (total length 65 to 74 cm). Adults in the white phase closely resemble the larger adult Gannet, but the *black of the wing involves some of the secondaries*; tarsus and toes yellow to red. It differs from the adult Blue-faced Booby in having a *white tail*. The brown phase is chiefly gray brown, but the lower back, belly, and tail are whitish; color of soft parts variable. Immature: dark brown, lighter underneath except for a *darker breast band*; skin at base of bill bluish; tarsus and toes yellowish.

DISTRIBUTION AND VARIATION: Breeds on islands in tropical and subtropical parts of Atlantic, Pacific, and Indian Oceans. Strays northward to Cuba and, accidentally, to Louisiana (one record) and Florida (St. Petersburg and Dry Tortugas; *S. s. sula*; Woolfenden, 1965).

372. Gannet, *Morus bassanus* (L.)

IDENTIFICATION: Larger than most boobies (total length 75 to 100 cm). Adult: *white with black primaries*; head and neck tinged with straw yellow. Bill gray blue; iris pale gray. Immature: *dark gray with white spots above*; white with dark flecks below. Intermediates frequent in spring.

DISTRIBUTION: Breeds on islands in north Atlantic Ocean, southward in North America to Newfoundland (formerly Nova Scotia and New Brunswick) and in the Old World to Great Britain and northern Norway. Winters from the coast of Virginia (rarely farther north) southward to the Florida Keys and westward to the coast of Mississippi, casually to Texas; also along the coast of western Europe and northwestern Africa. Florida: November to April, rarely into early summer.

ORDER PELECANIFORMES: LXXXII. FAMILY PHALACROCORACIDAE, Cormorants. Large water birds with a totipalmate foot, blackish plumage, and a *strongly hooked bill*. Bill not depressed or serrate; gular pouch moderate.

373. Great Cormorant, *Phalacrocorax carbo* (L.)

IDENTIFICATION: Larger than *P. auritus* (total length 86 to 100 cm). Adult: mostly iridescent black, with *white patches on flanks* and under side of head. Gular pouch yellowish and black; iris blue green. Immature: essentially grayish brown, but with the *white chin and abdomen* separated by a *brownish breast*; a light area also extends upward behind eye. At all ages the *pouch is cleft medially by feathers*, and there are 14 rectrices.

DISTRIBUTION AND VARIATION: Breeds from coast of eastern Canada through Greenland and Iceland to northern Europe (*P. c. carbo*); also other races in Africa, Asia, the East Indies, Australia, Tasmania, and New Zealand. *P. c. carbo* winters from its southern breeding range southward on the Atlantic Coast of the United States to New Jersey, casually farther. Represented in Florida by 2 specimens and a few sight records, late October to May (Woolfenden and Lohrer, 1968; Stevenson, 1970b).

374. Double-crested Cormorant, *Phalacrocorax auritus* (Lesson)

IDENTIFICATION: Total length 73 to 84 cm. Adult: *chiefly iridescent black*, with an orange gular pouch (no white in plumage); iris green. Immature: dark gray brown above; whitish underneath, *but not posteriad*; gular pouch more yellowish. Newly hatched young are naked. At all ages the *gular pouch is not divided medially by feathers*, and there are 12 rectrices. At a distance a cormorant may be confused with a loon or merganser, but the *gular pouch* and *hooked bill* will distinguish it.

DISTRIBUTION AND VARIATION: Breeds from southwestern Alaska and Canada (Pacific Coast and southern portions) southward locally to Baja California, southern Mexico, northern Gulf Coast, Florida, and Cuba. Winters from the southern half of breeding range southward to British Honduras. The race breeding in Florida is *P. a. floridanus*; it is rare or absent during summer in northwestern Florida and most common on the coast at that season. *P. a. auritus* ranges south in winter throughout the mainland.

ORDER PELECANIFORMES: LXXXIII. FAMILY ANHINGIDAE, Darters. *Head and neck more slender* than in cormorants, *bill* longer and *spear-shaped, tail longer*. Plumage blackish or brownish dorsally.

375. Anhinga, *Anhinga anhinga* (L.)

IDENTIFICATION: Total length 84 to 91 cm (with a disproportionately long tail). Adult male: iridescent black, with white streaks on back and a large white patch on wing. Adult female: with less white dorsally, and with the head, neck, and breast vinaceous brown. Immature: similar to female, but more brownish on body and brownish gray on throat and breast. Iris red; gular pouch and webbing yellowish to orange; skin around eye blue in breeding adults. Young are naked when hatched. The *proportions of the head and neck, shape of the bill, and length of the tail* (see description of family) will separate it from cormorants or other species.

DISTRIBUTION AND VARIATION: Breeds from northwestern Mexico to Ecuador near Pacific Coast; also from southern Texas, western Tennessee, central Alabama, southern Georgia, and central North Carolina southward to Cuba, eastern Mexico, and Central America (*A. a. leucogaster*); another race is in South America (Palmer, 1962). Winters from the northern Gulf Coast and the coast of South Carolina southward; rarely strays northward and westward to the northern United States, southern Canada, Kansas, and Arizona. Virtually confined to fresh water except in south Florida; resident over most of Florida, but usually withdraws from northwestern corner in winter.

ORDER PELECANIFORMES: LXXXIV. FAMILY FREGATIDAE, Frigatebirds. Compared with other members of the order, wings longer, *bill long and hooked*, legs short, *tail very long and deeply forked*, and webs incised.

376. Magnificent Frigatebird, *Fregata magnificens* Mathews

IDENTIFICATION: Total length 95 to 105 cm, the *deeply forked tail* making up about 40 percent of this. Adult male: black, with iridescence above; gular pouch and feet reddish; iris blue gray. Adult female: upperparts duller; breast and sides of abdomen white, otherwise blackish below; gular pouch dark. Immature: similar to adult female, but with entire head and neck white. Newly hatched young are naked. (See description of family.)

DISTRIBUTION AND VARIATION: Breeds on islands and coastal mainland from Baja California southward to the Equator and from the Bahama Islands and eastern Mexico to southern Brazil; also off coast of west Africa (Murphy, 1967). Wanders northward to northern California, the northern Gulf Coast, and the Carolinas; often carried much farther by tropical storms, even into the interior. Common resident in southern Florida, decreasing northward, where chiefly May to September; accidental inland. Began nesting on Marquesas Keys by 1969. (Robertson and Ogden, personal communication; *F. m. rothschildi*).

ORDER CICONIIFORMES: LXXXV. FAMILY ARDEIDAE, Herons, Egrets, and Bitterns. Long-legged wading birds, with the lower tibia bare and the webbing vestigial. *Bill long, compressed, and spear-shaped* in this family; lores bare. Neck long. Plumes often prominent in breeding plumage, when much powder down present.

377. Great White Heron, *Ardea occidentalis* Audubon[1]

IDENTIFICATION: One of Florida's largest birds (total length 115 to 135 cm). Plumage entirely white; bill, tarsus, and toes greenish yellow (to orange in breeding season). Iris yellow.

DISTRIBUTION: Southern Florida, Cuba, the Isle of Pines (formerly Jamaica), and the coast of Yucatan and Quintana Roo. Wanders northward rarely to coastal Texas, Alabama, and Mississippi, and to northern Florida, the Carolinas, and Pennsylvania. Most common on the Keys and in Florida Bay.

378. Great Blue Heron, *Ardea herodias* L.[1]

IDENTIFICATION: Similar in size and form to the Great White Heron (total length 105 to 130 cm). Adult: head white with black plumes; otherwise *chiefly gray above and streaked underneath*. Bill and feet yellowish to greenish; iris yellow. Immature: similar, but darker on the head, especially the crown. "Wurdemann's Heron": similar, but head and neck mostly white (a hybrid between Great White and Great Blue Herons).

DISTRIBUTION AND VARIATION: Breeds from southeastern Alaska, northern Alberta, central Saskatchewan, southern Manitoba, central Ontario, and southern Quebec southward to southern Mexico and the West Indies (rarely); also on the Galapagos Islands; wanders farther north. Winters in southern half of breeding range southward to Panama and northern South America and on the Bermuda Islands. *A. h. wardi* breeds throughout Florida; in winter *A. h. herodias* ranges south to central Florida.

379. Green Heron, *Butorides virescens* (L.)

IDENTIFICATION: One of the smaller members of the family (total length 38 to 56 cm). Adult: *crown and back mostly dark green*; *neck chiefly maroon*, streaked underneath; underparts mostly ashy gray; *tarsus and toes orange*; iris yellow. Immature: more strongly streaked on neck and underparts and with more greenish tarsus and toes.

DISTRIBUTION AND VARIATION: Breeds from western Washington, southern Nevada, central Arizona, north-central Texas, central Minnesota and Michigan, southeastern Canada, and coastal Maine to northwestern South America and the West Indies, including all of Florida. Winters from the extreme southern United States to northern

---

1. These two herons are now regarded as a single species; see p. xv.

South America (perhaps not in interior of northwestern Florida). The eastern race is *B. v. virescens*.

380. Cattle Egret, *Bubulcus ibis* (L.)

IDENTIFICATION: Stockier than the Little Blue Heron, thus appearing smaller (total length about 50 to 64 cm). Plumage entirely white, except for *buffy plumes anteriad in breeding adults*; tarsus and toes yellow in summer adults, mostly dark in winter and in immatures; *bill yellow* (to red in breeding plumage); iris yellow.

DISTRIBUTION AND VARIATION: Breeds locally throughout southeastern United States and up Atlantic Coast to New Jersey; also southward to Bolivia and Brazil. Casually strays north to eastern Canada and Illinois and westward to southern California. Also breeds throughout most of the Old World, first becoming established in the United States at Lake Okeechobee, Florida, around 1950; now statewide in summer and virtually so in winter. Only *B. i. ibis* is known in the New World. (See Davis, 1960.)

381. Little Blue Heron, *Egretta caerulea* (L.)

IDENTIFICATION: About the size of the Snowy Egret and Louisiana Heron (total length 50 to 73 cm). Adults: *dark slaty blue, except for the maroon head and neck*, a color pattern confusible with that of a dark-phase Reddish Egret. Bill bluish with a dark tip; tarsus and toes dark greenish; iris yellow. Immature: almost wholly white, but with dark wing tips; later with scattered patches of slaty blue; tarsus and toes grayish green.

DISTRIBUTION: Breeds from central Oklahoma, southeastern Missouri, northwestern Tennessee, Alabama, and coastal Massachusetts (casually) southward through eastern Mexico and the West Indies to Peru and Uruguay, wandering after the breeding season to Ontario, Quebec, and Argentina. Winters from the southeastern United States southward.

382. Reddish Egret, *Egretta rufescens* (Gmelin)

IDENTIFICATION: Larger than Little Blue Heron (total length 68 to 81 cm). Dark phase: dark slate-colored, except for rufous head and neck; basal half of bill light (pinkish in life), distal half dark; iris white or pale yellow. Coloration lighter than in adult Little Blue Heron and *feet blackish*. White phase: plumage entirely white; *bill two-toned*. Individuals without the two-toned bill or rufous head and neck (if in dark phase) may represent either immatures or a nonbreeding plumage. White individuals may be distinguished from the smaller white herons and egrets by the *uniformly dark tarsus and toes*, and from the Great Egret by the *dark bill*; dark immatures have *flecks of rusty* on the body.

DISTRIBUTION AND VARIATION: Resident from Baja California, coastal Texas, and southern Florida to Guatemala, El Salvador, Cuba, and the Bahamas, wandering to the northern Gulf Coast; also wanders southward to Venezuela in winter. *E. r. rufescens* is uncommon northward along the coast of Florida to Merritt Island and Taylor County and rare farther north or inland; breeds north to Tampa Bay.

383. Great Egret, *Egretta alba* (L.)

IDENTIFICATION: Plumage entirely white, as in Great White Heron, but size smaller (total length 90 to 105 cm). The combination of a *yellow to orange bill and black tarsus and toes* will set it apart from other herons. Iris yellow.

DISTRIBUTION AND VARIATION: Breeds from southern Oregon and Idaho, northern Texas, central Oklahoma, southern Minnesota, western Wisconsin, central Illinois, southern Indiana, northern Ohio, and Long Island southward to southern South America, wandering northward into southern Canada after the breeding season (*E. a. egretta*; also *E. a. alba* occurs widely in the Old World).

384. Snowy Egret, *Egretta thula* (Molina)

IDENTIFICATION: An entirely white wader about the size of a Little Blue Heron (total length 50 to 68 cm), but differing from other waders in its combination of *black bill, black tarsi, and yellow to orange toes*; lores yellow to reddish; iris yellow. Plumes also more conspicuous than in Little Blue Heron. In immatures the *greenish tarsi* also contrast with the yellow toes.

DISTRIBUTION AND VARIATION: Breeds from central California, northern Nevada and Utah, and eastern Colorado southward to extreme northern Mexico (*E. t. brewsteri*); also from southern Texas and Oklahoma eastward near the coast (and up Mississippi River) to Florida, northward to Long Island, and southward to northern Chile and Argentina (*E. t. thula*). After the breeding season the species wanders to the northern United States and southern Canada. Winters from California, the Gulf Coast, Florida Peninsula, and coastal North Carolina southward throughout most of tropical America.

385. Louisiana Heron, *Egretta tricolor* (Müller)

IDENTIFICATION: About the size of the Little Blue Heron (total length 58 to 66 cm). Adults: *mostly dark slate above* (except for buff on lower back) *and white on under side* of body, wings, and tail. Bill bluish to greenish; tarsus and toes yellow to orange; iris red. Immature: similar, but with considerable rusty on head and neck.

DISTRIBUTION AND VARIATION: Breeds from central Baja California, the Gulf Coast, and New Jersey southward to Ecuador, Colombia, Venezuela, northern Brazil, the Bahamas, and the Greater Antilles. Some stray northward in summer. *E. t. ruficollis* occurs over most of Florida, but in the interior of the Panhandle is rare in summer and absent in winter.

386. Black-crowned Night Heron, *Nycticorax nycticorax* (L.)

IDENTIFICATION: About the size of the Little Blue Heron, but stockier (total length 58 to 66 cm). Adult: *glossy greenish black on crown and back*; wings light gray; underparts (including forehead and sides of head) white. Iris orange red; lores greenish in fall and winter; tarsus and toes greenish to reddish. Immature: brownish gray, with white flecks dorsally; whitish with dark streaks ventrally, progressing toward adult plumage in spring. The *legs are shorter than in the Yellow-crowned Night Heron*, and only about 25 mm of tarsus protrudes beyond the tail in flight.

DISTRIBUTION AND VARIATION: Breeds from Oregon, southern Washington, southern Idaho, southeastern Wyoming, northeastern Montana, southern Saskatchewan, southwestern Manitoba, central Minnesota and Wisconsin, southern Michigan, Ontario, and Quebec, and northeastern New Brunswick southward (locally) to the Hawaiian Islands, Peru, and Argentina (*N. n. hoactli*); also other races in Chile, Europe, Asia, and Africa. Withdraws in winter from most of northern and central United States, Texas (except near coast), and the Mexican plateau.

387. Yellow-crowned Night Heron, *Nyctanassa violacea* (L.)

IDENTIFICATION: Similar in size and form to Black-crowned Night Heron (total length 55 to 71 cm). Adults: dark slate-colored (varied with white above), except for creamy white crown and a patch of white (surrounded by black) on side of head. Darker than the Great Blue Heron and not streaked below. Tarsus and toes yellow to greenish; iris orange to reddish. Immature: similar to that of Black-crowned Night Heron, but dark markings in plumage are more gray than brown, and *legs longer* (about 50 mm of tarsus projecting beyond tail in flight).

DISTRIBUTION AND VARIATION: Breeds from Baja California, Sonora, Texas, most of Oklahoma, southern Missouri, Illinois, and Indiana, western Tennessee, central Alabama, and along coastal plain to Massachusetts, southward to Ecuador and Brazil (at least near coast) and the West Indies; also locally on oceanic islands. Wanders northward after breeding season to Colorado, Nebraska, southeastern Wisconsin, southern Ontario, New Brunswick, and Newfoundland. Winters from Baja California, Sonora, the Gulf Coast, and South Carolina southward. *N. v. violacea* occurs throughout the state in summer, but withdraws from much of northern Florida in winter.

388. Least Bittern, *Ixobrychus exilis* (Gmelin)

IDENTIFICATION: Much smaller than other herons (total length 28 to 36 cm). Male: essentially glossy greenish black above, rufous brown on neck and wings, also some buff on wings; underparts whitish or light buff, lightly streaked in immature. Bill yellowish, except dark dorsally and distally; lores, iris, and feet yellow to greenish. Adult female: similar to male, except dark brown above. Immature female: rufous with buffy markings on back. A darker phase is seen occasionally. Hatchling: covered with buffy down, lighter below.

DISTRIBUTION AND VARIATION: Breeds from Oregon, Montana (casually), southern Manitoba, eastern North and South Dakota, Minnesota, northern Michigan, southern Ontario, New York, coastal Maine, and southern New Brunswick southward to Paraguay and Brazil, wandering occasionally to southern Canada. Winters chiefly from southern California, southern Texas, and central Florida southward. *I. e. exilis* breeds locally throughout the state, but most individuals retire from the northern parts in winter.

389. American Bittern, *Botaurus lentiginosus* (Rackett)

IDENTIFICATION: Similar in size and form to the night herons, averaging slightly larger (total length 60 to 85 cm). *Mostly variegated brown and buffy dorsally*, but underparts light buff, the throat streaked with brown; a black (or dull gray) patch on side of head and neck; chin white. Bill and iris yellowish; feet greenish.

DISTRIBUTION: Breeds from the middle latitudes of Canada southward to southern California, central Arizona, southern Colorado, central Kansas, central Missouri, western Tennessee, southwestern Ohio, Pennsylvania, and eastern Maryland and Virginia (casually to northern Texas, Louisiana, and Florida). Winters from southwestern British Columbia, Utah, Arizona, central New Mexico and Oklahoma, northern Arkansas, southern Indiana and Ohio, and Delaware southward to Panama, Grand Cayman, and Puerto Rico. Casual records far outside range in summer, winter, and migration.

ORDER CICONIIFORMES: LXXXVI. FAMILY CICONIIDAE, Storks. Large wading birds with (usually) black and white plumage and much bare skin on head; bill long and heavy, not so compressed as in Ardeidae; middle claw not pectinate.

390. Wood Stork, *Mycteria americana* L.

IDENTIFICATION: About the size of the Great Blue Heron (total length 90 to 120 cm). Adult: head and neck blackish (bare); plumage white, except for the black remiges and rectrices. Immature: similar, but head and neck partly covered with small, grayish feathers. *Bill heavy, decurved*, light-colored in immatures; toes partly pinkish in life, more intensely pink in breeding season.

DISTRIBUTION: Breeds widely in South America, up both coasts of Central America and Mexico, in Cuba, Hispaniola, and peninsular Florida (possibly west to Jackson County). Nonbreeding birds wander northward, especially in summer and fall, to most parts of the United States, at least rarely. Uncommon to rare in Florida Panhandle.

ORDER CICONIIFORMES: LXXXVII. FAMILY THRESKIORNITHIDAE, Ibises and Spoonbills. Size and form as in Ardeidae, but *bill either decurved* (ibises) *or spatulate* (spoonbills); middle claw not pectinate; bare skin confined to face in ibises.

391. Glossy Ibis, *Plegadis falcinellus* (L.)

IDENTIFICATION: About size of Little Blue Heron (total length 56 to 64 cm), but *bill strongly decurved*. Adult: *head, neck, back, and underparts maroon*; plumage otherwise dark, with greenish reflections. Bill, tarsus, and toes dull; *bare skin around eye a bright, chalky blue* in breeding season, but *no white feathering around eye*. Immature: plumage duller, with white streaks on head and neck.

DISTRIBUTION: Breeds in eastern Texas (casually), in Louisiana and Arkansas, and from Long Island southward to the Greater Antilles; winters from the Florida Peninsula southward. Wanders northward to the northern United States and southern Canada. Also ranges over much of Europe, southern Asia, the East Indies, Australia, New Zealand, Africa, and Madagascar.

392. White-faced Ibis, *Plegadis chihi* (Vieillot)

IDENTIFICATION: Strongly similar to the Glossy Ibis in size, form, and coloration (total length 50 to 64 cm), adults differing in the *white feathering surrounding the eye* (that is, above, below, and behind); bill, tarsus, and toes reddish. Immatures are probably not separable in the two forms, sometimes regarded as a single species.

DISTRIBUTION: Formerly nested from southern Oregon and Idaho, western Colorado, central Kansas, eastern Nebraska, and southern

Minnesota southward to southern Mexico, though present breeding range much reduced; also in South America (southern Peru and Brazil to central Argentina); one old Florida breeding record. Winters chiefly in Mexico; wanders to Florida, northwestern United States, Baja California, and Pacific islands off Mexico.

393. White Ibis, *Eudocimus albus* (L.)

IDENTIFICATION: Size and form like that of the Glossy Ibis (total length 56 to 78 cm). Adult: *white with black wing tips*. Immature: brownish with white underparts, lower back, and rump; head and neck streaked. Many are in intermediate plumage in summer. At all ages the bill, lores, tarsus, and toes tend to be pinkish; *iris grayish blue*.

DISTRIBUTION: Breeds in the Coastal Plain from North Carolina to Texas and through eastern Mexico and the western Antilles to Venezuela; also along Pacific Coast from northwestern Mexico southward to northwestern Peru; strays northward occasionally to Colorado, Nebraska, Missouri, Vermont, and Quebec. Withdraws from northernmost parts of breeding range in winter.

394. Scarlet Ibis, *Eudocimus ruber* (L.)

IDENTIFICATION: Identical in form to the White Ibis, but slightly larger (total length 71 to 76 cm). Adult: *scarlet* with black wing tips. Immature: similar to that of White Ibis, but *bill entirely dark and head and neck not streaked*; *iris dark* (Palmer, 1962). Hybrids between this species and the White Ibis are *pink*.

DISTRIBUTION: Northern South America, from Venezuela to Brazil, rarely straggling to the southeastern United States (but feral origin of some open to question); also introduced and breeding at Greynolds Park (Miami) and wandering over south Florida (Bundy, 1962 and 1965; Map 39).

395. Roseate Spoonbill, *Ajaia ajaja* (L.)

IDENTIFICATION: Somewhat larger than the Little Blue Heron (total length 70 to 90 cm). Adult: bill and bare head greenish; neck and back white, sometimes tinged with pinkish; tail and sides of breast buffy; otherwise *mostly pink*; tarsus and toes pinkish; iris reddish. Immature: similar, but paler (sometimes entirely white), buff of adults replaced by pinkish; head and neck feathered, and tarsus and toes yellowish to dark. At all ages the *roughened, spatulate bill* should render it unmistakable.

DISTRIBUTION: Breeds from western Mexico, coastal Texas, southwestern Louisiana, and southern Florida southward to most of South America, withdrawing from Louisiana, Texas, and northern Mex-

Map 39. Distribution of the Scarlet Ibis (*Eudocimus ruber*)

ico in winter; rarely wanders northward in summer to northern Baja California, southern California, Utah, Colorado, Nebraska, Kansas, Indiana, Pennsylvania, South Carolina and Northern Florida (to Duval and Wakulla Counties).

ORDER PHOENICOPTERIFORMES: LXXXVIII. FAMILY PHOENI-COPTERIDAE, Flamingos. Long-legged wading birds having some characteristics of the Ciconiiformes and some of the Anseriformes. Toes webbed. Bill proportionately short and thick, *abruptly decurved at midpoint*, and the mandibles lamellate. Tongue thick and fleshy. Coloration white to red, with blackish remiges.

396. American Flamingo, *Phoenicopterus ruber* L.

IDENTIFICATION: (See characteristics of the order.) Total length 105 to 125 cm. The *heavy, bent bill*, extremely long neck and legs, and *reddish plumage with black remiges* combine to make this bird unmistakable to a careful observer. Bill black distally, yellow to orange basally; tarsus and toes pinkish; iris yellow. Immatures differ in being paler underneath and brownish gray above.

DISTRIBUTION AND VARIATION: Resident in Yucatan, the Bahamas, the Greater Antilles, and on islands off Venezuela, wandering to Oklahoma, the Southeastern States, the Lesser Antilles, and northern South America; the few that have taken up residence in Florida Bay, along with most other recent Florida records, probably are of captive origin (*P. r. ruber*); accidentally wanders to northwestern Florida (Gaither and Gaither, 1966; Hallman, 1962). Another race breeds in Africa, southern Europe, and southern Asia.

ORDER ANSERIFORMES: LXXXIX. FAMILY ANATIDAE, Swans, Geese, and Ducks. Moderate to large waterfowl with webbed front toes, a depressed, *membranous-covered bill provided with a nail*, and, in most species, with lamellate tomia. Legs laterally attached and short; neck long or moderate.

397. Whistling Swan, *Olor columbianus* (Ord)

IDENTIFICATION: The largest of Florida waterfowl (total length 120 to 150 cm). Adult: plumage *entirely white*; bill, tarsus, and toes black, except yellow spot at base of bill. Immature: plumage ashy gray, with brownish cast on head and neck; bill pinkish. The Snow Goose, also a large white waterfowl, has black remiges.

DISTRIBUTION: Breeds in Alaska and northern Canada. Winters on the Pacific Coast from British Columbia to southern California, and on the Atlantic Coast from Maryland (rarely farther north) to North Carolina; also on the Gulf Coast, at least locally, from Texas to northwestern Florida; occasionally inland in winter; outlying records from central Florida (Nicholson, 1955), southern Florida, Mexico, Bermuda, Cuba, Puerto Rico, and Newfoundland.

398. Canada Goose, *Branta canadensis* (L.)

IDENTIFICATION: Larger than most waterfowl (total length 64 to 100 cm). *Chin, upper throat, and side of head white*; *remainder of head and neck black*; upperparts otherwise brownish gray; underparts somewhat lighter.

DISTRIBUTION AND VARIATION: Breeds from Alaska, northern Canada, and western Greenland southward to northeastern California, Utah, Kansas, southern Canada, and Massachusetts; descendants of captive birds nest locally in northern Florida (*B. c. maxima*). Winters from southern Canada and the Great Lakes southward to northern Mexico and the Gulf States, in Florida south to Lake Okeechobee; also in Japan. Only *B. c. interior* is frequently seen in Florida, but there is a specimen of the much smaller *hutchinsi* from Wakulla County, Dec. 24, 1956 (TT 501).

399. Brant, *Branta bernicla* (L.)

IDENTIFICATION: Color pattern roughly similar to that of *B. canadensis*, but size smaller (total length 55 to 70 cm). It differs also in having the *white patch* (streaked with black) *on the side of the neck* and the *black extending onto the breast*.

DISTRIBUTION AND VARIATION: Breeds in Arctic regions of the world, wintering southward to Europe, Egypt, southern Asia, and in the United States (*B. b. hrota*) on the Atlantic Coast from Massachusetts to North Carolina, rarely to Florida (occasionally to the Gulf Coast and inland); also on the Pacific Coast from British Columbia to California.

400. White-fronted Goose, *Anser albifrons* (Scopoli)

IDENTIFICATION: Total length 66 to 76 cm. Adult: plumage brownish gray, except for *white area around bill* and *black splotches on underside*; bill pink; tarsus and toes yellow. Immature: similar, but without the white forehead or black markings ventrally; differs from immature Blue Goose in having the *bill, tarsus, and toes yellowish*. At all ages there is a *white stripe on the body at the edge of the wing*.

DISTRIBUTION AND VARIATION: Breeding range nearly circumpolar. Winters southward to northern Africa, southern Asia, and from southern British Columbia, Montana, the Dakotas, and Illinois southward to southern Mexico and the Gulf States. The few sight records (and 2 photographs) in northern and central Florida (mostly adults) probably pertain to *A. a. gambelli*, but *no specimens have been taken*.

401. Snow Goose, *Chen hyperborea* (Pallas)[1]

IDENTIFICATION: Total length 60 to 75 cm. Adults: *plumage entirely white, but for black primaries*; bill, tarsus, and toes pink. Immature: similar, but washed with pale gray, especially dorsally. (Considered by some only a color phase of the Blue Goose.)

DISTRIBUTION AND VARIATION: Arctic Coast from northeastern Siberia to northern Greenland in breeding season. Winters south to Japan, northern Baja California, the Gulf Coast, and on the Atlantic Coast from New Jersey to North Carolina, rarely through Florida to the West Indies. Florida specimens represent *C. h. hyperborea*.

402. Blue Goose, *Chen caerulescens* (L.)[1]

IDENTIFICATION: Size of Snow Goose, but plumage chiefly dark. Adult: *body bluish gray*, with white feather edgings on upperparts; *abdomen, head, and neck white* (sometimes stained with rust); bill, tarsus, and toes pink. Immature: similar, but head and neck dark. Unlike the similar immature White-fronted Goose, it has the *bill and feet dark*.

DISTRIBUTION: Breeds chiefly on large islands of northeastern Canada. Winters chiefly along Gulf Coast from northwestern Florida to Veracruz; also in peninsular Florida and on the Atlantic Coast from Maine to Georgia in small numbers; regular and sometimes common as a fall migrant in Florida Panhandle, October-December.

403. Fulvous Tree Duck, *Dendrocygna bicolor* (Vieillot)

IDENTIFICATION: A medium-sized duck with a disproportionately long neck (total length 45 to 54 cm). Upperparts mostly brownish black, but with a *white bar across the base of the dark tail*; head, neck, and underparts rich buffy or reddish brown (paler in immature); a white stripe on side of body (or hidden by wings). Bill, tarsus, and toes bluish gray. Downy young: upperparts gray with white patches on back and wings; superciliary stripe white and line through eye gray; underparts grayish white.

DISTRIBUTION AND VARIATION: Breeds from central California, southeastern Texas, and southern Arkansas southward to southern Mexico; also in two areas in northern and eastern South America, Cuba, eastern Africa, Madagascar, India, and Ceylon. *D. b. helva* strays eastward to the Atlantic Coast and peninsular Florida in irregular numbers and has nested locally in south Florida, September to May, rarely summering.

---

1. These geese are now considered conspecific.

404. Mallard, *Anas platyrhynchos* L.

IDENTIFICATION: One of the larger species of ducks in Florida (total length 50 to 65 cm). Adult male: head and neck dark glossy green, separated from color of body by a conspicuous white ring; breast chestnut; remainder of body, wings, and tail chiefly gray to brown; *speculum purplish blue, bordered with white*; bill yellow; tarsus and toes orange. Female and immature male in fall: chiefly dark brown above and lighter brown below (feathers dark with light edgings); bill orange and dark brown in female, yellow in male; *speculum*, tarsus, and toes *as in adult male*. Downy young: very similar to downy Mottled Duck, but upperparts darker.

DISTRIBUTION AND VARIATION: Breeds throughout Europe and Asia (except southern parts), and northern and central North America, moving southward in winter to Africa, southern Asia, Borneo, southern Mexico, and, rarely, Panama and the West Indies. *A. p. platyrhynchos* breeds in North America from Alaska and northwestern to southeastern Canada southward to southern Texas, Illinois, northern Alabama, and Virginia, and casually farther south. (Escapees, or their descendants, have nested in north Florida.) Winters from southern Canada southward, including most of Florida, but rare or absent in extreme southern parts (October-June).

405. Black Duck, *Anas rubripes* Brewster

IDENTIFICATION: Size of Mallard (total length 53 to 64 cm). Dark brown (fuscous) above, with *lighter color confined to edges of the back feathers*; sides of head and throat lighter, but *with dark streaks, even on chin*; speculum purple, bordered with black only. (No great change of plumage with age or sex.) Bill yellow to greenish; tarsus and toes red to yellowish olive.

DISTRIBUTION: Breeds from the northern parts of Manitoba, Ontario, and Quebec, and from Labrador and Newfoundland southward to North Dakota, northern Minnesota, Wisconsin, northern Illinois, Indiana, Ohio, West Virginia, and eastern Virginia (rarely North Carolina coast). Winters from northern Indiana, southern Michigan, southern Ontario, New York, and along the coast from New Brunswick to Connecticut, southward to southern Texas, the Gulf Coast, and Florida, except very rare in southern parts (October-May).

406. Mottled Duck, *Anas fulvigula* Ridgway

IDENTIFICATION: Averaging slightly smaller than the Black Duck (total length 50 to 56 cm) and very similar in appearance; *head and neck more buffy, chin unstreaked, secondaries usually not tipped with white, and with longitudinal buffy bands in feathers of the back.* Downy young: upperparts brownish, with yellowish buff on back edge of wing and as 2 spots on middle and lower back; dark streak through eye and a smaller malar streak; underparts and feet yellowish buff to straw-colored; bill bluish gray.

DISTRIBUTION AND VARIATION: Resident along the Gulf Coast from southern Texas to Dauphin Island, Alabama; accidental in south Florida and, doubtfully, near Pensacola (*A.f. maculosa*; Weston, 1965); also (*A.f. fulvigula*) resident in the Florida Peninsula from the Chassahowitzka River and Alachua and Seminole Counties southward to Florida Bay. Recently to Franklin County (TT 2792), but most sight records outside stated range are not satisfactory.[1]

407. Gadwall, *Anas strepera* L.

IDENTIFICATION: Smaller than the Mallard, Black Duck, and Mottled Duck (total length 45 to 58 cm). Adult male: mostly grayish brown, with a white belly and *black lower back and tail coverts; speculum white, partly surrounded by black*; chestnut on bend of wing; bill dark; tarsus and toes yellowish. Female and immature male: similar but browner, lacking the black around tail; *bill greenish yellow*.

DISTRIBUTION: Breeds from southern Alaska, southern British Columbia, north-central Alberta, central Manitoba, and Quebec southward to southwestern California, Nevada, northern Arizona, northwestern New Mexico, northern Texas, southwestern Kansas, northern Iowa, central Minnesota, southern Wisconsin, northwestern Pennsylvania, New Jersey, Delaware, Maryland, and northeastern North Carolina; also in Iceland, the British Isles, and continental Eurasia. Winters from southern Alaska, southern British Columbia, Washington, northeastern Colorado, northern Arkansas, southern Illinois, West Virginia, Long Island, and eastern Maryland southward to southern Baja California, southern Mexico, the Gulf Coast, and southern Florida; also into Africa and southern Asia. Florida: October to May, occasionally in summer.

408. Pintail, *Anas acuta* L.

IDENTIFICATION: About size of Mallard (total length 53 to 76 cm). Adult male: *head and neck brown, except for an extension of the white from the underparts up each side of neck*; body and wings otherwise chiefly vermiculated gray, but with white edgings on the longer wing coverts; tail mostly black and the *central rectrices greatly elongated*. Female and immature male: chiefly brown, streaked with dark on head and neck, and mottled on wings and body; underparts whitish (streaked in immature). *Bill, tarsus, and toes bluish gray* in both sexes.

DISTRIBUTION: Breeds from northern Alaska and Canada south-

---

1. More recent breeding near Milton probably involves *A.f. maculosa*.

ward to southern California, northwestern Nevada, northern Arizona, southern Colorado, central Nebraska, Iowa, and Illinois, southern Michigan, northern Ohio, and northwestern Pennsylvania; also in western Greenland, in Iceland, and most of Eurasia. Winters from southeastern Alaska, northern California, southern Nevada, Colorado, Oklahoma, Iowa, southern Illinois, southern Ohio, Massachusetts, Long Island, and Chesapeake Bay (rarely farther north) southward to southern Mexico, Central America, Colombia, the Greater Antilles (rarely Lesser Antilles), and the Bahamas; also from Europe (except northeastern) and central Asia southward to Africa, southern Asia, Borneo, the Philippines, and the Hawaiian Islands. Florida: September to April.

409. Bahama Duck, *Anas bahamensis* L.

IDENTIFICATION: Small to medium-sized (total length 38 to 48 cm). General coloration gray brown, but with *white cheeks and throat*; tail fawn-colored, not especially elongate even in male. Sexes alike.

DISTRIBUTION AND VARIATION: Greater Antilles (except Jamaica), northern Lesser Antilles, Bahama Islands, and South America (Guianas to eastern Argentina and Peru to Bolivia and Chile); also the Galapagos Islands; *A. b. bahamensis* is accidental in the Florida Peninsula, Virginia, and Wisconsin.

410. Green-winged Teal, *Anas crecca* L.

IDENTIFICATION: One of the smallest of Florida ducks (total length 33 to 38 cm). Male: *head chestnut with a lateral green stripe*, terminating in black at nape; body vermiculated gray to brownish, lighter underneath; *speculum dark green*; feet blue gray. Female: browner and without rich head color of male; bill pinkish, spotted with black.

DISTRIBUTION AND VARIATION: Breeds in Europe, Asia (except southern), Alaska, Canada, and the northern United States, moving southward in winter. *A. c. carolinensis* breeds from north-central Alaska and northern and southeastern Canada southward to southern California, eastern Arizona, northern New Mexico, northern Nebraska, southern Minnesota, northern Ohio, northwestern Pennsylvania, western New York, and Maine (casually farther south); and winters from southern British Columbia, central Montana, northern Nebraska, Iowa, Wisconsin, southern Illinois, Ohio, northwestern Pennsylvania, Massachusetts, New Brunswick, and Nova Scotia southward to Honduras, the northern Gulf Coast, southern Florida, the Bahamas, and rarely into the Greater Antilles. Florida: October-April.

411. Blue-winged Teal, *Anas discors* L.

IDENTIFICATION: Slightly larger than the Green-winged Teal (total length 35 to 41 cm). Male in breeding plumage: head dark gray to black, with a *white crescent in front of eye*; back blackish with white feather edgings; underparts brown with spots and bars of black; a large chalky blue patch and a smaller green patch *on wing*; bill dark; tarsus and toes yellowish. Female and fall male: similar, but chiefly gray brown on body and head and without the white crescent; less blue on wing; underparts light. Bill relatively large.

DISTRIBUTION AND VARIATION: Breeds from British Columbia, northern Saskatchewan, southern Quebec, and Nova Scotia southward to southern California, southern New Mexico, central Texas, Louisiana, Tennessee, and North Carolina (casually south to Florida). Winters from southern California, Texas, and North Carolina (casually farther north) southward to Ecuador and Brazil. Season in Florida August to June, occasionally throughout summer. *A. d. discors* and *A. d. orphna* both occur in the state.

412. Cinnamon Teal, *Anas cyanoptera* Vieillot

IDENTIFICATION: About the size of the Blue-winged Teal (total length 38 to 43 cm). Male: *head, neck, and most of body dark cinnamon*; color of wings, tail (including coverts), bill, tarsus, and toes as in Blue-winged Teal. Female: probably not distinguishable by plumage from female Blue-winged Teal.

DISTRIBUTION AND VARIATION: Breeds from southwestern Canada and Wyoming southward to California, west-central Texas, and central Mexico (casually in Central America); also in Colombia and from southern Peru, Chile, and southeastern Brazil to the southern tip of South America. The North American race (*A. c. septentrionalium*) has strayed eastward in winter to New York, North and South Carolina, Georgia, Florida (October-April), and Cuba. (See Owre, 1962.)

413. European Wigeon, *Anas penelope* (L.)

IDENTIFICATION: Medium-sized (total length 43 to 51 cm). Adult male: *crown creamy buff, rest of head and neck cinnamon red*, shading to chestnut on breast; upperparts and flanks chiefly vermiculated gray; remainder of underparts white; large white patches on wings; tail and under coverts black; bill blue, with black tip; feet grayish blue. Female and immature male in fall: head and neck gray, tinged with cinnamon and streaked with dusky; upperparts brown, with lighter feather edgings and a large green patch (but little white) on wing; sides of body reddish brown; remainder of underparts white; distinguished only with difficulty (not at all in the field?) from female of the common American Wigeon.

DISTRIBUTION: Breeds in the northern parts of Europe and Asia, ranging southward in winter to Africa, southern Asia, and the Philippines; also rarely winters near the Atlantic Coast of North America from Massachusetts to Florida (November-May) and near the Pacific Coast in California to northern Baja California; recorded in fall farther north and in spring in the interior. Suspected of breeding in Canada.

414. American Wigeon, *Anas americana* (Gmelin)

IDENTIFICATION: About size of European Wigeon (total length 45 to 55 cm). Adult male: pattern as in the European Wigeon, but colors differing; *sides of head green and crown white to creamy* (not buffy); flanks more brownish; face and neck gray-streaked; bill and feet bluish. Female and immature: very similar to corresponding plumage of European Wigeon, but head and neck without a trace of cinnamon. Hybridizes with *A. penelope*.

DISTRIBUTION: Breeds from western and central Alaska, Yukon and Mackenzie River basins, Great Slave Lake, southern Manitoba (rarely to Churchill), western Minnesota, and Wisconsin southward to northeastern California and the northern parts of Nevada, Arizona, Colorado, and Nebraska; formerly or rarely eastward to southern Ontario and northwestern Pennsylvania. Winters from southern Alaska, southern British Columbia, Oregon, southern Nevada, southwestern Utah, northeastern Colorado, southern Illinois, Wisconsin, Ontario, and Maryland (rarely New England) southward to Central America and the West Indies. Florida: September to May, occasionally into summer.

415. Northern Shoveler, *Anas clypeata* (L.)

IDENTIFICATION: About size of American Wigeon (total length 43 to 53 cm). Adult male: *head and upper neck metallic green; lower neck, breast, and part of wing white*; rest of wing dusky distally, blue (anteriad), and green (posteriad) near body; remainder of upperparts chiefly dark, with white feather edgings; *remainder of underparts chestnut brown*; bill dark; tarsus and toes orange. Female and immature male: mostly brownish, with streaks of dusky on head and neck and lighter feather edgings on body; wing pattern similar to that of male, but with less white; bill greenish. Adult males in fall are usually in eclipse plumage—similar to that of female, but darker and with more reddish brown below. In any plumage the *monstrous bill* will separate it from other ducks.

DISTRIBUTION: Breeds from western Alaska, Great Slave Lake, central Alberta and Saskatchewan, west side of Hudson Bay, southern Ontario, northwestern Pennsylvania, and Delaware (occasionally farther north) southward to southern California, central Arizona (formerly), south-central New Mexico, Kansas, Nebraska, western Iowa, northern Alabama, and northeastern North Carolina; also over most of Europe and Asia. Winters from southwestern British Columbia, southeastern

Washington, central Arizona, southwestern New Mexico, east-central Texas, northern Alabama, and South Carolina (casually farther north) southward to Central America and the West Indies; also southward to Africa and southern Asia. Florida: September-early June.

416. Wood Duck, *Aix sponsa* (L.)

IDENTIFICATION: Medium-sized (total length 43 to 51 cm). The gaudily colored male in breeding plumage is unmistakable among North American ducks, with *green, violet, buff, black, and white in a patch-work arrangement*, but is similar to the female during the eclipse plumage of late summer; feet yellowish; iris red; bill (except nail) reddish. Female and immature male in fall: *head crested* (as in adult male) and lead gray, with the *eye-ring*, chin, and throat *white*; upperparts (including breast and sides) chiefly brown; underparts white. Downy young: upperparts dark grayish, darker on head, with white patch on elbow; dark streak through eye; underparts yellowish; bill dark with light tip; feet yellow.

DISTRIBUTION: Breeds from southwestern British Columbia, Washington, northern Idaho, and northwestern Montana southward to central California; also from southern Manitoba, northern Minnesota, and southeastern Canada southward to southern Texas, the Gulf Coast, southern Florida (not Keys), and Cuba, and westward to the eastern parts of the Dakotas, Nebraska, Kansas, Oklahoma, and Texas; casual in Rocky Mountain states. Winters in all but northern parts of breeding range and casually to Jamaica and Bermuda.

417. Redhead, *Aythya americana* (Eyton)

IDENTIFICATION: About size of American Wigeon (total length 43 to 56 cm). Male: *head and most of neck chestnut red*; lower neck and breast black; remainder of body mostly gray, but lighter ventrally; feet gray blue; iris yellowish. Female and immature male in fall: plain brownish above and white underneath, the *area around the bill lighter*. In both sexes the bill is gray blue, this color separated from the black tip by an indistinct light ring, and the *wing stripe appears gray* in flight.

DISTRIBUTION: Breeds from central British Columbia, northern Alberta, Mackenzie, southern Saskatchewan, southern Manitoba, and northwestern Minnesota southward to southern California, central Nevada, central Arizona, northwestern New Mexico, southern Colorado, western Nebraska, northern Iowa, southern Wisconsin, and northwestern Pennsylvania. Winters from southern British Columbia, Utah, Nevada, northeastern Colorado, northern Arkansas, southern Illinois, and eastern Maryland southward to Baja California, central Mexico, the northern Gulf Coast, and (rarely) to southern Florida (October-May, occasionally later).

418. Ring-necked Duck, *Aythya collaris* (Donovan)

IDENTIFICATION: Medium-sized (total length 38 to 41 cm). Male: head (slightly crested) and neck purplish, the latter with a maroon ring at its base; breast and upperparts black; sides and underparts whitish, *conspicuously white on side of breast*; feet gray green; iris yellowish. Female: brownish gray above, including breast; underparts, area in front of eye, and eye-ring whitish; feet light gray. Both sexes have a *narrow white ring near the tip of the bluish bill*, and the *wing stripe appears gray* in flight.

DISTRIBUTION: Breeds from southern Alaska, Mackenzie, and the remainder of southern Canada southward to eastern Oregon, eastern California, Arizona, central Colorado, northern Nebraska, northern Iowa, northwestern Pennsylvania, and Maine. Winters from southern British Columbia, Nevada, New Mexico, northern Texas, northeastern Arkansas, the Ohio Valley, southern Pennsylvania, and Massachusetts southward to the Canal Zone, the northern Gulf Coast, the Bahamas, the Greater Antilles, and Bermuda. Florida: October to April, but occasional in summer.

419. Canvasback, *Aythya valisineria* (Wilson)

IDENTIFICATION: One of Florida's larger ducks (total length 50 to 60 cm). Male: head and neck reddish brown; breast blackish; body white, except posterior end and tail black. Female: similar, but with the head and neck duller brown and the body grayish. Although the color of the male easily distinguishes it from all other ducks except the Redhead, the species is best identified by the *continuous, gently sloping profile of the forehead and long bill*. Feet gray blue; iris reddish.

DISTRIBUTION: Breeds from central Alaska, northern Mackenzie, Great Slave Lake, and southeastern Manitoba southward to southern Oregon, northern California, western Nevada, northern Utah (casually central), northern Colorado, western Nebraska, and northern Minnesota. Winters from southern British Columbia, northwestern Montana, northern Colorado, Iowa, Illinois, Indiana, Lakes Erie and Ontario, and eastern Massachusetts (rarely Quebec) southward to southern Mexico, the Gulf Coast, and southern Florida (rarely farther); November-April, rarely into summer.

420. Greater Scaup, *Aythya marila* (L.)

IDENTIFICATION: About size of Ring-necked Duck (total length 40 to 46 cm) and of similar color pattern. Male: head, neck, and breast black, the *head showing entirely greenish reflections in strong light*; body light gray above, blackish at posterior end, and white underneath. Female: upperparts brownish except for a *prominent white area around base of bill*. In both sexes the *white wing stripe* (as seen in flight) *extends onto the innermost primaries*. Feet gray blue; iris yellowish.

DISTRIBUTION AND VARIATION: Breeds in Alaska, Canada, and the northern parts of Europe and Asia, moving southward in winter. *A. m. nearctica* breeds southward to southern Canada, North Dakota, and southeastern Michigan (rarely); occasionally summers south to Florida (Stevenson, 1965). Winters, chiefly coastwise, from southeastern Alaska to southern California, and from Quebec to Florida, western Cuba, and Texas (casually the Bahamas). Florida: November-June.

421. Lesser Scaup, *Aythya affinis* (Eyton)

IDENTIFICATION: Slightly smaller than the Greater Scaup (total length 38 to 46 cm) and of almost identical color. The male's head in strong light, however, *reflects both purple and green*, and in both sexes the *white wing stripe* of the secondaries *changes to gray on the innermost primaries*. (Also see bill characters in Key.) Summering males of both species resemble females, and both lack the white at the base of the bill at that season. Downy young: upperparts dark brown with 2 pairs of light spots on back; sides of head light brown; underparts mostly buffy yellow.

DISTRIBUTION: Breeds from central Alaska and northern Canada southward to central British Columbia, northern Idaho, northeastern Colorado, Nebraska, and northeastern Iowa (casually farther east). Winters from southern British Columbia, northeastern Colorado, Iowa, southern Great Lakes, and eastern Maryland (rarely northeastward) southward to Central America, Colombia, Ecuador, Trinidad, and the West Indies. Regularly summers in small numbers within its winter range, and has nested in Florida.

422. Common Goldeneye, *Bucephala clangula* (L.)

IDENTIFICATION: About size of Redhead (total length 43 to 53 cm). Adult male: *head dark metallic green, with a rounded white spot in front of eye*; back, tail, and posterior end of body blackish; neck, breast, and underparts white, and much white also on wings. Female and immature male: *head chocolate brown, abruptly separated from light underparts*; upperparts otherwise grayish brown; breast and sides gray; remainder of underparts white. The head appears large in comparison with most other ducks. Feet orange yellow; iris yellow.

DISTRIBUTION AND VARIATION: Breeds in northern parts of Eurasia and North America, wintering farther south. *B. c. americana* breeds from western and central Alaska, and the northern parts of Mackenzie, Manitoba, Ontario, and Quebec, central Labrador, and Newfoundland southward to southern British Columbia, northwestern Montana, eastern North Dakota, northern Minnesota, northern Michigan, southeastern Canada, northeastern New York, and northern New England. Winters from southeastern Alaska, southern British Columbia, central Montana, northeastern Wyoming, central Nebraska, northern Iowa, Minnesota, central Wisconsin, northern Michigan, and southeastern Canada southward to Baja California, northern Mexico, and the Gulf Coast, very rarely to southern Florida (November-April).

423. Bufflehead, *Bucephala albeola* (L.)

IDENTIFICATION: Size of the Blue-winged Teal (total length 33 to 38 cm). Adult male: *head black* with iridescent reflections, *except for a large white patch from crown to eye*; back black; sides and underparts white. Feet pinkish yellow; bill dark with a light tip. Female and immature male: dark gray above and whitish underneath, with a *small white spot below and behind eye*; feet dusky bluish. Head appears relatively large.

DISTRIBUTION: Breeds from southern Alaska, Great Slave Lake, and western and northwestern Ontario southward to southern British Columbia, northern Montana, southern Saskatchewan, and central Manitoba; also in mountains of Oregon and northeastern California, and, formerly, farther southeast. Winters from coastal Alaska, southern British Columbia, northwestern Montana, the Great Lakes, coastal Maine, and southwestern New Brunswick southward to Baja California, south-central Mexico, and the Gulf Coast, rarely to southern Florida (November-April) and the northern Antilles.

424. Oldsquaw, *Clangula hyemalis* (L.)

IDENTIFICATION: Moderately small (total length about 38 to 46 cm, except for elongate rectrices of adult male). Adult male in winter: *chiefly white, but with patches of chocolate brown on sides of neck, breast* (entirely brown), *wings, and back*; much more brown in plumage in late spring and summer, but seldom in Florida then. Feet dark gray blue; bill dark *with wide pinkish band* near middle. Female and immature male: mostly brown above, on sides of neck, and at base of throat (top of head white in female); otherwise white; bill darker. All winter birds differ from other ducks in showing a *dark patch on the side of an otherwise white head* and *no whitish areas in the outstretched wing*.

DISTRIBUTION: Breeds in northern parts of Alaska, Canada, and Eurasia, moving southward in winter to middle latitudes of Eurasia and in North America to California, the Great Lakes, and the Gulf Coast (rarely inland); also occasionally to central and southern Florida (November-March, rarely later).

425. Harlequin Duck, *Histrionicus histrionicus* (L.)

IDENTIFICATION: Size of Oldsquaw (total length 38 to 46 cm), but without elongate tail feathers. Adult male: *upperparts bluish gray, with prominent white markings, especially anteriad, bordered by black*; also a *reddish brown stripe over and behind the eye*, and a large patch of the same color on each side of the body; underparts grayish brown. Feet dark grayish blue. Female and immature male: upperparts dull brown, with *2 or 3 white patches on side of head*; underparts whitish.

DISTRIBUTION: Breeds in central and northeastern Asia, and from western Alaska southward in the mountains to British Columbia, California, and Colorado; also from northeastern Quebec, Greenland, and Iceland to eastern Quebec. Winters southward to Korea, Japan, central California (rarely southern), and, on the Atlantic Coast, from Massachusetts to Long Island; casually in the interior and in South Carolina and Florida (Williams, 1968).

426. Common Eider, *Somateria mollissima* (L.)

IDENTIFICATION: Larger than any common Florida duck (total length 56 to 66 cm). Adult male: the *black underparts and lower sides* are in striking contrast to the *white breast and upperparts*; crown black; feet yellowish green. Female and immature male: essentially brown, with *dark bars on breast*. The *gently sloping profile* in both sexes is similar to that of the Canvasback or male Surf Scoter. (Also see Key.)

DISTRIBUTION AND VARIATION: Breeds in northwestern Europe, northeastern Siberia, and Arctic North America, ranging southward in winter to the Azores, the Mediterranean Sea, Washington, southern Ontario, New Jersey, and (casually) North Carolina. Four Florida records, 2 representing collected specimens of undetermined race (Sprunt, 1958; Robertson, 1967).

427. King Eider, *Somateria spectabilis* (L.)

IDENTIFICATION: Slightly smaller than the Common Eider (total length 53 to 61 cm). Adult male: body and wings black (except for white patches in wings best seen in flight and white breast and upper back); head and neck also white, thus giving the effect of *white in front and black behind*; feet yellowish. Female: similar to female Common Eider, including black bars (but see Key). Immature male: dull brown, with lighter feather edgings; bill flesh-colored. Profile about as in Common Eider.

DISTRIBUTION: Breeds along Arctic and north Atlantic coasts from Russia through North America to Labrador and Greenland. Winters southward to southern Alaska and New York, rarely to California, the Central States, the Great Lakes, and Florida (2 specimens; Owre, 1962; TT 2809).

428. White-winged Scoter, *Melanitta deglandi* (Bonaparte)

IDENTIFICATION: Slightly larger than other scoters (total length 48 to 58 cm). Regardless of age or sex, the *white wing patch* (2 in male) will distinguish this species from other scoters, but it may not be visible when bird is at rest. Adult male: black above, with a white patch around eye; dark brown ventrally; bill yellow with a dark cere; feet pink and black. Adult female: entirely dark, except feet light brownish red. Immature: dark brown, with 2 *light spots on each side of head*, often tending to merge, this plumage being almost identical with corresponding plumage of Surf Scoter. Bill deep at base.

DISTRIBUTION AND VARIATION: Breeds from northwestern Alaska, northwestern, central, and eastern Canada southward to northeastern Washington, southern Manitoba, and central North Dakota; casual in summer farther south. Winters southward to southern British Columbia, Colorado, Nebraska, the Gulf Coast, and rarely southern Florida (*M. d. deglandi*; November-April; one June specimen).

429. Surf Scoter, *Melanitta perspicillata* (L.)

IDENTIFICATION: Slightly smaller than White-winged Scoter (total length 43 to 56 cm). Adult male: *entirely black except for white spots on the forehead and nape* (sometimes only one); feet reddish; bill orange, blue, and white. Adult female: similar to corresponding plumage of White-winged Scoter, but with light areas on side of head and crown darker and more contrasting. Immature: similar to immature White-winged Scoter. *Bill deep at base; no white in wings.*

DISTRIBUTION: Breeds from western Alaska and northwestern Canada southward to northern British Columbia, Great Bear Lake, and northwestern Saskatchewan; also in James Bay and central Labrador; present in summer but not breeding in other parts of Canada and Alaska. Winters along the Pacific Coast from the Aleutian Islands to the Gulf of California; on the Great Lakes; and on the Atlantic and Gulf Coasts from the Bay of Fundy to Florida; also occasionally in the interior. Chiefly north Florida, November to March, rarely into summer.

430. Black Scoter, *Melanitta nigra* (L.)

IDENTIFICATION: Size of Surf Scoter (total length 43 to 56 cm). Adult male: entirely black, but with a gibbous *yellow to orange cere* above a black bill (bill itself not so thick as in other scoters); feet dark. Female and immature male: uniformly dark brown, except for a large light gray patch on each side of head (whiter in immature); can be confused with male Ruddy Duck in this plumage.

DISTRIBUTION AND VARIATION: Breeds from Iceland, northern Russia, northern Siberia, and northern Alaska southward to Scotland, north-central Siberia, and the Aleutian Islands; summers but may not breed in Canada. Winters southward to western Europe, the Mediterranean, Black, and Caspian Seas, Japan, eastern China, southern California, the eastern Gulf Coast, and Florida (especially northern, rarely remaining into summer; *M. n. americana*).

431. Ruddy Duck, *Oxyura jamaicensis* (Gmelin)

IDENTIFICATION: A small grayish duck (total length 35 to 41 cm) with a dark crown and a *light gray* (white in male) *cheek patch*, thus resembling a female Black Scoter. In breeding plumage (spring), the male has a blue bill, a bright reddish brown back, bluish feet, and reddish irides. It has the distinctive habit of often *carrying the tail erect*. The female has one dark cheek stripe. Downy young: dark gray above with a white spot on each side of back; side of head white with a dark malar stripe; breast light gray; remainder of underparts whitish.

DISTRIBUTION AND VARIATION: Breeds from British Columbia, Great Slave Lake, Alberta, northern Manitoba, and Nova Scotia southward to Guatemala, northern New Mexico, central Texas, northern Iowa, and northern Illinois; also in the Greater Antilles; occasionally (recent records) in Ohio, Pennsylvania, and Florida (*O. j. rubida*); also in the Andes of Colombia (*O. j. andina* and *ferruginea*). North American population winters from southern British Columbia, southern Nevada, central Arizona, southern New Mexico, central and northeastern Texas, Missouri, southern Illinois, northwestern Pennsylvania, and Massachusetts south to Baja California, Costa Rica, southern Florida, and the Bahamas. Also many summer records in Florida.

432. Masked Duck, *Oxyura dominica* (L.)

IDENTIFICATION: Smaller than most Florida ducks (total length 30 to 36 cm). Similar in form to the Ruddy Duck, but with a *white speculum*. Male: cinnamon brown, with fore part of head black, and back and sides spotted with black; bill bright blue. Female: duller, and with 2 black stripes on side of head. Feet reddish gray.

DISTRIBUTION: Resident in the Greater Antilles and from western and northeastern Mexico southward to Ecuador, Peru, Bolivia, Chile, Brazil, Uruguay, and eastern Argentina. Casual in southern Texas and on some of the Lesser Antilles; accidental in Wisconsin, Vermont,

Massachusetts, Maryland, Louisiana, and Florida (Hames, 1959; Langridge, 1962; Owre, 1962).

433. Hooded Merganser, *Lophodytes cucullatus* (L.)

IDENTIFICATION: Total length 40 to 48 cm. Adult male: upperparts black, except for the *prominent crest on the head, which is white with a black rim*; underparts white; sides brown; feet and iris yellow. Female and immature male: dark gray above, whitish underneath, with a *grayish brown crest*; feet and bill yellowish green. Bill with *vertical* serrations. Downy young: upperparts dark brown with white spots on wings and back; sides of head and neck buffy brown; throat dusky; remainder of underparts mostly white.

DISTRIBUTION: Breeds from southeastern Alaska, northern British Columbia, Great Slave Lake, central Alberta, Manitoba, central Ontario, southern Quebec, and New Brunswick southward to southwestern Oregon, central Idaho, south-central Wyoming, central Nebraska, south-central Iowa, eastern Missouri, eastern Arkansas, and western Tennessee; also casually to Louisiana, Alabama, South Carolina, and Florida (south to Hillsborough River). Winters from southern British Columbia, Utah, Colorado, Nebraska, the Great Lakes, Pennsylvania, New York, and Massachusetts southward to Baja California, Mexico City, Veracruz, the northern Gulf Coast, and southern Florida (November-April), and the northern Antilles.

434. Common Merganser, *Mergus merganser* L.

IDENTIFICATION: A very large duck (total length 53 to 68 cm). Adult male: head and most of upperparts black, the head with greenish iridescence, *abruptly contrasting with the white underparts at the upper neck*; iris red. Female and immature male: head, upper neck, and crest reddish brown, *in abrupt contrast to white of entire underparts*; remainder of upperparts gray; chin white; iris yellowish. Serrations *slanted backward*. Bill and feet orange red in both sexes.

DISTRIBUTION AND VARIATION: Breeds across northern and central Eurasia and from southeastern Alaska, British Columbia, central Alberta, Manitoba, northern Ontario, central Quebec, and Newfoundland southward to central California, Arizona, northern Mexico, South Dakota, Michigan, and New York. Winters southward to the Mediterranean, Black, and Caspian Seas, northern Africa, and southern Asia; also (*M. m. americanus*) from Alaska coast, southern Canada, and the northern United States southward to Baja California, northern Mexico, Louisiana, Alabama, and northern Florida (rarely—fewer than 1 percent of the mergansers in Florida are of this species; late November to mid-April).

435. Red-breasted Merganser, *Mergus serrator* L.

IDENTIFICATION: Slightly smaller than the Common Merganser (total length 48 to 66 cm). Adult male: head (crested) dark metallic green, which color is separated by an incomplete white ring from the *brownish lower neck and breast*; upperparts chiefly black, but with considerable white on wings; underparts whitish; bill and iris red. Female and immature male: similar to female Common Merganser, but with the *gray brown throat gradually merging into the whitish underparts* without the abrupt change of that species; lower mandible orange; iris yellowish. Both sexes have orange red feet; serrations on tomia *slanted backward*.

DISTRIBUTION AND VARIATION: Breeds across northern (and central) Eurasia and from northern Alaska, Mackenzie, and northeastern Labrador southward to southeastern Alaska, northwestern British Columbia, central Alberta, southern Manitoba, the central parts of Minnesota, Wisconsin, and Michigan, southern Ontario, northern New York, and coastal Massachusetts. Winters southward to northern Africa, southern Europe, southern Asia, Japan, and (*M. s. serrator*) from southeastern Alaska, the Great Lakes, and southeastern Canada southward to southern Baja California, Sonora, central Arizona and New Mexico, the Gulf Coast, and southern Florida (late October to May, occasionally in summer).

ORDER FALCONIFORMES: XC. FAMILY CATHARTIDAE, New World Vultures. Bill, feet, and claws strong, sharp, and curved throughout the order, but not so well developed in this family. The facial disk of the otherwise-similar owls is lacking, and only the Osprey in this order has a reversible front toe. Cere present at base of bill. In the vultures the *head is naked and the nostrils perforate*.

436. Turkey Vulture, *Cathartes aura* (L.)

IDENTIFICATION: One of Florida's larger soaring birds (total length 66 to 81 cm). Plumage fuscous brown, *lighter on remiges*; *head naked and red* (grayish in immature); bill whitish; feet flesh-colored.

DISTRIBUTION AND VARIATION: Breeds from southern British Columbia, central Alberta, western Ontario, southern Michigan, southern Ontario, central New York, southwestern Massachusetts, and Connecticut southward to the tip of South America; withdraws from southern Canada and the northern United States in winter. *C. a. septentrionalis* occurs throughout Florida at all seasons. (See Wetmore, 1964).

437. Black Vulture, *Coragyps atratus* (Bechstein)

IDENTIFICATION: Slightly smaller than the Turkey Vulture (total length 58 to 69 cm). Plumage *entirely black, but for a white area in the primaries*; *head naked and dark gray*; bill dark with a whitish tip; feet bluish gray.

DISTRIBUTION: Resident from southern Arizona, northern Mexico, western Texas, eastern Oklahoma, southeastern Kansas, Missouri, the southern parts of Illinois, Indiana, and Ohio, eastern West Virginia, and Maryland (casually farther north) southward to Argentina and Chile. Absent or casual on the Florida Keys.

ORDER FALCONIFORMES: XCI. FAMILY ACCIPITRIDAE, Kites, Hawks, and Eagles. Claws and bill more strongly developed than in vultures; *head feathered* (North American species) and *nostrils imperforate*. The lack of a facial disk separates this family from the owls, the lack of a reversible toe from the ospreys, and the usually scutellate tarsi from the falcons.

438. White-tailed Kite, *Elanus caeruleus* (Vieillot)[1]

IDENTIFICATION: Slightly smaller than the Red-shouldered Hawk (total length 38 to 43 cm). Adult: *head and tail mostly white*; upperparts otherwise pearl gray, except darker toward anterior edge of wing near body; underparts chiefly white, but with dark wing tips; feet yellowish; iris orange red. Immature: more brownish above with a rusty breast-band. *Tarsus reticulate*.

DISTRIBUTION AND VARIATION: Resident in central California; also from Oklahoma and Nevada (rarely) and southern Texas southward to Central America; also (at least formerly) in south-central Florida where very rare (*E. c. majusculus*). (See Kale, *Fla. Field Nat.* 2:4–7.)

439. Swallow-tailed Kite, *Elanoides forficatus* (L.)

IDENTIFICATION: Size of the Mississippi Kite, but tail much longer (total length 50 to 64 cm). The glossy black upperparts, white head and underparts (overlaid with rust in nestlings), and black flight feathers, coupled with the *long, deeply forked tail*, make this bird almost unmistakable; feet and base of bill light bluish gray.

DISTRIBUTION AND VARIATION: Breeds from central and southern Texas, the northern Gulf Coast, and the lowlands of South Carolina and Georgia southward to Bolivia, Brazil, and northern Argentina; formerly nested northward to eastern Nebraska and Minnesota and apparently extending breeding range northward now; breeds over most of Florida, but not on outer Keys nor extreme northwestern parts (*E. f. forficatus*). Winters from southern Mexico to northwestern South America (accidentally in southeastern United States); late February–early September.

---

1. For use of *caeruleus*, see Parkes, 1958.

440. Mississippi Kite, *Ictinia misisippiensis* (Wilson)

IDENTIFICATION: Near size of Sharp-shinned Hawk (total length 33 to 38 cm). Adult: dark gray above, *lighter gray on head* and underparts; *tail* and wing tips *almost black*; feet yellowish; iris red. Immature: mostly blackish above; underparts buffy to white with rich brown streaks; tail barred. The *long, pointed wings* help to distinguish it from other small hawks.

DISTRIBUTION: Breeds from Arizona, New Mexico, southeastern Kansas, southeastern Missouri, western Tennessee, northern Mississippi, central Alabama and Georgia, and South Carolina southward to Texas, the Gulf Coast, and northern Florida, but expanding northward and westward in recent years. Winters from southern Texas and peninsular Florida (casually) southward through Mexico to Guatemala; also in Paraguay. Occasionally recorded north of breeding range. Florida: April to early September, with a few sight records in winter.

441. Everglade Kite, *Rostrhamus sociabilis* (Vieillot)

IDENTIFICATION: Slightly smaller than the Red-shouldered Hawk (total length 40 to 46 cm). Adult male: *dark slate*, except for the tail coverts, basal portions of outer rectrices, and narrow tip of tail, *all of which are white*; iris, tarsus, and toes red. Adult female: chiefly fuscous above and heavily streaked with same below; white areas approximately as in adult male; cere, tarsus, and toes dull orange. Immature: similar to adult female, but with more bright brown above. In all plumages, the *white basal part of the tail* separates it from all other hawks of comparable size.

DISTRIBUTION AND VARIATION: Resident from peninsular Florida, Cuba, the Isle of Pines, and Veracruz southward to northern and eastern South America (to Argentina); presently occurs locally in Florida in the southern half of the Peninsula (*R. s. plumbeus*), rarely straying northward.

442. Goshawk, *Accipiter gentilis* (L.)

IDENTIFICATION: One of the largest North American hawks (total length 50 to 66 cm). Like other hawks of this genus, it has comparatively *short, rounded wings and a long tail*. Adult: upperparts light blue gray, the flight feathers of the wing darker; *underparts and superciliary stripe white, with fine dark bars below*; iris reddish brown; bill blue gray. Immature: upperparts grayish brown, except the head and neck light, with blackish streaks; tail distinctly barred; underparts white, with longitudinal rows of dark spots; *superciliary stripe white*; iris yellow. At all ages the cere, tarsus, and toes are yellow and the bill grayish blue; upper *half* of tarsus feathered.

DISTRIBUTION AND VARIATION: Breeds from northern Eurasia, northwestern Alaska, northwestern Mackenzie, northern Alberta, northern Saskatchewan, Ontario, Quebec, Labrador, and Newfoundland southward to northern California, Nevada, southeastern Arizona, Colorado, northern Minnesota, Michigan, Pennsylvania, and western Maryland (possibly to Tennessee and North Carolina in high mountains); also in western Mexico (Jalisco); southward in Old World to southern Eurasia, Burma, and Japan. *A . g . atricapillus* winters southward to southern California, western Mexico, Texas, Louisiana, Tennessee, Kentucky, West Virginia, Virginia, and (casually) Florida (4 records).

443. Sharp-shinned Hawk, *Accipiter striatus* Vieillot

IDENTIFICATION: About size of Kestrel or Merlin (total length 25 to 35 cm). Adult: *upperparts dark blue gray*; underparts white, with *light reddish brown crossbars*; tail barred; iris red. Immature: dark brown (fuscous) above; underparts white, with longitudinal dark streaks; tail more heavily barred; superciliary area streaked; iris yellow. In any plumage the cere, tarsus, and toes are yellowish, the bill grayish blue, and the *tip of the tail truncate*.

DISTRIBUTION AND VARIATION: Breeds from northwestern Alaska and northern Canada southward to southern California (and southern Mexico, *A . s . suttoni*), southern New Mexico, Texas, Louisiana, Alabama, South Carolina, and probably northern Florida; also in Greater Antilles. Winters from southern British Columbia, western Montana, Nebraska, southern Minnesota, Illinois, southern Michigan, southern Ontario, New York, southern Vermont, New Hampshire, Maine and Nova Scotia southward to Panama, the northern Gulf Coast, and southern Florida (*A . s . velox*; September-April).

444. Cooper's Hawk, *Accipiter cooperii* (Bonaparte)

IDENTIFICATION: Color pattern as in corresponding ages of the Sharp-shinned, but size larger (total length 35 to 51 cm) and *tip of tail slightly rounded*. Upper *third* of tarsus feathered.

DISTRIBUTION: Breeds from southern British Columbia, central Alberta, northwestern Montana, Wyoming, eastern North Dakota, southern Manitoba, western Ontario, northern Michigan, southern Ontario, southern Quebec, Maine, New Brunswick, and Nova Scotia southward to Baja California, northern Mexico, south-central Texas, Louisiana, central Mississippi, and northern and central Florida. Winters from Washington, Colorado, Nebraska, Iowa, the southern parts of Wisconsin, Minnesota, Michigan, and Ontario, New York, Vermont, southern Maine, and Massachusetts southward to Costa Rica and southern Florida; September to April, rarely summering.

445. Red-tailed Hawk, *Buteo jamaicensis* (Gmelin)

IDENTIFICATION: Hawks of this genus are heavy-bodied, with long, broad wings and relatively short tails; this is the largest of the common Florida species (total length 48 to 63 cm). Adult: mottled dark brownish gray above; underparts mostly white, except for dark ends of remiges; *tail reddish brown above*, paler below. Immature: similar, but with more dark markings on belly and flanks and tail not red (alternating bands of light and dark gray); iris yellow. Most Red-tails give the impression of white underparts *separated by a dark band across the abdomen*. Feet yellowish to straw-colored; bill grayish blue. Considerable variation of color occurs, and sight records of very pale individuals in Florida have been referred to the race *kriderii*.

DISTRIBUTION AND VARIATION: Breeds from central Alaska and central and eastern Canada southward to Panama, the northern Gulf Coast, and the northern Antilles. *B. j. borealis* breeds throughout northern Florida and *B. j. umbrinus* the central and southern parts of the mainland. The species winters from western Canada and the northern United States to the southern edge of its breeding range.

446. Red-shouldered Hawk, *Buteo lineatus* (Gmelin)

IDENTIFICATION: Somewhat smaller than Red-tailed Hawk (total length 43 to 57 cm). Adult: upperparts varying from brown to gray, but with white flecks scattered throughout; *bend of wing reddish brown*; underparts light, with *crossbars of reddish brown*; tail black, broken by 4 narrow white crossbars. Immature: upperparts similar to those of adult, but with little or no reddish on bend of wing; underparts white with dark spots in longitudinal rows; tail with many narrow light and dark bars; iris yellow. At all ages the *wings* (in flight) *show a light area toward the tip*, the cere, tarsus, and toes are yellow, and the bill is grayish blue.

DISTRIBUTION AND VARIATION: Breeds from northern California to northern Baja California (*B. l. elegans*); also from eastern Nebraska, central Minnesota, Wisconsin, northern Michigan, southern Ontario, and southern Quebec southward (through eastern parts of Kansas and Texas) to Veracruz, the Gulf Coast, and the Florida Keys; *B. l. alleni* breeds across northern and central Florida, *B. l. extimus* in southern Florida, including the Keys. The species withdraws from southern Canada and the extreme northern United States in winter.

447. Broad-winged Hawk, *Buteo platypterus* (Vieillot)

IDENTIFICATION: Smaller than the Red-shouldered Hawk (total length 33 to 46 cm). Adult: upperparts dark gray brown; underparts white, with brown crossbars; *tail with broad, alternating bands of black and white* (3 each). Immature: upperparts more flecked with whitish; underparts light, with dark spots in longitudinal rows; bars on tail more numerous and less contrasting; iris yellowish. Cere, tarsus, and toes yellow at all ages, and bill grayish blue.

DISTRIBUTION AND VARIATION: Breeds from Alberta, Saskatchewan, Manitoba, Ontario, Quebec, Nova Scotia, and New Brunswick southward to Texas, the Gulf Coast, and northern Florida (Levy and Alachua Counties; *B. p. platypterus*); winters from southern Florida and Guatemala to northern Peru and western Brazil. Other races populate Cuba, Puerto Rico, the Lesser Antilles, and Tobago. A melanistic phase is seen occasionally.

448. Swainson's Hawk, *Buteo swainsoni* Bonaparte

IDENTIFICATION: About size of Red-tailed Hawk (total length 48 to 56 cm). Adult, light phase: dark gray brown above; underparts chiefly buffy, with a *brown band across upper breast* (gray in female); tail with several alternating light and dark bands; throat white. Immature: with more white or light markings dorsally, and underparts streaked and spotted with dark (brown breastband lacking); iris yellowish. Dark phase: *uniformly dark gray brown*, except for lighter areas in wings and faint barring on tail. Intermediate color phases also occur. The light phase is best identified in flight by the *dark remiges contrasting with light wing lining*. Cere, tarsus, and toes yellow; bill grayish blue.

DISTRIBUTION: Breeds from Alaska, northwestern Mackenzie, Saskatchewan, Manitoba, western Minnesota, and Illinois southward to Baja California, northern Mexico, south-central Texas, and southwestern Missouri. Winters in Argentina and extreme southern Florida, October-April; rarely in migration in north and central Florida.

449. Short-tailed Hawk, *Buteo brachyurus* Vieillot

IDENTIFICATION: Close to size of Red-shouldered Hawk (total length 43 to 46 cm). Light phase: dark brownish gray above and almost *immaculately white below*. Dark phase: *entire plumage dark brownish gray*. In both phases the tail may be finely barred (only rarely in Florida birds), and the cere, tarsus, and toes are yellow. The underparts of immatures are mottled on belly and throat in the dark phase and rufous buff in the light phase.

DISTRIBUTION AND VARIATION: Resident from Tamaulipas southward to Argentina; also breeds in Florida Peninsula, probably wintering only in the southern half (*B. b. fuliginosus*; Rand, 1960).

450. Rough-legged Hawk, *Buteo lagopus* (Pontoppidan)

IDENTIFICATION: Size of Red-tailed Hawk (total length 48 to 61 cm). Light phase: upperparts light, with dark longitudinal streaks or blotches; underparts mostly light, but with dark tips on primaries, dark spot on bend of wing, and a *broad, dark band across belly*; *basal part of tail white, tip dark*; iris yellow. Dark phase: chiefly blackish above and below, except for light areas in wings; *tail as in light phase*. Many intermediate plumages occur, but the tail character holds constant (though it may be barred distally).

DISTRIBUTION AND VARIATION: Breeds in northern Eurasia and North America, reaching central Eurasia and the southern United States in winter; *B. l. sanctijohannis* winters to southern California, southern Arizona, New Mexico, northeastern Texas, Missouri, Tennessee, and Virginia; casually to southern Texas, the Gulf Coast, and Florida (numerous sight records but *no specimen*).

451. Golden Eagle, *Aquila chrysaetos* (L.)

IDENTIFICATION: Size of Bald Eagle (total length 76 to 105 cm). Adult: chiefly sooty brown, but with *golden brown head and neck*, a light area at base of tail, and remiges lighter than remainder of wing. In this plumage a soaring bird *can be mistaken for a Turkey Vulture*. Immature: uniformly dark except for a *large white patch on each wing* and the *white basal half of the tail*. At all ages the *tarsus is entirely feathered*, and the cere and toes are yellow.

DISTRIBUTION AND VARIATION: Breeds widely in Eurasia (especially northern and mountainous parts) and North America. *A. c. canadensis* breeds from northern Alaska, British Columbia, Mackenzie, northern Saskatchewan, northern Manitoba, and Quebec southward to northern Baja California, north-central Mexico, the western parts of Texas, Oklahoma, Kansas, Nebraska, and South Dakota, northern Ontario, New York, northern New Hampshire, and Maine; summers also in mountains (and may nest) southward to North Carolina and Tennessee. South in winter rarely to the Gulf Coast and Florida (November to February).

452. Bald Eagle, *Haliaeetus leucocephalus* (L.)

IDENTIFICATION: The largest raptore regularly present in Florida (total length 76 to 110 cm). Adult: *head and tail entirely white*; body and wings dark brown; bill, cere, iris, tarsus, and toes yellow, *only two-thirds of the tarsus feathered*. Immature: mostly dark, but with *light areas in wings, belly, and tail*, increasingly so with age (not so distinct as those of immature Golden Eagle); bill and iris dark; feet yellow.

DISTRIBUTION AND VARIATION: Breeds from northwestern Alaska, Mackenzie, Manitoba, central Ontario, southeastern Quebec, and Newfoundland southward (locally) to Baja California, central Arizona, New Mexico, the Gulf Coast, and all of Florida (*H. l. leucocephalus*); withdraws from northern Canada in winter. Florida birds nest in winter and many leave the state by summer.

453. Marsh Hawk, *Circus cyaneus* (L.)

IDENTIFICATION: Size of Red-shouldered Hawk, but tail longer (total length 45 to 61 cm). Adult male: upperparts, head, and neck pearl gray; underparts otherwise essentially white; tail barred light and dark; iris yellow; bill bluish gray. Adult female and immature: dark brown above; *underparts reddish brown or buffy* (streaked in adult female); iris dark. Regardless of age or sex, the *rump is white*, and the cere, tarsus, and toes are yellow.

DISTRIBUTION AND VARIATION: Breeds in western and northern Europe, northern and central Asia, and northern and central North America. *C. c. hudsonius* breeds from northern Alaska, northwestern Mackenzie, northern Manitoba, northern Ontario, central Quebec, Labrador, and Newfoundland southward to northern Baja California, southern Arizona, southern New Mexico, northern Texas, western Oklahoma, Kansas, Missouri, southern Illinois, southern Indiana, Ohio, West Virginia, and southeastern Virginia; also casually in Florida. Winters from southern Canada and the northern United States southward to northern South America and the Greater Antilles. Florida: August-May.

ORDER FALCONIFORMES: XCII. FAMILY PANDIONIDAE, Ospreys. Two external characters separating this family from other hawks are the *reversible outer toe* and the fact that the *claws are rounded in cross section*. The tarsus is reticulate. Monotypic.

454. Osprey, *Pandion haliaetus* (L.)

IDENTIFICATION: Size of Red-tailed Hawk (total length 50 to 64 cm). Dark grayish brown above, except for *considerable white in crown*; *underparts chiefly white*; tail finely barred; cere, tarsus, and toes gray; iris red or yellow.

DISTRIBUTION AND VARIATION: Virtually cosmopolitan. *P. h. haliaetus* breeds from northwestern Alaska, Great Slave Lake, northern Manitoba, northern Ontario, central Quebec, southern Labrador, and Newfoundland southward (locally) to northwestern Mexico, central Arizona, central New Mexico, southern Texas, the Gulf Coast, and the Florida Keys; most common coastwise. Winters from central California, southern Texas, Louisiana, the northern Gulf Coast, coastal South Carolina, and the Bahamas southward to Peru and Brazil (casually farther).

ORDER FALCONIFORMES: XCIII. FAMILY FALCONIDAE, Falcons and Caracaras. Diurnal birds of prey that usually have pointed wings, a toothed upper mandible, rounded scales on front of tarsus, and a rounded nostril with a central tubercle.

455. Caracara, *Caracara cheriway* (Jacquin)

IDENTIFICATION: Size of Red-tailed Hawk (total length 50 to 64 cm). Adult: *crown (with crest) black*; remainder of upperparts and belly blackish; throat and upper breast white, transversely barred on lower breast; *tail and coverts chiefly white*, but former *with a dark tip*; cere and bare face orange red; bill white; tarsus and toes yellowish. Immatures are browner and have longitudinal streaks on breast.

DISTRIBUTION AND VARIATION: Resident from Baja California and the southern parts of Arizona, New Mexico, and Texas southward to northern and western South America; also central Florida, Cuba, and the Isle of Pines (*C. c. audubonii*).

456. Peregrine Falcon, *Falco peregrinus* Tunstall

IDENTIFICATION: About size of Red-shouldered Hawk, but with *narrow, pointed wings* (total length 38 to 51 cm). Adult: dark blue gray with darker bars above; underparts light with dark crossbars; *a black vertical bar below each eye*. Immature: browner above, and with longitudinal streaks beneath. At all ages the cere, tarsus, and toes are yellow.

DISTRIBUTION AND VARIATION: Cosmopolitan except for New Zealand and small islands of eastern Pacific. In North America, has bred from northern Alaska, northern Canada, and southern Greenland southward to northwestern Mexico, central Arizona, southwestern Texas, Kansas, Arkansas, northeastern Louisiana, northern Alabama, and northwestern Georgia; winters from Vancouver Island, the Pacific Coast of the United States, western and southern Arizona, southern New Mexico, Colorado, southern Nebraska, Missouri, Illinois, Indiana, Ohio, southern Ontario, Pennsylvania, New York, Massachusetts, and New Brunswick southward to northern Chile, central Argentina, and Uruguay. Both *F. p. anatum* and *F. p. tundrius* have been collected in Florida (White, 1968); September-May.

457. Merlin, *Falco columbarius* L.

IDENTIFICATION: About size of American Kestrel (total length 25 to 33 cm). Adult male: dark grayish blue above; light underneath, with *bold, dark markings arranged longitudinally*; tail blackish, with narrow white crossbars. Adult female and immature: more brownish above. Cere, tarsus, and toes yellow.

DISTRIBUTION AND VARIATION: Breeds in northern Eurasia, Alaska, Canada, and into northeastern North Dakota, the northern parts of Minnesota, Wisconsin, Michigan, Ohio, and New York, and in New Hampshire and Maine. Winters from coastal Washington, Wyoming, Colorado, southern Texas, southern Louisiana, Alabama, and South Carolina southward to northern Peru, Colombia, and northern Venezuela; casually north to Nebraska, Iowa, Illinois, Indiana, Ohio, southern Ontario, southern Quebec, and Maine. Most Florida specimens are referred to *F. c. columbarius*, but *bendirei* has been collected on Key West and Merritt Island; September-May.

458. American Kestrel, *Falco sparverius* L.

IDENTIFICATION: The smallest diurnal raptore in Florida (total length 23 to 31 cm). Male: upperparts mainly reddish brown, but with grayish blue on top of head and on wings, and *black spots or bars on wings, head, and back*; underparts lighter, with rounded black spots; tip of tail with a broad black band and a narrow white one. Female: color of upperparts duller (blue almost wanting) and with more dark bars; several dark crossbars also on tail; underparts similar to the male in ground color, but with medium brown markings arranged in longitudinal rows. *Black markings on the face are diagnostic in both sexes.* Bill gray blue; cere, tarsus, and toes yellow.

DISTRIBUTION AND VARIATION: Breeds from northern Alaska, Mackenzie, Alberta, northern Manitoba, northern Ontario, southern Quebec, and Nova Scotia southward through Mexico and the West Indies (except extreme southern Florida) to the tip of South America. Winters from southern British Columbia, Nevada, Utah, Colorado, Nebraska, Iowa, Minnesota, Illinois, southern Michigan, southern Ontario, New York, central parts of Vermont and New Hampshire, southern Maine, and Nova Scotia southward to southern limit of breeding range. *F. s. sparverius* occurs throughout Florida in winter; *F. s. paulus* breeds over all but the southern tip, but is locally rare or absent.

ORDER GALLIFORMES: XCIV. FAMILY PHASIANIDAE, Pheasants and Quail. Upland game birds with fowl-like bills, strong feet, and strong, blunt claws. The tarsi and toes are bare and the head usually feathered.

459. Bobwhite, *Colinus virginianus* (L.)

IDENTIFICATION: Total length 23 to 26 cm, but tail very short. Male: upperparts and breast chiefly brown or reddish brown, with lighter and darker markings; a broad white stripe over each eye, bordered by black above; sides of neck with longitudinal rows of white spots; chin white, bordered by black below; flanks with alternating stripes of white and reddish brown; remainder of underparts white, with wavy narrow black bars running transversely; feet yellowish brown.

Female: similar, but without black on face, and with white on head and chin replaced by buff; bill and feet yellowish brown. Downy young: upperparts reddish brown medially, bounded on back by 2 whitish stripes and on head by 2 broad ochraceous stripes; remainder of upperparts mixed with brownish and blackish; underparts grayish, except yellowish white on chin and throat.

DISTRIBUTION AND VARIATION: Resident from southwestern Wyoming, South Dakota, Minnesota, Wisconsin, Michigan, southern Ontario, central New York, southern Vermont and New Hampshire, and southwestern Maine southward (through eastern Colorado, eastern New Mexico, and eastern Mexico) to southern Mexico, the northern Gulf Coast, and the Greater Antilles (some). The breeding form in north Florida is *C. v. virginianus*, which is replaced south of Gainesville and Anastasia Island by *C. v. floridanus*; absent from Keys. *C. v. mexicanus* is frequently released, but does not maintain its integrity.

ORDER GALLIFORMES: XCV. FAMILY MELEAGRIDIDAE, Turkeys. Large upland game birds (total length 75 to 125 cm) with much *bare*, *wrinkled*, *red to blue skin on the head*, and highly iridescent plumage; contour feathers nearly truncate.

460. Turkey, *Meleagris gallopavo* L.

IDENTIFICATION: The largest of upland game birds (total length 90 to 125 cm). Upperparts with metallic coppery and bronze reflections, the feathers tipped with black; primaries and secondaries barred gray and white; tail brown, with a subterminal dark band and a terminal one of *rust* color (usually white in domestic turkeys); underparts more coppery; *head and neck bare, the skin red to blue and wrinkled*; feet reddish to brownish. Males have a beard of coarse, black bristles hanging from center of breast, and spurs on the tarsi. Downy young: upperparts cinnamon, heavily spotted with dark brown; underparts and sides of head yellowish white to buffy.

DISTRIBUTION AND VARIATION: Resident from Arizona, New Mexico, central Colorado, southeastern Oklahoma, southern Missouri, Kentucky, West Virginia, and northern Pennsylvania southward to southern Mexico, the northern Gulf Coast, and southern Florida (except Keys); also introduced into parts of California, Utah, Wyoming, and South Dakota (Map 40). The breeding birds of Florida are usually referred to *M. g. osceola*, but there is some possibility of *M. g. silvestris* in north Florida.

ORDER GRUIFORMES: XCVI. FAMILY GRUIDAE, Cranes. Birds in this order are often long-legged and have the lower tibia devoid of feathers; typically they inhabit wet places. The cranes (Gruidae) are the largest members of the order and bear a superficial resemblance to herons and

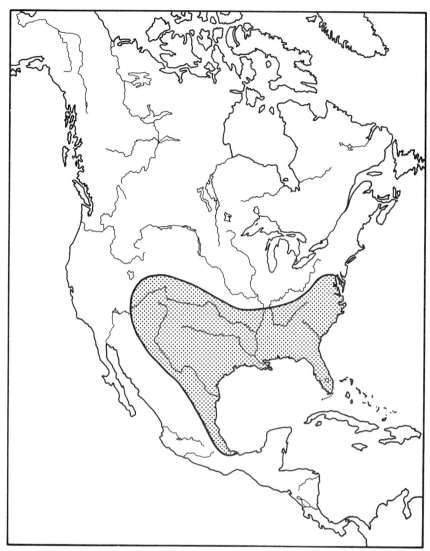

Map 40. Distribution of the Turkey (*Meleagris gallopavo*)

egrets. An observable point of difference is the presence of feathers or bristles on the lores; also the bill is relatively shorter, the tertials are very long, and the hallux is usually reduced.

461. Whooping Crane, *Grus americana* (L.)

IDENTIFICATION: A giant white bird with a long neck and long legs (total length 125 to 140 cm). Adult: *plumage white*, except for black primaries and a *bare red face and crown*. Immature: white overlaid with patches of rust, especially above; head feathered. The color of the adult's plumage is similar to that of a Wood Stork, but the latter bird has a heavier, decurved bill, more black in the wings, and a blackish or gray head and neck.

DISTRIBUTION: Breeds in south-central Mackenzie (northwestern Canada) and winters on coast of Texas and northern Mexico, chiefly on the Aransas Wildlife Refuge; formerly wintered in other Southern States, including one record for Florida (Hallman, 1965).

462. Sandhill Crane, *Grus canadensis* (L.)

IDENTIFICATION: About size of Great Blue Heron (total length 100 to 120 cm). Adult: plumage evenly medium gray, except chin and throat lighter and primaries darker; *a patch of bare, red skin toward front of head* (but with minute bristles). Immature: plumage more brownish; head entirely feathered. Downy young: reddish brown, with white spot in front of wing (Sutton, 1946).

DISTRIBUTION AND VARIATION: Breeds from northeastern Siberia, northern Alaska, and northern Canada southward (locally) to Oregon, northeastern California, eastern Nevada, northwestern Utah, Wyoming, Colorado, and the Dakotas; also locally from southern Mississippi (*G. c. pulla*) to southern Georgia and southern Florida (*G. c. pratensis*). Northern-breeding birds move southward in winter, ranging from central California and southern Texas southward to Baja California and north-central Mexico; also eastward to Florida (*G. c. tabida*; John Aldrich, personal communication). Records in extreme West Florida may pertain to the newly described *G. c. pulla* (Proc. Biol. Soc. Wash., 85:63–70).

ORDER GRUIFORMES: XCVII. FAMILY ARAMIDAE, Limpkins. Intermediate in size between cranes and rails. They differ from each of those groups in the *shape of the outermost primary*, which is narrow but widens near the tip. However, as the family is monotypic, any good character for the species also distinguishes the family.

463. Limpkin, *Aramus guarauna* (L.)

IDENTIFICATION: Total length 63 to 71 cm. Olive brownish, with *white flecks arranged in longitudinal rows* on head, neck, underparts, upper back, and wing coverts; chin mostly white; bill reddish at base; feet dark greenish. Downy young: mostly dark brownish, but chin and much of underparts lighter.

DISTRIBUTION AND VARIATION: Resident from southern Mexico through Central America and the Amazon Basin to central Argentina and Uruguay; also (*A. g. pictus*) from Florida, especially peninsular, and southern Georgia (rare) to Cuba, the Isle of Pines, and Jamaica; accidental farther north.

ORDER GRUIFORMES: XCVIII. FAMILY RALLIDAE, Rails, Gallinules, and Coots. Smaller than limpkins and cranes (total length less than 50 cm). Body compressed, but other external structures variable. As in the case of cranes and limpkins, the wing tip is rounded.

464. King Rail, *Rallus elegans* Audubon

IDENTIFICATION: The largest of Florida's rails (total length 38 to 48 cm). Top of head and stripe through eye dark brown; remainder of upperparts reddish brown, with longitudinal streaks or rows of darker brown spots; *wing much more reddish*; chin and line over eye white; *throat and breast bright reddish brown*; abdomen barred with dark gray and white; bill yellowish to reddish at base; iris reddish brown; feet greenish olive. Downy young are entirely black.

DISTRIBUTION AND VARIATION: Breeds from eastern Nebraska, Iowa, central Minnesota, the southern parts of Wisconsin, Michigan, and Ontario, and from New York, Connecticut, and Massachusetts southward through east-central Kansas and central Oklahoma to western and southern Texas, the Gulf Coast, and southern Florida except the Keys (*R. e. elegans*); also in central Mexico. Winters from southeastern Texas, Louisiana, Mississippi, southern Alabama, and eastern North Carolina southward to eastern Mexico, the Gulf Coast, and southern Florida (casually somewhat farther north, and reaching the northern United States and southern Canada rarely in fall). May be conspecific with the Clapper Rail.

465. Clapper Rail, *Rallus longirostris* Boddaert

IDENTIFICATION: Slightly smaller than the King Rail (total length 30 to 38 cm). Color pattern similar to that of the King Rail, except much more grayish throughout; also much darker in some subspecies than in others. Downy young entirely black, except for small white areas on breast (Wetherbee and Meanley, 1965).

DISTRIBUTION AND VARIATION: Resident on the Pacific Coast from San Francisco Bay (locally) to northwestern Peru; on the Atlantic and Gulf Coasts from Connecticut to Brazil; accidental inland. The following races breed in Florida: *R. l. waynei*, Atlantic Coast southward to Merritt Island; *R. l. scottii*, from Palm Beach County around southern tip of mainland and up Gulf Coast to Pensacola; *R. l. insularum*, Florida Keys. In addition, *R. l. crepitans* has been taken at Amelia Island in winter, and *R. l. saturatus* has occurred on the Gulf Coast south to Tampa Bay and at Tallahassee (TT 2810 and 2811).

466. Virginia Rail, *Rallus limicola* Vieillot

IDENTIFICATION:Total length 20 to 27 cm. Coloration practically identical to that of the King Rail, except that side of head is gray. Tarsus, toes, iris, and base of bill reddish.

DISTRIBUTION AND VARIATION: Breeds from British Columbia, Alberta, central Saskatchewan, central Manitoba, western and southern Ontario, southern Quebec, New Brunswick, and Nova Scotia southward to northwestern Baja California, east-central Arizona, northern New Mexico, western Oklahoma, Missouri, Illinois, northern Alabama, West Virginia, northern Virginia, coastal North Carolina, and in southern Mexico (*R. l. limicola*); also south to Guatemala, in the high mountains of Ecuador, and from central Chile and near Buenos Aires to the southern tip of South America. Winters near the Pacific Coast part of breeding range and from the southern United States southward; Florida records range from early September to May.

467. Sora, *Porzana carolina* (L.)

IDENTIFICATION: Size of Virginia Rail, but *bill short* (total length 20 to 25 cm). Adult: upperparts medium brown, with dark markings on head and neck, and white spots and streaks on wings and back; *face and throat black*; breast and sides of neck light blue gray; abdomen barred with dark brown and white; bill yellowish green; tarsus and toes greenish. Immature: differs from adults chiefly in absence of black in plumage; chin white; breast and sides of head and neck buffy brown. (The rare Yellow Rail is similar to this plumage, but is buffier, much smaller, and has a white wing patch.)

DISTRIBUTION: Breeds from British Columbia, Mackenzie, Saskatchewan, northern Manitoba, northern Ontario, central and southern Quebec, and New Brunswick southward to northwestern Baja California, Nevada, Arizona, southern New Mexico, Oklahoma, Missouri, Illinois, Indiana, central Ohio, West Virginia, and Pennsylvania. Winters from northern coastal California, northeastern Texas, the southern parts of Arizona, Louisiana, Mississippi, and South Carolina southward to Peru and British Guiana. Florida: late August-late May.

468. Yellow Rail, *Coturnicops noveboracensis* (Gmelin)

IDENTIFICATION: Total length 15 to 20 cm. Upperparts reddish brown, with white crescentic markings on wings and back, and blackish markings on wings; *secondaries forming a big white patch* conspicuous in flight; chin and line over eye light tan; breast buffy brown; flanks barred brown and white; belly whitish; *bill short and greenish*; tarsus and toes yellowish brown.

DISTRIBUTION AND VARIATION: Summers, breeding at least locally, from Mackenzie, Manitoba, Ontario, Quebec, New Brunswick, and Maine southward to Alberta, Saskatchewan, North Dakota, Minnesota, Wisconsin, Ohio, Massachusetts, and Connecticut (*C. n. noveboracensis*). Winters from Oregon to southern California and in southern Louisiana and Mississippi, southern and central Alabama, and Florida (September-May); also a resident race in south-central Mexico.

469. Black Rail, *Laterallus jamaicensis* (Gmelin)

IDENTIFICATION: Sparrow-sized (total length 13 to 15 cm). Adult: *nape, back, and wings brownish black, the back and wings with rounded white spots*; head and most of underparts dark slate, but belly with white markings; *bill short and dark gray*; iris red; tarsus and toes greenish. Immature: similar, but chin, throat, and breast lighter and top of head browner. Downy young: bluish black. (Note that downy young of the large rails are entirely black and close to the size of adult Black Rails.)

DISTRIBUTION AND VARIATION: Breeds locally from Kansas, Indiana, Ohio, New York, Connecticut, and Massachusetts southward to western North Carolina and Florida; also in southern coastal California and probably adjacent Baja California, the coast of Peru, and central Chile. The eastern race, *L. j. jamaicensis*, winters from southern Louisiana and southern Georgia southward to Cuba, Puerto Rico, and Jamaica.

470. Purple Gallinule, *Porphyrula martinica* (L.)

IDENTIFICATION: Slightly smaller than the Coot (total length 30 to 36 cm). Adult: *head, neck, and underparts brilliant purplish blue; back, wings, and tail greenish*; under side of tail and its coverts white; bill red, tipped with yellowish; frontal shield chalky blue; tarsus and toes greenish yellow. Immature: head, neck, and underparts light grayish brown (almost white on chin and belly); *back and wings darker and tinged with greenish*; frontal shield greenish yellow. Downy young are coal black.

DISTRIBUTION: Breeds from southern Illinois, western and south-central Tennessee, central Alabama, and South Carolina southward to the Gulf Coast (from Louisiana eastward) and the Lesser Antilles; and from southern Texas southward through eastern Mexico to Peru and northern Argentina. Winters from western Mexico, southern Texas, southern Louisiana, and northern Florida southward to southern limit of breeding range. (Many records of wandering birds outside range northward to Canada and westward to Arizona and Utah.) Florida: late March to early November, wintering in Peninsula; chiefly a transient on the Keys.

471. Common Gallinule, *Gallinula chloropus* (L.)

IDENTIFICATION: About size of Coot (total length 36 to 38 cm). Adult: head, neck, and most of underparts slate gray; wings and back grayish brown: *each flank with a longitudinal white bar*; under side of tail mostly white; *frontal shield*, base of bill, and lower tibia *red*; tip of bill yellow; tarsus and toes gray green. Immature: like adults except for dull-colored bill and frontal shield, white above eye and on throat, *browner upperparts*, and pale underparts. Downy young: mostly black, but throat white and crown nearly bare.

DISTRIBUTION AND VARIATION: Throughout much of Eurasia, Africa, North and Central America, and all but southern South America. *G. c. cachinnans* breeds locally from northern California, central Arizona, central Texas, Oklahoma, Kansas, Nebraska, Iowa, central Minnesota, southern Wisconsin, Michigan, southern Ontario, southern Quebec, Vermont, and Massachusetts southward to southern Texas, the Gulf Coast, and southern Florida; also in Bermuda and the Galapagos Islands; winters from north-central California, southern Arizona, southern Texas, Louisiana, the southern parts of Mississippi and Alabama, and eastern North Carolina southward to Panama and southern Florida; another race is resident in the West Indies. Casual winter records occur much farther north.

472. American Coot, *Fulica americana* Gmelin

IDENTIFICATION: Total length 33 to 41 cm. Plumage dark gray, *almost black on head and neck*; frontal shield and 2 spots near tip of bill reddish; remainder of *bill white*; tarsus and toes gray green, the *toes lobed*. *Downy young* differ from those of gallinules in having *orange bristles anteriad and white hairlike bristles throughout*, giving a frosted effect.

DISTRIBUTION AND VARIATION: *F. a. americana* breeds from British Columbia, southern Mackenzie, northern Alberta, central Saskatchewan, Manitoba, western Ontario, northern Minnesota, Wisconsin, and Michigan, southern Ontario and Quebec, and New Brunswick southward (locally) to Baja California, Nicaragua, southern parts of Texas, Louisiana, and Mississippi, northern Alabama, west-central Georgia, and southern Florida, rarely westward to Panama City (Hallman, 1966a); also Cuba, the Isle of Pines, Jamaica, and Grand Cayman; withdraws in winter from most of western and northeastern United States and southern Canada; other races are resident in the Andes and on the Hawaiian Islands.

ORDER CHARADRIIFORMES: XCIX. FAMILY JAÇANIDAE, Jaçanas. Primitive shorebirds somewhat resembling gallinules, with *extremely long toes and claws* and, usually, a frontal shield. Wing tip rounded. Spurs often present on bend of wing (wrist). Florida's species appears in the Hypothetical List (769H).

ORDER CHARADRIIFORMES: C. FAMILY HAEMATOPODIDAE, Oyster-catchers. Although this order is diverse, all members have the lower tibia bare, and most have at least vestigial webbing between the front toes. The hallux, if present, is reduced and elevated. Most species have pointed wing tips. The nostrils are perforate and often slitlike. The Haematopodidae (oystercatchers) are highly distinctive, as all members have contrasting dark and white plumage, reddish legs, and a narrow, chisel-shaped bill. Hallux lacking.

473. American Oystercatcher, *Haematopus palliatus* Temminck

IDENTIFICATION: Total length 43 to 53 cm (larger than most shore-birds). Adult: head, neck, and breast sooty black; remainder of upper-parts grayish brown, but with a large white patch in each wing; under-parts otherwise white; *bill greatly compressed and red*; eye-ring and iris red; tarsus and toes light pinkish. Immature: similar, but head and neck paler; feathers of upperparts with buffy edgings; bill brownish. Downy young: entire head, neck, and upperparts mottled gray, with 2 black stripes running down back; remainder of underparts white.

DISTRIBUTION AND VARIATION: Breeds locally along the coast from New Jersey to northeastern Florida and from Texas to southwest-ern Florida (casual farther north in summer); ranges to southeast Florida (rarely to Cuba) in winter and perhaps in summer; also resident from eastern Mexico to Argentina and on the Pacific Coast from Colombia to Chile (*H. p. palliatus*; also *H. p. frazari* ranges from Baja California to southern Mexico).

ORDER CHARADRIIFORMES: CI. FAMILY CHARADRIIDAE, Plovers and Turnstones. Compared with other shorebirds, *most plovers have the upper mandible swollen toward the tip*, and in turnstones the culmen is flattened. Typically the *tarsus is reticulate* in front and the *hallux lacking*. Plumage bright or contrasting in breeding season.

474. Semipalmated Plover, *Charadrius semipalmatus* Bonaparte

IDENTIFICATION: Most plovers of this genus have a single dark ring around the throat, sometimes incomplete. Total length 165 to 190 mm. Adult in breeding plumage: upperparts chiefly dark brown; forehead white, bordered by black, which extends behind eye; *white ring around neck, a black ring immediately ventral to it*; remainder of underparts white; *tarsus, toes, and base of bill orange*; tip of bill dark; bill relatively short. Immature, adults in winter plumage: black replaced by dark brown; bill entirely dark.

DISTRIBUTION: Breeds from Alaska and northern Canada southward to middle latitudes of Canada and to Nova Scotia. Winters from central California, northern Mexico, the Gulf Coast, and coastal South Carolina (rarely farther north) southward to the southern tip of South America. Chiefly a transient inland. Nonbreeding birds remain all summer on Florida coasts and elsewhere far south of the breeding range.

475. Piping Plover, *Charadrius melodus* Ord

IDENTIFICATION: Size of Semipalmated Plover (total length 150 to 190 mm). *Upperparts mostly pale brownish gray*; mark above eye and the *usually incomplete ring around lower throat* black (brownish in winter); forehead, nape, and underparts white; *tarsus, toes, and base of bill orange*; *bill very short* and its tip dark.

DISTRIBUTION AND VARIATION: Breeds from central Alberta, southern Saskatchewan, southern Manitoba, northern Michigan, southern Ontario, and southwestern Newfoundland southward to central Nebraska, the southern edge of the Great Lakes, and coastal Virginia. Winters along the Atlantic and Gulf Coasts from South Carolina to Texas; also in the Bahamas and Greater Antilles to Puerto Rico; July to early May in Florida. Most Florida specimens are referable to *C. m. melodus*, but *circumcinctus* has also been collected (Stevenson and Baker, 1970).

476. Snowy Plover, *Charadrius alexandrinus* L.

IDENTIFICATION: About the size and color of the Piping Plover (total length 150 to 175 mm). Plumage differs from that of Piping Plover in *having the black band vestigial* and in having a dark spot behind the eye; *bill longer, black*; *tarsus and toes dusky*. Downy young: top of head, back, and wings pale grayish buff, mottled with black; underparts and collar around neck white.

DISTRIBUTION AND VARIATION: Occurs at least locally or in migration in nearly all parts of the world except far northern latitudes and the interior of some continents (for example, South America). *C. a. nivosus* breeds on the Pacific Coast from southern Washington to Baja California, and in the interior of western Nevada, Utah, eastern Colorado, southern New Mexico, southwestern Kansas, northwestern Oklahoma, and north-central and southern Texas; winters from the coast of Oregon to west-central Mexico and on the Gulf Coast from southern Texas to "western Florida" (A.O.U. *Checklist*).[1] *C. a. tenuirostris* is resident on

---

1. The basis for the inclusion of Florida is not known to me.

the Gulf Coast from Louisiana to southern Florida (only one summer record on Keys), and in the Bahamas and Greater Antilles.

477. Wilson's Plover, *Charadrius wilsonia* Ord

IDENTIFICATION: Somewhat larger than the Semipalmated Plover, which it resembles in color (total length 175 to 200 mm). It differs from that species in its paler upperparts, less black on top of the head and upper back, and in having a *white superciliary stripe* (with some buffy brown posteriad in adult male); also *bill longer, thicker, and black*; tarsus and toes pinkish. Downy young: mottled blackish and buff above, light buffy below; bare area on side of throat.

DISTRIBUTION AND VARIATION: Breeds on the Pacific Coast from central Baja California to Peru (*C. w. beldingi*) and on the Atlantic Coast from Virginia (rarely New Jersey) to British Honduras and the West Indies (*C. w. wilsonia*; also a third race in northern South America). Winters from Baja California and the northern Gulf Coast southward to Peru and Brazil.

478. Killdeer, *Charadrius vociferus* (L.)

IDENTIFICATION: Larger than other Florida plovers of this genus (total length 230 to 280 mm). Upperparts mostly grayish brown, but *rump bright cinnamon*; white ring around nape; underparts entirely white except for *2 black rings around throat and breast*; forehead and area around eye white, bordered by black below; also a black bar above forehead; tarsus and toes grayish brown; eyelids reddish. Downy young: differs from adult chiefly in having *only one black ring* and no cinnamon on rump.

DISTRIBUTION AND VARIATION: *C. v. vociferus* breeds from northwestern British Columbia, northern Alberta, southern Mackenzie, northeastern Manitoba, northern Ontario, southern Quebec, Maine, and New Brunswick southward to southern Baja California, central Mexico, the southern parts of Texas, Louisiana, Mississippi, and Alabama, and southern Florida (Dickie, 1965); another race is resident in Peru and northern Chile. North American birds winter from southern British Columbia, Oregon, northern Utah, Colorado, Oklahoma, southern Ohio, and Long Island southward to northern South America.

479. Mountain Plover, *Charadrius montana* (Townsend)

IDENTIFICATION: Slightly larger than Wilson's Plover (total length 200 to 230 mm), but bill smaller. Breeding plumage: upperparts mostly *plain* grayish brown (not mottled), but with fore parts of crown and line through eye black in summer; upper breast with brownish wash; forehead, superciliary line, throat, and remainder of underparts white; tarsus long and toes short, their color light gray brown. Winter: no black on head; underparts tinged with buff.

DISTRIBUTION: Breeds from northern Montana and northeastern North Dakota southward through eastern Wyoming, western Nebraska, Colorado, and western Kansas to central and southeastern New Mexico, western Texas, and western Oklahoma. Winters from central California, southern Arizona, and central and coastal Texas southward to southern Baja California and northern Mexico; casual in other western states and accidental in Massachusetts, Alabama, and Florida.

480. American Golden Plover, *Pluvialis dominica* (Müller)

IDENTIFICATION: About size of Killdeer (total length 240 to 280 mm). Adults in breeding plumage: *upperparts blackish, with rounded yellowish spots*; white stripe from forehead, over eye, to side of breast; underparts black; tarsus and toes bluish gray. Adults in winter, immature: upperparts somewhat duller; *underparts* (including sides of head and neck) *dusky*, the belly lighter. *Axillars not black in any plumage, and rump dark. Hallux lacking.*

DISTRIBUTION AND VARIATION: Breeds in northern parts of Siberia, Alaska, and Canada. Winters in southern Asia, Australia, Tasmania, New Zealand, and other islands in southwest Pacific (*P. d. fulva*); also from Bolivia, Paraguay, and southern Brazil south to east-central Argentina and Uruguay; spring and fall migrant through Florida, rarely wintering, September to April (*P. d. dominica*).

481. Black-bellied Plover, *Pluvialis squatarola* (L.)

IDENTIFICATION: Slightly larger than Killdeer (total length 265 to 330 mm). Adult in breeding plumage: wings and *back black, with crescentic white markings*; top of head and nape white, with fine black markings in longitudinal rows; a broad white stripe from forehead, over eye, to side of breast; face, throat, breast, and abdomen black; tarsus and toes bluish gray. Adult in winter plumage, immature: less contrast in color of upperparts; underparts mostly whitish, but with dark markings on throat and sides. In flight the *black axillars* are the best field mark, and the *rump is white*. Hallux present.

DISTRIBUTION: Breeds in northern parts of Russia, Siberia, Alaska, and Canada. Winters from the British Isles, southern Europe, southern Asia, and the Solomon Islands southward to southern Africa, Madagascar, Australia, and New Zealand; also from southwestern British Columbia, the Gulf Coast, and New Jersey southward along coasts to

Chile and Brazil. Frequent inland on migration. Non-breeding birds summer south of breeding range.

482. Surfbird, *Aphriza virgata* (Gmelin)[1]

IDENTIFICATION: Near size and shape of the Red Knot (total length about 250 mm). Breeding plumage: upperparts streaked black and white, with brownish on back; tip of tail black, but its *base and upper tail coverts white*. Underparts white, heavily spotted with dark except narrowly on throat. Base of lower mandible yellow; feet greenish yellow, the tarsus reticulate. Winter plumage and immature: less brownish above; breast washed with gray.

DISTRIBUTION: Breeds in south-central Alaska at high elevations, though nonbreeding birds occur on coast south to California. Winters on Pacific Coast from southeastern Alaska to southern South America; casual on Gulf of Mexico, including 2 records (one photograph) at Pensacola, Florida (Weston and Baxter, 1957; James, 1968).

483. Ruddy Turnstone, *Arenaria interpres* (L.)[1]

IDENTIFICATION: About size of Killdeer (total length 230 to 250 mm). Adult in breeding plumage: upper back and folded wing mostly reddish brown, but with black markings; remainder of *head and neck white, with black lines*; throat and breast black; abdomen white; *tarsus and toes orange*. Adult in winter: plumage somewhat duller. Immature: plumage much duller, with little reddish or black, but *extensive brown on breast is distinctive*.

DISTRIBUTION AND VARIATION: Breeds in Iceland and the northern parts of Alaska, Canada, Greenland, Scandinavia, Russia, and Siberia; nonbreeding birds summer on coasts far south of breeding range. Winters from central California, the Gulf Coast, coastal South Carolina, and the Bahamas southward along both coasts to central Chile and southern Brazil; also from the British Isles, the Mediterranean Sea, and southern Asia southward to South Africa, Madagascar, Australia, and New Zealand, and on smaller Pacific islands. *A. i. morinella* is regular on the coast of Florida and very rare inland; *A. i. interpres* is of accidental occurrence (Loftin and Olson, 1960).

ORDER CHARADRIIFORMES: CII. FAMILY SCOLOPACIDAE, Sandpipers and allies. Similar to the plovers (Charadriidae), but bill longer and not greatly enlarged near tip, hind toe usually present, tarsus usually scutellate in front, and plumage usually duller.

---

1. The Surfbird and Ruddy Turnstone have recently been transferred to the Family Scolopacidae (A.O.U., 1973), but the Surfbird can hardly be keyed to that family on the basis of external morphology.

484. American Woodcock, *Philohela minor* (Gmelin)

IDENTIFICATION: Near size of Common Snipe (total length 235 to 305 mm). Front of crown gray; *back of head with 3 broad, black crossbars*, alternating with 3 narrow, rusty ones; upperparts rusty brown, with numerous black and gray markings; underparts and sides of head rusty brown; bill long, pinkish toward base; tarsus and toes pinkish; *wings rounded in flight, the 3 outermost primaries quite narrow*; eyes placed high on head. Downy young: mottled brownish above, buffy below.

DISTRIBUTION: Breeds from southeastern Manitoba, western and central Ontario, southern Quebec, northern New Brunswick, Nova Scotia, and southern Newfoundland southward (through central Minnesota, central Iowa, west-central Missouri, and Arkansas) to eastern Texas, Louisiana, Mississippi, south-central Alabama, and southern Florida. Winters from eastern Oklahoma, southern Missouri, southwestern Tennessee, northern Mississippi, northern Alabama, western South Carolina, west-central North Carolina, and Virginia southward to southern edge of breeding range.

485. Common Snipe, *Capella gallinago* (L.)

IDENTIFICATION: Total length 255 to 305 mm. *Crown, neck, and back blackish, with longitudinal stripes of buff; a reddish brown band* and more narrow black one *near end of tail*; remainder of upperparts brownish gray, with buffy markings; breast buffy with dark markings in longitudinal rows; remainder of underparts whitish, marked with dark gray vertical bars on flanks; *bill long*, pinkish toward base; tarsus and toes greenish.

DISTRIBUTION AND VARIATION: Breeds widely in Eurasia, Alaska, Canada, and the northern United States (south to California, Arizona, Colorado, Nebraska, Iowa, northern Illinois, Michigan, Ohio, Pennsylvania, and New Jersey). Winters south in Old World to south-central Africa, Ceylon, Java, and the Philippines; North American birds (*C. g. delicata*) range from southern British Columbia, Utah, Colorado, Nebraska, western Kentucky, Tennessee, and Virginia southward to Colombia, Venezuela, and Brazil. Florida: September-May.

486. Long-billed Curlew, *Numenius americanus* Bechstein

IDENTIFICATION: One of Florida's largest shorebirds (total length 500 to 660 mm). Upperparts cinnamon buff, with blackish markings, the latter forming longitudinal rows on neck; sides of neck, throat, and flanks similarly marked; chin whitish; remainder of *underparts cinnamon buff, unmarked; bill very long (150 to 200 mm in adults), decurved*, and pinkish toward base; tarsus and toes bluish gray.

DISTRIBUTION AND VARIATION: Breeds from southern British Columbia, Alberta, Saskatchewan, and Manitoba southward to Utah, New Mexico, and Texas (formerly farther east). Winters from California, western Nevada, Arizona, Texas, Louisiana, and South Carolina southward to Baja California, Guatemala, and Panama. Florida specimens are assigned to *N. a. americanus*; chiefly August to May, rare.

487. Whimbrel, *Numenius phaeopus* (L.)

IDENTIFICATION: Similar in appearance to Long-billed Curlew, but smaller (total length 400 to 450 mm). Crown with *2 broad, blackish stripes*, separated by a narrow buffy gray one; a light stripe over eye; neck streaked with light gray and blackish; remainder of upperparts blackish brown, with buffy feather edgings; underparts light buffy to whitish, with *darker flecks on breast and bars on sides*; *bill long (75 to 100 mm), decurved, and devoid of pinkish*; tarsus and toes grayish.

DISTRIBUTION AND VARIATION: Breeds in the northern parts of Eurasia, Alaska, and Canada (except eastern). Nonbreeders summer much farther south. Winters from western Europe, southern Asia, and the Philippine Islands southward to South Africa, Madagascar, Ceylon, Australia, Tasmania, New Zealand, and smaller islands of southwest Pacific; also (*N. p. hudsonicus*) from central California to southern Chile and from the coasts of Texas, Louisiana, and South Carolina southward to Colombia and Brazil.

488. Upland Sandpiper, *Bartramia longicauda* (Bechstein)

IDENTIFICATION: Somewhat larger than Killdeer (total length 280 to 330 mm). *Crown blackish, separated medially by a narrow strip of buffy*; neck and back with alternating longitudinal dark and light buffy streaks; folded wings with blackish, brown, buffy, and white markings; chin white; sides of head ochraceous; line over eye light; remainder of underparts buffy to white, with blackish markings on throat and sides; *tail relatively long*, usually protruding beyond folded wings; *bill short, slender, greenish toward base*; tarsus and toes greenish.

DISTRIBUTION: Breeds from northern Alaska, western and southern Canada, the central parts of Minnesota, Wisconsin, Michigan, and Maine southward to eastern Washington, northeastern Oregon, Idaho, southern Montana, northern Utah, southeastern Wyoming, central Colorado, northwestern Oklahoma, north-central Texas, central Missouri, southern Illinois, southern Indiana, southern Ohio, eastern West Virginia, central Virginia, and Maryland. Winters from northern Argentina, Uruguay, and southern Brazil to south-central Argentina and Chile. Florida: July-September; March-May.

489. Spotted Sandpiper, *Actitis macularia* (L.)

IDENTIFICATION: About size of Wilson's Plover (total length 175 to 205 mm). Adult in breeding plumage: upperparts grayish brown; superciliary stripe and *underparts white*, the latter *with rounded black spots*; neck streaked with dusky; base of bill flesh-colored; feet fleshy gray. Adult in winter, immature: similar, but without spots underneath or streaks on neck; *dark of upperparts extending to sides of breast*.

DISTRIBUTION: Breeds from northwestern Alaska, western and central Canada, and Labrador southward to southern California, southern Nevada, central Arizona, northern New Mexico, central Texas, (northern Mississippi and Alabama?), eastern Tennessee, northern Georgia, western North Carolina, Virginia, and eastern Maryland. Winters from southern British Columbia, southwestern Arizona, the southern parts of New Mexico, Texas, Louisiana, Mississippi, and Alabama, and coastal South Carolina southward to northern Chile, central Bolivia, and southern Brazil. Florida: early July-early June.

490. Solitary Sandpiper, *Tringa solitaria* Wilson

IDENTIFICATION: Size of Spotted Sandpiper (total length 190 to 215 mm). *Upperparts dark gray, with rounded white spots*; superciliary line whitish; throat and upper breast with dark streaks; chin, eye-ring, and remainder of underparts white.

DISTRIBUTION AND VARIATION: Breeds from central Alaska and western and central Canada southward to east-central British Columbia, southern Alberta, central Saskatchewan, southern Manitoba, and eastward across central Canada to Labrador. Winters from Baja California, southern Texas, and southeastern Georgia southward to south-central Argentina. Only *T. s. solitaria* is known to Florida; early July-late May, though rare in winter.

491. Greater Yellowlegs, *Tringa melanoleucus* (Gmelin)

IDENTIFICATION: Slightly smaller than the Willet (total length 305 to 380 mm). Head and neck white, longitudinally streaked with blackish; remainder of upperparts blackish, with rounded white spots; underparts white, with crescentic black markings, those on the throat arranged in longitudinal rows; belly almost pure white; *bill almost as long as tarsus* (about twice length of head) *and often slightly upcurved*; tarsus and toes yellow. Plumage less contrasting in winter.

DISTRIBUTION: Breeds in central and southern Alaska and across the southern and central parts of Canada. Winters from southwestern British Columbia, Oregon, central California, southern Nevada, southwestern Arizona, central New Mexico, central Texas, the southern parts of Louisiana, Mississippi, and Alabama, and eastern South

Carolina southward to southern South America. Small numbers summer far south of breeding range, including Florida.

492. Lesser Yellowlegs, *Tringa flavipes* (Gmelin)

IDENTIFICATION: Smaller than the Greater' Yellowlegs (total length 230 to 280 mm) and identical to it in coloration; differs in having *bill straight, much shorter than tarsus, and only slightly longer than head.*

DISTRIBUTION: Breeds in Alaska and at middle latitudes of Canada. Winters from eastern Mexico, the northern Gulf Coast, and South Carolina southward to Chile and Argentina. Small numbers summer far south of the breeding range. Florida: late June to late May or later.

493. Willet, *Catoptrophorus semipalmatus* (Gmelin)

IDENTIFICATION: Near size of Whimbrel (total length 350 to 430 mm), but bill straight. Adult in breeding plumage: upperparts chiefly ashy gray, with dark, wavy crossbarring on back, and blackish longitudinal streaks on head and neck; a white patch on each wing, visible even when wing is folded; throat marked like head and neck; chin and remainder of underparts white, the breast and sides with dark, wavy crossbars. Adult in winter plumage: light gray above and on sides of breast; remainder of underparts, chin, and forehead whitish. Immature: similar to adults in winter, but more brownish, with buffy markings above. Downy young: gray above, with irregular black markings; white below. In all plumages the tarsus, toes, and base of bill are light blue gray.

DISTRIBUTION AND VARIATION: Breeds from eastern Oregon, Idaho, central Alberta, central Saskatchewan, southern Manitoba, and Minnesota southward to northeastern California, Nevada, northern Utah, northern Colorado, and eastern South Dakota; also along coast in southern Nova Scotia and from New Jersey to southern Texas, in the Bahamas and Greater Antilles, and on the coast of Venezuela. *C. s. semipalmatus* is the breeding bird in Florida, but is joined by *C. s. inornatus* for the winter.

494. Red Knot, *Calidris canutus* (L.)

IDENTIFICATION: About size of Killdeer (total length 255 to 280 mm), but more compact, *resembling a large Sanderling.* Adult in breeding plumage: upperparts light to medium gray, with blackish and reddish brown spots and light feather edgings; crown streaked with blackish; superciliary line and *underparts largely reddish brown.* Adult in winter, immature: upperparts light gray, without black or brown markings, except for streaking on crown; underparts whitish, with darker markings on breast and flanks.

DISTRIBUTION AND VARIATION: Breeds in Arctic Zone of Old and New Worlds. *C. c. rufa* winters on the Atlantic and Gulf Coasts from Massachusetts to Argentina; also on the Pacific Coast of South America. A few summer far south of breeding range, including Florida.

495. Purple Sandpiper, *Calidris maritima* (Brünnich)

IDENTIFICATION: About size of Ruddy Turnstone (total length 205 to 240 mm). Winter: *upperparts dark gray*, with some purplish on back; *throat and breast smoky gray*; superciliary line, chin, and remainder of underparts white, the sides with blackish spots; bill yellowish near base; *tarsus and toes orange*. In breeding plumage (rare in Florida) it is browner with streaking on the breast and rusty flecks dorsally.

DISTRIBUTION: Breeds in extreme northern parts of eastern Canada, in Europe, and in Siberia (except eastern). Winters on Atlantic Coast from southwestern Greenland and Newfoundland southward to Florida (westward to Panama City; Stevenson, 1960b); accidental in Texas; also, in the Old World, winters from Iceland, the British Isles, and the Baltic Sea to the Mediterranean Sea. Florida: November to May, rare.

496. Sharp-tailed Sandpiper, *Calidris acuminata* (Horsfield)

IDENTIFICATION: Very similar to the Pectoral Sandpiper, but bill shorter and shafts of all primaries partly white (only the outermost primaries white in *C. melanotos*).

DISTRIBUTION: Breeds in northern Siberia; winters on islands of southwest Pacific Ocean. Rare migrant on Pacific Coast of United States and accidental in southern Florida (Ogden, 1968).

497. Pectoral Sandpiper, *Calidris melanotos* (Vieillot)

IDENTIFICATION: About size of Ruddy Turnstone (total length 205 to 240 mm). Top of head, neck, and upper back buffy, heavily streaked with dusky; feathers of back and wings (except remiges) blackish, with rusty or whitish edgings; primaries dusky; central rectrices dark, outer ones white; *breast buffy, heavily streaked with dusky*; remainder of underparts white; *tarsus, toes*, and base of bill *greenish*.

DISTRIBUTION: Breeds in northern parts of eastern Siberia, Alaska, and Canada. Winters from Peru, southern Bolivia, northern Argentina, and Uruguay southward to southern South America; also in southwest Pacific to Australia and New Zealand; casual in winter in southern United States. Florida: July-November (winter); March-May.

498. White-rumped Sandpiper, *Calidris fuscicollis* (Vieillot)

IDENTIFICATION: Smaller than the Pectoral Sandpiper, which it resembles (total length 175 to 205 mm). Adult in breeding plumage: differs from the Pectoral Sandpiper in having broader black streaks on back and *entirely white upper tail coverts*. Adult in winter: upperparts light gray, intermixed with feathers (chiefly on wings) that are blackish and edged with rusty and white; sides of neck and breast whitish and only faintly streaked with dusky. Immature: similar to adults in breeding plumage, but feathers of upperparts edged with rusty and white, and breast less distinctly streaked. Tarsus, toes, and base of bill greenish.

DISTRIBUTION: Breeds on Arctic Coast of Alaska and Canada. Winters from Paraguay and southern Brazil to southern tip of South America. Florida: late April to June and August to October, but very rare in fall.

499. Baird's Sandpiper, *Calidris bairdii* (Coues)

IDENTIFICATION: Slightly smaller than Pectoral Sandpiper (total length 175 to 190 mm). Top of head, neck, and back buffy, with dusky streaks; feathers of back and wings (except remiges) dark, with *light grayish buff edging*, the latter *giving a scaly effect*; upper tail coverts with a dark center; breast light buffy, less heavily streaked than in Pectoral or White-rumped Sandpipers; *tarsus and toes blackish*.

DISTRIBUTION: Breeds in northeastern Siberia and the Arctic portions of Alaska, Canada, and Greenland. Winters in South America, often at high altitudes. Rare in Florida, July to October and April to May.

500. Least Sandpiper, *Calidris minutilla* (Vieillot)

IDENTIFICATION: One of the smallest of sandpipers (total length 130 to 165 mm). Breeding plumage: very similar to the Pectoral Sandpiper, but much smaller; also streaking on breast less extensive. Winter plumage: upperparts more grayish; breast only indistinctly streaked. At all seasons the *tarsus and toes are yellowish green*. (Compare with Semipalmated and Western Sandpipers.)

DISTRIBUTION: Breeds from Alaska across northern and central Canada, but nonbreeders summer much farther south. Winters from coastal Oregon, southern Nevada, western and central Arizona, southern Utah, central New Mexico, central and northeastern Texas, northern Oklahoma, Arkansas, northern Alabama, and coastal North Carolina southward to the Galapagos Islands, central Peru, and central Brazil. Florida: chiefly July to late May.

501. Dunlin, *Calidris alpina* (L.)

IDENTIFICATION: About size of Pectoral Sandpiper (total length 190 to 230 mm). Breeding plumage: crown and back rusty red, longitudinally streaked with blackish; remainder of upperparts whitish to light gray, the nape streaked with blackish; sides of head, throat, and breast white, with fine dark streaks; *large black patch on anterior part of abdomen*; remainder of underparts white. Winter plumage: upperparts and breast *evenly medium gray*; superciliary line, chin, and remainder of underparts whitish. At all seasons the *bill is black, long, and drooped at the tip*; tarsus and toes blackish.

DISTRIBUTION AND VARIATION: Breeds in northern parts of Eurasia and North America, wintering southward to northern Africa, southern Asia, northwestern Mexico, and the southern United States. *C. a. hudsonia*[1] winters on the Atlantic and Gulf Coasts from Massachusetts to Texas, rarely inland in Florida (chiefly October-May).

502. Semipalmated Sandpiper, *Calidris pusilla* L.

IDENTIFICATION: About size of Least Sandpiper (total length 140 to 165 mm). Breeding plumage: upperparts chiefly gray, with some buffy and with light feather edgings; nape and upper back streaked; underparts chiefly white, but with blackish streaks on throat and upper breast. Winter plumage: no buffy in upperparts; streaks on throat faint. *Bill, tarsus, and toes blackish at all seasons, the bill comparatively short and thick.*

DISTRIBUTION: Breeds in the northern and middle latitudes of Alaska and Canada. Winters from the northern Gulf Coast and coastal South Carolina (sight records only in U.S.?) southward to Peru, northern Chile, Paraguay, and southern Brazil. Transient in Florida, rarely wintering; rare inland except during migration.

503. Western Sandpiper, *Calidris mauri* Cabanis

IDENTIFICATION: Very similar in size and coloration to the Semipalmated Sandpiper (total length 140 to 165 mm), but with more rufous on back and wings in breeding plumage; *bill longer* (at least as long as tarsus) *and noticeably thicker at base.*

DISTRIBUTION: Breeds on coast of northern and western Alaska. Winters from coast of California, the northern Gulf Coast, and coastal North Carolina southward to Colombia, Ecuador, Peru, Venezuela, and the West Indies. Virtually resident in Florida; rare inland except during migration.

---

1. See MacLean and Holmes, *Auk* 88:896.

504. Sanderling, *Calidris alba* (Pallas)

IDENTIFICATION: Near size of Dunlin (total length 180 to 230 mm). Breeding plumage: upperparts and sides of breast largely black, with rusty or white feather edgings on wings and back; black arranged as spots on head and neck; remainder of underparts white. Winter plumage: upperparts mottled gray and white; underparts mostly white (the *palest of Florida sandpipers*). At all seasons the *short bill, tarsus, and toes are black.*

DISTRIBUTION: Breeds in northernmost parts of Canada, Greenland, and Eurasia. Winters on the Pacific Coast from southern British Columbia to southern Chile, on the Atlantic and Gulf Coasts from Massachusetts to southern Argentina, and in the Old World on all coasts except northernmost and New Zealand. Some remain all summer far south of the breeding range, thus the species is resident in Florida, though only a rare transient inland.

505. Buff-breasted Sandpiper, *Tryngites subruficollis* (Vieillot)

IDENTIFICATION: About size of Dunlin (total length 190 to 215 mm). Breeding plumage: head, neck, and upper back alternately streaked with buffy and blackish; feathers of back and wing coverts blackish with buffy edgings; remiges dark gray, marbled with white; sides of head and *underparts generally cinnamon buff*, with darker speckling on throat and sides of breast. Winter plumage: *underparts* largely whitish, *extensively washed with buff. Bill short and slender; tarsus and toes yellowish green.*

DISTRIBUTION: Breeds in northern parts of Alaska and Canada. Winters in central Argentina. Rare transient in Florida, August to September and April to May.

506. Ruff, *Philomachus pugnax* (L.)

IDENTIFICATION: Near size and form of Upland Sandpiper (total length 255 to 315 mm, the female smaller than the male). Breeding plumage: black with purplish iridescence or rusty with dark crossbars; ruff on neck of male. Winter plumage: upperparts grayish; breast lighter gray; remainder of underparts white; ruff absent. The *tarsus and toes are yellowish green*, as in the Pectoral Sandpiper, but the *tarsus is longer*; bill yellowish basally.

DISTRIBUTION: Breeds over Eurasia except in southern parts. Winters from western and southern Europe and southern Asia southward to South Africa, Ceylon, Thailand, and Borneo. Casual in eastern North America and the West Indies. *Sight records only* in Florida, some supported by photographs (Langridge, 1964; Stitt, 1964; Olson, 1965).

507. Stilt Sandpiper, *Micropalama himantopus* (Bonaparte)[1]

IDENTIFICATION: About size of Dunlin, but legs longer (total length 190 to 230 mm). Breeding plumage: top of head, neck, and upper back with alternating light and dark longitudinal streaks; superciliary stripe whitish, bordered above and below with lighter feather edgings; chin and throat longitudinally streaked with white and dark; remainder of underparts crossbarred with blackish and white. Winter plumage, immature: less black in upperparts; underparts mostly white, but with some dark markings on the throat, breast, and sides. In all plumages the *rump and upper tail coverts are essentially white, and the long tarsus and the toes are greenish*.

DISTRIBUTION: Breeds in northern parts of Alaska and Canada. Winters from Bolivia, western Brazil, and Paraguay southward to Uruguay and east-central Argentina, and rarely northward to southern Texas and peninsular Florida. Mainly a transient in Florida, July to October and March to May.

508. Short-billed Dowitcher, *Limnodromus griseus* (Gmelin)

IDENTIFICATION: Similar in size and form to the Common Snipe (total length 240 to 280 mm). Breeding plumage: upperparts buffy, with longitudinal blackish streaks on top of head, neck, and upper back, and large black markings on wings; underparts cinnamon, with blackish rounded spots on breast and belly, and bars on flanks. Winter plumage: medium gray above and on breast; superciliary stripe, chin, and remainder of underparts white, with dark markings on upper breast and sides. At all seasons the *rump and lower back are entirely white*, the long straight bill is greenish toward the base, and the tarsus and toes are greenish.

DISTRIBUTION AND VARIATION: Breeds in southern Alaska, southern Mackenzie, and the northern parts of Alberta, Saskatchewan, and Manitoba (probably farther east). Winters from central California, western Nevada, southern Arizona, southern New Mexico, western Texas, the northern Gulf Coast, and coastal South Carolina southward to northwestern Peru and east-central Brazil. Many remain far south of breeding range in summer. *L. g. hendersoni* and *L. g. griseus* may occur at any time of the year in Florida, most abundantly on the coast.

509. Long-billed Dowitcher, *Limnodromus scolopaceus* (Say)

IDENTIFICATION: Slightly larger than the Short-billed Dowitcher (total length 280 to 305 mm), but not certainly distinguishable from it in the field except by call notes. Extremes may be separated by bill length.

---

1. Submerging this genus in *Calidris* has been proposed recently.

DISTRIBUTION: Breeds along the Arctic Coast from northeastern Siberia to Mackenzie. Winters from central California, western Nevada, southern Arizona, southern New Mexico, west-central Texas, the northern Gulf Coast, and Florida southward to Guatemala. Most frequent around fresh water, September to April.

510. Marbled Godwit, *Limosa fedoa* (L.)

IDENTIFICATION: Slightly larger than the Willet (total length 430 to 530 mm). Head and neck buffy, longitudinally streaked with blackish; feathers of back and wing coverts blackish with buffy edgings; *underparts generally buffy*, with wavy crossbars of dusky (longitudinal streaks on throat); chin and superciliary stripe whitish; *bill very long, upcurved, and flesh-colored toward base*; feet bluish gray. In winter plumage the buffy ground color of the underparts is virtually unmarked.

DISTRIBUTION: Breeds from central Alberta, southern Saskatchewan, and southern Manitoba southward to central Montana, central North Dakota, northeastern South Dakota, and west-central Minnesota. Winters from central California, western Nevada, and southeastern Texas southward to Ecuador, northern Peru, and Chile; also on the coasts of South Carolina, Georgia, and Florida, where virtually resident.

511. Hudsonian Godwit, *Limosa haemastica* (L.)

IDENTIFICATION: Very similar in form to the Marbled Godwit, but smaller (total length 355 to 430 mm). Breeding plumage: similar to the Marbled Godwit, but much darker above and more reddish underneath. Winter plumage: upperparts, line through eye, and breast brownish gray; superciliary line, chin, and abdomen whitish. In all plumages the *tail is white at the base and tip*, unlike that of the Marbled Godwit. Bill and feet as in that species.

DISTRIBUTION: Breeds from northwestern Mackenzie to northern Manitoba. Winters on the coast of Chile, and from Paraguay, southern Brazil, and Uruguay southward to the tip of South America. Very rare transient in Florida (location of specimens unknown).

ORDER CHARADRIIFORMES: CIII. FAMILY RECURVIROSTRIDAE, Avocets and Stilts. Large shorebirds with a *long, slender bill, upcurved in some species*. Legs also very long. Plumage contrasting. *Toes webbed at base*.

512. American Avocet, *Recurvirostra americana* Gmelin

IDENTIFICATION: Size of Willet, but bill and legs longer (total length 380 to 510 mm). Breeding plumage: head, neck, and breast light cinnamon, except for white eye-ring; remainder of upperparts black contrasted with white; abdomen and flanks white; bill long and recurved; lower legs and feet gray blue; iris red. Winter plumage: head, neck, and breast white, tinged dorsally with light gray. Immature; similar to winter adult, but with a rufous tinge on neck and buffy feather edgings dorsally. Hallux present.

DISTRIBUTION: Breeds from south-central Oregon, east-central Washington, southern Idaho, northern Montana, central Alberta, southern Saskatchewan, and southern Manitoba southward to southern California, southern Nevada, northern Utah, southern parts of New Mexico and Texas, and eastward to eastern parts of North and South Dakota, western Nebraska, eastern Colorado, and north-central Oklahoma. Winters from north-central California and southern Texas southward to Guatemala and eastward to Florida (August to May, occasionally remaining throughout summer).

513. Black-necked Stilt, *Himantopus mexicanus* (Müller)

IDENTIFICATION: Similar in form to American Avocet, but smaller (total length 330 to 380 mm). Adult: *upperparts more or less black* (darker and glossier in male); forehead, incomplete eye-ring, and *entire underparts white*; bill long, slender, and straight, or only slightly upcurved; *legs very long and pinkish*. Immature: similar, but with rusty feather edgings on crown, wings, and back. *Hallux lacking*. Downy young: upperparts brownish gray, mottled with dusky and with several large black spots; head, neck, and underparts buffy.

DISTRIBUTION: Breeds from southern Oregon, Idaho, southern Saskatchewan, northern Utah, southern Colorado, eastern New Mexico, and coastal Texas and Louisiana southward to northeastern Baja California and (locally) to Nicaragua; also on the Galapagos Islands, on the coast of Ecuador (and Peru?), and from southern New Jersey southward through peninsular Florida to the Bahama Islands, Greater Antilles, northern Lesser Antilles, and northern South America to Brazil. Winters from central California, northwestern Mexico, southern Texas, the Louisiana coast, and (rarely) southern Florida southward to limit of breeding range. Occurs in northwest Florida only as a rare transient.

ORDER CHARADRIIFORMES: CIV. FAMILY PHALAROPODIDAE, Phalaropes. Small shorebirds with colorful breeding plumage and *lobate toes*. Wings long and pointed. Tarsus scutellate, serrate behind.

514. Red Phalarope, *Phalaropus fulicarius* (L.)

IDENTIFICATION: Size of Sanderling, which it resembles in winter plumage (total length 190 to 230 mm). Female in breeding season: top of head and back of neck mostly blackish; remainder of upperparts black-ish with buffy feather edgings; remiges dark gray, with white patch; sides of head white; *underparts brick red*; *bill comparatively short and thick, yellow toward base*; *tarsus and toes yellowish in life*. Male in breeding plumage: similar to female, but differs in having top of head and back of neck streaked with blackish and buff, and the underparts paler. Winter plumage, immature: upperparts chiefly dark gray to medium gray; a dusky patch on side of head; forehead, sides of neck, and entire underparts white (immatures have some buff in plumage); bill, tarsus, and toes duller than in breeding adults.

DISTRIBUTION: Breeds in Arctic Zone of North America and Eurasia. Winters in Atlantic and Pacific Oceans off Africa and South America, rarely farther north; also in Gulf of Mexico off Pensacola. Otherwise rare transient in Florida.

515. Wilson's Phalarope, *Steganopus tricolor* Vieillot

IDENTIFICATION: Slightly larger than Sanderling (total length 205 to 255 mm). Female in breeding plumage: top of head, lower neck, and back light gray; occiput, nape, *superciliary stripe*, and chin *white*; stripe on *side of head and neck blackish, becoming reddish as it continues down neck* and onto back; stripe across wing reddish, the remainder of wing pearl gray; throat and breast buffy; upper tail coverts and abdomen white; bill, tarsus, and toes dull-colored. Male in breeding plumage; similar, but duller, completely lacking the chestnut and gray stripes on wings and back; top of head brown. Winter plumage: upperparts mainly light gray; face and underparts white; feet greenish yellow. *Bill long and slender* (as long as tarsus).

DISTRIBUTION: Breeds from the interior of British Columbia, cen-tral Alberta, central Saskatchewan, northeastern Manitoba, central Minnesota, and southern Wisconsin, Michigan, and Ontario southward to south-central California, central Nevada, western and northern Utah, northeastern Colorado, central Kansas, western Nebraska, eastern South Dakota, and northern Indiana. Winters in Chile and Argentina. Accidental in western Europe. Florida: rare, August to September and April to June.

516. Northern Phalarope, *Lobipes lobatus* (L.)

IDENTIFICATION: About size of Sanderling (total length 175 to 205 mm). Female in breeding plumage: upperparts mostly dark gray, striped with buffy on back; throat and incomplete eye-ring white; *lower part of throat and sides of neck reddish brown*; remainder of underparts chiefly white, with dark markings on sides. Male in breeding plumage: similar to female, but with less contrasting color above; face mostly white. Winter plumage: upperparts mostly dark gray, with white feather edgings (buffy in immature); auriculars blackish, surrounding an incomplete white eye-ring; usually some buff on sides of neck; underparts white, with grayish flecks along sides. In all plumages the tarsus, toes, and *slender bill* are dull-colored.

DISTRIBUTION: Breeds in northern Eurasia, Alaska, the northern half of Canada, and on the coast of Greenland. Winters in the Pacific Ocean off Ecuador, Peru, and Chile; in the Atlantic Ocean off southern Argentina and off north Africa (rarely farther north); also off Arabia, India, Malaya, southern Japan, China, the East Indies, and New Zealand. Rare migrant in Florida, August to December and April to May.

ORDER CHARADRIIFORMES: CV. FAMILY STERCORARIIDAE, Jaegers and Skuas. Gull-like birds with hooked bills, white in the primaries, and (usually) elongate central rectrices. Front toes webbed. *Nostrils far out on bill. Cere and 2 other sheaths covering upper mandible.*

517. Pomarine Jaeger, *Stercorarius pomarinus* (Temminck)

IDENTIFICATION: *Size of Ring-billed Gull* (total length 51 to 58 cm). Adult: head blackish; nape white; remainder of upperparts dusky; *central rectrices* 75 to 100 mm longer than others, *blunt, and twisted*; auriculars tinged with yellow; underparts varying from white (light phase) to dusky (dark phase). All degrees of intermediacy between the two extremes are known to occur. Immature: similar to adults, but underparts more or less barred with buffy, and *central rectrices scarcely longer than others*. In all (?) plumages the *tarsus is yellowish*, the toes dusky, and several outer primaries have white shafts.

DISTRIBUTION: Breeds on Arctic Coast and offshore islands of North America and Eurasia. Winters in Atlantic Ocean from Cape Hatteras to the West Indies and off the coast of Africa; in the Pacific Ocean from southern California to Peru and off eastern Australia; casual inland. Florida: September-May.

518. Parasitic Jaeger, *Stercorarius parasiticus* (L.)

IDENTIFICATION: *Size of Laughing Gull* (total length 41 to 53 cm). Adult: coloration similar to corresponding phases of Pomarine Jaeger, but *central rectrices pointed*, projecting 75 to 100 mm beyond others. Immature: considerably more buffy on head, neck, and back than young of Pomarine Jaeger; central rectrices short, but *pointed*. In all plumages, tarsus and toes dusky, and several outer primaries with white shafts.

DISTRIBUTION: Breeds in Arctic portions of North America and Eurasia. Winters in the Pacific Ocean from southern California to southern Chile, northern and eastern Australia, and New Zealand southward; in the Atlantic Ocean from Maine and the British Isles south to northern

Argentina and the west coast of Africa. Frequent in summer far south of breeding range. Casual inland. Florida: September-May (summer).

519. Long-tailed Jaeger, *Stercorarius longicaudus* Vieillot

IDENTIFICATION: About size of Parasitic Jaeger, but *central rectrices longer* (up to half the total length of 43 to 58 cm in adults). Adult: similar to light-phase adults of other jaegers, but entire neck and sides of head white (except where tinged with yellow), back lighter gray, and *central rectrices pointed and projecting 20 to 25 cm beyond other tail feathers*; *tarsus light blue gray in life.* (Dark phase unknown.) Immature: upperparts chiefly brownish gray, streaked with whitish on head, neck, and throat, and with light feather edgings on back; breast mostly brownish gray; abdomen light, with crossbars of dusky; central rectrices scarcely longer than other tail feathers, but pointed. In all plumages *only 2 outer primary shafts are white* (others light).

DISTRIBUTION: Breeds in extreme northern North America and Eurasia. Winters in the Atlantic Ocean from latitude of New York and Spain southward to that of southern South America; also in Pacific Ocean from Peru to southern Chile. Very rare transient in Florida.

ORDER CHARADRIIFORMES: CVI. FAMILY LARIDAE, Gulls and Terns. [1] Bill somewhat less hooked (if at all), but relatively longer, than in jaegers; nostrils placed far out on bill, especially in gulls. Plumage usually with black or gray and white. Front toes webbed. Central rectrices not elongate.

520. Glaucous Gull, *Larus hyperboreus* Gunnerus

IDENTIFICATION: Larger than Herring Gull (total length 66 to 81 cm). Adult: entire plumage apparently white (the mantle being very light gray); bill and iris yellow; tarsus and toes flesh-colored. Immature: *pale puffy or creamy* flecked with dusky in first winter, except for *white primaries*; in second winter, *entirely white*; bill, tarsus, and toes flesh colored, the bill with a dusky tip. *Bill longer and heavier than in Iceland Gull.*

DISTRIBUTION AND VARIATION: Breeds in northern parts of North America and Eurasia, mainly within the Arctic Circle; winters southward to 30 or 35 degrees North latitude, *L. h. hyperboreus* occasionally reaching Florida; December to May, rarely summering.

521. Iceland Gull, *Larus glaucoides* Meyer

IDENTIFICATION: Slightly smaller than Herring Gull (total length 51 to 66 cm). Coloration almost identical to corresponding ages of the

---

1. A Band-tailed Gull (*Larus belcheri*) was photographed near Naples in the winters of 1974–75 and 1975–76.

Glaucous Gull, but *bill noticeably smaller*; also *wing tips usually extending beyond tail tip at rest*.

DISTRIBUTION AND VARIATION: Breeds in Iceland, Greenland, and extreme northern Canada, wintering southward to Europe, southern California, and Virginia (a few to Florida; *L. g. glaucoides*; December to April, rarely summering).

522. Great Black-backed Gull, *Larus marinus* L.

IDENTIFICATION: About size of Glaucous Gull (total length 70 to 80 cm). Adult: *mantle very dark gray*; remainder of plumage entirely white; bill and iris yellow; tarsus and toes light flesh-colored. Immature: upperparts whitish with brown mottling; remiges dark brown; underparts whiter, but also with faint flecks of brownish; bill light at base, remainder dark. In this plumage it is *whiter above and below than first-year Herring Gull, and the bill is noticeably heavier*.

DISTRIBUTION: Breeds along Atlantic Coast from central Greenland to North Carolina; also on northern coasts of Europe. Winters from Newfoundland to Florida and Alabama, but rare in Gulf of Mexico (Williams, 1963); also in Europe southward to the Mediterranean, Black, and Caspian Seas; occasional in summer far south of breeding range. Florida: chiefly October to May, but also summering.

523. Herring Gull, *Larus argentatus* Pontoppidan

IDENTIFICATION: The largest common Florida gull (total length 58 to 66 cm). Adult: mantle pearl gray, the wing tips black with white spots; remainder of plumage white; *bill yellow*, with red spot near tip. Immature: *entire plumage dark brownish gray*, mottled with whitish; flight feathers darker; in this plumage much darker, especially below, than immature of *L. marinus*; bill flesh-colored at base, dusky at tip. Second-year birds are whiter except for the *dark rectrices*. At all ages the *tarsus and toes are flesh-colored*.

DISTRIBUTION AND VARIATION: Breeds in Alaska, Canada, the northern United States (Montana to New York), south on Atlantic Coast to North Carolina, and in northern Eurasia. Winters from southern edge of breeding range to Panama, Bermuda, and Barbados (*L. a. smithsonianus*); also to northern and central Africa and southern Asia. Frequent in summer far south of breeding range, including Florida.

524. Ring-billed Gull, *Larus delawarensis* Ord

IDENTIFICATION: Smaller than Herring Gull (total length 45 to 51 cm). Adult: plumage almost identical to that of adult Herring Gull, but the yellowish *bill has a black ring near its tip*; *tarsus and toes usually yellowish to greenish*; iris yellow. Immature: upperparts white, mottled with brownish gray on wings, back, neck, and head; primaries mostly dark; *tail* chiefly white toward base, but *with a narrow, blackish, subterminal band*; underparts mostly whitish; bill light toward base, dusky at tip; legs greenish or pinkish.

DISTRIBUTION: Breeds from south-central Oregon, central Washington, Alberta, north-central Saskatchewan, and south-central Manitoba southward to northeastern California, south-central Idaho, south-central Colorado, southeastern Wyoming, and northeastern South Dakota; also from central Quebec and northeastern Newfoundland southward to northern Michigan; southern Ontario, and northern New York. Winters from Oregon along Pacific Coast to southern Mexico (casually El Salvador), around larger lakes in the interior of Mexico and the United States, and along the Atlantic and Gulf Coasts from the Gulf of St. Lawrence to Cuba. Many summer far south of breeding range, including Florida.

525. Laughing Gull, *Larus atricilla* L.

IDENTIFICATION: Smaller than Ring-billed Gull (total length 38 to 43 cm). Adult in breeding plumage: *head black*, but for incomplete, white eye-ring; *mantle dark gray*, except wing tips black; remainder of plumage white; bill, tarsus, and toes reddish. Adult in winter: *bill black*; head white with dusky markings on sides and occiput; otherwise as in breeding plumage. Immature: upperparts chiefly *dark* brownish gray, shading to blackish on primaries and near tips of secondaries; tail whitish with a subterminal blackish band; underparts largely white, but grayish on throat and breast; *bill, tarsus, and toes blackish*. Downy young: upperparts whitish, mottled with blackish; black spots large and distinct on top and sides of head and throat; rest of throat and belly smoky buff; breast buffy white.

DISTRIBUTION: Breeds locally on Atlantic and Gulf Coasts from Nova Scotia to Venezuela; also on the Pacific Coast of northwestern Mexico and in southeastern California (inland). Winters on the Pacific Coast from southern Mexico to northern Peru, and on the Atlantic and Gulf Coasts from North Carolina to Brazil. Occasional inland.

526. Franklin's Gull, *Larus pipixcan* Wagler

IDENTIFICATION: Slightly smaller than the Laughing Gull (total length 35 to 38 cm). Adult in breeding plumage: similar to corresponding plumage of the Laughing Gull, but *with a white band between gray and black of primaries* (which are also tipped with white); secondaries more extensively tipped with white; middle rectrices pearl gray, outer ones white; *bill dark red*. Adult in winter: similar, but front of head white and hind part dusky; less white in wings; bill and feet dark. Immature: upperparts chiefly grayish brown; *upper tail coverts, forehead, and underparts white*; bill and feet dusky.

DISTRIBUTION: Breeds from southeastern Alberta, central and southern Saskatchewan, and southwestern Manitoba southward to east-central Oregon, south-central Montana, northwestern Utah, eastern North Dakota, southwestern Minnesota, northeastern South Dakota, and northwestern Iowa. Winters on the Pacific Coast from Guatemala to the Gulf of Panama, the Galapagos Islands, and southern Chile; also on Gulf of Mexico from Texas to Louisiana, casually to Florida (where chiefly a rare transient in late fall) and Puerto Rico.

527. Bonaparte's Gull, *Larus philadelphia* (Ord)

IDENTIFICATION: The smallest of the common Florida gulls (total length 30 to 36 cm). Adult in breeding plumage: *head black; mantle pale gray*, but primaries white with black tips; remainder of plumage white; bill black; tarsus and toes reddish. Adult in winter: similar, but *head white with a dark spot behind* eye; tarsus and toes flesh-colored. Immature: similar to winter adults, but mantle darker, and a subterminal blackish tail band present; legs yellowish. Adults in spring may be confused with the larger Laughing Gull, but have a lighter back and show *much white in wings* in flight.

DISTRIBUTION: Breeds in western and central Alaska and western Canada. Winters on the Pacific Coast from Washington to southern Baja California and Jalisco; on the Atlantic and Gulf Coasts from Maine to Texas and Yucatan; also in the interior at Lakes Erie and Ontario and occasionally elsewhere; more frequent inland during migration. Florida: chiefly November to May.

528. Little Gull, *Larus minutus* Pallas

IDENTIFICATION: *Even smaller than Bonaparte's Gull* (total length 26 to 29 cm). Adult in winter: mantle medium gray; occiput and nape somewhat darker; forehead, tail, and *tips of remiges white*; underparts white, except for *contrastingly dark wings*. Immature: similar, but with a subterminal black tail band, and the *axis of the outstretched wing is black above* (as in Kittiwake); also under surface of wing white. *Bill and feet black in winter*.

DISTRIBUTION: Breeds on mainland of northern Europe and into central Siberia; one breeding record in Ontario. Winters from western part of breeding range, Iceland, and the British Isles southward to the Mediterranean, Black, and Caspian Seas. Casual in Canada, Greenland, and the eastern United States (few Florida records and one specimen; Monroe, 1959b).

529. Black-legged Kittiwake, *Rissa tridactyla* (L.)

IDENTIFICATION: Near size of Laughing Gull (total length 38 to 43 cm). Adult in breeding plumage (unknown in Florida?): similar to corresponding plumage of the larger Ring-billed Gull, but with less black on wing tips, *no black band on bill*, and the *tarsus and the 3 toes (rarely 4) black*. Adult in winter: similar, but *with dusky coloration on back of head and neck*; base of bill dusky, tip yellowish. Immature: similar to winter adult, but darker on back of head and neck, *wing axis black* as in Little Gull, and a black terminal tail band present (tail notched); bill black. *Feet always dark.*

DISTRIBUTION AND VARIATION: Breeds on the coasts of Alaska, northern Canada (south to Newfoundland), Greenland, Iceland, the British Isles, and northern Eurasia. Winters from southern parts of breeding range southward to New Jersey, Florida, and the northern Antilles, but very rare on the west coast (*R. t. tridactyla*) (rarely inland); also from British Columbia southward to northwestern Baja California; and in the Old World to western Africa, the Mediterranean Sea, and Japan. Florida: chiefly November to March, rare.

530. Sabine's Gull, *Xema sabini* (Sabine)

IDENTIFICATION: About size of Bonaparte's Gull (total length 33 to 36 cm). Adult in breeding plumage (unknown in Florida): head and throat slate-colored to black; neck, tail, secondaries, tertiaries distally, and entire underparts white; primaries chiefly black; *bill black with yellow tip*. Adult in winter: differs from breeding plumage in its white head and neck. Immature: forehead, lores, base of tail, and entire underparts white; wing pattern as in adult; crown, nape, and back ashy brown, with light edgings on back feathers. The best field marks are the *white triangle on the upper wing surface* (when spread) and the *decidedly forked tail*.

DISTRIBUTION AND VARIATION: Breeds on Arctic Coast of Alaska, Canada, and Eurasia. Winters chiefly in the southern parts of the Atlantic and Pacific Oceans (one Florida specimen, probably *X. s. sabini*).

531. Gull-billed Tern, *Gelochelidon nilotica* (Gmelin)

IDENTIFICATION: Near size of Bonaparte's Gull (total length 33 to 38 cm). Adult in breeding plumage: top of head and nape black; *remainder of upperparts evenly pearl gray*; underparts white; bill, tarsus, and toes black; *bill thick and short; tail short and only slightly forked*. Adult in winter: similar, but without black on head and neck; dark spots in front of and behind eye. Immature: similar to winter adult, but upperparts with buff, and bill yellowish basally. Downy young: upperparts gray buff, with large dusky spots on head, wings, rump, and flanks, and 2 dark stripes down neck and back; underparts whitish; feet and base of bill orange.

DISTRIBUTION AND VARIATION: Southern United States to South America; also western and southern Europe and southern Asia to central Africa, Australia, and Tasmania. *G. n. aranea* breeds from the coast of Maryland and the northern Gulf Coast (also interior of Florida Peninsula) southward to the Bahama and Virgin Islands, and winters from the coast of Texas and Louisiana, Tampa Bay, and northeastern Florida southward to northern South America. Uncommon in northwest Florida (coast only), April to August.

532. Forster's Tern, *Sterna forsteri* Nuttall

IDENTIFICATION: Near size of Gull-billed Tern, but tail longer (total length 35 to 38 cm). Adult in breeding plumage: similar to corresponding plumage of the Gull-billed Tern, but mantle somewhat darker, primaries whiter, and *tail longer and more deeply forked* (tail *may* protrude well beyond folded wings); bill orange toward base, dusky near tip; tarsus and toes reddish. Adult in winter: similar, but top of head white, *nape dusky*, and stripe across eye black. Immature: washed with brownish on back; remiges dusky. In all plumages the *inner web of the outermost tail feather is dark*. The most abundant small tern in winter, but compare with Common and Roseate Terns.

DISTRIBUTION: Breeds from south-central Alberta, southern Saskatchewan, and southern Manitoba southward to south-central California, north-central Utah, eastern Colorado, western Nebraska, eastern South Dakota, north-central Iowa, southern Minnesota, and southern Wisconsin; also from northeastern Mexico to the Alabama coast, and on the coasts of Maryland and Virginia. Winters from central California to Guatemala, and from the northern Gulf Coast, the interior of northern Florida, and the coast of Virginia southward to eastern Mexico and the Greater Antilles. Small numbers summer far south of breeding range, therefore resident in Florida.

533. Common Tern, *Sterna hirundo* L.

IDENTIFICATION: About size of Forster's Tern (total length 33 to 38 cm). Similar to corresponding plumages of Forster's Tern, but differs in having *outer* web of the outermost tail feather darker than its inner web; *primaries never pure white*; *immatures have the leading edge of wing dark near body*. Foot color as in Forster's Tern, but base of bill redder in breeding plumage and more yellow in winter. In winter all show a *dark streak on side of head beginning at the eye and continuing around the nape*. Throat and breast often light gray in breeding adults. Downy young: upperparts buffy with darker mottling; some distinct, rather large blackish spots on back of head, nape, and rump; sides of throat smoky gray; remaining underparts whitish.

DISTRIBUTION AND VARIATION: Breeds over most of Eurasia, in southern Canada, and the northern and eastern United States. *S. h. hirundo* breeds from south-central Mackenzie, northern Saskatchewan, central Manitoba, southeastern Ontario, southern and eastern Quebec, southeastern Labrador, Newfoundland, New Brunswick, and Nova Scotia southward to southeastern Alberta, northeastern Montana, North Dakota, northeastern South Dakota, central Minnesota, northeastern Illinois, northwestern Indiana, northern Ohio, northwestern Pennsylvania, central New York, and northwestern Vermont; also along the Atlantic Coast from Maine to North Carolina and on the coast of Texas; recent Florida nestings on coast of Panhandle (Hallman, 1961) and at Pompano Beach (McGowan, 1969); may have nested formerly on the Dry Tortugas (but see Robertson, 1964), and nests at Bermuda. Winters from Baja California, the northern Gulf Coast, and South Carolina along the coasts southward to southern Ecuador and southern South America.

534. Roseate Tern, *Sterna dougallii* Montagu

IDENTIFICATION: Slightly larger than Least Tern, but tail much longer (total length 33 to 38 cm). Similar to corresponding plumages of Forster's Tern, but *upperparts paler, tail relatively longer and entirely white*, and underparts tinged with pale pinkish in breeding plumage; bill red basally and black toward tip in breeding season. Immature: streaked on top of head; feathers on back margined with buff. Downy young: similar to downy Common Tern, but dorsal spotting finer (as in Royal Tern); general coloration may be grayish or buffy; feet more brownish.

DISTRIBUTION AND VARIATION: Breeds at scattered coastal points over much of the world, but not in South America. *S. d. dougallii* breeds locally along the Atlantic Coast from Nova Scotia to Virginia, and on the Dry Tortugas and occasionally the Florida Keys (Bonney and Johnston, 1964); also in the Bahamas, the West Indies, western Europe, and South Africa. Winters from the West Indies to Brazil and from the Azores to South Africa; casual in the southeastern United States in winter. Florida: chiefly April to September, but generally rare.

535. Sooty Tern, *Sterna fuscata* L.

IDENTIFICATION: About size of Forster's Tern (total length 38 to 43 cm). Adult: *upperparts sooty black* (darker on top of head), except for white on forehead and outer rectrices; *underparts entirely white*; bill, tarsus, and toes black. Immature: *mostly sooty brown*, but somewhat grayer underneath, and with white under wing coverts; feathers of back tipped with white. Downy young: upperparts, head, and throat dark gray with silvery tinge, finely spotted with whitish; remaining underparts white.

DISTRIBUTION AND VARIATION: Breeds chiefly on remote oceanic islands of the Atlantic, Pacific, and Indian Oceans. *S. f. fuscata* breeds on islands in the Gulf of Mexico (including the Dry Tortugas) and in the West Indies. Adults winter in Gulf and Caribbean areas, but juvenals migrate to the west coast of Africa, remaining there for several years (Robertson, 1964 and 1969). Storm-borne individuals may appear anywhere in the eastern United States in summer and early fall. Florida: mostly April to September, but accidental in winter.

536. Bridled Tern, *Sterna anaethetus* Scopoli

IDENTIFICATION: Similar in size and coloration to the Sooty Tern (total length 35 to 38 cm). Adult in summer: like adult Sooty Tern, but upperparts less dark and *interrupted by white on nape*, and the white of forehead extending to a point behind eye. Adult in winter, immature: upperparts mostly blackish, with lighter feather edgings, but top of head white with blackish streaks; underparts and *nape whitish*.

DISTRIBUTION AND VARIATION: Breeds on scattered oceanic islands of the world. *S. a. recognita* breeds off the east coast of British Honduras and in the West Indies; storm-borne individuals in spring and summer have appeared in Alabama, Florida, and South Carolina; also probably regular in summer off south Florida, and accidental in the state in winter (TT 2818); normal winter range imperfectly known. Florida: April-September.

537. Least Tern, *Sterna albifrons* Pallas

IDENTIFICATION: The smallest of Florida terns (total length 21 to 26 cm). Adult in breeding plumage: crown, nape, and line through eye black; upperparts pale pearl gray; *forehead white* (extending to a point above eye); underparts white; *tarsus, toes, and bill mostly yellow*. Adult in winter (August to February): black of head largely replaced by gray; bill, tarsus, and toes duller. Immature: similar to adult in winter, but with dark markings on wings, back, and upper side of tail; remiges buffy at first. (In this plumage, similar to immature Common Tern.) Downy young: upperparts mostly white, spotted with dusky.

DISTRIBUTION AND VARIATION: Breeds from central California southward to Peru; in the interior along the Colorado, Red, Missouri, and Mississippi Rivers as far north as Nebraska and Iowa; and along the Atlantic and Gulf Coast from Massachusetts to eastern Brazil, including the West Indies; also from Europe and southern Asia southward to central Africa and Australia. *S. a. antillarum* breeds inland in small numbers in the Florida Peninsula and abundantly on all coasts, wintering off northeastern Brazil, casually northward to Florida. *S. a. athalassos* may be expected occasionally in the state in migration. Florida; late March to early October.

538. Royal Tern, *Thalasseus maximus* (Boddaert)

IDENTIFICATION: A gull-sized tern (total length 46 to 54 cm, at least one-third of which is the deeply forked tail). Adult in breeding plumage: top of head black and crested; upperparts pale gray (appearing white at a distance); underparts white; *bill and feet orange to yellow*. Adult in winter: forehead white, and crown streaked with same. Immature: similar to adults in winter, but with dark markings on wings. Downy young: highly variable; ground color light buffy, overlaid with no, little, or much dark coloration; color of bill and feet variable.

DISTRIBUTION AND VARIATION: Breeds locally from Baja California to west-central Mexico and from Maryland and the northern Gulf Coast southward to the southern Caribbean; also on the west coast of Africa. Winters from central California and North Carolina southward to Peru and Argentina; casually strays farther north or inland. Recent Florida breeding at 5 localities on Atlantic and Gulf coasts.

539. Sandwich Tern, *Thalasseus sandvicensis* (Latham)

IDENTIFICATION: Slightly larger than Forster's Tern (total length 35 to 41 cm) and *slightly crested*. Similar to corresponding plumages of the Royal Tern, but *bill of adult black with yellow tip*; in immature, bill largely yellow, later almost entirely black. Downy young: mostly white or buffy white, but sometimes with fine, dark mottling.

DISTRIBUTION AND VARIATION: Breeds from Virginia to Duval County, Florida (R. Loftin, personal communication), from southern Texas to the Florida Panhandle (Port St. Joe, Stevenson, 1972b; formerly in the Peninsula), in the Bahamas, on islands off Yucatan, and in western and southern Europe. The American race, *T. s. acuflavidus*, winters from southern Mexico to Panama, and from the Louisiana coast and Florida to Argentina. Resident over most of Florida coastline, but uncommon on Atlantic and not wintering in Panhandle.

540. Caspian Tern, *Hydroprogne caspia* (Pallas)

IDENTIFICATION: The largest of Florida terns (total length 48 to 58 cm). Similar to corresponding plumages of the Royal Tern, but differs in the following respects: *bill thicker and red*; *tail shorter and less deeply forked, usually not extending to tip of folded wing*; crest less apparent;

primaries darker, especially on under side. Downy young: mostly grayish white or buffy white, usually heavily mottled or spotted with dusky.

DISTRIBUTION: Breeds locally from central Mackenzie and Manitoba southward to eastern Washington and Oregon, western Nevada, southern California, northern Utah, and northwestern Wyoming; also in Baja California, and in northeastern Wisconsin, Michigan, southeastern Canada, northwestern Pennsylvania, and on the coast of Virginia, South Carolina, Louisiana, southern Texas, and Florida (Hillsborough and Pinellas Counties; Meyerriecks, 1963; Rohwer, 1968; Schreiber and Dinsmore, 1972; and Brevard County; Robert Barber, personal communication). Also in much of Europe and southern Asia, Australia, New Zealand, and Africa. Winters from central California southward to southern Baja California; from North Carolina along the Atlantic and Gulf Coasts to Texas and the Greater Antilles; and from the Mediterranean Sea to southern Africa, Australia, and New Zealand. Sometimes seen inland in Florida.

541. Black Tern, *Chlidonias niger* (L.)

IDENTIFICATION: Scarcely larger than Least Tern (total length 23 to 26 cm). Adult in breeding plumage: *upperparts chiefly dark gray; head, neck, and underparts black*, except under tail coverts white. Adult in winter: upperparts as in summer; head, neck, and underparts white (many are in mottled intermediate plumage in late summer). Immature: similar to winter adult, but with brownish feather edgings above and gray on sides. Feet purplish red.

DISTRIBUTION AND VARIATION: Breeds from southeastern British Columbia, Alberta, Saskatchewan, Manitoba, Ontario, New Brunswick, and Maine southward to California, northern Nevada, northern Utah, Colorado, Nebraska, Missouri, Kentucky, Ohio, Pennsylvania, and western New York; also over most of Europe. *C. n. surinamensis* winters on the Pacific Coast from Panama to Peru, and on the Atlantic from Panama to Surinam (also sight records in winter in southern United States). Regular transient in Florida, rare in spring (April-June), but common on Gulf Coast in fall, when regular throughout state (July-October).

542. Brown Noddy, *Anous stolidus* (L.)

IDENTIFICATION: Size of Sandwich Tern (total length 35 to 41 cm). *Crown white, shading through gray on nape and sides of head to brownish black over remainder of body, wings, and the unforked tail*; immatures are darker on top of head. Only the Black Noddy, accidental in Florida, is of similar color. Downy young: essentially whitish or dusky (2 color phases).

DISTRIBUTION AND VARIATION: Breeds on islands in warmer parts of the world. *A. s. stolidus* breeds on the Dry Tortugas, in the Bahamas, off the coasts of Yucatan, British Honduras, Jamaica, and Hispaniola, on the Virgin Islands, the Lesser Antilles, off the coast of Venezuela, and on scattered islands eastward to the west coast of Africa. Presumably winters at sea. Storm-borne birds have appeared along entire Atlantic Coast of Florida and northward to South Carolina and Bermuda, and westward to southern Louisiana and eastern Mexico. Florida: April-September.

543. Black Noddy, *Anous tenuirostris* (Temminck)

IDENTIFICATION: Distinguishable from the Brown Noddy by its *more blackish* (less brownish) *plumage and more slender bill*; crown and tail as in Brown Noddy.

DISTRIBUTION AND VARIATION: Breeds on small islands in the central and southwestern Pacific Ocean, the tropical southern Atlantic Ocean, and off the coasts of British Honduras and Venezuela; a few summer on the Dry Tortugas (*A. t. atlanticus*; Robertson *et al.*, 1961).

ORDER CHARADRIIFORMES: CVII. FAMILY RYNCHOPIDAE, Skimmers. Gull-sized water birds with dark upperparts and contrasting white underparts; some bright color on *bill* (yellow to red), which *is extremely compressed, with the lower mandible protruding far beyond the upper.* The long wings and forked tail are ternlike.

544. Black Skimmer, *Rynchops niger* L.

IDENTIFICATION: About size of Royal Tern (total length 43 to 51 cm). *See description of family.* Adult in breeding plumage: upperparts chiefly black, but with white tips on secondaries, white outer rectrices, and white forehead; underparts entirely white; bill red toward base, dark distally; feet red. Adult in winter: similar, but more brownish above and the nape white. Immature: similar, but with buff in upperparts and on side of head. Downy young: upperparts largely blackish, but with light buffy feather edgings; underparts white; bill not hypognathous.

DISTRIBUTION AND VARIATION: Breeds locally on the Atlantic and Gulf Coasts from Massachusetts to Yucatan, and in northern and eastern South America; also on the Pacific Coast from southern California to southern South America. *R. n. niger* winters from the northern Gulf of Mexico (casually up Atlantic Coast to North Carolina) southward to eastern Mexico, Cuba, and the Virgin Islands, and from southern California to Nicaragua. No Florida breeding records south of Fort Myers and Brevard County; regular inland at Orlando, Lakeland, and Lake Okeechobee; occasionally elsewhere, especially after storms.

ORDER CHARADRIIFORMES: CVIII. FAMILY ALCIDAE, Auks, Murres, and Puffins. Water birds with webbed front toes and the hallux vestigial or lacking. Bill laterally compressed in most and not noticeably hooked. Tail short; wings small, with short secondaries. Plumage dark, usually with contrastingly white underparts.

545. Razorbill, *Alca torda* L.

IDENTIFICATION: Near size of Ring-billed Gull (total length 38 to 45 cm). Winter adult: upperparts (except side of head and neck and tip of secondaries) black; remainder of plumage white. *Bill very high* (culmen arched) *and compressed, with vertical ridges and a light bar.* Immature: similar, but bill smaller and without ridges or a bar. Hallux lacking.

DISTRIBUTION AND VARIATION: Breeds in northernmost parts of North America and Europe (down Atlantic Coast to Maine). Winters southward to western Mediterranean and South Carolina (one Florida record of *A. t. torda*; Cruickshank, 1967).

546.   Dovekie, *Alle alle* (L.)

IDENTIFICATION: Total length 17 to 23 cm, *the tail being extremely short.* Adult in winter, immature: upperparts blackish; underparts and sides of neck white; neck short; *bill black, stubby, and cylindrical.* Breeding plumage (unlikely in Florida): head, neck, and throat entirely dark. Hallux lacking.

DISTRIBUTION AND VARIATION: Breeds in Greenland, Iceland, and on islands off northern Europe. Winters from the breeding range southward to New Jersey, the Azores, and the Mediterranean and Baltic Seas, casually farther. Florida records are predominantly in late November and early December and very erratic, sometimes involving thousands of birds; accidental in Gulf of Mexico.

ORDER COLUMBIFORMES: CIX. FAMILY COLUMBIDAE, Pigeons and Doves. Short-legged land birds with the *bill constricted near the middle and provided with a basal cere.* Neck and tarsus short; *head small*; wings pointed. Iridescent plumage frequent.

547.   White-crowned Pigeon, *Columba leucocephala* L.

IDENTIFICATION: Size of Rock Dove (total length 305 to 355 mm). *Top of head white* (light grayish buff in female); *remainder of plumage slaty gray*, with iridescent bronze green on lower neck and upper back; bill reddish with white tip; iris and orbital skin white; tarsus and toes orange red.

DISTRIBUTION: Breeds from the Florida Keys (southern mainland rarely?) and Bahama Islands southward to islands off the coast of Central America. Winters chiefly south of Florida. Accidental in summer or fall northward to Punta Rassa and Fort Pierce (Hubbard, 1965). Florida: late March-early November, a few wintering (Map 41).

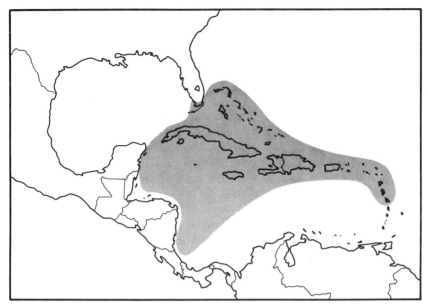

Map 41. Breeding range of the White-crowned Pigeon
(*Columba leucocephala*)

548. Scaly-naped Pigeon, *Columba squamosa* Bonnaterre

IDENTIFICATION: Larger than Rock Dove (total length 330 to 405 mm). Plumage mostly slate gray, but head, throat, and breast reddish, and back of neck chestnut to purplish, with brownish feather edgings. Orbital skin and iris yellow to red; tarsus and toes red; bill red with whitish tip.

DISTRIBUTION: Resident in the West Indies and islands off the coast of Venezuela. Accidental at Key West.

549. Rock Dove, *Columba livia* Gmelin

IDENTIFICATION: Total length 305 to 355 mm. Chiefly slate gray, but darker on foreparts; neck and breast with iridescence; 2 black bands on wing and one near end of tail. Orbital skin blue gray; iris orange red; cere whitish; feet reddish. (Various other plumages occur due to previous interbreeding of domestic stock.) The common domestic pigeon.

DISTRIBUTION: Originally in Eurasia and northern Africa; now widely introduced, at least locally, throughout the world. All Florida birds are probably domestic or semi-domestic.

550. Zenaida Dove, *Zenaida aurita* (Temminck)

IDENTIFICATION: About size of Mourning Dove (total length 255 to 305 mm), but tail shorter. Chiefly gray brown above and *cinnamon brown below*, with lavender reflections on breast and abdomen; wing coverts spotted with black; remiges mostly blackish, but *secondaries broadly tipped with white*; *rectrices* mostly dark, *with broad whitish tips*; purplish reflections on neck and sides of head. Tarsus and toes reddish; orbital skin grayish blue; iris orange to red.

DISTRIBUTION AND VARIATION: Breeds in the Bahamas and Greater Antilles southward to islands off the coast of Central America and Grenada; also on Yucatan Peninsula. Accidental on Florida Keys (*Z. a. zenaida*) and in Osceola County (Steffee and Mason, 1967).

551. White-winged Dove, *Zenaida asiatica* (L.)

IDENTIFICATION: About size of Mourning Dove, but tail shorter and not pointed (total length 280 to 305 mm). Upperparts grayish brown, tending to gray on head and neck; underparts lighter, almost white posteriad; reflections on neck greenish; most rectrices and secondaries with broad white tips; *outermost wing coverts white, forming a large patch*; bases of primaries blackish. Iris, tarsus, and toes reddish; orbital skin grayish blue.

DISTRIBUTION AND VARIATION: Resident from southern Nevada, southeastern California, central Arizona, southern New Mexico, and the lower Rio Grande Valley southward to western Panama; from the Bahama Islands to the Greater Antilles; and from southwestern Ecuador to northern Chile. In fall and winter (rarely at other seasons) some individuals of, presumably, *Z. a. asiatica* stray into the eastern United States, but most of the population moves southward or is resident. Recent breeding records in Florida near Homestead (Fisk, 1968) and Naples (Below, 1965) represent imported birds, thus possibly other subspecies, although importation "from Venezuela" seems in error.

552. Mourning Dove, *Zenaida macroura* (L.)

IDENTIFICATION: Total length 280 to 330 mm, *the pointed tail constituting about half of this*. Upperparts mostly grayish brown, the underparts lighter; top of head in male light blue gray, and neck with considerable iridescence; *rectrices graduated, broadly tipped with white*; tarsus and toes pinkish.

DISTRIBUTION AND VARIATION: Breeds from Alaska, British Columbia, Alberta, Saskatchewan, Manitoba, Ontario, Quebec, Maine, and New Brunswick southward to western Panama; also in the Bahamas, Cuba, the Isle of Pines, Hispaniola, and Puerto Rico. Winters from the United States southward within breeding range. *Z. m. carolinensis* is the common form over most of Florida, but the breeding birds of the Keys are *Z. m. macroura* (Aldrich and Duvall, 1958; TT 2812); *Z. m. marginella* is accidental in winter.

553. Ringed Turtle Dove, *Streptopelia risoria* (L.)[1]

IDENTIFICATION: About size of Mourning Dove, but tail not pointed (total length 255 to 305 mm). A *very light-colored dove* (rufous buff above, lighter below) *with a black band on the lower neck*; remiges and rectrices largely grayish; iris, tarsus, and toes reddish. Some varieties almost white.

DISTRIBUTION: Northern Africa and Arabia (*S. roseogrisea*). The domestic variety has escaped from captivity and become established locally over much of the world; established in Florida at St. Petersburg, and reported also from Gainesville (Austin, personal communication), Miami, Homestead (Heinzman and Heinzman, 1965), Cocoa, Tallahassee, Orlando (Nicholson, 1960), and perhaps elsewhere.

554. Ground Dove, *Columbina passerina* (L.)

IDENTIFICATION: *Florida's smallest dove* (total length 150 to 180 mm). Adult: upperparts grayish brown; underparts lighter gray (vinaceous in male), with dark spots on breast; *remiges reddish brown* (conspicuous in flight); tail dark, rounded; bill yellowish to reddish toward base, darker toward tip; iris reddish; tarsus and toes yellowish to pinkish. Immature: feathers tipped with white.

DISTRIBUTION AND VARIATION: Resident from southern California, central Arizona, southern Texas, western Mississippi, west-central Alabama, South Carolina, and Bermuda southward to Costa Rica and the West Indies; also from Colombia and Venezuela to Ecuador and Brazil. The eastern race, *C. p. passerina*, occurs throughout Florida, and *C. c. pallescens* is accidental on St. George Island (Franklin Co.) (TT 2814).

555. Inca Dove, *Scardafella inca* (Lesson)

IDENTIFICATION: Slightly larger than the Ground Dove (total length about 200 mm). Adult: underparts pale grayish vinaceous anteriorly (nearly white on chin), shading to buff on belly, flanks, and lower tail coverts; chin, throat, and upper part of chest immaculate, but *feathers of other portions tipped with blackish*, the edges widest on flanks (producing a barred appearance); upperparts, including all wing

---

1. This "species" is a domestic descendant of the African Collared Dove, *Streptopelia roseogrisea*; see Goodwin, 1967.

coverts, grayish brown, *each feather tipped with a crescentic bar of blackish*; feet pinkish; iris orange to red; orbital skin grayish. Immature: similar, but buffier below; also some buffy or white tips on wing coverts. Rufous in wing suggests Ground Dove, but *tail much longer, strongly graduated, and outer rectrices mostly white*. Iris straw-colored.

DISTRIBUTION: Resident from southern Arizona, New Mexico, southern Texas, and Tamaulipas southward to Costa Rica and Nicaragua. Present and breeding at Key West, Florida, since 1965 (TT 2815).

556. Key West Quail-Dove, *Geotrygon chrysia* Bonaparte

IDENTIFICATION: About size of Mourning Dove, but tail short and rounded (total length 280 to 305 mm). Upperparts mostly reddish brown, with greenish or purplish reflections on top of head, nape, and upper back; underparts largely white, but breast pale vinaceous; *malar stripe white*, bordered by chestnut below. Female somewhat duller. Bill reddish at base; iris yellow to red; orbital skin and feet reddish.

DISTRIBUTION: Resident in the Bahama Islands and Greater Antilles (except Jamaica); accidental in southern Florida (Inwood, 1965; Nelson, 1966).

557. Ruddy Quail-Dove, *Geotrygon montana* (L.)

IDENTIFICATION: Similar in form to Key West Quail-Dove, but smaller (total length 240 to 280 mm). Male: *upperparts mostly chestnut red*, with purplish reflections anteriad; underparts cinnamon to white, with a vinaceous tinge on breast; *2 dark stripes below eye*; orbital skin, bill, and feet reddish to purplish; iris variable.

DISTRIBUTION AND VARIATION: Resident from lowlands of central Mexico southward to Bolivia, Paraguay, and Brazil; also the Greater Antilles, Grenada, and Trinidad; *G. m. montana* occupies most of this range and has occurred accidentally in southern Florida (see Robertson and Mason, 1965).

ORDER PSITTACIFORMES: CX. FAMILY PSITTACIDAE, Parrots and allies. Tropical birds of variable size with a *stout, strongly hooked bill* and *zygodactylous toes*. The plumage is typically brilliant, with green most frequent, but red, yellow, and blue common.

558. Carolina Parakeet, *Conuropsis carolinensis* (L.)

IDENTIFICATION: Much larger than the Budgerigar (total length about 305 to 330 mm, about half of which is tail). Adult: *head and neck yellow to orange*; rest of plumage mostly green, but under surface of tail, feet, and wing lining yellowish; bill straw-colored. Immature: entirely green except *forehead and region in front of eye orange*; feet greenish. *Tail long and graduated*.

DISTRIBUTION AND VARIATION: Formerly ranged from North Dakota, eastern Nebraska, Iowa, southern Wisconsin, Ohio, and central New York southward to the Gulf Coast of eastern Texas, Louisiana, Mississippi, Alabama, and to central Florida. *Extinct*; no Florida record since 1920 (*C. c. carolinensis*).

559. Budgerigar, *Melopsittacus undulatus* (Shaw)

IDENTIFICATION: About size of House Sparrow, but *tail much longer and strongly graduated* (total length about 200 to 230 mm). Wild type: head and neck yellowish, shading to vivid green on body; *rounded black spots on throat*, black crossbars (feather tips) on neck and upper back, and large black markings on wing coverts; remiges largely dusky greenish; rectrices steel blue. White and blue phases also occur.

DISTRIBUTION: Australia; released and well established in Pinellas County, Florida, where largely semi-domestic; occasional in nearby counties.

ORDER CUCULIFORMES: CXI. FAMILY CUCULIDAE, Cuckoos, Anis, and allies. Small to medium-sized birds with decurved bills and *zygodactylous toes*. The tail is long and graduated, and the rectrices (only 8 to 10) are often tipped with white in American species. Nostrils comparatively low on bill.

560. Mangrove Cuckoo, *Coccyzus minor* (Gmelin)

IDENTIFICATION: About size of Yellow-billed Cuckoo (total length 280 to 330 mm), from which it differs mainly in having a *dark stripe through the eye*, *no reddish brown on wings*, and the underparts more or less buffy.

DISTRIBUTION AND VARIATION: Breeds (and largely resident) from western Mexico, southern Florida, and the Bahama Islands southward to northern South America. *C. m. maynardi* breeds chiefly on the Keys and islands off the west coast northward to Pinellas County, some remaining for the winter; rare in Dade County.

561. Yellow-billed Cuckoo, *Coccyzus americanus* (L.)

IDENTIFICATION: About size of Mourning Dove (total length 280 to 330 mm). Upperparts brownish olive; underparts white; *primaries largely bright reddish brown*; *rectrices graduated, with large white spots at tips* on ventral surface; *lower mandible mostly yellow*; feet grayish blue.

DISTRIBUTION AND VARIATION: Breeds from southern British Columbia, North Dakota. Minnesota, southern Ontario, Quebec, and New Brunswick southward to Baja California, Sinaloa, Tamaulipas, the northern Gulf Coast, the Greater Antilles, and the northern Lesser Antilles. Winters from northern South America to central Argentina and Uruguay. *C. a. americanus* breeds throughout Florida, but is accidental in winter; April-November.

562. Black-billed Cuckoo, *Coccyzus erythropthalmus* (Wilson)

IDENTIFICATION: Similar in size and coloration to the Yellow-billed Cuckoo, but with the *lower mandible blackish*, less reddish (or none) in the primaries, and *white spots on under side of tail much smaller*. Eyering of red skin; feet grayish blue.

DISTRIBUTION: Breeds from the southern parts of Saskatchewan, Manitoba, Ontario, and Quebec, and from New Brunswick and Nova Scotia southward to southeastern Wyoming, Nebraska, northwestern Arkansas, eastern Kansas, eastern and central Tennessee, North Carolina, northern Georgia, and probably northwestern South Carolina. Winters in Colombia, Venezuela, Ecuador, and northern Peru. Florida: April to May and September to November; rare.

563. Smooth-billed Ani, *Crotophaga ani* L.

IDENTIFICATION: Slightly smaller than a male Boat-tailed Grackle (total length 305 to 380 mm), which it resembles in general form and coloration. Plumage entirely black; *bill compressed and with a high dorsal ridge, devoid of longitudinal grooves*; rectrices graduated.

DISTRIBUTION: Resident from extreme southeastern Mexico, southern Florida, and the Bahama Islands southward to Ecuador and northern Argentina; regular Florida range from Brevard County to the Keys, but most numerous near lower east coast (Map 42).

564. Groove-billed Ani, *Crotophaga sulcirostris* Swainson

IDENTIFICATION: Virtually identical to Smooth-billed Ani in size and color, but *usually with longitudinal grooves on the upper mandible*, and the depth of the bill is less (under 20 mm). However, it is not always possible to distinguish the 2 species in the field.

DISTRIBUTION: Resident from southern Baja California, southern Sonora, and the lower Rio Grande southward and eastward to Peru and British Guiana. A few stray eastward to Florida (especially northwestern) in fall and winter (late September-March).

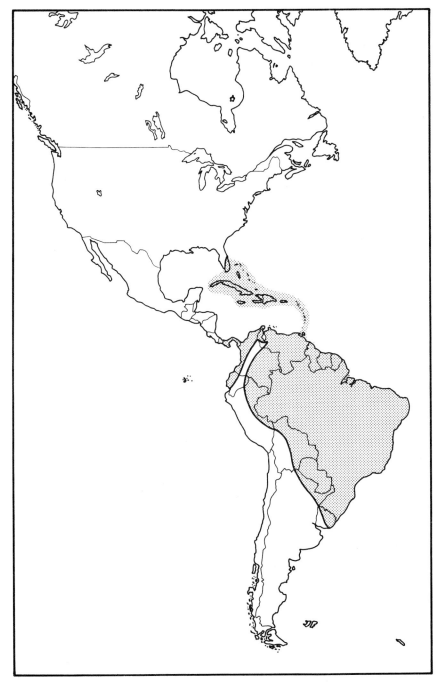

Map 42. Distribution of the Smooth-billed Ani (*Crotophaga ani*)

ORDER STRIGIFORMES: CXII. FAMILY TYTONIDAE, Barn Owls. Owls are nocturnal birds of prey, with bill and foot structure as in hawks except for the *reversible outer toe*. Unlike hawks, they have the *eyes directed forward and surrounded by a radial disk* (outwardly radiating slender feathers). Head large, appearing even larger because feathers do not lie close. In the family Tytonidae the legs are relatively long, extending beyond the tail in flight, and the *middle claw is pectinate*.

565. Barn Owl, *Tyto alba* (Scopoli)

IDENTIFICATION: (See description of order and family.) Somewhat smaller than the Barred Owl (total length 38 to 53 cm). Upperparts golden brown, mottled with gray and white; underparts white or buffy, with sparse darker spots; facial disk white; ear tufts lacking; feet and bill yellowish. *No other American owl is so white underneath unless also whitish above.*

DISTRIBUTION AND VARIATION: Almost cosmopolitan. *T. a. pratincola* breeds from southwestern British Columbia, North Dakota, the southern parts of Minnesota, Wisconsin, Michigan, Ontario, and Quebec, and from Massachusetts southward to eastern Guatemala, the northern Gulf Coast, and southern Florida (may not nest on Keys); probably withdraws from northern parts of breeding range in winter, when it also appears on Florida Keys.

ORDER STRIGIFORMES: CXIII. FAMILY STRIGIDAE, Typical Owls. As described for the order, but legs usually shorter than in Barn Owl and *middle claw not pectinate*.

566. Screech Owl, *Otus asio* (L.)

IDENTIFICATION: *Florida's only small owl with ear tufts* (total length 20 to 23 cm). Red phase: upperparts bright rufous, marked with longitudinal blackish streaks; underparts chiefly white, but marked with rufous on facial disk and breast and with black (sparse) on body. Gray phase: ground color brownish gray rather than rufous, which color does not appear anywhere. (Many Florida birds are in an intermediate brown phase.) Toes only sparsely feathered; legs shorter than in Burrowing Owl. Iris and feet yellowish.

DISTRIBUTION AND VARIATION: Mostly resident from southeastern Alaska, the southern parts of British Columbia, Manitoba, Ontario, and Maine southward to southern Mexico and Florida. The breeding form throughout Florida is *O. a. floridanus*, but *O. a. asio* reaches north Florida in winter (Stevenson and Baker, 1970).

567. Great Horned Owl, *Bubo virginianus* (Gmelin)

IDENTIFICATION: *Florida's only large owl with ear tufts*, but see also Short-eared and Long-eared Owls (total length 46 to 64 cm). Upperparts mottled with blackish, brown, gray, buff, and white; ear tufts long; facial disk reddish brown; *throat white*; *underparts* whitish or buffy, *with dark crossbars*; *tarsus and toes feathered*; iris yellow.

DISTRIBUTION AND VARIATION: Resident virtually throughout North, Central, and South America. *B. v. virginianus* is the breeding form in Florida, but it does not reach the Keys.

568. Burrowing Owl, *Speotyto cunicularia* (Molina)

IDENTIFICATION: About size of Screech Owl (total length 23 to 25 cm). *Upperparts grayish brown, interrupted by white, usually rounded spots*; throat and breast similar to upperparts (except evenly brownish in juvenals); chin and belly white; *no ear tufts*; tarsus long and only sparsely feathered; iris and bill yellow.

DISTRIBUTION AND VARIATION: Resident (?) from the southern parts of western Canada southward to central Mexico (to Central America in winter; *S. c. hypugaea*); resident in the Florida Peninsula, on Upper Matecumbe Key, Cuba and on the Bahama Islands (*S. c. floridana*); also (other races) locally in the Antilles, on other offshore islands, and in Central and South America. *S. c. hypugaea* is accidental in winter at Pensacola, Cedar Key, and in Dade County. (See Ligon, 1963.)

569. Barred Owl, *Strix varia* Barton

IDENTIFICATION: *A large owl with no ear tufts* (total length 43 to 61 cm). Upperparts brown, with white crossbars; breast similar, but with white predominating; abdomen whitish, with *longitudinal* streaks of brown; tarsus thickly feathered, but *toes only sparsely so*; bill and toes yellowish.

DISTRIBUTION AND VARIATION: Resident from northern British Columbia, northern Alberta, central Manitoba, Ontario, Quebec, and Nova Scotia southward (east of Rocky Mountains) to Texas, the Gulf Coast, and the Florida Keys. The Florida race is *S. v. georgica*.

570. Long-eared Owl, *Asio otus* (L.)

IDENTIFICATION: *A medium-sized owl with long ear tufts* (total length 33 to 41 cm). Upperparts mottled with brownish, buff, and light gray, with a tendency toward crossbarring; *facial disk mostly rufous*; *underparts* whitish to rufous, *with dark markings arranged in longitudi-*

nal rows (little evidence of crossbars); *ear tufts long and placed near middle of head*; tarsus and toes well feathered; iris yellow.

DISTRIBUTION AND VARIATION: Northern parts of Eurasia, southern Alaska, and southern Canada southward to southern Europe, northern Africa (Morocco, Tunisia), southern Asia (except southeastward), northern Baja California, southern Arizona, Oklahoma, Arkansas, and Virginia. Strays somewhat farther south in winter, with *A. o. wilsonianus* rarely reaching Florida (about 6 records, to southern tip and Keys).

571. Short-eared Owl, *Asio flammeus* (Pontoppidan)

IDENTIFICATION: Size of Long-eared Owl (total length 33 to 43 cm). Ground color buffy to whitish, longitudinally streaked with brown (streaks almost lacking on abdomen); ear tufts short and doubtfully visible in the field; *facial disk blackish near eyes*, bordered by white; tarsus and toes fully feathered; iris yellow. Also identified by its selection of *open habitats* and its *erratic flight*.

DISTRIBUTION AND VARIATION: Breeds over much of Eurasia, on the Hawaiian (and certain other oceanic) Islands, and from northern Alaska and Canada southward to California, Utah, Colorado, Kansas, Missouri, and Virginia; also in Hispaniola, Puerto Rico, the higher mountains of Colombia and Ecuador, and southern South America. The North American bird, *A. f. flammeus*, regularly reaches Florida in small numbers in winter (late October-April).

572. Saw-whet Owl, *Aegolius acadicus* (Gmelin)

IDENTIFICATION: The smallest owl likely to occur in Florida (total length 18 to 26 cm). Adult: upperparts dark cinnamon brown, the *head finely streaked* and the back spotted *with white*; tail incompletely barred; underparts white, heavily streaked with reddish brown; legs and feet feathered. Immature: upperparts brown (not reddish); whiter between and above eyes. At all ages the iris is yellow and the *ear tufts lacking*.

DISTRIBUTION AND VARIATION: Breeds from southern Alaska, central British Columbia and Alberta, southern Saskatchewan and Manitoba, northern Ontario, and central and eastern Quebec southward at higher elevations to southern California, southern Arizona, southern Mexico, Oklahoma, central Missouri, Ohio, Tennessee, and North Carolina. Southward in winter to lowlands of southwestern United States, Louisiana, Alabama, Georgia, and northeastern Florida (*A. a. acadicus*, one record; Lesser and Stickley, 1967).

ORDER CAPRIMULGIFORMES: CXIV. FAMILY CAPRIMULGIDAE, Nightjars. Mottled brownish or grayish birds with *small feet, a small bill, and a very large mouth*, usually surrounded by strong bristles. Tarsus partly feathered. Wings long and pointed.

573. Chuck-will's-widow, *Caprimulgus carolinensis* Gmelin

IDENTIFICATION: *The largest Florida nightjar* (total length 280 to 305 mm). Upperparts mottled with dark brown, golden brown, and black, the last color forming longitudinal stripes on the head and wide bars on the primaries; underparts similar, but lighter; large white patches in lateral rectrices of male (buffy in female); *rictal bristles long, pinnate near base. Throat brown anterior to band.*

DISTRIBUTION: Breeds from eastern Kansas, Missouri, the southern parts of Illinois, Indiana, Ohio, Maryland, and New Jersey southward to central Texas, the northern Gulf Coast, and the upper Florida Keys (Map 43). Winters from the northern Gulf Coast (rarely) and central Florida southward to eastern Mexico, Central America, the Greater Antilles, and Colombia.

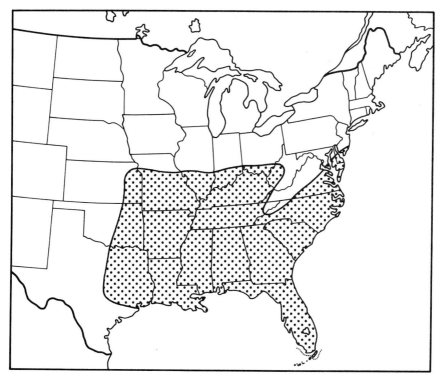

Map 43. Breeding range of the Chuck-will's-widow
(*Caprimulgus carolinensis*)

574. Whip-poor-will, *Caprimulgus vociferus* Wilson

IDENTIFICATION: *Smaller than Chuck-will's-widow* (total length 230 to 255 mm) and differing from it in color pattern by having more blackish and gray in plumage and *black anterior to the throat band*. The large tail patches are white in the male, buffy in the female. *Rictal bristles not pinnate*.

DISTRIBUTION AND VARIATION: Breeds from central Arizona, southern New Mexico, and southwestern Texas southward to Honduras (*C. v. arizonae*); also from central Saskatchewan, southern Manitoba, western and southern Ontario, southern Quebec, New Brunswick, and Nova Scotia southward (east of the Great Plains) to northeastern Texas, western Arkansas, the northern parts of Mississippi, Alabama, and Georgia, northwestern South Carolina, and the eastern parts of North Carolina and Virginia (*C. v. vociferus*). The latter race winters from southern Texas, the northern Gulf Coast, southern Georgia, and coastal South Carolina southward to Costa Rica and Cuba. Florida: early September–early May.

575. Common Nighthawk, *Chordeiles minor* (Forster)

IDENTIFICATION: About size of Whip-poor-will (total length 205 to 255 mm). Upperparts blackish, with buffy or light gray feather edgings; *underparts mostly crossbarred with white and dark gray*; *primaries dusky, with a large patch of white*; throat with a white band (buffy in female). (Compare with Lesser and Antillean Nighthawks.)

DISTRIBUTION AND VARIATION: Breeds from southern Yukon, lower Mackenzie Valley, northern parts of Saskatchewan, Manitoba, and Ontario, and from Quebec and Newfoundland southward to southern California, central Nevada, northwestern to southeastern Arizona, southeastern Mexico, Panama, the northern Gulf Coast, and the middle Florida Keys. Winters from Colombia and Venezuela southward to central Argentina (casually in Florida; sight records). The breeding race in Florida is *C. m. chapmani*, but the following others have been collected in migration: *minor*, *sennetti*, *henryi*, and *hesperis* (Selander, 1954; Sprunt, 1963; Stevenson and Baker, 1970).

576. Antillean Nighthawk, *Chordeiles gundlachii* Lawrence[1]

IDENTIFICATION: Slightly smaller than the Common Nighthawk (total length about 200 mm). Almost identical to the Common Nighthawk, but somewhat buffier below and tail shorter (less than 100 mm). Call note quite distinct.

DISTRIBUTION AND VARIATION: Breeds from the Greater Antilles,

---

1. Included under *Chordeiles minor* in A.O.U. *Check-list*, but see Eisenmann, 1962.

the Bahama Islands, and the Florida Keys (north to Key Biscayne) southeastward to the Virgin Islands (Map 44). The Florida race is *C. g. vicinus*.

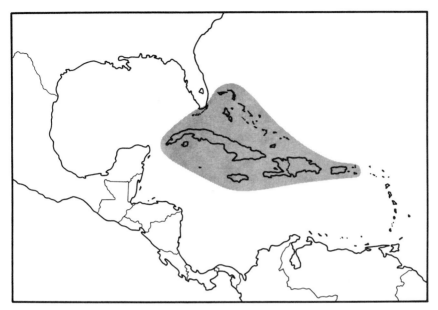

Map 44. Breeding range of the Antillean Nighthawk

ORDER APODIFORMES: CXV. FAMILY APODIDAE, Swifts. The order consists of small birds with weak feet and weak bills; the wings are long and pointed. The swifts (Apodidae) are especially long-winged, often short-tailed, and have remarkable powers of flight. The weak feet, *small bill, and large mouth* resemble those in the nightjars. Hallux reversible; tarsus not scaled. Plumage often dull. Rectrices only 10.

577. Chimney Swift, *Chaetura pelagica* (L.)

IDENTIFICATION: A swallow-sized bird with a short, rounded tail (total length about 125 mm). *Entire plumage dusky*, becoming darker on crown and wings; *wings long and pointed*; *rectrices spiny-tipped*; bill and feet reduced, but gape large.[1]

DISTRIBUTION: Breeds from southeastern Saskatchewan, southern Manitoba, central Ontario, southern Quebec, and Nova Scotia southward to southeastern Texas, the northern Gulf Coast, and south-central Florida. Winters in the upper Amazon drainage, at least in northeastern Peru and to Chile.

---

1. The extremely similar Vaux' Swift (*Chaetura vauxi*) is likely to occur in west Florida in fall.

ORDER APODIFORMES: CXVI. FAMILY TROCHILIDAE, Humming-birds. *Very small* birds with long, pointed wings, small feet, and a *long, tubular bill*. Tongue a long, protrusible double tube. Plumage often with metallic greens, blues, or reds.

578. Cuban Emerald, *Chlorostilbon ricordii* (Gervais)

IDENTIFICATION: Larger than the Ruby-throat (total length 100 to 115 mm). Male: *plumage mostly metallic green above and below*, the forked tail more dusky. Female: similar, but more brownish underneath. A *white spot behind eye* in both sexes. (Other Florida hummingbirds have mostly whitish underparts.)

DISTRIBUTION: Northern Bahama Islands, Cuba, and the Isle of Pines; accidental on east coast of Florida and the Keys (several sight records).

579. Ruby-throated Hummingbird, *Archilochus colubris* (L.)

IDENTIFICATION: The smallest bird of regular occurrence in Florida (total length slightly over 75 mm). Adult male: upperparts brilliant metallic green, shading to blackish on wings and tail; *throat metallic red*; remainder of underparts whitish, tinged laterally with green; bill long and slender. Adult female: similar, but throat whitish. Immature: upperparts duller (less metallic); underparts buffy white, the throat of males spotted with dusky. (Also see Key.)

DISTRIBUTION: Breeds from central Alberta and Saskatchewan, the southern parts of Manitoba, Ontario, and Quebec, and from New Brunswick and Nova Scotia southward (east of the Great Plains) to southeastern Texas, the Gulf Coast, and southern Florida. Winters from southern Texas and central (rarely northern) Florida southward to Central America and Cuba.

580. Black-chinned Hummingbird, *Archilochus alexandri* (Bourcier and Mulsant)

IDENTIFICATION: Very similar to the Ruby-throat in size and general coloration but *male with a black chin and purple throat*. Adult female: identical to female Ruby-throat, except *middle rectrices longer than outer ones*. Immatures of the two species are indistinguishable.

DISTRIBUTION: Breeds from southwestern British Columbia and northwestern Montana southward to northern Baja California, northwestern Mexico, and south-central to northeastern Texas. Winters from southeastern California and northwestern Mexico southward to south-central Mexico. Accidental in Louisiana and northwestern Florida (Hallman, 1963; Weston, 1965).

581. Rufous Hummingbird, *Selasphorus rufus* (Gmelin)

IDENTIFICATION: Slightly larger than the Ruby-throat (total length 88 to 100 mm). Adult male: *upperparts mostly rufous*, but with greenish on top of head; central rectrices, belly, and flanks also rufous; remainder of underparts whitish, except for the *orange red throat*. Adult female, immature: green largely replacing rufous on upperparts, but *some rufous present in tail*; little red on throat. (Also see Key.)

DISTRIBUTION: Breeds from southeastern Alaska, southern Yukon, east-central British Columbia, southwestern Alberta, and western Montana southward to northwestern California and southern Idaho. Winters in northern and central Mexico, rarely straying eastward to Florida (Pensacola to the east coast, late September to March).

ORDER CORACIIFORMES: CXVII. FAMILY ALCEDINIDAE, Kingfishers. Small to medium-sized birds with *small, syndactyl feet*, a large head, and a strong, spear-shaped bill. Plumage often with blues and greens.

582. Belted Kingfisher, *Megaceryle alcyon* (L.)

IDENTIFICATION: Total length 280 to 355 mm. Male: upperparts light slaty blue, streaked on head and crest with blackish; remiges with considerable black and white barring; white spots also on lateral rectrices; underparts white, with a bluish band across the upper breast and the same color on sides of abdomen (both areas tinged with brown in immature). Female: similar, but the breastband overlaid with brown, and an additional reddish brown band across abdomen; sides also rufous. Color of feet variable. (Also see description of family.)

DISTRIBUTION AND VARIATION: Breeds from northwestern Alaska, southern Yukon, southwestern Mackenzie, the central parts of Alberta, Saskatchewan, and Manitoba, northern Ontario, central Quebec, and Labrador southward to Panama, the northern Gulf Coast, and central Florida (perhaps southern Florida formerly). Winters from the northern United States southward to Panama, the West Indies, and Bermuda. The eastern form, *M. a. alcyon*, occurs in Florida.

ORDER PICIFORMES: CXVIII. FAMILY PICIDAE, Woodpeckers. Small to medium-sized birds with strong, *zygodactyl* (rarely 3-toed) *feet, stiff rectrices*, and a chisel-shaped bill.

583. Common Flicker, *Colaptes auratus* (L.)

IDENTIFICATION: About size of Belted Kingfisher (total length 305 to 330 mm). Male: upperparts mostly grayish brown, with blackish crossbars, but with more gray on crown, red on occiput, and a *white rump*; rectrices black; *remiges dusky, with yellow shafts*; underparts light tan gray to buffy, with *rounded black spots on lower breast and abdomen*, a black bar across upper breast, and a black malar stripe; under side of rectrices and remiges largely yellow. Female: similar, but lacking the black malar stripe. (Yellow is replaced by red in a western race.)

DISTRIBUTION AND VARIATION: Breeds from central Alaska, northwestern Mackenzie, northern Manitoba, James Bay, central Quebec, southern Labrador, and Newfoundland southward to southern Mexico, southern Texas, the northern Gulf Coast, the upper Florida Keys, Cuba, and Grand Cayman (see Short, 1965; Map 45). Withdraws from northernmost parts of breeding range in winter. *C. a. auratus* is the form breeding in Florida, but *C. a. luteus* joins it in the northern parts in winter.

584. Pileated Woodpecker, *Dryocopus pileatus* (L.)

IDENTIFICATION: The largest common Florida woodpecker (total length 380 to 430 mm). Male: *plumage mostly dull blackish*, but top of head (including crest) and malar stripe light red; white on chin, throat, under and behind eye, and on wings (the latter patch largely hidden when wing is folded); bill gray, lighter near base; iris yellowish. Female: differs only in having the malar stripe dark gray and the forehead gray or brown.

DISTRIBUTION AND VARIATION: Resident locally from the southern parts of Mackenzie and Manitoba, northern Ontario, central Quebec, New Brunswick, and Nova Scotia southward to central California, central Texas, the Gulf Coast, and the upper Florida Keys. The breeding form in northwestern Florida is *D. p. pileatus*, and the Peninsula is occupied by *D. p. floridanus*.

585. Red-bellied Woodpecker, *Centurus carolinus* (L.)

IDENTIFICATION: Smaller than the Flicker (total length 230 to 255 mm). Adult male: *crown entirely light red*; *remainder of upperparts transversely barred with black and white*; underparts smoky gray, with reddish tinge on belly; under side of tail barred black and white. Adult female: similar, but forehead and front of crown gray. Immature: similar to adults, but with little red on head or belly.

DISTRIBUTION AND VARIATION: Resident from southeastern Minnesota, southern parts of Wisconsin, Michigan, and Ontario, western New York, and Delaware southward to southern Texas, the Gulf Coast, and the Florida Keys. *C. c. carolinus* occupies the northern parts of the state and is replaced by *C. c. perplexus* from about Venice and Stuart southward.

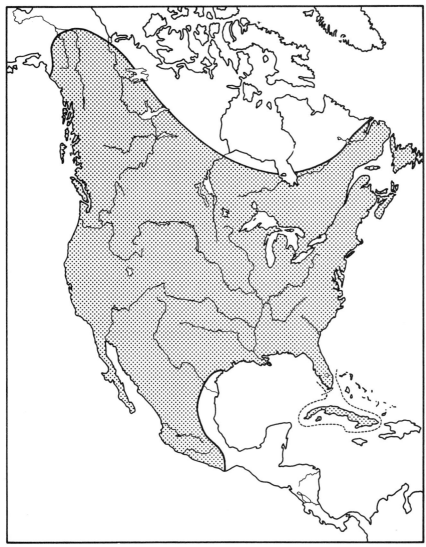

Map 45. Breeding range of the Common Flicker (*Colaptes auratus*)

586. Red-headed Woodpecker, *Melanerpes erythrocephalus* (L.)

IDENTIFICATION: Near size of Red-bellied Woodpecker (total length about 230 mm). Adult: *entire head, neck, and upper breast bright red*; remainder of upperparts black, except for the secondaries, which are mostly pure white; remainder of underparts white, except for the black tail. Immature: *head, neck, and upper breast mostly brownish gray* (later mixed with red); otherwise similar to adults, but feathers of upperparts tipped with brownish gray, and with some black in secondaries. Bill bluish gray.

DISTRIBUTION AND VARIATION: Breeds from southern Saskatchewan and Manitoba, western and southern Ontario, southern Quebec, New York, and southern New Hampshire southward to northern New Mexico, central Texas, the Gulf Coast, and southern Florida; casually west to California. Withdraws from northern parts of breeding range in winter. The eastern form, *M. e. erythrocephalus*, occurs in Florida.

587. Yellow-bellied Sapsucker, *Sphyrapicus varius* (L.)

IDENTIFICATION: Slightly smaller than Red-headed Woodpecker (total length about 205 mm). Adult male: top of head, chin, and throat entirely red; *side of head with 3 black stripes, separated by 2 white ones*; upperparts mostly blackish, with white or light spots, some of which are rounded; upper breast black (continuous with malar stripe); sides of body barred with gray and white; sides of breast, lower breast, and belly yellowish. (Entire head, neck, and breast red in 2 western races.) Adult female: similar, but nape blackish (partly white in male), and chin and throat white. Immature: similar to adult female, but with less contrast, especially on head and neck. In all plumages there is a *longitudinal white stripe on the wing*.

DISTRIBUTION AND VARIATION: Breeds from southeastern Alaska, southern Mackenzie, northern parts of Manitoba and Ontario, southern parts of Quebec and Labrador, and from Newfoundland southward at high elevations to southern California, central Arizona, northern New Mexico, southeastern South Dakota, eastern Missouri, central Illinois, northwestern Indiana, northern Ohio, western Pennsylvania, northern New York, central New England, and in the mountains to extreme eastern Tennessee and northern Georgia. Winters from the Pacific Coast of Alaska and Canada and the middle latitudes of the United States southward to Panama and the West Indies. *S. v. varius* and *S. v. appalachiensis* have been collected in Florida (Stevenson and Baker, 1970; early October-early May).

588. Hairy Woodpecker, *Dendrocopos villosus* (L.)

IDENTIFICATION: About size of Yellow-bellied Sapsucker (total length 190 to 215 mm). *Upperparts black, with rounded white spots*; underparts mostly whitish, but with black on sides of throat; *outer rectrices entirely white*; males have a red spot on nape.

DISTRIBUTION AND VARIATION: Resident from central Alaska, the Yukon, central Mackenzie, northern Manitoba, James Bay, south-central Quebec, and Newfoundland southward to northern Baja California, southeastern Mexico, the northern Gulf Coast, and southern Florida. The southeastern race, *D. v. audubonii*, occurs in Florida.

589. Downy Woodpecker, *Dendrocopos pubescens* (L.)

IDENTIFICATION: Color pattern almost identical to that of the Hairy Woodpecker, but size smaller (total length about 150 mm). The bill appears *relatively* smaller than the Hairy Woodpecker's, and the *outer rectrices are barred*.

DISTRIBUTION AND VARIATION: Resident from southeastern Alaska, southwestern Mackenzie, northern Alberta, central Saskatchewan, northern Manitoba, James Bay, southern Quebec, and Newfoundland southward to southern California, central Arizona, northern New Mexico, south-central Texas, the northern Gulf Coast, and southern Florida (*D. p. pubescens*).

590. Red-cockaded Woodpecker, *Dendrocopos borealis* (Vieillot)

IDENTIFICATION: Near size of Hairy Woodpecker (total length 175 to 205 mm). Top of head entirely black; *remainder of upperparts transversely barred with white and blackish*, except for absence of white from central rectrices; malar stripe black; *sides of head* (below and behind eyes) *white*; underparts mostly whitish, but with black markings along sides of body and, distally, on the rectrices. Males have a small red spot on each side of the head, and immatures a larger one on the crown.

DISTRIBUTION AND VARIATION: Resident from eastern Oklahoma, southern Missouri, the Cumberland Mountains of Kentucky and Tennessee, northern Georgia, South Carolina, and southeastern Virginia southward to eastern Texas, the Gulf Coast, and southern Florida (to northern parts of Monroe and Palm Beach Counties, formerly to southern Dade County; Map 46). Northern Florida is occupied by *D. b. borealis*, and the Peninsula south of Gainesville by *D. b. hylonomus*.

591. Ivory-billed Woodpecker, *Campephilus principalis* (L.)

IDENTIFICATION: Larger than the Pileated Woodpecker (total length 480 to 530 mm). Male: upperparts chiefly black, but with a large white patch on each wing and a white stripe running down side of neck and onto back; underparts black; back of crest red; bill ivory or white. Female: differs only in having *entire crest black*. In both sexes the *white wing patch is large and conspicuous even when the wing is folded*.

DISTRIBUTION: Formerly resident from northeastern Texas, southeastern Oklahoma, northeastern Arkansas, southeastern Missouri, southern Illinois, southern Indiana, and southeastern North Carolina southward to the Brazos River (Texas), the northern Gulf Coast, and southern Florida. Now nearly extinct, but a few still persisted in wilder sections of Florida in the 1960's.

ORDER PASSERIFORMES: CXIX. FAMILY TYRANNIDAE, Tyrant Flycatchers. The members of this large family have the *bill depressed at*

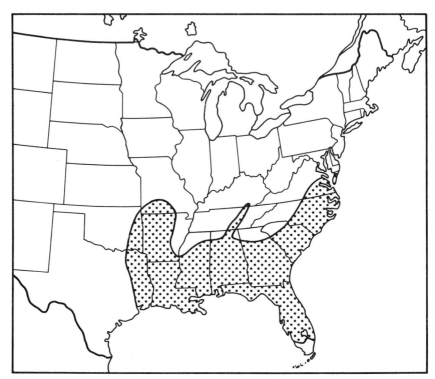

Map 46. Distribution of the Red-cockaded Woodpecker
(*Dendrocopos borealis*)

the base, hooked at the tip, and *its base surrounded by rictal bristles*. In life they are more easily recognized by habits than by structure.

592. Eastern Kingbird, *Tyrannus tyrannus* (L.)

IDENTIFICATION: Size of Mockingbird, but tail shorter (total length 205 to 230 mm). Upperparts dark gray (darker on head and neck), with white edgings on remiges and a *broad white band across tip of tail*; concealed crown patch orange (this mark and the white tail tip less conspicuous or lacking in immature).

DISTRIBUTION: Breeds from northern British Columbia, south-central Mackenzie, central Saskatchewan, central Manitoba, northern Ontario, and southeastern Quebec southward to western Washington, most of Oregon, northeastern California, northern Nevada, southern Idaho, northern Utah, Colorado, northeastern New Mexico, central Texas, the northern Gulf Coast and the southern Florida mainland (March-October). Winters in Peru and Bolivia.

593. Gray Kingbird, *Tyrannus dominicensis* (Gmelin)

IDENTIFICATION: Larger than the Eastern Kingbird (total length 230 to 255 mm). *Upperparts light gray, devoid of any true white*; concealed crown patch orange (lacking in immature); underparts whitish; wing lining tinged with yellow. *Bill noticeably larger* than in Eastern Kingbird.

DISTRIBUTION AND VARIATION: Breeds from the coast of South Carolina (casually) and Georgia around the coast of Florida to Alabama; casual inland, but more frequent in migration near Lake Okeechobee (*T. d. dominicensis*); also breeds in the Bahama Islands, Greater Antilles, northern Lesser Antilles, and southward to central Venezuela (Map 47). Winters chiefly from Hispaniola and Puerto Rico southward to Colombia and Venezuela. Casual much farther north along Atlantic Coast in migration. Florida: March to October, rarely into winter.

594. Western Kingbird, *Tyrannus verticalis* Say

IDENTIFICATION: Size of Gray Kingbird (total length 230 to 255 mm). Top of head and neck light gray, with concealed orange crown patch (lacking in immature); back grayish; wings and *tail* dark, the latter *edged with white*; chin and throat light gray or whitish; *remainder of underparts yellow*; bill smaller than Gray Kingbird's and light at base of lower mandible. As compared with *T. melancholicus* (Hypothetical List), the bill is smaller and not entirely black below, the back is greener, the tail blacker, and the chin less white. (Compare with Great Crested Flycatcher.)

DISTRIBUTION: Breeds from western Oregon and Washington, southern British Columbia, the southern parts of Alberta, Saskatchewan, and Manitoba, and from western Minnesota (rarely eastward to Ohio and southern Ontario) southward to northern Baja California, northern Mexico, southern New Mexico, northeastern Texas, northeastern Oklahoma, east-central Kansas, and north-central Missouri. Winters from southern Mexico to northern Nicaragua; also on the Florida Keys and northward (rarely) to Pensacola and the coast of South Carolina. Florida: September to May.

595. Scissor-tailed Flycatcher, *Muscivora forficata* (Gmelin)

IDENTIFICATION: Size of Eastern Kingbird, but tail much longer (total length 280 to 380 mm). Adult male: top of head, neck, and back pearl gray to whitish, with some partly concealed reddish on crown and back; wings brownish gray, with lighter feather edgings; *tail blackish, very long, and deeply forked*; lateral rectrices largely white; *underparts* chiefly white, but *tinged with salmon on breast* and with red on axillars; tip of tail dark. Adult female: axillars tinged with salmon, but otherwise like male. Immature: similar to adults, but crown patch lacking.

DISTRIBUTION: Breeds from eastern New Mexico, western Oklahoma, southeastern Colorado, Nebraska, central and southeastern Kansas, central Arkansas, and northwestern Louisiana southward to south-

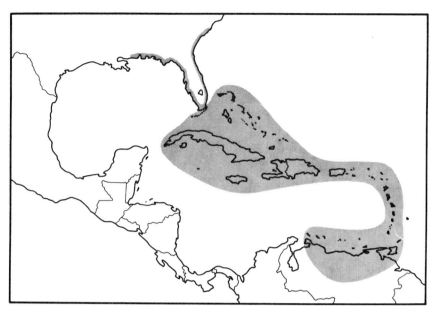

Map 47. Breeding range of the Gray Kingbird (*Tyrannus dominicensis*)

ern Texas and northern Mexico (Nuevo Leon). Winters from southeastern Mexico to western Panama; also in small numbers in southern Florida and casually northward to the northern Gulf Coast. Florida: October to May, rarely June.

596. Great Crested Flycatcher, *Myiarchus crinitus* (L.)

IDENTIFICATION: Size of Eastern Kingbird (total length 205 to 230 mm). Upperparts mostly brownish olive, but wings and tail brown, with *rufous on primaries and rectrices*, and white edgings on secondaries; *chin and throat light gray*; *belly bright yellow*, but merging with greenish on breast and sides; under side of tail mostly light cinnamon; crown patch wanting. Lower mandible mostly light.

DISTRIBUTION AND VARIATION: Breeds from southeastern Saskatchewan, southern Manitoba, central Ontario, southwestern Quebec, northern Maine, and New Brunswick southward to western Oklahoma, central Texas, the Gulf Coast, and the Florida Keys. Winters from eastern Mexico and central Florida to central Colombia; March to September in north Florida. The breeding race in Florida (resident in south Florida) is *M. c. crinitus*, but *M. c. boreus* occurs in migration.

597. Wied's Crested Flycatcher, *Myiarchus tyrannulus* (Müller)

IDENTIFICATION: Slightly larger than Great Crested Flycatcher (total length 215 to 250 mm) and similar in color pattern; differs in having *much less rufous in tail*, more brownish in the back, and paler under-parts. It differs from *M. cinerascens* in having the *outer web of the outermost tail feather whitish* and the *bill entirely black*.

DISTRIBUTION AND VARIATION: Breeds from southern Nevada, central Arizona, southwestern New Mexico, and southern Texas southward to Costa Rica; also from northern South America and the Lesser Antilles southward to Paraguay, northern Argentina, and south-central Brazil. Winters from central Mexico southward. Casual in Florida, December to April (*M. t. cooperi*).

598. Ash-throated Flycatcher, *Myiarchus cinerascens* (Lawrence)

IDENTIFICATION: Almost identical in coloration to Wied's Crested Flycatcher, but slightly smaller (total length 200 to 215 mm); they are *probably not separable in the field. Outer web of outer rectrix* light, but *not whitish*; *lower mandible brown at base*.

DISTRIBUTION AND VARIATION: Breeds from southwestern Ore-gon, eastern Washington, southern Idaho, southwestern Wyoming, Colorado, New Mexico, and northern and central Texas southward to northern Baja California, Guerrero, and southwestern Tamaulipas. Win-ters from southeastern California, central Arizona, and southern Tamaulipas southward to Guatemala, El Salvador, and Costa Rica; casual in fall (October-December) along Gulf Coast eastward to Pen-sacola, Florida (*M. c. cinerascens*).

599. Eastern Phoebe, *Sayornis phoebe* (Latham)

IDENTIFICATION: Smaller than Eastern Kingbird (total length 165 to 180 mm). *Upperparts chiefly grayish brown, but darker on head and neck*; underparts mostly whitish, but tinged with yellowish on belly and with grayish on sides; *bill entirely dark*.

DISTRIBUTION: Breeds from northeastern British Columbia, central Mackenzie, northern Saskatchewan and Manitoba, northwestern and central Ontario, southern Quebec, and New Brunswick southward to southern Alberta, southwestern South Dakota, southeastern Colorado, western Oklahoma, eastern New Mexico, central and northeastern Texas, Arkansas, northeastern Mississippi, east-central Alabama, northern (probably central) Georgia, western South Carolina, and most of North Carolina. Winters from central and northeastern Texas, Ar-kansas, Tennessee, and Virginia southward to southern Mexico, the northern Gulf Coast, and southern Florida; late September to April.

600. Yellow-bellied Flycatcher, *Empidonax flaviventris* (Baird and Baird)

IDENTIFICATION: *Almost identical to the Acadian Flycatcher in fall and not distinguishable from it in the field* unless calling. When museum skins are compared, this species has a grayer (less white) ground color on the throat and a less greenish crown; its smaller size is often evident. (Also see Key.)

DISTRIBUTION: Breeds from northern British Columbia, northern Alberta, southern Mackenzie, central Saskatchewan and Manitoba, northern Ontario, central Quebec, southern Labrador, and Newfoundland southward to the northern parts of North Dakota, Minnesota, Wisconsin, southern Ontario, northeastern Pennsylvania, and New York. Winters from eastern Mexico to eastern Panama. Very few Florida specimens, all taken in September and October and mostly in the Panhandle.

601. Acadian Flycatcher, *Empidonax virescens* (Vieillot)

IDENTIFICATION: Total length 140 to 155 mm. Upperparts mostly olive green, but crown more brownish and tail gray; remiges blackish, the secondaries edged with whitish; wing coverts blackish, tipped with 2 yellowish wing bars; eye-ring conspicuous; underparts whitish in spring, often very yellow in fall, except rectrices dusky and throat whitish. Compared in the hand with the rare Yellow-bellied Flycatcher, it differs in its *longer wing tip, greener crown, and* (usually) *less yellow underparts; not distinguishable from that species in the field in fall,* except by call notes.

DISTRIBUTION: Breeds from southeastern South Dakota, northern Iowa, the southern parts of Wisconsin, Michigan, Ontario, and New York, and from northeastern Pennsylvania and southwestern Connecticut southward (east of the Great Plains) to central and southeastern Texas, the Gulf Coast, and north-central Florida (April-October) (Map 48). Winters from Costa Rica to Ecuador and western Venezuela.

602. Traill's Flycatcher, *Empidonax traillii* (Audubon)[1]

IDENTIFICATION: Size of Acadian Flycatcher, but differing from it in having *more gray brown upperparts,* virtually no yellow below, and more buffy wing bars. Differs from *E. minimus* in having all rectrices about equally long: *Can be safely distinguished in the field from the Least Flycatcher only by call notes.*

DISTRIBUTION AND VARIATION: Breeds from central Alaska, central Yukon, northwestern Mackenzie, northeastern Alberta, the northern parts of Saskatchewan, Manitoba, and Ontario, and from Quebec, Newfoundland, and Nova Scotia southward to northern Baja California,

---

1. Now regarded as comprising two species (A.O.U., 1973). Of 17 Florida specimens I examined, one (TT 597) seems clearly referable to *E. alnorum* (the Alder Flycatcher) and another (TT 598) to *E. traillii* (the Willow Flycatcher).

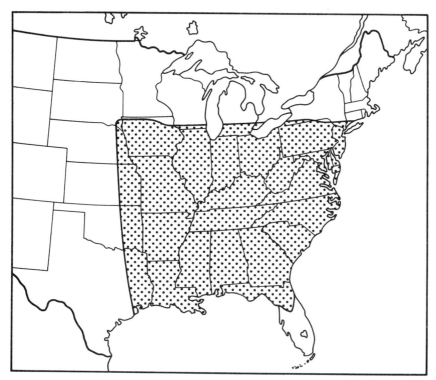

Map. 48. Breeding range of the Acadian Flycatcher
(*Empidonax virescens*)

southern Nevada and Arizona, southwestern New Mexico, western
Texas, Arkansas, northeastern Tennessee, and northern Georgia.
Known in Florida only as a rare (or uncommon) fall migrant, August to
October (Stevenson, 1966).

603. Least Flycatcher, *Empidonax minimus* (Baird and Baird)

IDENTIFICATION: Total length about 120 to 140 mm, averaging
smaller than any other flycatcher in Florida. *Indistinguishable in the
field*, except by voice, from the Traill's Flycatcher, but usually smaller,
wing bars whiter, bill shorter and narrower, and *with a notched tail*.

DISTRIBUTION: Breeds from southwestern Yukon, central Mackenzie, northeastern Alberta, northern Saskatchewan, central Manitoba,
northern Ontario, central Quebec, and Prince Edward and Cape Breton
Islands southward to northeastern British Columbia, Montana, northeastern Wyoming, southwestern South Dakota, northeastern Kansas,
southwestern Missouri, central Illinois, south-central Indiana, northern
Ohio, western Pennsylvania, West Virginia, and central New Jersey;
also, at higher elevations, southward to eastern Tennessee and northern

Georgia. Winters from central Mexico to Panama; also small numbers in south Florida; otherwise known only as a rare fall migrant in Florida, but more frequent in the Panhandle and southern parts, September and October.

604. Eastern Wood Pewee, *Contopus virens* (L.)

IDENTIFICATION: Size between that of Eastern Phoebe and Acadian Flycatcher (total length 150 to 165 mm). Coloration similar to that of Eastern Phoebe, but with *prominent whitish wing bars and edgings on secondaries*, lighter lower mandible, and a complete dusky band across breast; *eye-ring inconspicuous or lacking*; tarsus shorter than in other species of flycatchers in Florida.

DISTRIBUTION: Breeds from southern Manitoba, western and central Ontario, southern Quebec, northern Maine, central New Brunswick, and northern Nova Scotia southward (east of the Great Plains) to central and southeastern Texas, the Gulf Coast, and (rarely) central Florida (late March-November). Winters from central Costa Rica to northwestern South America (Venezuela to Peru); also possibly in south Florida (sight records).

605. Olive-sided Flycatcher, *Nuttallornis borealis* (Swainson)

IDENTIFICATION: Slightly smaller than Eastern Kingbird (total length 180 to 205 mm). Upperparts mostly medium to dark gray, but secondaries with whitish edges; *white patch on flank* (often concealed when wing is folded); *underparts white medially, medium gray laterally*; under tail coverts mottled gray and white. Resembles a Wood Pewee, but larger.

DISTRIBUTION: Breeds from northern Alaska, west-central and southern Yukon, west-central and southern Mackenzie, northeastern Alberta, northern Saskatchewan, north-central Manitoba, northern Ontario, central Quebec, and central Newfoundland southward to northern Baja California, central parts of Nevada and Arizona, northern New Mexico, central Saskatchewan, southern Manitoba, northeastern North Dakota, central Minnesota, the northern parts of Wisconsin and Michigan, southern Ontario, northeastern Ohio, and Massachusetts; also southward in the Appalachian Mountains to eastern Tennessee and western North Carolina. Winters mainly from Colombia and northern Venezuela to Peru. Florida: late April to May, and August to October; very rare.

606. Vermilion Flycatcher, *Pyrocephalus rubinus* (Boddaert)

IDENTIFICATION: Near size of Eastern Wood Pewee (total length 140 to 150 mm). Adult male: stripe through eye, nape, back, and both upper and lower surfaces of wings and tail brownish gray; *rest of plumage brilliant red*. Female: upperparts, including the crown, brownish gray; chin and throat white; *breast whitish, with dusky streaks; belly and under tail coverts pinkish*. Immature male: similar to female, but with some reddish feathers on crown and breast.

DISTRIBUTION AND VARIATION: Breeds from southwestern California, southern Nevada, southwestern Utah, northeastern Arizona, southwestern New Mexico, and western and central Texas southward to Guatemala and Honduras; also on the Galapagos Islands, and from Venezuela and Colombia to central Chile and southern Argentina. Withdraws from northernmost parts of breeding range in winter, at which time it ranges eastward to Florida (*P. r. mexicanus*, September-April).

ORDER PASSERIFORMES: CXX. FAMILY ALAUDIDAE, Larks. Small, largely terrestrial birds with *pointed wings and long tertials*. Tarsus scutellate and rounded behind. *Hind claw typically longer than its toe*.

607. Horned Lark, *Eremophila alpestris* (L.)

IDENTIFICATION: Slightly larger than House Sparrow (total length 165 to 180 mm). Male: upperparts mostly of varying shades of brown or buff, tinged with vinaceous on neck; remiges darker, some with lighter edgings; tail blackish, the outer rectrices edged with white laterally; upper tail coverts almost as long as rectrices; *forehead and line over eye whitish, bordered by black above; also a black stripe through eye and a black patch on throat and breast*; chin and upper throat white or yellowish; remainder of underparts white, with darker markings on lower breast and sides; *hind claw very long*; the ear tufts ("horns") are usually flattened in museum skins. Female: similar to male, but less boldly marked on head and throat.

DISTRIBUTION AND VARIATION: Breeds over most of Eurasia and North America, southward to southern Mexico, southern Texas, southwestern Louisiana, the northern parts of Mississippi, Alabama, and Georgia, and much of North Carolina. Winters in the United States casually southward to the northern Gulf Coast and peninsular Florida. All Florida specimens have represented *E. a. praticola*.

ORDER PASSERIFORMES: CXXI. FAMILY HIRUNDINIDAE, Swallows. Small birds with *long, pointed wings and a short, depressed bill*; gape wide; 12 rectrices: feet small.

608. Bahama Swallow, *Callichelidon cyaneoviridis* (Bryant)

IDENTIFICATION: Size and color pattern like the Tree Swallow's (total length 125 to 150 mm), but *tail deeply forked*. Adult: upperparts metallic green to blue, except for the duller remiges and rectrices; underparts white (with region just below eye mottled with dusky in female). Immature: more brownish above, and with brown on sides of breast.

DISTRIBUTION: Resident in the Bahama Islands, some moving to eastern Cuba in winter; accidental in south Florida.

609. Tree Swallow, *Iridoprocne bicolor* (Vieillot)

IDENTIFICATION: Total length 125 to 150 mm. Adult: *upperparts steel blue*; *underparts pure white*; remiges and rectrices dusky. Immature: more brownish above, but showing some iridescence in strong light; an incomplete breast band may be present. Tail not noticeably forked.

DISTRIBUTION: Breeds from north-central Alaska, southwestern Yukon, west-central and southern Mackenzie, the northern parts of Alberta, Saskatchewan, Manitoba, Ontario, and Quebec, southern Labrador, and Newfoundland southward to the coast of southern California, west-central Nevada, east-central Oregon, southeastern Washington, Idaho, west-central Utah, south-central New Mexico, southeastern Wyoming, southern North Dakota, the eastern parts of South Dakota and Nebraska, northeastern Kansas, south-central Missouri, northwestern Tennessee, south-central Indiana, central Ohio, northern West Virginia and Virginia, central Maryland, northeastern Pennsylvania, eastern New York, northern Connecticut, and Rhode Island (casually farther south). Winters from southern California, southwestern Arizona, northern Mexico, the northern Gulf Coast, and southern New Jersey (rarely Massachusetts coast) southward to Honduras and Cuba. Florida: late July to early June.

610. Bank Swallow, *Riparia riparia* (L.)

IDENTIFICATION: Slightly smaller than other Florida swallows (total length about 130 mm). Upperparts grayish brown; *chin, throat, and abdomen white, interrupted at breast by a clear-cut, broad, brownish collar*.

DISTRIBUTION AND VARIATION: Breeds from the British Isles and northernmost Eurasia southward to northwestern Africa and southern Asia; also from north-central Alaska, southern Yukon, northwestern and south-central Mackenzie, northeastern Alberta, east-central Saskatchewan, northeastern Manitoba, northern Ontario, southern Quebec, southern Labrador, and southwestern Newfoundland southward to southern Alaska, British Columbia, southern California, western Nevada, northern Utah, Colorado, Oklahoma, Texas, Arkansas, northern Alabama, central West Virginia, and eastern Virginia (casually farther south). Winters in southern and eastern Africa and southern Asia; also from Colombia and British Guiana southward to Peru, Bolivia, central Brazil, Paraguay, and northern Argentina. *R. r. riparia* occurs as a spring and fall transient throughout Florida, late July to October and April to May.

611. Rough-winged Swallow, *Stelgidopteryx ruficollis* (Vieillot)

IDENTIFICATION: Size of Tree Swallow (total length about 140 mm). Similar to the Bank Swallow above, but *underparts shading gradually* from light brownish gray on chin, throat, and breast to white of abdomen and under tail coverts.

DISTRIBUTION AND VARIATION: Breeds from British Columbia, southern Alberta, southwestern Saskatchewan, southeastern Manitoba, western and southern Ontario, southwestern Quebec, central Vermont, and New Hampshire southward to Peru, Bolivia, Paraguay, and Argentina. *S. r. serripennis*, the breeding form in the East, nests in Florida south to northern Collier County and Lake Okeechobee and winters from Mexico (probably northern), coastal Texas, the northern Gulf Coast (irregular), and Florida (casually South Carolina) southward through Mexico to western Panama.

612. Barn Swallow, *Hirundo rustica* L.

IDENTIFICATION: Size of Tree Swallow, *but tail long and deeply forked* (total length 150 to 190 mm). Adult: upperparts steel blue, except remiges and rectrices duller, and forehead reddish brown; *chin and throat reddish brown, separated from the cinnamon abdomen by a more or less complete blackish collar* (females somewhat paler). Immature: much paler below, and the forehead practically white.

DISTRIBUTION AND VARIATION: Breeds from north-central Alaska, southern Yukon, western Mackenzie, Saskatchewan, southern Manitoba, central Ontario, southeastern Quebec, Labrador, and southwestern Newfoundland southward to northwestern Baja California, central Mexico, central Texas, northern Louisiana, central Alabama, northern Georgia, northwestern South Carolina, and eastern North Carolina; also along the Gulf Coast from Louisiana to northern Florida (to Franklin and Alachua counties) (*H. r. erythrogaster*); also throughout most of Europe, northern and central Asia, and northwestern Africa. *H. r. erythrogaster* winters chiefly in Panama and South America (except southern parts), casually northward to Florida, where it is most common as a transient (July-November; March-June).

613. Cliff Swallow, *Petrochelidon pyrrhonota* (Vieillot)

IDENTIFICATION: Size of Tree Swallow (total length 125 to 155 mm). Crown and back metallic blue black, the latter somewhat streaked with white; *forehead light brown or whitish*; remiges and rectrices dusky; throat, sides of head, and nape chestnut; *some blackish markings on lower throat*; remainder of underparts shading from light gray brown on breast to white on abdomen; *rump patch usually cinnamon buff*, but sometimes darker brown; *tail not forked or notched.* (Compare with Cave Swallow.)

DISTRIBUTION AND VARIATION: Breeds from central parts of Alaska and Yukon, western Mackenzie, central Saskatchewan, southern Manitoba, central Ontario, and southern Quebec southward to central Mexico, southern and northeastern Texas, central Missouri, western Kentucky, west-central Tennessee, northern Alabama and Georgia, and western North Carolina.[1] Winters from southern Brazil to central Chile and central Argentina, accidentally in Florida. *P. p. pyrrhonota*, the breeding form in the East, is mainly a transient in Florida, rare in spring; August-October, April-May.

614. Cave Swallow, *Petrochelidon fulva* (Vieillot)

IDENTIFICATION: Resembles a small Cliff Swallow (total length about 130 mm), but differs as follows: *darker on forehead*, rump, and lower tail coverts; *lighter on throat* and sides of head; *no black markings on lower throat.*

DISTRIBUTION AND VARIATION: Breeds from southeastern New Mexico, south-central Texas, and the Greater Antilles southward to southeastern Mexico. Winters in the Greater Antilles and, probably, South America. Accidental in southern Florida, though possibly regular in spring on the Dry Tortugas (*P.f. cavicola*).

615. Purple Martin, *Progne subis* L.

IDENTIFICATION: The largest common Florida swallow (total length 180 to 205 mm). Adult male: plumage mostly metallic blue black, but rectrices and remiges duller. (One-year-old birds may be more similar to adult female.) Adult female, immature: upperparts similar, but duller; underparts and sides of head and neck fading gradually from gray anteriad to whitish on abdomen and under tail coverts, the *latter not tipped with dark.*

DISTRIBUTION AND VARIATION: Breeds from southwestern British Columbia southward to Baja California, Sonora, and Arizona; also (east of Rocky Mountains) from northeastern British Columbia, the central parts of Alberta and Saskatchewan, southern Manitoba, western On-

---

1. Began nesting at Lake Okeechobee in 1974.

tario, the northern parts of Minnesota and Wisconsin, the southern parts of Ontario and Quebec, New Brunswick, and central Nova Scotia southward to south-central Texas, the Gulf Coast, and southern Florida (January-October). Winters in northern South America, casually north to Peninsular Florida (*P. s. subis*).

616. Cuban Martin, *Progne dominicensis* (Gmelin)

IDENTIFICATION: Size and general color of the Purple Martin. Adult male: identical to Purple Martin but for *white on abdomen* (sometimes concealed) and sometimes on breast and under tail coverts. Adult female: differs from female Purple Martin in having the gray breast, throat, and sides *abruptly contrasted* with the white abdomen. *Under tail coverts tipped with dusky.*

DISTRIBUTION AND VARIATION: *P. d. cryptoleuca* breeds in Cuba and the Isle of Pines; probably winters in South America. Accidental in southern half of Florida Peninsula.

617. Southern Martin, *Progne modesta* Gould

IDENTIFICATION: Smaller than Purple Martin (total length about 165 mm), from which it differs also in its *more deeply forked tail* and the female's *dark belly, with broad, pale feather edgings on the lower abdomen.*

DISTRIBUTION AND VARIATION: Breeds from Bolivia and Argentina southward. Migrates northward in summer to northern South America and Panama, and accidentally to Key West, Florida (Robertson, personal communication; *P. m. elegans*).

ORDER PASSERIFORMES: CXXII. FAMILY CORVIDAE, Crows, Jays and Magpies. Medium-small to medium-large birds with rounded wing tips and rounded or graduated tails. *Nostrils usually concealed.* North American species usually show a combination of only black, blue, gray, and white in the plumage.

618. Blue Jay, *Cyanocitta cristata* (L.)

IDENTIFICATION: Slightly larger than the Mockingbird (total length 255 to 315 mm). Top of head and back grayish lavender, interrupted on nape by a black collar; wings and tail bright blue, the secondaries and rectrices barred with black and tipped with white; underparts smoky, with *whitish on throat separated by a black crescent below*; *head crested.*

DISTRIBUTION AND VARIATION: Breeds from central parts of Alberta and Saskatchewan, southern Manitoba, central Ontario, southern Quebec, and Newfoundland southward (east of the Rocky Mountains) to the Texas Panhandle, the Gulf Coast, and the tip of the Florida mainland. Locally withdraws from northern parts of breeding range in winter. *C. c. cristata* breeds south to central Florida, and *C. c. semplei* from Osceola and Hillsborough Counties southward; a specimen of *C. c. bromia* was collected on St. George Island in December, 1972 (TT 3157).

619. Scrub Jay, *Aphelocoma coerulescens* (Bosc)

IDENTIFICATION: Size of Blue Jay, but *head not crested* (total length 265 to 315 mm). *Back gray; remainder of upperparts entirely blue*, somewhat tinged with gray, the primaries being largely gray; underparts whitish to light gray, except for medium gray on under side of rectrices and breast, blue on sides of throat, and black on sides of head. In juvenals the crown, nape, back, and breastband are brownish gray.

DISTRIBUTION AND VARIATION: Resident from southwestern Washington, Oregon, northern Nevada, southeastern Idaho, northern Utah, southwestern Wyoming, all but northeastern Colorado, and central Texas southward to southeastern Mexico; also (*A. c. coerulescens*) in Florida from Jacksonville Beach (rarely) and Levy County southward almost to Miami and Naples, locally distributed (Map 49).

620. Common Crow, *Corvus brachyrhynchos* Brehm

IDENTIFICATION: About size of Red-shouldered Hawk (total length 430 to 505 mm). Plumage entirely brownish black, with slight iridescence. (See Fish Crow.)

DISTRIBUTION AND VARIATION: Breeds from interior of British Columbia, southwestern Mackenzie, the northern parts of Saskatchewan, Manitoba, and Ontario, central Quebec, and southern Newfoundland southward to northern Baja California, central Arizona, north-central New Mexico, central Texas, the Gulf Coast, and the southern tip of the Florida mainland. Withdraws from Canada in winter, at which time a few reach the Florida Keys (UM 438). The breeding form in the Florida Peninsula is *C. b. pascuus*, and *brachyrhynchos* occupies the Panhandle (Johnston, 1961).

621. Fish Crow, *Corvus ossifragus* Wilson

IDENTIFICATION: Smaller than the Common Crow (total length 380 to 430 mm), from which it also differs in its *glossier, more bluish black plumage*, and its *shorter tarsus* (less than 50 mm). The *more nasal, often double, call note* is the most certain means of distinguishing it in the field, though juvenal Common Crows have a similar call.

DISTRIBUTION: Breeds along the Atlantic Coast from Rhode Island southward and along the Gulf Coast from southeastern Texas eastward,

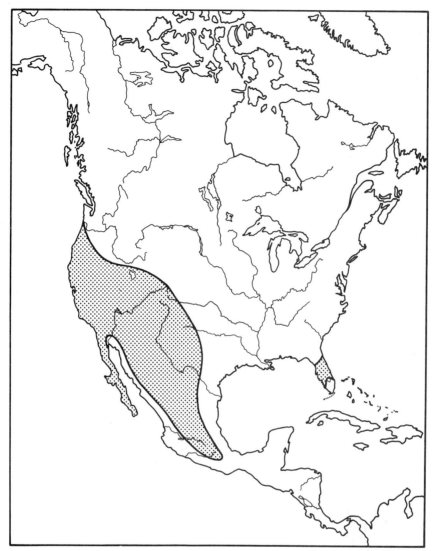

Map 49. Distribution of the Scrub Jay (*Aphelocoma coerulescens*)

penetrating inland to northwestern Louisiana, eastern Oklahoma, Arkansas, extreme western Tennessee, the central parts of Mississippi, Alabama, and Georgia, western parts of the Carolinas, and central Virginia; withdraws from northern parts of inland range in winter. Absent from the Florida Keys and extreme southern tip of mainland (Map 50).

ORDER PASSERIFORMES: CXXIII. FAMILY PARIDAE, Titmice. Small birds with gray, white, and black in the soft, lax plumage, occasionally brown. The bill is short and conical, and the nostrils hidden. The wings are rounded and scarcely, if any, longer than the tail.

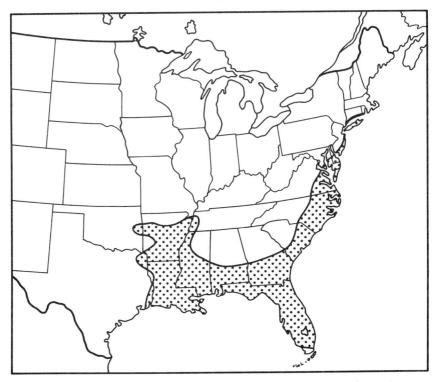

Map 50. Distribution of the Fish Crow (*Corvus ossifragus*)

622. Carolina Chickadee, *Parus carolinensis* Audubon

IDENTIFICATION: The size of a small warbler (total length 100 to 115 mm). *Crown, nape, chin, and throat black*; remainder of upperparts medium gray, except somewhat darker on remiges and rectrices, and secondaries edged with white; sides of head and neck and remainder of underparts whitish, *the former in contrast to adjacent black*.

DISTRIBUTION AND VARIATION: Resident from southeastern Kansas, southwestern and east-central Missouri, central parts of Illinois, Indiana, and Ohio, southern Pennsylvania (except in mountains), and central New Jersey southward to south-central Texas, upper Texas coast, the Gulf Coast, and central Florida (Map 51). The race in the Panhandle is *P. c. carolinensis*, in the Peninsula *P. c. impiger*.

623. Tufted Titmouse, *Parus bicolor* L.

IDENTIFICATION: Slightly smaller than the House Sparrow (total length 140 to 165 mm). Forehead black; remainder of upperparts medium gray to brownish gray, the *crested head* somewhat darker gray; underparts whitish, except for *rusty brown on flanks*.

DISTRIBUTION: Resident from southeastern Nebraska, central and eastern Iowa, southeastern Minnesota, southern parts of Wisconsin, Michigan, and Ontario, northwestern Pennsylvania, the New York City area, and southwestern Connecticut southward to eastern Texas, the Gulf Coast, and southern Florida (northern Monroe and western Broward Counties).

ORDER PASSERIFORMES: CXXIV. FAMILY SITTIDAE, Nuthatches. Small land birds with plumage combinations of chiefly black, gray, and white, but sometimes with brown. Toes strong (longest toe plus its claw at least as long as tarsus). *Bill long and straight*. Tail rather short.

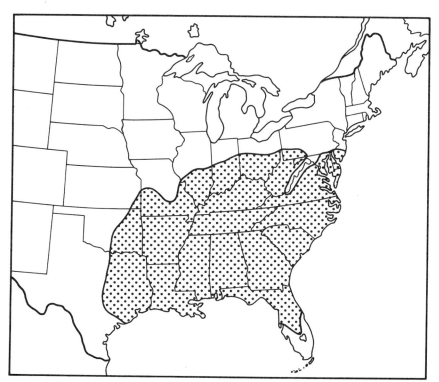

Map 51. Distribution of the Carolina Chickadee (*Parus carolinensis*)

624. White-breasted Nuthatch, *Sitta carolinensis* Latham

IDENTIFICATION: Size of a small sparrow (total length 130 to 155 mm). *Top of head, nape, upper back, and much of wings and tail black*; remainder of upperparts bluish gray, except for white patches in rectrices; underparts, including sides of head and neck, whitish, interrupted by rufous brown posteriad.

DISTRIBUTION AND VARIATION: Resident from southern British Columbia, southeastern Alberta, northwestern and central Montana, the southern parts of Manitoba, Ontario, and Quebec, northern Maine, north-central New Brunswick, and central Nova Scotia southward to southern Baja California, southern Mexico, the northern Gulf Coast, and, at least formerly, central Florida (*S. c. carolinensis*); apparently disappearing from most of its Florida range.

625. Red-breasted Nuthatch, *Sitta canadensis* L.

IDENTIFICATION: Smaller than White-breasted Nuthatch (total length 105 to 120 mm). Crown, nape, and most of rectrices black; back bluish gray; wings dusky; *superciliary stripe white*; a black stripe running through eye; chin and throat whitish; *remainder of underparts tinged with rusty*; females have a blue-gray crown.

DISTRIBUTION: Breeds from southeastern Alaska, southern Yukon, southwestern Mackenzie, central Saskatchewan, southern Manitoba, James Bay, western and northern Ontario, southern and eastern Quebec, and Newfoundland southward (mostly in the Canadian Life Zone) to San Francisco Bay, mountains of southern California, southeastern Arizona, south-central Colorado, southwestern South Dakota, southeastern Manitoba, central Minnesota, Wisconsin, northern Michigan, southern Ontario, southern New York, western Connecticut, Massachusetts, and (at high elevations) to eastern Tennessee and western North Carolina; also on Guadelupe Island. Withdraws from extreme northern parts of breeding range in winter, ranging southward then to the southern parts of Arizona and New Mexico, south-central Texas, the northern Gulf Coast, and central Florida (to Sarasota and northern Volusia Counties; France, 1969; Merrill, 1969).

626. Brown-headed Nuthatch, *Sitta pusilla* Latham

IDENTIFICATION: One of Florida's smallest birds (total length about 100 mm). *Top of head and neck mostly light brown* (grayer in immature) with lighter feather edgings, but interrupted by a white spot on lower nape; back bluish gray; remiges and rectrices dusky; underparts whitish or light buffy.

DISTRIBUTION AND VARIATION: Resident from southeastern Oklahoma, central Arkansas, Louisiana, northern Mississippi, north-central Alabama, northern Georgia, central North Carolina, southeastern Virginia, and southern Maryland and Delaware southward to southeastern Texas, the Gulf Coast, and southern Florida. *S. p. pusilla* is the breeding form in the Panhandle and *S. p. caniceps* in the Peninsula (to northern Monroe County and Fort Pierce; Map 52).

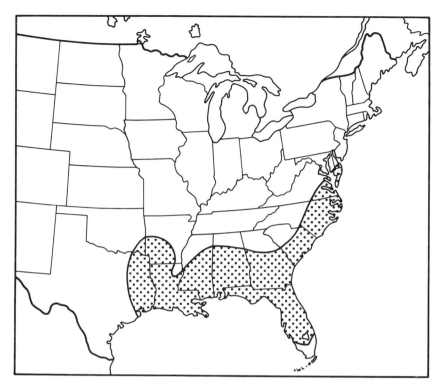

Map 52. Distribution of the Brown-headed Nuthatch (*Sitta pusilla*)

ORDER PASSERIFORMES: CXXV. FAMILY PYCNONOTIDAE, Bulbuls. Small to medium-small birds with soft, lax plumage and, usually, *hairlike bristles on the nape*, intermixed with feathers. Wings rounded, usually not longer than tail. Tarsus short, not longer than bill plus rictus.

627. Red-whiskered Bulbul, *Pycnonotus jocosus* (L.)

IDENTIFICATION: Total length about 180 mm. *Top of head black and crested*; remainder of upperparts grayish brown; *auriculars and under tail coverts reddish*; malar streak black; remainder of underparts (except most of tail) white. Juvenals are paler above and without red.

DISTRIBUTION AND VARIATION: India and southeastern Asia; introduced locally elsewhere. The Florida population (Kendall, Dade County) represents *P. j. emeriae*. (See Stimson, 1962; also Banks and Laybourne, 1968.)

ORDER PASSERIFORMES: CXXVI. FAMILY CERTHIIDAE, Creepers. Small birds usually with gray, brown, black, and white in the plumage and with *stiffened rectrices. Bill long and slender, often decurved*. Claws long, slender, and curved.

628. Brown Creeper, *Certhia familiaris* L.

IDENTIFICATION: Slightly smaller than the Carolina Wren (total length 125 to 145 mm). Upperparts mostly brown, with short whitish streaks, but *rump bright rusty*, and the *pointed, graduated rectrices* entirely grayish brown; underparts light gray or whitish. Its *woodpecker-like posture* in the field will separate it from species of similar color.

DISTRIBUTION AND VARIATION: Breeds over much of Eurasia; also from southeastern Alaska, British Columbia, the central parts of Alberta and Saskatchewan, southern Manitoba, northern Ontario, southern Quebec, and Newfoundland southward at high elevations to Nicaragua, eastern Tennessee, and western North Carolina; at lower elevations to west-central California, southeastern Nebraska, southern Iowa, southern Wisconsin, central Michigan, northeastern Ohio, southern Ontario, and Massachusetts. Withdraws from northernmost parts of breeding range (and from higher altitudes?) in winter, ranging southward at that season to southeastern California, southern parts of Arizona, New Mexico, and Texas, the northern Gulf Coast, and central (rarely southern) Florida. Most Florida specimens have been assigned to *C.f. americana*, but *nigrescens* has been taken once in the Panhandle (Weston, 1965).

ORDER PASSERIFORMES: CXXVII. FAMILY TROGLODYTIDAE, Wrens. Small birds of brownish to white plumage, *barred with blackish on wings and tail*. Bill long, slender, and decurved. Rictal bristles reduced or lacking. Wings rounded and short. Tail tip rounded.

629. House Wren, *Troglodytes aedon* Vieillot[1]

IDENTIFICATION: Much smaller than Carolina Wren (total length 110 to 130 mm). Upperparts brown to reddish brown, transversely barred with dusky; underparts light grayish brown to whitish, *faintly barred with dusky on flanks*; *superciliary line faint*; tail of moderate length.

DISTRIBUTION AND VARIATION: Breeds from southern and east-central British Columbia, central Alberta, southern parts of Saskatchewan and Manitoba, central Ontario, southern Quebec, Maine, and New Brunswick southward to northern Texas, Arkansas, Tennessee, northern Georgia and probably northern Alabama (Harriett Wright, personal communication); also through the mountains of Mexico and Central America to the southern tip of South America; also in the central and southern Lesser Antilles and the Falkland Islands (Map 53). Winters from the southwestern United States, northeastern Texas, central

---

1. Includes *Troglodytes brunneicollis* and *T. musculus*, but the description applies to the North American population only.

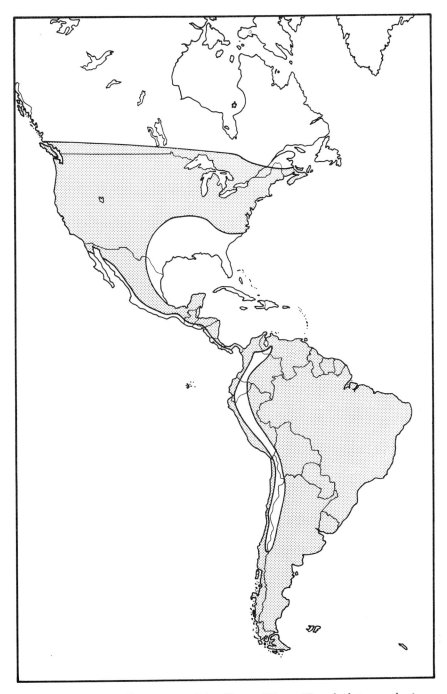

Map 53. Breeding range of the House Wren (*Troglodytes aedon*)

Arkansas, the northern parts of Mississippi, Alabama, and Georgia, and eastern North Carolina southward to the southern limit of breeding range, the northern Gulf Coast, and the Florida Keys. Subspecies known to occur in Florida are *aedon*, *baldwini*, and *parkmanii* (September-May).

630. Winter Wren, *Troglodytes troglodytes* (L.)

IDENTIFICATION: About size of House Wren, but *tail very short* (total length 90 to 105 mm). Coloration of upperparts similar to that of House Wren, but less distinctly barred; underparts buffy anteriad, *heavily spotted and barred with blackish posteriad*; lower mandible and feet light brown.

DISTRIBUTION AND VARIATION: Breeds from Alaska, southern Yukon, south-central Mackenzie, central Saskatchewan, southern Manitoba, northern Ontario, central Quebec, Newfoundland, Iceland, and the northern parts of Europe and Asia southward to the central parts of California and Idaho, the Great Lakes, northern New York, Massachusetts (south in mountains to northern Georgia), and southern Eurasia. Winters from southern Alaska and the Aleutian Islands, northeastern Colorado, southeastern Nebraska, central Iowa, Missouri, central Illinois, the southern parts of Michigan and Ontario, Ohio, central New York, and Massachusetts southward to southern California, Arizona, New Mexico, central Texas, the Gulf Coast, and central (rarely southern) Florida; also withdraws in winter from extreme northern Eurasia. Both *T. t. hiemalis* and *T. t. pullus* are known to occur in Florida (Stevenson and Baker, 1970).

631. Bewick's Wren, *Thryomanes bewickii* (Audubon)

IDENTIFICATION: Slightly smaller than the Carolina Wren (total length 125 to 140 mm). Upperparts chiefly reddish brown, but remiges and rectrices duller brown, barred with blackish; *outer rectrices spotted with white*; underparts and superciliary stripe whitish, the tail coverts barred with blackish; *tail proportionately very long*.

DISTRIBUTION AND VARIATION: Breeds from southwestern British Columbia, central Washington, Nevada, southern Utah, southwestern Wyoming, central Colorado, southeastern Nebraska, the southern parts of Iowa, Wisconsin, Michigan, and Ontario, and from Ohio, central Pennsylvania, and Virginia southward to southern Baja California, southern Mexico, southern and eastern Texas, Arkansas, and the central parts of Mississippi, Alabama, Georgia, and South Carolina. Resident in the West, but eastern population withdraws from Canada and the northern United States in winter, some reaching the Gulf Coast, rarely southeastward to central Florida, but becoming scarcer in state (*T. b. bewickii* and *T. b. altus*; September to March).

632. Carolina Wren, *Thryothorus ludovicianus* (Latham)

IDENTIFICATION: Size of a small sparrow (total length 125 to 150 mm). *Upperparts more reddish brown* than in other Florida wrens, especially on rump; remiges and rectrices barred with dusky; superciliary stripe, chin, and throat whitish; stripe through eye reddish brown; remainder of *underparts ranging from buffy white to cinnamon brown*; lower mandible and feet flesh-colored.

DISTRIBUTION AND VARIATION: Resident from southeastern Nebraska, the southern parts of Iowa and Illinois, central Indiana, southeastern Michigan, the southern parts of Ontario and Pennsylvania, southeastern New York, southern Connecticut, and southeastern Massachusetts southward to central Mexico, the northern Gulf Coast, and upper Key Largo, Florida. The Panhandle is occupied by *T. l. ludovicianus*, the Peninsula by *T. l. miamensis*, and Dog Island (Franklin County) by *T. l. nesophilus* (Stevenson, 1973).

633. Long-billed Marsh Wren, *Telmatodytes palustris* (Wilson)

IDENTIFICATION: Suggests a small Carolina Wren (total length 100 to 140 mm). Coloration of upperparts like that of Carolina Wren, but less reddish and with *white streaks on upper back*; underparts mostly white, but light brown on flanks; lower mandible and feet flesh-colored to light brown.

DISTRIBUTION AND VARIATION: Breeds from central British Columbia, northern Alberta, south-central Saskatchewan, southern Manitoba, southern Ontario, southwestern Quebec, southern Maine, and eastern New Brunswick southward to northern Baja California, south-central Mexico, the northern Gulf Coast, and Florida (to Tampa Bay and New Smyrna); may not breed in interior of eastern United States south of Kansas, Missouri, Illinois, Indiana, Ohio, and Virginia. Winters in southern parts of breeding range northward to central parts of Gulf States in the East, but to British Columbia in the West; also southward to southern Mexico and southern Florida. The Florida breeding races are *T. p. griseus*, New Smyrna northward, and *T. p. marianae*, Gulf Coast from Pensacola to Tampa Bay (formerly to Charlotte Harbor). They are joined in winter by *T. p. palustris*, *T. p. dissaeptus*, and *T. p. iliacus* (Stevenson and Baker, 1970).

634. Short-billed Marsh Wren, *Cistothorus platensis* (Latham)

IDENTIFICATION: About size of Winter Wren, but tail not as short (total length 100 to 115 mm). Upperparts with about equal amounts of brown and black, *broken on crown* and back *by longitudinal streaks of white*; black on wings and tail forming transverse bars; underparts varying from white to buffy brown, the latter on upper breast, sides, and under tail coverts; lower mandible and feet flesh-colored; light superciliary stripe not conspicuous.

DISTRIBUTION AND VARIATION: Breeds from southeastern Saskatchewan, southern Manitoba, western and southern Ontario, southern Quebec, southern Maine, and eastern New Brunswick southward (east of Great Plains) to east-central Arkansas, southern Illinois, central Indiana, south-central Ohio, eastern West Virginia, Maryland, and southeastern Virginia; nonbreeding birds may summer southward to northern Florida. Winters from south-central and northeastern Texas, western Tennessee, northern Alabama, and central Virginia southward to central Mexico, the northern Gulf Coast, and southern Florida (*C. p. stellaris*, late September to May). Other races are resident from southern Mexico to western Panama, and from northern South America southward through the Andes to the southern tip of South America.

ORDER PASSERIFORMES: CXXVIII. FAMILY MIMIDAE, Mimic Thrushes. Medium-small birds usually with much gray or brown in the plumage, occasionally bluish or blackish. Structurally similar to wrens, but *rictal bristles more developed and plumage usually not barred*.

635. Mockingbird, *Mimus polyglottos* (L.)

IDENTIFICATION: Total length 230 to 280 mm. Adult: upperparts chiefly medium gray, except darker on wings and tail; *large white patches in primaries and outer rectrices*; underparts whitish (often stained gray by smoke in cities); iris variable, often pale. Juvenal: similar to adults, but back with a brownish cast, remiges tipped with white, and *throat and breast spotted with dusky*.

DISTRIBUTION AND VARIATION: Breeds from San Francisco Bay, northern inland California, central Nevada, northern Utah, southeastern Indiana, north-central Ohio, eastern West Virginia, Maryland, and central New Jersey (casually northward to southern Canada) southward to southern Mexico, the northern Gulf Coast, the Greater Antilles, and the Virgin Islands. Withdraws from extreme northern parts of breeding range in winter. The eastern race, *M. p. polyglottos*, is resident throughout Florida.

636. Gray Catbird, *Dumetella carolinensis* (L.)

IDENTIFICATION: Resembles a small Mockingbird in form (total length 200 to 230 mm). *Upperparts dark gray, shading to blackish on tail and top of head*; underparts and sides of head medium dark gray, except for *chestnut under tail coverts* (paler in immature).

DISTRIBUTION: Breeds from southern British Columbia, the central parts of Alberta and Saskatchewan, southern Manitoba, western Ontario, the southern parts of Ontario and Quebec, and from New Brunswick and Nova Scotia southward through northern and eastern Washington and eastern Oregon to north-central Utah, east-central Arizona, north-central New Mexico, western Oklahoma, eastern Texas, and the southern interior of Louisiana, Mississippi, Alabama, and Georgia (formerly to Florida). Winters from southeastern Texas, northeastern Louisiana, southeastern Arkansas, the central parts of Mississippi, Alabama, Georgia, and South Carolina, eastern North Carolina, and southeastern Virginia southward to Panama and the Lesser Antilles; casually much farther north.

637. Brown Thrasher, *Toxostoma rufum* (L.)

IDENTIFICATION: Slightly larger than Mockingbird (total length 250 to 305 mm). *Upperparts bright reddish brown*, except generally duller on remiges; *underparts white*, tinged with buffy, and *with dark brown or blackish spots arranged in longitudinal rows*; *tail much longer than in thrushes*; *iris yellow*; feet light brown.

DISTRIBUTION AND VARIATION: Breeds from southeastern Alberta, the southern parts of Saskatchewan and Manitoba, western Ontario, northern Minnesota and Michigan, southern Ontario, southwestern Quebec, northern Vermont, central New Hampshire, and southwestern Maine southward to central Montana, eastern Wyoming and Colorado, northern and eastern Texas, the Gulf Coast, and the upper Florida Keys. Winters chiefly south of latitude 40° N., occasionally reaching Cuba. The breeding race in Florida is *T. r. rufum*, and *T. r. longicauda* has occurred in northwest Florida in fall (Stevenson and Baker, 1970).

638. Curve-billed Thrasher, *Toxostoma curvirostre* (Swainson)

IDENTIFICATION: About size of Brown Thrasher, but *bill averaging longer and much more decurved. Upperparts light gray brown*, the wing coverts and other rectrices narrowly tipped with white; underparts buffy white, sparsely spotted with brownish gray, except white in center of throat; iris orange.

DISTRIBUTION AND VARIATION: Resident from northwestern and central Arizona, New Mexico, and western and southern Texas southward to Oaxaca; accidental in Nebraska and near Pensacola, Florida (*T. c. palmeri* apparently "hybridized" with a Brown Thrasher in 1932; Weston, 1965).

639. Sage Thrasher, *Oreoscoptes montanus* (Townsend)

IDENTIFICATION: Smaller than Mockingbird and with the *tail square-tipped* (total length about 200 mm). Brownish gray above, whitish below, with longitudinally arranged rows of dark spots; white spots at tips of outer rectrices; iris yellow.

DISTRIBUTION: Breeds from south-central British Columbia, central Idaho, south-central Montana, and northern and southeastern Wyoming southward to south-central California, southern Nevada, Utah, north-central New Mexico, northwestern Texas, and western Oklahoma; also in southwestern Saskatchewan. Winters from central California, northern Arizona, southern New Mexico, and central and southern Texas southward to northern Mexico; also casually farther north. Accidental eastward to New York, coastal Alabama, and Florida (Gilchrist County; Johnston, 1969).

ORDER PASSERIFORMES: CXXIX. FAMILY TURDIDAE, Thrushes. Small to medium-sized birds with the *front of the tarsus booted*. Plumage frequently with spots in adults or young. Tail tip truncate in North American species.

640. American Robin, *Turdus migratorius* L.

IDENTIFICATION: Size of Blue Jay, but tail shorter (total length 230 to 265 mm). Adult: upperparts chiefly medium gray, but darker on remiges (with light edgings), rectrices, and top of head; *underparts mostly rufous brown*, but lower belly and chin white, and throat streaked with black and white; sides of head blackish. Immature: similar, but underparts spotted with blackish. Bill yellowish in adults.

DISTRIBUTION AND VARIATION: Breeds from the northern parts of Alaska and Canada southward to southern Mexico, eastern Texas, the southern parts of Louisiana, Mississippi, and Alabama, and northern Florida (*T. m. achrusterus*; Olson, 1961). Winters chiefly in the United States, but ranges at that season southward to Guatemala, the Florida Keys (including *T. m. migratorius* and *T. m. nigrideus*; Stevenson and Baker, 1970), the Bahama Islands, and western Cuba.

641. Wood Thrush, *Hylocichla mustelina* (Gmelin)

IDENTIFICATION: Smaller than the Brown Thrasher and tail much shorter (total length 190 to 210 mm). Upperparts shading gradually from *bright reddish brown on head, neck, and upper back* to olive brown on tail; sides of head grayish, streaked with white; underparts white with *large, rounded, blackish spots on throat, breast, and sides*; feet flesh-colored.

DISTRIBUTION: Breeds from southeastern South Dakota, the central parts of Minnesota and Wisconsin, northern Michigan, southern Ontario and Quebec, northern Vermont, central New Hampshire, and southwestern Maine southward to southeastern Texas, the southern parts of Louisiana, Mississippi, and Alabama, and northern Florida (mostly north of 30 degrees latitude, but see Goin and Goin, 1969). Winters from southern Texas through eastern Mexico to Panama, casually north to central Texas and northern Florida.

642. Hermit Thrush, *Catharus guttatus (Pallas)*

IDENTIFICATION: Smaller than the Wood Thrush (total length 165 to 180 mm). Upperparts shading gradually from olive brown on head to *bright reddish brown on rump and tail*; underparts white, tinged with buff or brownish gray, and spotted with dark brown anteriad (except on chin and midline of throat); tarsus and toes flesh-colored. *Easily confused with Gray-cheeked and Swainson's Thrushes* except in direct sunlight.

DISTRIBUTION AND VARIATION: Breeds from central Alaska, southern Yukon, southern Mackenzie, northwestern Saskatchewan, southern Manitoba, northern Ontario, central Quebec, southern Labrador, and Newfoundland southward to southern California, southeastern Arizona, southern New Mexico, western Texas, southern Saskatchewan, central Minnesota and Wisconsin, north-central Michigan, central Ontario, northeastern Ohio, central Pennsylvania, eastern West Virginia, western Maryland, northeastern Pennsylvania, southern New York, Connecticut, Massachusetts, and, locally, Long Island. Winters from northwestern Washington, southwestern British Columbia, and the middle latitudes of the United States southward to Guatemala, the northern Gulf Coast, and southern Florida (October to early May). *C. g. faxoni* and *C. g. crymophilus* have been collected in Florida (Stevenson and Baker, 1970).

643. Swainson's Thrush, *Catharus ustulatus* (Nuttall)

IDENTIFICATION: About size of Hermit Thrush (total length 165 to 190 mm). Upperparts generally olive brown; underparts white posteriad, but *heavily tinged with buff on chin, throat, and breast, some of which color extends upward on head to include the eye-ring*; breast and sides of throat spotted (sometimes also streaked) with dusky; tarsus and toes flesh-colored. *Easily confused with Hermit or Gray-cheeked Thrush in mediocre light conditions.*

DISTRIBUTION AND VARIATION: Breeds from central Alaska, northern Yukon, western and southern Mackenzie, northern Manitoba and Ontario, central Quebec, southern Labrador, and Newfoundland southward to southwestern California, the central parts of Nevada, Utah, and Colorado, southern Saskatchewan and Manitoba, the northern parts of Minnesota, Wisconsin, and Michigan, the southern parts of Ontario, Quebec, and Vermont, central New Hampshire, and Maine; also, in the mountains, to eastern West Virginia. All Florida specimens have been collected in spring and fall (*C. u. incanus*, *clarescens*, and *swainsoni*; September-October and April-May). (See Stevenson and Baker, 1970.)

644. Gray-cheeked Thrush, *Catharus minimus* (Lafresnaye)

IDENTIFICATION: About size of Hermit Thrush (total length 165 to 200 mm). Coloration almost identical to that of Swainson's Thrush, but usually *with less buff on throat and breast and with none on sides of head*; tarsus and toes flesh-colored. *Easily confused with Swainson's or Hermit Thrush in less than optimal light conditions.*

DISTRIBUTION AND VARIATION: Breeds from northeastern Siberia, northern parts of Alaska, Mackenzie, Manitoba, and Quebec, central Labrador, and Newfoundland southward to southwestern Alaska, northeastern British Columbia, central Saskatchewan, southeastern New York, and northwestern Massachusetts. Winters in Hispaniola and from Nicaragua to northern Peru and northwestern Brazil. Both *C. m. aliciae* (*sensu* Todd, 1958) and *C. m. bicknelli* occur as transients in Florida, September to November and April to May. (See Stevenson and Baker, 1970.)

645. Veery, *Catharus fuscescens* (Stephens)

IDENTIFICATION: About size of Hermit Thrush (total length 165 to 190 mm). *Upperparts chiefly reddish brown*, but remiges and rectrices somewhat duller; chin and abdomen white; *throat and upper breast buffy, finely speckled with dusky* (spots smaller than in other thrushes); lower breast tinged with light gray; tarsus and toes flesh-colored.

DISTRIBUTION AND VARIATION: Breeds from eastern British Columbia, north-central Alberta, southern parts of Saskatchewan, Manitoba, Ontario, and Quebec, and from central Newfoundland southward to northeastern Arizona, northeastern South Dakota, southeastern Minnesota, northeastern Ohio, eastern Pennsylvania, central New Jersey, and (in the mountains) to eastern Tennessee and northern Georgia. Winters from Central America to central and northeastern Brazil. Regular transient in Florida, September to October and April to May (*C. f. fuscescens*, *salicicola*, and *fuliginosus*; Stevenson and Baker, 1970).

646. Eastern Bluebird, *Sialia sialis* (L.)

IDENTIFICATION: Slightly larger than House Sparrow (total length

150 to 180 mm). Adult male: *upperparts brilliant dark blue*, except for dusky on remiges; *chin, throat, and breast cinnamon brown*; abdomen and under tail coverts white. Adult female: *much paler throughout*, being more gray than blue on head and upper back. Juvenal: *upperparts chiefly gray, spotted with white on back*, but with considerable blue on remiges and rectrices; *chin, throat, and breast mottled gray and white*; abdomen and under tail coverts white.

DISTRIBUTION AND VARIATION: Breeds from the southern parts of Saskatchewan, Manitoba, Ontario, and Quebec, and from New Brunswick and southern Nova Scotia southward (east of the Rocky Mountains) to southeastern Arizona, Honduras, the northern Gulf Coast, southern Florida, and Bermuda. Withdraws from northernmost parts of breeding range in winter, when it rarely reaches the Keys, western Cuba, and the Bahamas (Bond, 1961). *S. s. sialis* breeds southward to central Florida, and *S. s. grata* breeds from about Lake County to southern Collier County and northern Palm Beach (formerly Dade) County.

647. Wheatear, *Oenanthe oenanthe* (L.)

IDENTIFICATION: About size of House Sparrow (total length 140 to 165 mm). Winter plumage: upperparts olive brown, except wings and distal half of tail black, and the *rump and basal half of tail white*; *a black line through eye*; underparts cinnamon brown.

DISTRIBUTION AND VARIATION: Breeds throughout most of Eurasia; also in Alaska, northern Canada, northern Greenland, and Iceland. Winters in Africa, southern Arabia, Mongolia, northern China, and casually to eastern China and the Philippines; accidental in Colorado (*O. o. oenanthe*), the eastern United States, Cuba, and Florida (2 fall records, presumably *O. o. leucorhoa*; Sprunt, 1963).

ORDER PASSERIFORMES: CXXX. FAMILY SYLVIIDAE, Gnatcatchers, Kinglets, and Old World Warblers. Mostly small, rather plainly colored birds with brown, gray, or olive green in the plumage. Bill slender and notched in North American species; nostrils slitlike, operculate. Wing tip rounded, but wing not especially short. Tarsus typically rather long.

648. Blue-gray Gnatcatcher, *Polioptila caerulea* (L.)

IDENTIFICATION: Very small, but tail long (total length 100 to 130 mm). Top of head, neck, and back light bluish gray; wings dusky; *tail nearly half of total length, black, becoming white on lateral rectrices*; underparts white; black line over eye in male. Reminiscent of miniature Mockingbird.

DISTRIBUTION AND VARIATION: Breeds from northern California, central Nevada, southern Utah, Colorado, and eastern Nebraska, central Minnesota, southern parts of Wisconsin, Michigan, and Ontario, western New York, and northern New Jersey southward to Baja California, Guatemala, the northern Gulf Coast, and southern Florida (northern Monroe and northwestern Dade Counties). Winters from southern California, southern Nevada, central Arizona, south-central Texas, Louisiana, the southern parts of Mississippi, Alabama, and Georgia, eastern parts of the Carolinas, and southeastern Virginia southward to Guatemala and Cuba; resident on the Bahama Islands. The eastern race, *P. c. caerulea*, occurs in Florida.

649. Golden-crowned Kinglet, *Regulus satrapa* Lichtenstein

IDENTIFICATION: One of Florida's smallest birds (total length 88 to 100 mm). *Crown yellow* (orange medially in male), *bordered by black*; a black streak through eye, bordered above and below by whitish; neck and upper back olive gray, shading to yellowish green on rump and lateral edges of the dusky rectrices; remiges similar to rectrices, but blackish toward base; a whitish wing bar on coverts; underparts whitish; *no eye-ring*.

DISTRIBUTION AND VARIATION: Breeds from southern Alaska, British Columbia, northern Manitoba and Ontario, southern Quebec, and Newfoundland southward to southern California, Guatemala (at high elevations), central Minnesota, northern Michigan, southern Ontario, northern New York, central Massachusetts, and southern Maine; also in the Appalachian Mountains to eastern Tennessee and western North Carolina. Winters from British Columbia, Alberta, the southern parts of Minnesota, Wisconsin, and Ontario, New York, New Brunswick, and Newfoundland southward to Guatemala (mountains), south-central Texas, the northern Gulf Coast, and northern (rarely central) Florida (*R. s. satrapa*, October-March).

650. Ruby-crowned Kinglet, *Regulus calendula* (L.)

IDENTIFICATION: Slightly larger than the Golden-crowned Kinglet (total length 100 to 115 mm). Color pattern of body, wings, and tail similar to that of Golden-crowned Kinglet, but *head olive brown except for a white eye-ring*; also a red crown patch in the male, usually concealed.

DISTRIBUTION AND VARIATION: Breeds from the northwestern parts of Alaska and Mackenzie, northern Manitoba, northern Ontario, central Quebec, southern Labrador, and Newfoundland southward (often at high elevations) to southern California (and Guadalupe Island), central Arizona and New Mexico, northern Michigan, southern Ontario, northern Maine, and Nova Scotia. Winters from southern British Columbia, Idaho, southern Utah, Nebraska, southern Iowa, northern Illinois, southern Ontario and Ohio, West Virginia, Maryland, and New Jersey southward to Baja California, Guatemala, the northern Gulf Coast, Cuba, and Jamaica (*R. c. calendula*, late September-early May).

ORDER PASSERIFORMES: CXXXI. FAMILY MOTACILLIDAE, Pipits and Wagtails. Small, terrestrial birds, most with *tail-bobbing habits*. Bill typically long and slender. Tail medium to long. *Hallux and its claw usually very long*. Wing pointed, with only 9 primaries (none rudimentary) and *long inner secondaries*.

651. Water Pipit, *Anthus spinoletta* (L.)

IDENTIFICATION: About size of House Sparrow, but tail longer (total length 150 to 180 mm). Upperparts grayish brown, *unstreaked*, becoming darker on remiges and most rectrices; *lateral rectrices largely white*; underparts buffy, with dusky streaks on breast. In late spring a few may be seen in breeding plumage, when they are more grayish above and tinted with cinnamon below; *tarsus and toes dark, the hind claw long*.

DISTRIBUTION AND VARIATION: Breeds over much of Europe and near middle latitudes of Asia; also from northern Alaska, northernmost Canada, and western Greenland southward to northern Oregon, northern Arizona and New Mexico, southern Hudson Bay, southern Labrador, and Newfoundland. In the New World, winters from southwestern British Columbia, Oregon, west-central Nevada, southern Utah, northern Texas, Arkansas, Tennessee, West Virginia, and southern New Jersey southward to Baja California, Oaxaca, Guatemala, the northern Gulf Coast, and southern Florida (*A. s. rubescens*, October-May).

652. Sprague's Pipit, *Anthus spragueii* (Audubon)

IDENTIFICATION: About size of the Water Pipit, differing from it in the following respects: *upperparts broadly streaked* with dusky and light buffy gray; underparts buffy anteriad to whitish on abdomen, lightly streaked with dusky on breast; *tarsus and toes straw-colored*; hind claw longer.

DISTRIBUTION: Breeds from northern Alberta, central Saskatchewan, and central Manitoba southward to Montana, North Dakota, and northwestern Minnesota. Winters from southern Arizona, central Texas, northwestern Louisiana, northwestern Mississippi, central Alabama, and northern Georgia southward to southern Mexico and the northern Gulf Coast; casually to south-central Florida and the Bahamas (October to April).

ORDER PASSERIFORMES: CXXXII. FAMILY BOMBYCILLIDAE, Waxwings. *Crested birds with soft, lax, plumage* of brown, yellow, and red, the *secondaries wax-tipped in adults*. Bill broad, slightly depressed at base, the nostrils concealed. Wings long and pointed, with only 9 functional primaries. *Terminal tail band yellow or red*.

653. Cedar Waxwing, *Bombycilla cedrorum* Vieillot

IDENTIFICATION: About size of a bluebird (total length 165 to 190 mm). Head, neck, breast, and upper back vinaceous gray brown, shading gradually to gray on rump and wings, and to yellow on belly; chin, forehead, and line through eye black; *head crested*; *rectrices pearl gray basally*, shading gradually to blackish, then *abruptly tipped with yellow*; under tail coverts very long and white; in older birds the *secondaries have red waxy tips*.

DISTRIBUTION: Breeds from southeastern Alaska, north-central British Columbia, northern Alberta, northwestern Saskatchewan, central Manitoba, northern Ontario, central and southeastern Quebec, and Newfoundland southward to northern California, northern Utah, Colorado, Oklahoma, central Missouri, southern Illinois, central Kentucky, eastern Tennessee (rarely northern Alabama), northern Georgia, and North Carolina. Winters from southern British Columbia, northern Idaho, central Arizona, north-central New Mexico, central Missouri, southern parts of Illinois, Michigan, and Ontario, and from Massachusetts southward to Panama, the northern Gulf Coast, and southern Florida (late October to early June); casually to Bermuda, the Bahama Islands, the West Indies, Colombia, and Venezuela.

ORDER PASSERIFORMES: CXXXIII. FAMILY LANIIDAE, Shrikes. Small to medium-small birds with gray, brown, black, and white often present in the plumage. *Bill relatively strong and hooked*, sometimes toothed. Tail long, often graduated. Wing typically short and rounded, with 10 primaries.

654. Loggerhead Shrike, *Lanius ludovicianus* L.

IDENTIFICATION: Close to size of Mockingbird, but tail shorter (total length 200 to 240 mm). Adult: upperparts mostly gray, becoming bluish gray on rump; remiges blackish, contrasting with white tips of secondaries and greater coverts; central rectrices black, lateral ones largely white; *stripe through eye black*; underparts white to light gray. Immature: upperparts more brownish than in adults, with fine, wavy crossbars of dusky; breast and sides differ in same manner from those of adult. *Bill black and hooked*.

DISTRIBUTION AND VARIATION: Breeds from southern British Columbia, central Alberta and Saskatchewan, the southern parts of Manitoba, Ontario, and Quebec, south-central Maine, and southwestern New Brunswick southward to southern Baja California, southern Mexico, the northern Gulf Coast, and the southern Florida mainland. Withdraws from Canada and the northern United States in winter; occasional at that season on Florida Keys. The breeding form in northern and central Florida is *L. l. ludovicianus*; *L. l. miamensis* ranges from Jupiter and Fort Myers southward (Rand, 1957). *L. l. migrans* has been collected in northern and central Florida in winter.

ORDER PASSERIFORMES: CXXXIV. FAMILY STURNIDAE, Starlings. Small to medium-sized birds with dark, often *iridescent plumage*, and frequently with *lanceolate contour feathers*. Legs and feet well developed. Bill relatively long, the nostrils operculate and rather low on bill. Wings usually pointed, with 10 primaries.

655. Starling, *Sturnus vulgaris* L.

IDENTIFICATION: About size and shape of a meadowlark (total length 190 to 215 mm). Adult in breeding plumage: *many contour feathers narrow and pointed*; *general coloration blackish, with purplish and greenish iridescence*, but with many feathers of back and abdomen tipped with whitish; remiges and rectrices dusky, with fine whitish edgings; *bill yellow*; feet flesh-colored. Adult in winter: spots much more extensive and larger, white and rounded on underside, light brown and often triangular on back; bill dark. Immature: dark gray above, lighter gray and streaked with white below; chin and throat largely white; no spotting or pointed feathers (except when acquiring adult plumage); bill dark.

DISTRIBUTION AND VARIATION: Resident over most of Europe and the middle latitudes of Asia; also, by introduction, in southern Canada and all of the United States except the southern parts of California, Arizona, and New Mexico. Winters southward through remainder of United States and into northeastern Mexico. Only occasional in extreme southwestern Florida, on the Keys (*S. v. vulgaris*), and the northern Antilles.

ORDER PASSERIFORMES: CXXXV. FAMILY VIREONIDAE, Vireos. Small arboreal birds with a tendency toward an olive green back and whitish to yellowish underparts. *Bill somewhat hooked at tip*. Outermost primary short to rudimentary. Rictal bristles present.

656. White-eyed Vireo, *Vireo griseus* (Boddaert)

IDENTIFICATION: Warbler-sized (total length 115 to 140 mm). Adult: upperparts grayish green, becoming greener on rump; *eye-ring and lores bright yellow*; wings mostly dusky, with *2 whitish bars on coverts*; rectrices dusky, with greenish lateral edgings; underparts white, tinged with yellowish on sides; feet bluish gray; *iris white*. Immature: similar, but back more grayish and *iris brown*.

DISTRIBUTION AND VARIATION: Breeds from eastern Nebraska, southern Indiana, Iowa, southern Wisconsin, and New York southward to northern Mexico, the northern Gulf Coast, and the Florida Keys; also in Bermuda. Winters from southern Texas, Louisiana, the central parts of Mississippi, Alabama, and Georgia, eastern parts of the Carolinas, and southeastern Virginia southward to Honduras, Panama, and Cuba. *V. g. griseus* breeds over northern Florida, and *V. g. maynardi* from Tampa Bay and Anastasia Island to the Keys; *V. g. noveboracensis* also occurs in migration and in winter (Weston, 1965).

657. Bell's Vireo, *Vireo bellii* Audubon

IDENTIFICATION: Florida's smallest vireo (total length 110 to 130 mm). Like White-eyed Vireo, but more grayish above, *eye-ring and lores white, and iris brown* at all ages. Also similar to Ruby-crowned Kinglet, but *bill larger and heavier*.

DISTRIBUTION AND VARIATION: Breeds from northern interior of California, southern Nevada, central Arizona, southwestern New Mexico, western Texas, eastern Colorado, central Nebraska, southeastern South Dakota, Iowa, southwestern Wisconsin, and northeastern Illinois southward to central Mexico, southern Texas, and northwestern Louisiana. Winters from northern Mexico southward to Nicaragua; also casually in southern Louisiana and the Florida Peninsula (probably *V. b. bellii*); transient in Panhandle. Florida: September to April, rare.

658. Yellow-throated Vireo, *Vireo flavifrons* Vieillot

IDENTIFICATION: Larger than White-eyed Vireo (total length 125 to 155 mm). Top of head, neck, and upper back decidedly greenish; *lower back and scapulars gray*; remiges and rectrices blackish, edged with white; *2 white wing bars on coverts*; *chin, throat, and breast canary yellow*; belly and under tail coverts white; feet bluish gray.

DISTRIBUTION: Breeds from southern Manitoba, north-central Minnesota, central Wisconsin, central Michigan, southern Ontario and Quebec, northern New Hampshire, and southwestern Maine southward (east of the Great Plains) to central and eastern Texas, the Gulf Coast, and central Florida. Winters from southeastern Mexico to Panama, in Cuba, the Isle of Pines, the Bahamas, and southern Florida; also rarely to Colombia and western Venezuela and northward to southern Texas and northern Florida. Florida: March to October (winter).

659. Solitary Vireo, *Vireo solitarius* (Wilson)

IDENTIFICATION: Size of Yellow-throated Vireo (total length 125 to 155 mm). *Head gray*, sometimes with a greenish cast; *eye-ring and lores white*; neck and back olive green; wings and tail dusky, with *2 whitish bars on wing coverts*; underparts white, tinged with yellowish on breast and sides; feet bluish gray.

DISTRIBUTION AND VARIATION: Breeds from central British Columbia, southwestern Mackenzie, central Saskatchewan and Manitoba, northern Ontario, southern Quebec, Newfoundland, and Nova Scotia southward (often at high elevations) to southern Baja California, Guatemala, El Salvador, north-central North Dakota, central Minnesota, southeastern Wisconsin, central Michigan, northeastern Ohio, northern New Jersey, central Connecticut, and Massachusetts; also, in the mountains, to eastern Tennessee, northern Georgia, northwestern South Carolina, and most of North Carolina. Winters from southern Arizona, northern Mexico, northern Louisiana, the central parts of Mississippi, Alabama, Georgia, and the Carolinas southward to Costa Rica and Cuba. Both *V. s. solitarius* and *V. s. alticola* winter in Florida, October to April.

660. Black-whiskered Vireo, *Vireo altiloquus* (Vieillot)

IDENTIFICATION: Size of Red-eyed Vireo (total length 140 to 165 mm). Top of head brownish gray; remainder of upperparts chiefly brownish olive; *a light superciliary stripe* and a dark stripe through eye; also a *dark malar streak*; underparts whitish; no wing bars; feet bluish gray; *iris red*.

DISTRIBUTION AND VARIATION: Breeds along the Florida coast (also a few inland in south Florida) from Cedar Key and New Smyrna Beach southward to the Bahamas, Greater Antilles, and Lesser Antilles; casually strays northward and westward to the Panhandle and to Louisiana (breeding?); late March to late September (Map 54). Winters mainly in northern South America, but also in the Lesser Antilles and Hispaniola (*V. a. barbatulus*).

661. Yellow-green Vireo, *Vireo flavoviridis* (Cassin)

IDENTIFICATION: Almost identical to the Red-eyed Vireo in size and coloration, but differs in having *more yellow on sides and under tail coverts*. Probably best regarded as conspecific with that species.

DISTRIBUTION AND VARIATION: Breeds from northern Mexico and extreme southern Texas southward to Bolivia, Paraguay, and Argentina. Winters in South America. *V. f. flavoviridis* is accidental at Pensacola, Florida (Monroe, 1959a), and in Quebec.

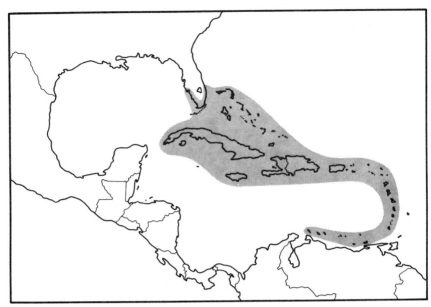

Map 54. Breeding range of the Black-whiskered Vireo (*Vireo altiloquus*)

662. Red-eyed Vireo, *Vireo olivaceus* (L.)

IDENTIFICATION: Size of Black-whiskered Vireo (total length 140 to 165 mm). Similar to Black-whiskered Vireo, but differs in its gray crown and *lack of a dark malar streak*. (See also Yellow-green Vireo.)

DISTRIBUTION: Breeds from British Columbia, southwestern Mackenzie, northeastern Alberta, central Saskatchewan and Manitoba, northern Ontario, central Quebec, and Nova Scotia southward to northern interior of Oregon, northern Idaho, southwestern and central Montana, Wyoming, eastern Colorado, Kansas, western Oklahoma, central Texas, the northern Gulf Coast, and the interior of southern Florida (northern Monroe County); March to early November. Winters from eastern Ecuador and southwestern Venezuela to eastern Peru and western Brazil.

663. Philadelphia Vireo, *Vireo philadelphicus* (Cassin)

IDENTIFICATION: Smaller than Red-eyed Vireo (total length 115 to 125 mm). Upperparts as in Red-eyed Vireo, but superciliary stripe less prominent; *underparts generally washed with yellow*, the breast being decidedly yellow; feet bluish gray.

DISTRIBUTION: Breeds from northeastern British Columbia, central Alberta, southern Saskatchewan and Manitoba, northern Ontario, central Quebec, and southwestern Newfoundland southward to north-central North Dakota, southern Ontario and Quebec, northern New Hampshire, central Maine, and northern New Brunswick. Winters from central Guatemala to Panama and northwestern Colombia. Rare transient in Florida, April to May and September to October.

664. Warbling Vireo, *Vireo gilvus* (Vieillot)

IDENTIFICATION: Smaller than Red-eyed Vireo (total length 120 to 150 mm), from which it differs in its *more brownish crown*, less prominent superciliary stripe, and more yellowish flanks; feet bluish gray. (See also Tennessee Warbler.)

DISTRIBUTION AND VARIATION: Breeds from northern British Columbia, southern Mackenzie, central Saskatchewan, southern Manitoba, western Ontario, northern Minnesota and Michigan, southern parts of Ontario, Quebec, Maine, and New Brunswick, and Nova Scotia southward to Baja California, north-central Mexico, central Texas, southern Louisiana, northwestern Alabama, western North Carolina, and eastern Virginia. Winters from southern Sonora and Vera Cruz to Guatemala and El Salvador. The eastern race, *V. g. gilvus*, is a rare transient in Florida, April to May and September to October.

ORDER PASSERIFORMES: CXXXVI. FAMILY COEREBIDAE, Honeycreepers. A heterogeneous group of small birds, usually of brilliant plumage. The bill is usually rather long and slender, but may be either downcurved, hooked, or pointed at the tip. As in hummingbirds, the tongue is structurally adapted for nectar feeding. Many authorities do not recognize the family, but divide its members among the warblers and tanagers.

665. Bananaquit, *Coereba flaveola* (Cabanis)

IDENTIFICATION: Warbler-sized (total length 100 to 130 mm). *Upperparts dark gray to black, interrupted by a white superciliary stripe*, a white patch on the primaries, and a yellow rump; chin and throat medium gray; *breast bright yellow*; abdomen whitish, tinged with yellow; under tail coverts white and under side of tail largely so.

DISTRIBUTION AND VARIATION: Resident in the West Indies (except Cuba); also on Cozumel, Quintana Roo, Grand Cayman, and Cayman Brac, and from southeastern Mexico to western Ecuador and Argentina. *C. f. bahamensis*, of the Bahamas, is accidental in southern and eastern Florida (Winter Park to Key West; Mason, 1960; Hames, 1960).

ORDER PASSERIFORMES: CXXXVII. FAMILY PARULIDAE, Wood Warblers. Chiefly small birds usually having greenish or olive, along with yellow, in the plumage. Bill straight and usually slender. Nostrils exposed and operculate. Only 9 primaries.

666. Black-and-white Warbler, *Mniotilta varia* (L.)

IDENTIFICATION: Total length 115 to 140 mm. *Head and back broadly streaked with black and white*; wings black to dusky, with 2 white bars on coverts; tail contrastingly marked with black, gray, and white; underparts white, heavily streaked with black in male, lightly so in female. Unlike any other North American bird of similar color pattern, this bird *creeps on the trunks or large limbs of trees like a nuthatch*.

DISTRIBUTION: Breeds from northeastern British Columbia, southwestern Mackenzie, central Saskatchewan, central Manitoba, southern Ontario and Quebec, and northern Newfoundland southward to central Alberta, eastern Montana, southwestern South Dakota, central Texas, northern Mississippi, central Alabama, northern Georgia, central South Carolina, and southeastern North Carolina. Winters from southern Baja California, northern Mexico, southern and southeastern Texas, the northern Gulf Coast, southern Georgia, and southeastern South Carolina southward to Ecuador, northern Venezuela, and the West Indies. Florida: July-late May.

667. Prothonotary Warbler, *Protonotaria citrea* (Boddaert)

IDENTIFICATION: Total length 125 to 140 mm. Adult male: *top of head and nape orange yellow*, sometimes interrupted on occiput by a dusky band; back greenish; *rump, wings, and tail bluish gray* to blackish, but some of the rectrices with large white patches; *underparts orange yellow anteriad*, gradually changing to yellow on abdomen; under tail coverts white and almost as long as rectrices. Adult female, immature: similar, but head and neck heavily overlaid with dusky greenish, and with no hint of orange in plumage. Feet flesh-colored. Juvenal: head and back grayish with a greenish cast; wings and tail as in adult; breast and throat yellowish gray; flanks slaty; abdomen whitish.

DISTRIBUTION: Breeds from east-central Minnesota, south-central Wisconsin, the southern parts of Michigan and Ontario, central New York, and New Jersey southward (east of the Great Plains) to eastern Texas, the Gulf Coast, and southern Florida (to northern Monroe County); March-early October. Winters from Yucatan to central Colombia and northern Venezuela, rarely to southern Florida (UM 4724).

668. Swainson's Warbler, *Limnothlypis swainsonii* (Audubon)

IDENTIFICATION: A rather large warbler (total length 140 to 150 mm). *Upperparts shading gradually from reddish brown on crown* through olive brown on back and rump *to dusky on wings and tail*; *dark line through the eye and a light one above it*; underparts whitish, lightly tinged with yellow and with dusky on breast and flanks; under tail coverts long and whitish; tarsus and toes flesh-colored. (Pattern suggestive of Red-eyed Vireo's.)

DISTRIBUTION: Breeds from northeastern Oklahoma, southeastern Missouri, southern Illinois, southwestern Indiana, southern Ohio, western West Virginia, southern Virginia, and southeastern Maryland southward to southeastern Louisiana, southern Mississippi and Alabama, and northern Florida; March to October (transient in Peninsula; Map 55). Winters in central and eastern Cuba, Jamaica, the Peninsula of Yucatan, in British Honduras, and the Bahamas.

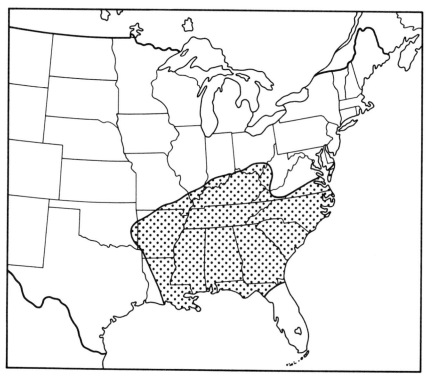

Map 55. Breeding range of the Swainson's Warbler
(*Limnothlypis swainsoni*)

669. Worm-eating Warbler, *Helmitheros vermivorus* (Gmelin)

IDENTIFICATION: Total length 125 to 140 mm. *Top of head marked with 3 longitudinal stripes of yellowish buff, alternating with 4 black stripes*; remainder of upperparts olive green, except for dusky on remiges and rectrices; underparts whitish, heavily tinged with buff on throat and breast, and with light gray posteriad; tarsus and toes flesh-colored.

DISTRIBUTION: Breeds from northeastern Kansas, southeastern Iowa, northern Illinois, southern Indiana, southern and east-central Ohio, southwestern and central Pennsylvania, central and southeastern New York, southern Connecticut, and western Massachusetts southward to northeastern Texas, central Arkansas, central and southern Alabama, northwestern Florida (Stevenson, 1961b and 1962a), northern Georgia, northwestern South Carolina, and east-central North Carolina (Map 56). Winters from southeastern Mexico, southern Florida, and the Bahama Islands southward to Panama; rarely farther north. Most Florida records range from late March to early May and from August to October.

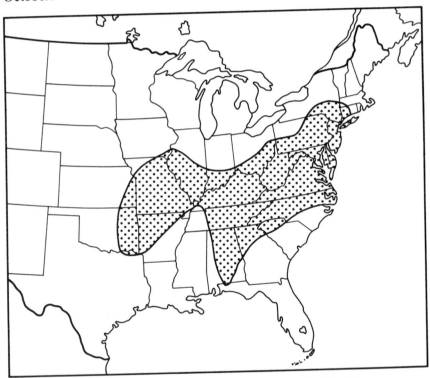

Map 56. Breeding range of the Worm-eating Warbler

670. Golden-winged Warbler, *Vermivora chrysoptera* (L.)

IDENTIFICATION: A rather small warbler (total length 100 to 130 mm). Adult male: *forehead, fore part of crown, and 2 wing bars* (usually united) *bright yellow*; remainder of upperparts bluish gray, sometimes marked with blackish on the back, remiges, and rectrices (back and wings tinged with greenish in fall); stripe through eye black, bordered above and below by white; *throat and upper breast black*; remainder of underparts white, tinged at times with yellowish and light gray. Adult female and immature: similar, but with *gray replacing black on head and throat*.

DISTRIBUTION: Breeds from southeastern Manitoba, east-central Minnesota, north-central Wisconsin, northern Michigan, southern Ontario, western and east-central New York, southern Connecticut, and eastern Massachusetts southward to southeastern Iowa, northern Illinois and Indiana, southern Ohio, eastern Tennessee, northern Georgia, northwestern South Carolina, western Virginia, north-central Maryland, and southeastern Pennsylvania. Winters from Guatemala to central Colombia and northern Venezuela and in the northern Antilles. Florida: early April to early May; August to October.

671. Blue-winged Warbler, *Vermivora pinus* (L.)

IDENTIFICATION: Size of Golden-winged Warbler (total length 100 to 130 mm). *Forehead and fore part of crown bright yellow* (more greenish in female); occiput, nape, and back olive green; rump yellowish green; wings and tail bluish gray to dusky, the *wings with 2 white bars* and the tail with large white patches on lateral rectrices; *stripe through eye black*; *underparts entirely bright yellow*, except for the white under tail coverts and the white and gray rectrices.

DISTRIBUTION: Breeds from east-central Nebraska, central Iowa, southeastern Minnesota, southern Wisconsin and Michigan, northern Ohio, northwestern Pennsylvania, western and southeastern New York, and southeastern Massachusetts southward to northeastern Texas, east-central Missouri, southern Illinois, central Tennessee, northern Alabama and Georgia, western North Carolina, western and northern Virginia, central Maryland, and Delaware. Winters from southeastern Mexico to Panama (casually to northern Colombia, Jamaica, Hispaniola, and in southern Florida). Florida: late March to early May; August to October.

672. Bachman's Warbler, *Vermivora bachmanii* (Audubon)

IDENTIFICATION: One of the smallest warblers (total length 100 to 115 mm). Adult male: *forehead, sides of head, and underparts mostly yellow, interrupted by a large black patch on throat and breast*; anal region and under tail coverts white; *crown blackish, fading to gray on nape*; back yellowish green; wing and tail dusky, the latter with white patches laterally. Adult female (and immature?): differs from male in the usual absence of black on crown and underparts, the smaller white patches in tail, and the less intense yellow underneath (sometimes virtually lacking).

DISTRIBUTION: Formerly bred locally from northeastern Arkansas, southeastern Missouri, and south-central Kentucky southeastward to west-central and central Alabama and southeastern South Carolina; perhaps also westward to Louisiana and northward to North Carolina and Virginia. Known to have wintered only in Cuba, the Isle of Pines, and Florida (one record). Probably nearing extinction (Stevenson, 1972c). Many *old* Florida records: late February to early April and July to early September.

673. Tennessee Warbler, *Vermivora peregrina* (Wilson)

IDENTIFICATION: Total length 115 to 130 mm. Adult in breeding plumage: top of head bluish gray (tinged with greenish in female); back greenish, shading to yellowish green on rump; remiges and rectrices dusky; a dusky streak through eye; *superciliary stripe and underparts almost entirely white*. Adult in fall: similar, but head, neck, and breast greenish. Immature in fall: differs from adult in having entire upperparts olive green, and the superciliary stripe and underparts yellowish green, except for the *white under tail coverts*. (See also Red-eyed and Warbling Vireos and Orange-crowned Warbler.)

DISTRIBUTION: Breeds from southern Yukon, central Mackenzie, the northern parts of Manitoba, Ontario, and Quebec, central Labrador, and western Newfoundland southward to south-central British Columbia, northwestern Montana, central Alberta and Saskatchewan, southern Manitoba, northern Minnesota, Wisconsin, and Michigan, south-central Ontario, northeastern New York, southern Vermont, central New Hampshire, southern Maine and New Brunswick, and central Nova Scotia. Winters from southeastern Mexico to Colombia and northern Ecuador; accidental in Florida (Stevenson, 1963a) and elsewhere in the United States in December. Most Florida records range from April to mid-May and September to early November.

674. Orange-crowned Warbler, *Vermivora celata* (Say)

IDENTIFICATION: Total length about 115 to 130 mm. Upperparts mostly olive green, but yellowish green on rump; wings and tail dusky; concealed crown patch rufous; *underparts (including tail coverts) greenish yellow, faintly streaked on breast with dusky*. Immature: similar, but more grayish on head. At any age the *yellowish under tail coverts* and *whitish patch near the bend of the wing are good field marks*.

DISTRIBUTION AND VARIATION: Breeds from central Alaska, northwestern and central Mackenzie, northern Manitoba and Ontario, and northwestern Quebec southward to northwestern Baja California, southeastern Arizona, western Texas, southeastern Saskatchewan, southern Manitoba, and western and central Ontario. Winters from northern California, southern Nevada, central Arizona, northeastern Texas, the central parts of Mississippi, Alabama, Georgia, and South Carolina, and from North Carolina (mainly eastern) southward to southern Baja California, Guatemala, the northern Gulf Coast, the Bahamas, and the Florida Keys (October-April; *V. c. celata* and *V. c. orestera*; Stevenson and Baker, 1970); *sight* records at Key West in August and September.

675. Nashville Warbler, *Vermivora ruficapilla* (Wilson)

IDENTIFICATION: Size of Tennessee Warbler (total length 115 to 130 mm). Adult: *head and neck gray*, admixed with chestnut on crown (concealed in fall) and *interrupted by a white eye-ring*; back olive green, shading to yellowish green on rump; remiges and rectrices mostly dusky; *underparts mostly bright yellow*, but somewhat buffy on belly. Immature in fall: similar, but head less gray (more dusky) and underparts more buffy than yellow.

DISTRIBUTION AND VARIATION: Breeds from the southern parts of British Columbia, Saskatchewan, and Manitoba, central Ontario, southern Quebec, and Nova Scotia southward to central California, northern Utah, southern Minnesota, northern Illinois, southern Michigan, northern Ohio, northeastern West Virginia, western Maryland, southeastern Pennsylvania, Connecticut, and Rhode Island. Winters from northern Mexico and southern Texas southward to Guatemala; also occasionally in Florida, especially southern (*V. r. ruficapilla*, September-May; very rare).

676. Northern Parula, *Parula americana* (L.)

IDENTIFICATION: A small warbler (total length 90 to 115 mm). Adult male: *upperparts mostly grayish blue, interrupted on upper back by a large greenish patch*; wings and tail mostly blackish, but with 2 white bars on wing coverts; streak through eye dark, and incomplete eye-ring white; *chin, throat, and lower breast yellow, separated on upper breast by a band of tawny behind a band of blue black*; abdomen whitish, more or less tinged with grayish. Adult female: similar to male, but with breast band almost wanting. Immature in fall: similar to adult female, but more greenish above and paler yellow below.

DISTRIBUTION: Breeds from southeastern Manitoba, western and central Ontario, southern Quebec, and the northern parts of Maine, New Brunswick, and Nova Scotia southward to eastern Texas, the northern Gulf Coast, and the interior of southern Florida (Collier

County). Winters from eastern Mexico southward to Nicaragua, casually Costa Rica; also from peninsular Florida and the Bahamas to the West Indies and Barbados; February to May (summer); July to November (winter).

677. Yellow Warbler, *Dendroica petechia* (L.)

IDENTIFICATION: Total length 115 to 140 mm. Adult: upperparts yellowish green (duller in female), except for the dusky remiges and rectrices, the latter also with considerable yellow medially; *underparts bright yellow, with reddish streaks on breast*; females are duller underneath, with the reddish streaks faint or lacking. Immature: similar to adult female, but even duller. There is considerable variation in fall birds, some of which are dark olive green above and similar to the Orange-crowned Warbler. *All, however, show some yellow in the tail*.

DISTRIBUTION AND VARIATION: Breeds from north-central Alaska, northern Yukon, northwestern and central Mackenzie, northern Manitoba and Ontario, north-central Quebec, central Labrador and Newfoundland southward to southern Baja California, central Peru, the coast of Venezuela, Trinidad, the West Indies, the Florida Keys, and the Galapagos Islands, exclusive of eastern and central Texas, Louisiana, Mississippi, most of Florida, and the southern parts of Arkansas, Alabama, Georgia, and South Carolina. Winters from southeastern Mexico southward to Peru and northern Brazil. *D. p. gundlachi* is apparently resident on the Florida Keys; *D. p. aestiva, amnicola,* and *rubiginosa* occur in migration probably throughout Florida (Stevenson and Baker, 1970); April-May; July-October.

678. Magnolia Warbler, *Dendroica magnolia* (Wilson)

IDENTIFICATION: Total length 115 to 130 mm. Adult male in spring: top of head and nape pale gray; back black, mixed with greenish posteriad; rump yellow; upper tail coverts and much of tail black; *lateral rectrices with considerable white basally; wings dusky, with a large white patch on coverts*; a broad black stripe through the eye, bordered above by white and below by the yellow chin and throat; remainder of underparts mostly yellow (white posteriad), *heavily streaked on breast and sides with black*. Adult female: similar to male, but back more greenish, stripe through eye gray, and underparts less heavily streaked with black. Immature in fall: similar to adult female, but head and nape more brownish olive, 2 separate white wing bars present, and black streaks of underparts faint.

DISTRIBUTION: Breeds from southwestern and south-central Mackenzie, northeastern Alberta, northern Saskatchewan, central Manitoba, northern Ontario, central and eastern Quebec, and southwestern Newfoundland southward to east-central British Columbia, south-central Alberta and Saskatchewan, southern Manitoba, northeastern Minnesota, central Wisconsin and Michigan, and southern Ontario; also (locally) to northeastern Ohio, east-central West Virginia, western Virginia, northwestern New Jersey, and northern Massachusetts. Winters from central Mexico southward to Panama; also in the Bahamas, the Greater Antilles, and southern Florida, casually farther north; April-May and September-early November (winter).

679. Cape May Warbler, *Dendroica tigrina* (Gmelin)

IDENTIFICATION: Total length 125 to 140 mm. Adult male: top of head blackish, sometimes mixed with dark chestnut; back olive green, spotted heavily with blackish; rump yellowish; wings blackish, with a large white patch on coverts; upper tail coverts and rectrices blackish, the latter with white on medial edges; *sides of head chestnut*; *underparts and sides of neck mostly yellow, heavily streaked with black on breast and sides*; under tail coverts white. Adult female: head, neck, and upper back light greenish gray, the crown spotted with blackish; lower back greenish, becoming yellowish on rump; no distinct white wing bars or patch; wings and tail mostly dusky; chin, throat, *sides of neck*, and breast *buffy to yellowish*, streaked with blackish; belly and under tail coverts white.

DISTRIBUTION: Breeds from southwestern and south-central Mackenzie, northeastern British Columbia, northern Alberta, central Saskatchewan, Manitoba, northern Ontario, and southern Quebec southward to northeastern North Dakota, northwestern and east-central Minnesota, northern Wisconsin and Michigan, northeastern New York, east-central Vermont, southern and eastern Maine, southern New Brunswick, and central Nova Scotia. Winters chiefly in the Bahama Islands, the Greater Antilles, and the Virgin Islands, but also in the Lesser Antilles and southern Florida; March-May and September-November (winter).

680. Black-throated Blue Warbler, *Dendroica caerulescens* (Gmelin)

IDENTIFICATION: Total length about 115 to 140 mm. Adult male: *upperparts mostly dark grayish blue*, often spotted with blackish; wings mostly blue on coverts and blackish on remiges, but with *a prominent white patch at base of primaries*; rectrices blackish, but outer ones with white patches medially; *sides of head, chin, and throat black*; remainder of underparts mostly white, but with black along sides. Adult female and immature: upperparts mostly olive green or brownish, but *usually with the white wing patch of the male*; *superciliary line whitish*; *patch behind eye dark*; underparts whitish, tinged with pale greenish, yellowish, or buffy.

DISTRIBUTION AND VARIATION: Breeds from north-central and northeastern Minnesota, western Ontario, northern Michigan, east-central Ontario, southwestern Quebec, northern Maine and New Brunswick, and central Nova Scotia southward to east-central Minnesota, northern Wisconsin, central Michigan, southern Ontario, northeastern Ohio, western and northeastern Pennsylvania, northern New Jersey, southeastern New York, Connecticut, Rhode Island, and Massachusetts; also southward in the mountains to eastern Tennessee, northeastern Georgia, and northwestern South Carolina. Winters in the Bahama Islands, the Greater Antilles, and the Virgin Islands; also in southern (rarely to northern) Florida. Both *D. c. caerulescens* and *D. c. cairnsi* occur in the state; April-May and late August-early November (winter).

681. Yellow-rumped Warbler, *Dendroica coronata* (L.)[1]

IDENTIFICATION: Total length 125 to 150 mm. Adult in breeding plumage: *crown and rump bright yellow*; remainder of upperparts mostly gray, longitudinally streaked with black; 2 white bars on wing coverts; chin white (yellow in *D. c. auduboni*); underparts chiefly white, but with heavy black streaks on breast and sides; *a patch of yellow on each side of breast*; females are somewhat duller. Winter plumage: similar to breeding plumage, but *much duller*; upperparts with considerable brownish; yellow patches on crown and breast obscure; underparts whitish, tinged with light grayish brown, also with obscure streaks on sides.

DISTRIBUTION AND VARIATION: Breeds from northern Alaska and Yukon, western and central Mackenzie, northern Manitoba and Ontario, central Quebec, north-central Labrador, and Newfoundland southward to northern Mexico, southern Arizona and New Mexico, western Texas, northern Minnesota and Michigan, central Ontario, northeastern New York, Massachusetts, and Maine. Winters from southwestern British Columbia, southeastern Colorado, Kansas, central Missouri, southern Wisconsin, northeastern Illinois, central Indiana, northern Ohio, western Pennsylvania, southern New York, and the coast of New England and Nova Scotia southward to central Panama, the northern Gulf Coast, Bermuda, the Bahama Islands, the Greater Antilles, and the Virgin Islands. *D. c. coronata* is the common Florida bird, but *hooveri* has been collected near Pensacola (Weston, 1965); also 2 sight records of *D. c. auduboni*.

682. Black-throated Gray Warbler, *Dendroica nigrescens* (Townsend)

IDENTIFICATION: Total length about 130 mm. Adult: top of head mostly black; back bluish gray, with black markings; remiges and rectrices blackish, with white edgings; 2 white bars on wing coverts;

---

1. Formerly called the Myrtle Warbler, but see A.O.U., 1973.

superciliary stripe narrow and yellow toward bill, broad and white toward neck; a broad black stripe through and below eye, bordered below by an equally broad white stripe; *chin and throat black* (mixed with white in female); remainder of underparts white, with black stripes along sides. Immature: top of head gray; back tinged with brown and without black markings; throat with very little or no black; otherwise like adults. The appearance from below is of a *Black-and-white Warbler with the head pattern of a Golden-winged Warbler*.

DISTRIBUTION: Breeds from the mainland of southwestern British Columbia, western Washington, central Oregon, southwestern Idaho, northern Utah, southwestern Wyoming, and northwestern and central Colorado southward at high elevations to northern Baja California, northwestern, central, and southeastern Arizona, and eastern and southern New Mexico. Winters from coastal and southern California and southern Arizona southward to southern Baja California and Guatemala (casually); also rarely eastward along the Gulf Coast to Florida (late September-early May).

683. Black-throated Green Warbler, *Dendroica virens* (Gmelin)

IDENTIFICATION: Total length 125 to 140 mm. Adult male: upperparts mostly yellowish green, more or less spotted with blackish; wings dusky, with 2 white bars on coverts; rectrices dusky, the lateral ones with considerable white; *sides of head and neck mostly bright yellow*, interrupted by a dark streak through the eye to side of neck; *chin, throat, breast, and sides black*; remainder of underparts white. Adult female: black patch of underparts reduced in size and mixed with yellow or white; otherwise similar to male. Immature: similar to adult female, but black even more diffuse.

DISTRIBUTION AND VARIATION: Breeds from south-central Mackenzie, north-central Saskatchewan and Manitoba, central Ontario, and the southern parts of Quebec, Labrador, Newfoundland, and Nova Scotia southward to central Alberta, southern Manitoba, east-central Minnesota, central Wisconsin and Michigan, eastern and south-central Ohio, Pennsylvania, and northern New Jersey; also southward, mostly at high elevations, to central Alabama, northern Georgia, and eastern South Carolina. Winters from southern Texas and southern Florida southward to Panama and the Greater Antilles, casually farther north and farther south. Both *D. v. virens* and *D. v. waynei* have been collected in Florida; March-May and September-November (winter).

684. Golden-cheeked Warbler, *Dendroica chrysoparia* Sclater and Salvin

IDENTIFICATION: Total length 115 to 130 mm. Male in breeding plumage: *upperparts, throat, and breast mostly black; sides of head bright yellow*, except, for black streak through eye; 2 broad white wing bars; much white on outer rectrices; lower breast, abdomen, and under tail coverts white; sides and flanks heavily streaked with black. Breeding female: similar to male, but black of *upperparts* replaced by olive green, *finely streaked with black*; throat yellow; usually some black on breast. Immature, adults in fall: similar to breeding adult, but *with white on throat and breast*.

DISTRIBUTION: Breeds in several counties in central Texas. Winters from southeastern Mexico to Nicaragua. Accidental near St. Petersburg, Florida, in August (Woolfenden, 1967).

685. Cerulean Warbler, *Dendroica cerulea* (Wilson)

IDENTIFICATION: Total length 100 to 130 mm. Adult male: *upperparts mostly grayish blue, longitudinally streaked with black* on head and back; 2 white wing bars present; remiges and rectrices dusky, edged with white; *underparts white, with a dusky or bluish band across breast*, and streaked with same color along sides. Adult female: upperparts similar to those of male, but more greenish; light superciliary stripe conspicuous; underparts tinged with yellow, and without breast band. Immature: similar to adult female, but more greenish above and yellowish below.

DISTRIBUTION: Breeds from southeastern Nebraska, northern Iowa, southeastern Minnesota, southern Wisconsin, Michigan, and Ontario, western and southeastern New York, eastern Pennsylvania, and northern New Jersey southward (east of the Great Plains) to eastern Texas, southeastern Louisiana, north-central Alabama, central North Carolina and Virginia, southern Maryland, and Delaware. Florida: April-early May and August-September, rare.

686. Blackburnian Warbler, *Dendroica fusca* (Müller)

IDENTIFICATION: Total length 115 to 140 mm. Adult male: upperparts largely black, but with orange on crown, superciliary stripe, and side of neck; *wing blackish, except for a large white patch on coverts*; lateral rectrices with large white patches, but tail otherwise black; spot under eye, *chin, throat, and upper breast brilliant orange*, fading through pale yellow to whitish posteriad; sides streaked with black; under tail coverts white. Adult female: similar to male, but upperparts more grayish brown, with dusky streaks; orange of underparts more restricted and less intense. Immature male: differs from adult female in being darker above, in having 2 white wing bars, and in having yellow (rather than orange) on breast. Immature female: upperparts much more brownish than in any other plumage; yellow almost lacking on crown and sides of head; breast only pale or buffy yellow.

DISTRIBUTION: Breeds from south-central Saskatchewan, southern Manitoba, northern Ontario, central Quebec, and the northern parts of Maine, New Brunswick, and Nova Scotia southward to the central parts of Minnesota, Wisconsin, and Michigan, southern Ontario, northeastern Ohio, western and central Pennsylvania, southeastern New York, and Massachusetts; also in the mountains to eastern Tennessee, north-central Georgia, and northwestern South Carolina. Winters from Guatemala to Venezuela and central Peru. Florida: April-May and August-early November.

687. Yellow-throated Warbler, *Dendroica dominica* (L.)

IDENTIFICATION: Total length 115 to 140 mm. Adult: *crown blackish, fading through medium gray to bluish gray on lower back*; wings dusky to blackish, with 2 white bars on coverts; tail dusky, with white patches on lateral rectrices; superciliary stripe white (usually yellow near bill), continuing to side of neck; a black patch on side of head, continuing as a narrow streak to side of breast; *chin and throat bright yellow*; remainder of underparts white, streaked with black on sides. Juvenal: very little, if any, yellow on throat or black on head.

DISTRIBUTION AND VARIATION: Breeds from central Oklahoma and Missouri, southern Illinois, central Indiana, southern Ohio, southwestern West Virginia, eastern Maryland, and southern Delaware southward to southeastern Texas, the Gulf Coast, and central Florida; frequently strays much farther north in spring. Winters from southern Texas, central Louisiana, southern Mississippi, central Alabama and Georgia, and southern South Carolina southward to Costa Rica, and Colombia (sight record), the Greater Antilles, and the Bahamas. The breeding form in most of the Florida Panhandle is *D. d. stoddardi*, and *D. d. dominica* occupies the Peninsula southward to Hillsborough and Osceola Counties; *D. d. albilora* has occurred in migration and in winter; July to April (summer).

688. Chestnut-sided Warbler, *Dendroica pensylvanica* (L.)

IDENTIFICATION: Total length 115 to 130 mm. Adult male: *crown yellow*, bordered by black; back streaked with black and light gray, tinged with yellow in center; wings dusky, with 2 white bars on coverts; tail dusky, with white on lateral rectrices; a black patch in front of and below eye, and a white patch behind eye; underparts mostly pure white, but *with a broad stripe of chestnut running down each side*. Adult female: similar, but with less black above and less chestnut on sides. Immature: wings and tail similar to those of adults, except for greenish feather edgings on wings; remainder of *upperparts evenly bright olive green; underparts white*, tinged with gray on sides; *white eye-ring conspicuous*.

DISTRIBUTION: Breeds from east-central Saskatchewan, west-central Manitoba, central Ontario, southern Quebec, central New Brunswick, and northern Nova Scotia southward to north-central North Dakota, eastern Nebraska, northwestern and southeastern Minnesota, southern Wisconsin, southern Michigan, northern Ohio, central and western Maryland, southeastern Pennsylvania, central New Jersey, New York, Massachusetts, and Maine; also in the mountains to south-eastern Tennessee (northern Alabama?), north-central Georgia, and northwestern South Carolina. Winters from southern Nicaragua to central Panama; also a few sight records at that season in southern Florida. Most Florida records fall in April and May, and from late August to late October.

689. Bay-breasted Warbler, *Dendroica castanea* (Wilson)

IDENTIFICATION: Total length 125 to 155 mm. Adult male in spring: *forehead and sides of head black*; *crown and nape chestnut*; *sides of neck buffy*; back black, with light gray feather edgings; wings dusky to black, with 2 white bars on coverts; rump gray; tail black, with white patches on lateral rectrices; chin, throat, and sides chestnut (lighter than crown); abdomen white; under tail coverts buffy. Adult female in spring: similar to male, but everywhere less brilliantly colored; head mostly gray, streaked with blackish; chestnut of underparts more restricted. Adult in fall and immature: wings and tail as described above; remainder of upperparts olive greenish streaked with dusky, at least in male; underparts whitish, washed with buff or yellowish; male with some rusty color on flanks; *under tail coverts buffy or yellowish. In all plumages the feet are dark.* (Compare with fall Blackpoll Warbler.)

DISTRIBUTION: Breeds from central Manitoba (possibly farther north and west), northern Ontario, central Quebec, New Brunswick, and central Nova Scotia southward to southern Manitoba, northeastern Minnesota, northern Wisconsin, southern Ontario and Quebec, north-eastern New York, central Vermont, New Hampshire, southern Maine, and southern Nova Scotia. Winters from central and eastern Panama to northern Colombia and western Venezuela. Florida: late April to May and late September to early November.

690. Blackpoll Warbler, *Dendroica striata* (Forster)

IDENTIFICATION: Total length 125 to 155 mm. Adult male in spring: *top of head black*; back streaked with black and gray; wings dusky to black, with 2 white bars on coverts; tail dusky, with white patches near ends of lateral rectrices; *sides of head white*; underparts white, streaked on sides of throat, breast and abdomen with black. Adult female in spring: *upperparts streaked with black and olive gray* (lacking the male's black crown and white side of head); underparts less heavily streaked than in male, and ground color usually tinged with yellowish. Adult in fall, immature: similar to adult female in spring, except more greenish above and below; *under tail coverts white*, as in other plumages. This field mark, along with the usually *pinkish legs*, will distinguish it from the Bay-breasted Warbler in fall.

DISTRIBUTION: Breeds from north-central Alaska, northern Yukon, northwestern and central Mackenzie, northeastern Saskatchewan, the northern parts of Manitoba, Ontario, Quebec, and Labrador, and from most of Newfoundland southward to central British Columbia, Alberta, and Manitoba, north-central Ontario, southern Quebec, eastern New York, northwestern Massachusetts, central New Hampshire, and the southern parts of Maine, New Brunswick, and Nova Scotia. Winters from eastern Ecuador, southeastern and central Colombia, and Venezuela southward to Chile and west-central Brazil. Florida: late April to early June and late September to early November; most common in spring.

691. Pine Warbler, *Dendroica pinus* (Wilson)

IDENTIFICATION: Total length 125 to 140 mm. Adult male: top of head, nape, and back bright olive green; wings and tail dusky, the former with 2 white bars on coverts; *underparts mostly bright yellow, usually streaked indistinctly on sides with dusky*; lower belly and under tail coverts white. Adult female: similar to male, but underparts paler and less extensively yellow. Immature: similar to adult female, but more olive brown above and more whitish below, *sometimes with no trace of yellow*.

DISTRIBUTION AND VARIATION: Breeds locally from southern Manitoba, western Ontario, northeastern Minnesota, northern Wisconsin and Michigan, central Ontario, southern Quebec, and central Maine southward to southeastern Texas, south-central Louisiana, the coast of Mississippi and Alabama, and southern Florida. Winters within breeding range as far north as Arkansas, Tennessee, western South Carolina, and coastal New Hampshire (casually). *D. p. pinus* breeds in the Panhandle and *D. p. florida* in the Peninsula; accidental on the Keys.

692. Kirtland's Warbler, *Dendroica kirtlandii* (Baird)

IDENTIFICATION: A rather large warbler (total length 140 to 155 mm). Adult male: *upperparts mostly gray to bluish gray, the back streaked with black*; an incomplete white eye-ring, surrounded by a *black stripe across eye*; wings and tail dusky, the former with 2 white bars on coverts; *underparts mostly yellow, streaked along sides with blackish*; under tail coverts white. Adult female: similar to male, but paler. Immature: similar to adult female, but browner above, sides tinged with brown, and breast with dusky specks. The *tail-bobbing habit* of this bird is shared by only the Palm and Prairie Warblers.

DISTRIBUTION: Breeds in several counties in northern Lower Michigan. Winters in the Bahamas. A very rare transient in Florida, April to May and late September to early November.

693. Prairie Warbler, *Dendroica discolor* (Vieillot)

IDENTIFICATION: Total length 100 to 130 mm. Adult male: top of head, nape, and *back yellowish green, the upper back striped with chestnut*; wings and tail dusky, the former with 2 yellowish bars on coverts; lateral rectrices with large white patches; ground color of underparts (except the white under tail coverts) entirely yellow, streaked with black along sides of neck and body; *also a black streak through the eye and one below it*. Adult female: duller above, and with little chestnut on back; streak under eye broader, and more grayish than black; underparts less heavily streaked. Immature: more grayish above than female; chin and throat whitish; remainder of underparts pale yellowish, streaked with dusky. The *tail-bobbing habit* aids in field identification.

DISTRIBUTION AND VARIATION: Breeds from southeastern South Dakota, eastern Nebraska and Kansas, central Missouri, Illinois, southern Wisconsin, northern Michigan, southern Ontario and Pennsylvania, northern New Jersey, southeastern New York, Massachusetts, and southern New Hampshire southward to northeastern Texas, northern and southeastern Louisiana, northern Mississippi, southwestern Alabama, much of Florida, southern Georgia, central South Carolina, and eastern North Carolina. Winters in the Florida Peninsula, the West Indies (except southern Lesser Antilles), and islands of the western Caribbean and southern Gulf of Mexico. The nominate race, *D. d. discolor*, breeds in northern Florida as far east as Leon (possibly Columbia) County, and *D. d. paludicola* breeds on the Keys and the coast of the Peninsula. (See Stevenson, 1956, and Gaither, 1964.) Statewide in migration, mid-March to mid-May and mid-July to early November.

694. Palm Warbler, *Dendroica palmarum* (Gmelin)

IDENTIFICATION: Total length 125 to 140 mm. Crown reddish brown (spring) or streaked with same color (fall); nape and back olive brown, with short dusky streaks; rump yellowish green; wings and tail dusky, with white patches on lateral rectrices; ground color of underparts entirely yellow (*D. p. hypochrysea*) or tinged with same (*D. p. palmarum*), but *under tail coverts always yellow*; throat and breast streaked with dusky in fall, the *streaks becoming reddish brown in spring*. A highly *terrestrial, tail-bobbing* warbler.

DISTRIBUTION AND VARIATION: Breeds from northeastern British Columbia, southwestern Mackenzie, northwestern Saskatchewan, northern Manitoba, northern Ontario, central Quebec, and southern Newfoundland southward to central Alberta and Saskatchewan, southeastern Manitoba, northeastern Minnesota, central Michigan, southern Ontario and Quebec, east-central New Hampshire, eastern Maine, New Brunswick, and Nova Scotia. Winters from southeastern Texas, central Louisiana, northern Mississippi, north-central and northeastern Tennessee, and North Carolina southward to the Yucatan Peninsula, northern Honduras, Panama, the Greater Antilles, the Virgin Islands, and Bermuda. Both *D. p. palmarum* and *hypochrysea* winter in Florida, the latter chiefly in the Panhandle; mid-September to mid-May.

695. Ovenbird, *Seiurus aurocapillus* (L.)

IDENTIFICATION: Members of this genus are somewhat larger than most warblers (total length 140 to 165 mm) and have brownish upperparts and streaked underparts. *Crown orange brown, bordered laterally by blackish*; remainder of upperparts olive brown to olive green; underparts white, streaked with dark brown on sides of throat, and heavily spotted with same on breast and sides of body; legs flesh-colored. Differs from the waterthrushes in having an *eye-ring* rather than a superciliary stripe.

DISTRIBUTION AND VARIATION: Breeds from northeastern British Columbia, south-central Mackenzie, central Saskatchewan and Manitoba, northern Ontario, southern Quebec, and Newfoundland southward to southern Alberta, eastern Colorado, southeastern Oklahoma, northern Arkansas, western Tennessee, northern Alabama and Georgia, western South Carolina, and central and northeastern North Carolina. Winters from northeastern Mexico and the southern parts of Texas, Georgia, and South Carolina (rarely on Louisiana coast and in Florida Panhandle) southward to northern Colombia and Venezuela and the Lesser Antilles. Both *S. a. aurocapillus* and *S. a. furvior* occur in Florida, August to May (Stevenson and Baker, 1970).

696. Northern Waterthrush, *Seiurus noveboracensis* (Gmelin)

IDENTIFICATION: Total length 125 to 160 mm. Upperparts dark brown; superciliary stripe and underparts whitish, tinged with yellowish or buffy, the *chin, throat, and sides streaked with dark brown*; legs flesh-colored. (Compare with Louisiana Waterthrush.)

DISTRIBUTION AND VARIATION: Breeds from north-central Alaska, northern Yukon, northwestern and south-central Mackenzie, northern parts of Saskatchewan, Manitoba, Ontario, and Quebec, central Labrador, and Newfoundland southward to central British Columbia, northern Idaho, western Montana, central Saskatchewan, northern parts of North Dakota, Minnesota, Wisconsin, and Michigan, northeastern Ohio, east-central West Virginia (mountains), northern Pennsylvania, and Massachusetts. Winters from southern Baja California, north-central Mexico, and central Florida southward to northern Ecuador, northeastern Peru, southern Venezuela, and the Guianas. Both *S. n. noveboracensis* and *notabilis* occur in Florida, and *limnaeus* may be expected; late March to May and early August to early November (winter).

697. Louisiana Waterthrush, *Seiurus motacilla* (Vieillot)

IDENTIFICATION: Total length 140 to 160 mm. Strongly similar to the Northern Waterthrush, but bill usually longer (more than 10 mm from nostril), *chin usually pure white, and flanks buffy* (contrasting with otherwise-white ground color of underparts). Juvenal: similar, but streaking much reduced.

DISTRIBUTION: Breeds from eastern Nebraska, north-central Iowa, east-central Minnesota, central Wisconsin, southern Michigan and Ontario, central New York and Vermont, southwestern New Hampshire, and Rhode Island southward to eastern Oklahoma, eastern Texas, central Louisiana, southern Mississippi and Alabama, northern Florida (Stevenson, 1961c; Austin, 1965), east-central Georgia, central South Carolina, and northeastern North Carolina. Winters from northern Mexico, southern Florida (rarely), and the Bahamas southward to central Colombia, western Venezuela, and Trinidad; also on Bermuda. Florida: chiefly late February to early May and early July to early October.

698. Kentucky Warbler, *Oporornis formosus* (Wilson)

IDENTIFICATION: Total length 125 to 145 mm. Crown largely black, but becoming gray toward occiput; remainder of upperparts olive green (more brownish on primaries); *lore and eye-ring yellow*; *a black stripe below eye to side of neck*; *entire underparts* (except rectrices) *bright yellow*; tarsus and toes flesh-colored.

DISTRIBUTION: Breeds from southeastern Nebraska, central Iowa, southwestern Wisconsin, northeastern Illinois, central Indiana, central and eastern Ohio, southern Pennsylvania, northern New Jersey, southeastern New York, and southwestern Connecticut southward to central and eastern Texas, southern parts of Louisiana, Mississippi, and Alabama, northwestern Florida, southwestern and central Georgia, and South Carolina. Winters from southeastern Mexico to northern Colombia and northwestern Venezuela; casually also in southern Florida.

Florida: late March to early May (summer); late July-early October (winter).

699. Connecticut Warbler, *Oporornis agilis* (Wilson)

IDENTIFICATION: One of the larger warblers (total length 135 to 155 mm). Adult male: upperparts mostly olive green, but more grayish on crown; eye-ring white; remainder of head, chin, throat, and upper breast gray; remainder of underparts pale to buffy yellow. Adult female: gray on chin and throat of male replaced by buff, but some gray present on breast; eye-ring duller. Immature: more brownish than adult female. Feet flesh-colored.

DISTRIBUTION: Breeds from east-central British Columbia, central Alberta and Manitoba, northern Ontario, and northwestern Quebec southward to the northern parts of Minnesota, Wisconsin, and Michigan, and central Ontario. Winters from northwestern Venezuela to northwestern and central Brazil. Rare transient in Florida, chiefly in May.

700. Mourning Warbler, *Oporornis philadelphia* (Wilson)

IDENTIFICATION: Total length 125 to 140 mm. Adult male: *entire head and throat dark slate gray, becoming black on breast and around eye*; *no white eye-ring*; plumage otherwise like that of Connecticut Warbler. Adult female: similar to male, but lighter gray on head and neck (without black), becoming whitish on throat; also with an *incomplete white eye-ring* in fall. Immature: similar to female, but more brownish above, more yellowish or buffy on throat and breast, and with an *incomplete white eye-ring (compare with immature Yellowthroat). Feet always flesh-colored.*

DISTRIBUTION: Breeds from the central parts of Alberta, Saskatchewan, and Manitoba, northern Ontario, southern Quebec, and Newfoundland southward to northeastern North Dakota, northwestern and east-central Minnesota, central Wisconsin, northeastern Illinois, southern Michigan, northern Ohio, northeastern Pennsylvania, southeastern New York, northwestern and central Massachusetts, central New Hampshire, southern Maine, and central Nova Scotia; also in the mountains to eastern West Virginia and adjacent parts of Virginia. Winters from southern Nicaragua to northern Ecuador, central Colombia, and western Venezuela. Very rare in Florida, September to May, but no specimens taken in winter.

701. Common Yellowthroat, *Geothlypis trichas* (L.)

IDENTIFICATION: Total length 115 to 140 mm. Adult male: *a black mask running from forehead through eye to side of neck, bordered above by a stripe of gray*; remainder of upperparts olive green to olive brown; chin, throat, breast, and under tail coverts brilliant yellow; belly buffy. Adult female: no black or gray on head; *eye-ring and lore whitish*; otherwise similar to male. Immature: similar to adult female, but quite buffy underneath, with little or no true yellow. Feet flesh-colored. *Fall birds*, especially immatures, *may be very confusing and mistaken for much rarer species.*

DISTRIBUTION AND VARIATION: Breeds from extreme southeastern Alaska, southern Yukon, northern Alberta, central Saskatchewan and Manitoba, central and northeastern Ontario, central Quebec, and southwestern Newfoundland southward to northern Baja California, southeastern Mexico, the northern Gulf Coast, and the southern tip of the Florida mainland. Winters from northern California, southern Arizona, southern and northeastern Texas, the central parts of Louisiana, Mississippi, Alabama, Georgia, and South Carolina, and eastern North Carolina southward to the Canal Zone and Trinidad. The breeding race in Florida is *G. t. ignota*; *G. t. typhicola* may breed in the northern tier of counties and does occur at other times of the year, along with *G. t. trichas* and *G. t. brachidactyla*.

702. Yellow-breasted Chat, *Icteria virens* (L.)

IDENTIFICATION: A *very large*, aberrant warbler (total length 175 to 190 mm). *Bill thicker* than in other warblers, and *culmen more decurved*. Upperparts olive green to olive gray; *lore and incomplete eye-ring white*; *chin, throat, and breast bright yellow*; belly and under tail coverts whitish; tail long, its tip rounded.

DISTRIBUTION AND VARIATION: Breeds from southern British Columbia, Alberta, and Saskatchewan, northwestern North Dakota, northeastern South Dakota, the southern parts of Minnesota, Wisconsin, Michigan, and Ontario, central New York, and southern Vermont and New Hampshire southward to south-central Baja California, south-central Mexico, the northern Gulf Coast, and northern Florida. Winters from southern Baja California, west-central Mexico, and southern Texas southward to western Panama; also rarely to northern and eastern United States, including Florida (*I. v. virens*); April-May (summer) and August-October (winter).

703. Hooded Warbler, *Wilsonia citrina* (Boddaert)

IDENTIFICATION: Total length 125 to 145 mm. Adult male: *forehead and sides of head bright yellow*; *remainder of head, neck, and throat black*; upperparts otherwise chiefly olive green, but dusky on wings and tail; remainder of underparts (except rectrices) bright yellow. Adult female, immature: with little or no black in plumage; otherwise like adult male. In any plumage the *lateral rectrices are extensively edged with white*, and the feet are flesh-colored.

DISTRIBUTION: Breeds from southeastern Nebraska, central Iowa, northern Illinois, southern Michigan and Ontario, northwestern Pennsylvania, western and southeastern New York, southern Connecticut, and Rhode Island southward to southeastern Texas, the Gulf Coast, and northern Florida (to Levy, Alachua, and Clay Counties). Winters from eastern Mexico to Panama and in the northern Antilles (a few sight records in Florida). Florida: chiefly March-April (summer) and July-October.

704. Wilson's Warbler, *Wilsonia pusilla* (Wilson)

IDENTIFICATION: Total length 115 to 125 mm. Adult male: *forehead bright yellow*; *crown black*; nape and back olive green; wings and tail dusky; *underparts* (except rectrices) *entirely brilliant yellow*. Adult female, immature: similar to adult male, but with little or no black in crown. In all plumages the feet are flesh-colored.

DISTRIBUTION AND VARIATION: Breeds from northern Alaska and Yukon, northwestern and central Mackenzie, northeastern Manitoba, northern Ontario, southern Labrador, and Newfoundland southward to southern California, central Nevada, northern Utah, northern New Mexico, central Saskatchewan, southern Manitoba, northern Minnesota, southern Ontario, northern Vermont, and central Maine and Nova Scotia. Winters from southern Baja California, northern Mexico, and southern Texas southward to western Panama, and casually eastward along the Gulf Coast to Florida (*W. p. pusilla*; September-April, very rare), Cuba, and Jamaica.

705. Canada Warbler, *Wilsonia canadensis* (L.)

IDENTIFICATION: Total length 125 to 145 mm. Upperparts entirely gray (top of head streaked with black in male); *underparts, eye-ring, and lores bright yellow, with longitudinal blackish streaks on breast, forming a transverse necklace* (bold in male, faint in female and immature); under tail coverts white; feet flesh-colored.

DISTRIBUTION: Breeds from north-central Alberta, central Saskatchewan and Manitoba, northern Ontario, and southern Quebec southward to southern Manitoba, central Minnesota, northern Wisconsin, central Michigan, northern Ohio, east-central Pennsylvania, northern New Jersey, Connecticut, Rhode Island, Massachusetts, Maine, and New Brunswick; also southward in the mountains to eastern Tennessee, northern Georgia, and western North Carolina. Florida: late April to May and August to October, very rare.

706. American Redstart, *Setophaga ruticilla* (L.)

IDENTIFICATION: Total length 125 to 140 mm. Adult male: *upperparts entirely black, except for salmon patches on wings* (bases of remiges) *and tail* (bases of lateral rectrices); black of head continuous with that of throat and upper breast; remainder of underparts white, except for salmon patches on each side of breast (and base of tail). Adult female and immature: distribution of colors like that of adult male, but black replaced by gray on head and by olive gray on rest of upperparts; salmon patches replaced by yellow (or partly salmon in immature male); otherwise underparts whitish.

DISTRIBUTION AND VARIATION: Breeds from southeastern Alaska, northern British Columbia, south-central Mackenzie, central Saskatchewan and Manitoba, northern Ontario, central Quebec, and Newfoundland southward to eastern Oregon, northern Utah and Colorado, southeastern Oklahoma, northeastern Texas, northwestern and southeastern Louisiana, central Mississippi, northwestern Florida (Stevenson, 1961a and 1962a), central Georgia, northwestern South Carolina, east-central North Carolina, and southeastern Virginia. Winters from southern and coastal California, southeastern Mexico, and southern Florida southward to Ecuador, northern Brazil, and British Guiana (occasional farther north). The breeding bird of the Southeast is *S. r. ruticilla*, but *S. r. tricolora* passes through Florida in migration (Stevenson and Baker, 1970); chiefly late March to late May and mid-July to early November.

ORDER PASSERIFORMES: CXXXVIII. FAMILY ICTERIDAE, Blackbirds, Orioles, and allies: Medium-small to medium-large birds frequently having black and yellow or orange in the plumage. The bill is typically rather slender, but occasionally stout, with operculate nostrils, a smooth tomium, and *no rictal bristles*. Typically the culmen parts the feathers of the forehead to a noticeable extent. Outermost primary almost as long as any.

707. Bobolink, *Dolichonyx oryzivorus* (L.)

IDENTIFICATION: Slightly larger than a House Sparrow (total length 165 to 190 mm). Adult male in breeding plumage: *entire head and underparts black; large patch of yellowish buff on nape and upper back*; remainder of upperparts mostly blackish to dark brown, but *scapulars and rump white* and secondaries and rectrices tipped with whitish. Adult female and immature (and adult male in fall): upperparts blackish, with buffy feather edgings; underparts mostly buffy to yellowish, lightly streaked with blackish on sides; *head striped with black and yellowish*. In all, the *bill is thick and conical and the rectrices pointed*.

DISTRIBUTION: Breeds from south-central British Columbia, the southern parts of Alberta, Saskatchewan, and Manitoba, central and south-central Ontario and Quebec, New Brunswick, and northern Nova Scotia southward to eastern Washington and Oregon, northeastern California, northern Nevada and Utah, central Colorado and Nebraska, northeastern Kansas, northern Missouri, central Illinois, south-central Indiana, southwestern and east-central Ohio, northern West Virginia, western Maryland, and central New Jersey (occasional summer records outside this range). Winters in eastern Bolivia, western Brazil, Paraguay, and northern Argentina. Abundant transient in Florida Peninsula; less common westward, especially in fall (April-June; August-early November).

708. Eastern Meadowlark, *Sturnella magna* (L.)

IDENTIFICATION: Near size and form of Starling (total length 230 to 255 mm). Top of head blackish, mottled with dark reddish brown, with 3 longitudinal whitish stripes (one medial and 2 lateral); remainder of upperparts varied with blackish, buffy, and reddish brown, *with a tendency for successive dark bars across the secondaries and rectrices to be united along the shaft of the feather*; side of head and superciliary stripe whitish to light brownish gray; sides of neck and body similar to upperparts, but with light colors predominating; under tail coverts similar; remainder of *underparts bright yellow, except for a black crescent on breast and sides of throat* (more restricted and mixed with light gray buff in immature and winter adults). Feet light brown.

DISTRIBUTION AND VARIATION: Breeds from southwestern South Dakota, northern Minnesota, Wisconsin, and Michigan, southeastern Ontario, southwestern and south-central Quebec, and central Nova Scotia southward to the southern tip of the Florida mainland, Cuba, the northern Gulf Coast, from Texas westward to northwestern Arizona, and southward to Colombia, Venezuela, and northern Brazil. The breeding race in Florida is *S. m. argutula*, but *S. m. magna* reaches the Panhandle in winter.

709. Western Meadowlark, *Sturnella neglecta* Audubon

IDENTIFICATION: Almost identical in size, form, and color pattern to the Eastern Meadowlark, but appearing lighter above (with less blackish and dark brown, and *with the blackish bars on secondaries and rectrices usually not connected along shafts*); also side of head more buffy. These characteristics are not entirely reliable, therefore field identifications of silent birds are suspect.

DISTRIBUTION AND VARIATION: Breeds from central British Columbia, Alberta, and Saskatchewan, southern Manitoba, western and southern Ontario, northern Michigan, and northwestern Ohio southward to Baja California, northern Mexico, central Texas, northwestern Louisiana, and northwestern Mississippi. Winters through most of breeding range and southward to south-central Mexico; a few wander eastward to the Panhandle of Florida (presumably *S. n. neglecta*; Monroe, 1958a); sight records farther east need substantiation.

710. Yellow-headed Blackbird, *Xanthocephalus xanthocephalus* (Bonaparte)

IDENTIFICATION: Size of meadowlark (total length 230 to 265 mm). Male: region around bill and eyes black; otherwise *entire head, neck, and upper breast bright yellow* (top of head mixed with brown in winter); remainder of plumage dull black, except for a white patch on wing coverts. Female: similar to adult male, but upperparts mostly dusky brown, and breast with white streaks.

DISTRIBUTION: Breeds from western Oregon, central Washington, central British Columbia, northeastern Alberta, north-central Saskatchewan, central and southeastern Manitoba, northern Minnesota, north-central Wisconsin, northeastern Illinois, and northwestern Ohio southward to southern California, southwestern Arizona, northeastern Baja California, south-central Nevada, southwestern Utah, east-central Arizona, southern New Mexico, northern Texas, northwestern Oklahoma, southern Kansas, northwestern Arkansas, southwestern to northeastern Missouri, and central Illinois; occasional summer records farther east. Winters from central California and Arizona, southern New Mexico, southwestern and central Texas, and southern Louisiana southward to central Mexico; casual eastward to Atlantic Coast, including all parts of Florida (late August to early May), Cuba, and the Bahamas.

711. Red-winged Blackbird, *Agelaius phoeniceus* (L.)

IDENTIFICATION: Total length 190 to 240 mm (males larger than females). Adult male: *bend of wing orange red, bordered by buff*; *rest of plumage black* (only slightly iridescent). Adult female and immature: upperparts blackish, with white feather edgings giving a streaked effect; superciliary line and ground color of underparts white to buffy, lightly streaked on throat, and heavily on body, with blackish; often some reddish on bend of wing. Certain species of sparrows have the general color pattern of the female Redwing, but are smaller.

DISTRIBUTION AND VARIATION: Breeds from northwestern British Columbia, southeastern Yukon, central Mackenzie, northern Saskatchewan, north-central Manitoba, northern Ontario, southern Quebec, and central Nova Scotia southward to Costa Rica, western Cuba, the Isle of Pines, and the northern Bahamas. Winters within breeding range as far north as southern Canada and the northeastern United States. The breeding races in Florida are: *A. p. littoralis*, coast of the Panhandle; *A. p. phoeniceus*, near Alabama and Georgia lines; *A. p. mearnsi*, most of

Peninsula; and *A. p. floridanus*, Keys and southern tip of mainland. *A. p. fortis* has been taken in the Panhandle in winter, and *A. p. phoeniceus* strays into the Peninsula at that season.

712. Tawny-shouldered Blackbird, *Agelaius humeralis* (Vigors)

IDENTIFICATION: Similar to the male Red-winged Blackbird, but *shoulder patches tawny rather than red*, more restricted in female than in male. Total length 190 to 215 mm.

DISTRIBUTION: Cuba and west-central Haiti; accidental on the Florida Keys.

713. Orchard Oriole, *Icterus spurius* (L.)

IDENTIFICATION: About size of House Sparrow, but tail longer (total length 150 to 180 mm). Adult male: entire head, upper breast, nape, upper back, tail, and most of wings black; *lower back and rump dark chestnut*; some of wing coverts and *remainder of underparts dark chestnut*; *almost a unique color pattern*. Adult female, immature: upper back and wings grayish, the coverts and remiges edged with whitish; remainder of upperparts olive green to yellowish green (rump); underparts mostly bright yellow; feet blue gray. One-year-old male: similar to adult female, but *chin and throat black*, and sometimes patches of black or chestnut appearing elsewhere in plumage.

DISTRIBUTION: Breeds from southern Manitoba, central and southeastern Minnesota, central Wisconsin, southern Michigan and Ontario, north-central Pennsylvania, central New York, and northeastern Massachusetts southward (east of the Rockies) to northern Mexico, the northern Gulf Coast, and the northern Florida Peninsula; transient throughout state, March-May (summer) and July-September. Winters from southern Mexico to Colombia and Venezuela, casually northward to southern Texas and southern Florida (UM 1436 plus sight records).

714. Spotted-breasted Oriole, *Icterus pectoralis* (Wagler)

IDENTIFICATION: Larger than other Florida orioles (total length 205 to 215 mm). Adult: lores, chin, throat, and tail black; wings and back mostly so; remainder of plumage mostly deep orange, but *sides of breast spotted with black*; also white on primaries and secondaries. Immature: head and underparts yellowish orange to yellow; back olive; wings and tail grayish, with white as on wings of adult; *this plumage suggests a large Prothonotary Warbler*.

DISTRIBUTION AND VARIATION: Resident from southeastern Mexico to eastern Nicaragua. *I. p. pectoralis* appeared in Miami in 1949 and is now resident near the coast as far north as Vero Beach and southward to Florida City.

715. Baltimore Oriole, *Icterus galbula* (L.)[1]

IDENTIFICATION: Total length 180 to 205 mm. Adult male: *entire head, throat, neck, and upper back black*; lower back yellowish orange; wings mostly black, but with orange and white on coverts, and white on edges of remiges; tail black with large orange patches on lateral rectrices; *remainder of underparts* (including tail) *deep yellow to orange*. Adult female and immature: similar, but with little or no black on head and neck, streaked and more olive on back, 2 white bars on wing coverts, and paler yellow underneath. Feet bluish gray.

DISTRIBUTION: Breeds from central Alberta and Saskatchewan, southern Manitoba, western Ontario, northern Michigan, southern Ontario and Quebec, central Maine, New Brunswick, and Nova Scotia southward to west-central Oklahoma, northeastern Texas, northwestern to southeastern Louisiana, central Mississippi, west-central and northern Alabama, north-central Georgia, western parts of the Carolinas, central Virginia, northern Maryland, and Delaware; casually to northeastern Colorado; apparently withdrawing from southeastern parts of breeding range, but nested at Key West, Florida, in 1972 (Ogden, 1972). Winters mainly from southeastern Mexico to central Colombia and northwestern Venezuela, but increasingly in the eastern United States, especially Florida (late August-early May).

716. Bullock's Oriole, *Icterus bullockii* (Swainson)

IDENTIFICATION: Size of Baltimore Oriole. Adult male: similar to corresponding plumage of Baltimore Oriole, but side of head orange, with a *black streak through eye*, and more white (no orange) on wing. Adult female and immature (including one-year-old male): top of head, nape, and tail olive green; *back grayish*; wings dark gray, with white edges on remiges and 2 white bars on coverts; chin, throat, and upper breast yellow to orange yellow, often with varying amounts of black; *abdomen whitish*. Individuals intermediate between the above description and that of the Baltimore Oriole are probably hybrids.

DISTRIBUTION AND VARIATION: Breeds from southern British Columbia and Alberta, southwestern Saskatchewan and North Dakota, and central South Dakota southward to northern Baja California, Mexico City, and northern Veracruz, and eastward in the United States to about 100 degrees longitude. Winters from west-central Mexico to northwestern Costa Rica, casually northward to central California and rarely eastward to Florida (probably *I. b. bullockii*); September to April.

---

1. Now lumped with *I. bullockii* as the Northern Oriole (A.O.U., 1973). The enlarged species is therefore more variable in color, and ranges throughout the western United States and into Mexico.

717. Rusty Blackbird, *Euphagus carolinus* (Müller)

IDENTIFICATION: Near size of Red-winged Blackbird (total length 215 to 240 mm, with male larger than female). Adult male: plumage entirely black, with bluish green reflections. Adult female: plumage entirely slate gray, somewhat darker on wings and tail. In both sexes the *feathers are tipped with rust above and buff below* in fall and much of winter, and the *iris is whitish*.

DISTRIBUTION AND VARIATION: Breeds from northern Alaska, northern Yukon, northwestern and central Mackenzie, northern Manitoba, Ontario, and Quebec, central Labrador, and Newfoundland southward to central Alaska, central British Columbia, south-central Alberta, central Saskatchewan and Manitoba, southern Ontario, northeastern New York, northern Vermont and New Hampshire, central Maine, southern New Brunswick, and Nova Scotia. Winters from southern Canada and the northeastern United States southward to central Colorado, southeastern Texas, the Gulf Coast, and southern Florida. *E. c. carolinus* has been collected in the state and *E. c. nigrans* just outside (late October-late April).

718. Brewer's Blackbird, *Euphagus cyanocephalus* (Wagler)

IDENTIFICATION: Slightly larger than the Rusty Blackbird (total length 230 to 255 mm), with males larger than females. Adult male: coloration similar to that of Rusty Blackbird, but *head reflections more purplish*. Adult female: almost identical to the female Rusty Blackbird, but distinguished in life by its *dark iris*. *In no plumage are the feathers ever edged with rust or buff*.

DISTRIBUTION: Breeds from the central parts of British Columbia, Alberta, and Saskatchewan, southern Manitoba, northern Minnesota, western Ontario, and northern Wisconsin southward to central California, southern Nevada, southwestern and central Utah, central Arizona, western and central New Mexico, northern Texas, Oklahoma, northern Iowa, southern Wisconsin, northeastern Illinois, northwestern Indiana, and southwestern Michigan. Winters from southwestern British Columbia, northern Washington, central Alberta, east-central Montana, central Oklahoma, Kansas, Arkansas, southwestern Tennessee, northeastern Mississippi, Alabama, Georgia, and the western parts of the Carolinas southward to Baja California, Oaxaca, central Veracruz, the northern Gulf Coast, and northern (rarely southern) Florida; November to April.

719. Boat-tailed Grackle, *Cassidix major* (Vieillot)

IDENTIFICATION: Larger than the Common Grackle (total length 305 to 430 mm, the male considerably larger than the female). Adult male: plumage entirely black, with bluish or purplish reflections, most pronounced on head and body; *tail strongly graduated and keeled*. Adult female and juvenal: upperparts dark grayish brown, blacker on wings and tail; lower belly and under tail coverts dark gray; remainder of underparts buffy brown (female) or whitish with dusky streaks (juvenal). Iris yellowish in adults of *C. m. torreyi*.

DISTRIBUTION AND VARIATION: Resident along the Atlantic and Gulf Coasts from southern New Jersey (except only in summer north of Virginia) to southeastern Texas, but largely absent in the Florida Panhandle; also breeds virtually throughout peninsular Florida. The breeding bird over most of Florida is *C. m. westoni* (see Chapman, 1967), but *C. m. alabamensis* may occur in Panhandle (Stevenson, in prep.), and *C. m. torreyi* occurs in northeastern Florida.

720. Common Grackle, *Quiscalus quiscula* (Linnaeus)

IDENTIFICATION: Total length 255 to 330 mm, the male being larger than the female. Male: Almost identical in form and plumage to the male Boat-tailed Grackle, but much smaller; head reflections greenish to purplish in most Florida birds. Female similar, but duller. In both sexes the *iris is yellowish*.

DISTRIBUTION AND VARIATION: Breeds from northeastern British Columbia, south-central Mackenzie, central Saskatchewan, northeastern Manitoba, western to northeastern Ontario, southern Quebec, southwestern Newfoundland, and northern Nova Scotia southward (east of the Rockies) to south-central Colorado, southeastern Texas, the Gulf Coast, and the Florida Keys. Withdraws from Canada and the extreme northeastern United States in winter; also from the Florida Keys. The breeding race in Florida is *Q. q. quiscula*, but both *stonei* and *versicolor* have been collected in north Florida in winter (Monroe, 1958c).

721. Brown-headed Cowbird, *Molothrus ater* (Boddaert)

IDENTIFICATION: Slightly larger than the House Sparrow (total length 180 to 205 mm). Adult male: *entire head, neck, and upper breast dark brown; plumage otherwise black*, with greenish or purplish reflections. Adult female: plumage mostly slate-gray, almost brownish on head, generally lighter and with short, dusky streaks underneath, and whitish on chin. Immature: similar to female, but more brownish, with buffy feather edgings, above; underparts lighter with indistinct brownish streaks. Bill thick at base.

DISTRIBUTION AND VARIATION: Breeds from central and northeastern British Columbia, south-central Mackenzie, central Saskatchewan, southern Manitoba, central Ontario, southwestern and east-central Quebec, New Brunswick, and southern Nova Scotia southward to northern Mexico, the northern Gulf Coast, and northern Florida (Monroe, 1957; summer records south to Levy and Brevard Counties). Winters from north-central California, southern Arizona, central Oklahoma and Missouri, southern Michigan, southern Ontario, New York, and Connecticut southward to southern Mexico, the northern Gulf Coast, southern Florida, and (casually) Cuba. Only *M. a. ater* is known to occur in Florida; July-April (summer).

722. Bronzed Cowbird, *Tangavius aeneus* (Wagler)

IDENTIFICATION: Slightly larger than Brown-headed Cowbird (total length 180 to 215 mm). Differs from that species in the *presence of a neck ruff* and of *red or orange irides*; also the bill is more slender (see Key). There is no brown on the head of the male, and the female is also blackish, though less iridescent than the male.

DISTRIBUTION AND VARIATION: Resident and breeding from central Arizona, southwestern New Mexico, and south-central Texas southward to western Panama. *T. a. aeneus* is accidental at Sarasota and Gainesville, Florida (Truchot, 1962; Robertson and Ogden, 1969; Roxie Laybourne, personal communication).

ORDER PASSERIFORMES: CXXXIX. FAMILY THRAUPIDAE, Tanagers. Mostly medium-small birds of brilliant plumage (males), predominating in tropical America. Typically the bill is rather stout and of finchlike proportions, but with the tomium either straight or gently curving and *often notched and toothed*.

723. Blue-gray Tanager, *Thraupis virens* (Swainson)

IDENTIFICATION: About the size of a bluebird (total length 160 to 190 mm). *Entire plumage pale grayish blue*, except for a large whitish patch on the wing coverts (Florida subspecies). No other Florida bird is so colored.

DISTRIBUTION AND VARIATION: Resident from southern Mexico southward to northwestern Peru, northwestern Bolivia, and the Amazon Basin; also on Trinidad and Tobago and (by introduction) from Hollywood, Florida, to Miami and spreading (*T. v. coelestis*; Arnold, 1961; Mason, 1964b; UM 4973) (Map 57).

724. Stripe-headed Tanager, *Spindalis zena* L.

IDENTIFICATION: About size of House Sparrow (total length 150 to 190 mm). Male: *head with broad stripes of black and white*; upperparts largely olive greenish otherwise, varying with subspecies; wings and tail blackish with broad, white edgings on flight feathers and greater coverts;

Map 57. Distribution of the Blue-gray Tanager (*Thraupis virens*)

breast and most of abdomen yellowish orange; lower abdomen and under tail coverts white. Female: races most likely in Florida are chiefly plain olive gray with the wing feathers edged with white.

DISTRIBUTION AND VARIATION: Resident in the Bahama Islands, the Greater Antilles, and on Cozumel Island; casual along the coast of southeastern Florida, including the Keys; *S. z. zena* (TT 2815) and possibly *S. z. townsendi* (Langridge, 1963).

725. Western Tanager, *Piranga ludoviciana* (Wilson)

IDENTIFICATION: About size of Summer Tanager (total length 180 to 190 mm). Adult male: *entire head red to orange* (yellow in fall and winter), fading to yellow on nape; bend of wing, lower back, and rump also yellow; remainder of upperparts dull black, but for a *white bar on greater coverts*; breast, abdomen, and under tail coverts bright yellow. Adult female: top of head, nape, and rump olive green; back olive gray; *wings and tail dark grayish, the former with 2 whitish bars on coverts* and whitish edgings on secondaries; underparts bright yellow, tinged with gray on breast and sides, and tending toward white on chin and throat.

DISTRIBUTION: Breeds from southern Alaska, northern British Columbia, south-central Mackenzie, northeastern Alberta, and central Saskatchewan southward (west of the Great Plains) to northern Baja California, southern Nevada, southwestern Utah, central and southeastern Arizona, southwestern New Mexico, and western Texas. Winters from southern Baja California and central Mexico southward to northwestern Costa Rica; casually north to California, southeastern Arizona, and southern Texas, and eastward along Gulf Coast to Florida (Sprunt, 1963; September-May).

726. Scarlet Tanager, *Piranga olivacea* (Gmelin)

IDENTIFICATION: About size of Summer Tanager (total length 165 to 190 mm). Adult male: *wings and tail jet black*; *remainder* of plumage brilliant light red; in fall, red replaced by greenish above and yellowish below. Adult female: similar to adult male in fall plumage, but wings and tail grayish; more greenish than corresponding plumage of Summer Tanager, and bill smaller. Wing bars rarely present in immatures.

DISTRIBUTION: Breeds from central Nebraska, eastern North Dakota, southeastern Manitoba, west-central and southern Ontario, southern Quebec, New Brunswick, and central Maine southward to north-central and southeastern Oklahoma, central Arkansas, west-central Tennessee, northeastern Mississippi, northwestern and east-central Alabama, northern Georgia, northwestern South Carolina, central North Carolina and Virginia, and Maryland. Winters from northwestern and central Colombia southward to central Peru and west-central Bolivia.

727. Summer Tanager, *Piranga rubra* (L.)

IDENTIFICATION: Size of Cardinal, but tail shorter (total length 175 to 205 mm). Adult male: *entire plumage red*, except dusky remiges; upperparts duller than underparts; *young males* in spring often *show patches of red and yellow plumage*; *bill yellow*. Adult female: upperparts olive green, except for dusky on remiges; underparts bright yellow, except duller on throat, breast, and sides. Immature: similar to adult female, but duller above, *more buffy below*, and bill horn-colored. No wing bars in any plumage. Bill heavier and longer than in other tanagers.

DISTRIBUTION AND VARIATION: Breeds from southeastern California, southern Nevada, central Arizona, New Mexico, Texas, and Oklahoma, southeastern Nebraska, southern Iowa, central Illinois, Indiana, and Ohio, West Virginia, Maryland, and Delaware southward to northern Mexico, the northern Gulf Coast, and southern Florida (northern Collier and Dade Counties). Winters from southern Baja California and Veracruz southward to south-central Peru, western Bolivia, west-central Brazil, western British Guiana, Cuba, and Jamaica (casually in Florida). Only the eastern form, *P. r. rubra*, is known in Florida; March to late October (winter).

ORDER PASSERIFORMES: CXL. FAMILY PLOCEIDAE, Weaver Finches. Small to medium-small birds with a rather heavy, cone-shaped bill and, usually, a low tarsus:wing ratio. Tail usually proportionately shorter than in Fringillidae, never long in North American species. Tomium angled or abruptly decurved toward base.

728. House Sparrow, *Passer domesticus* (L.)

IDENTIFICATION: Total length about 140 to 160 mm. Adult male: top of head and nape dark slaty gray; upperparts mostly dark grayish brown, broadly streaked on back with blackish and showing reddish brown on bend of wing (and sometimes on upper back); a white bar on wing coverts; remiges and rectrices more blackish; underparts chiefly light gray, but with *a large black patch on chin*, *throat*, *and upper breast* (feathers edged with gray in winter); *a patch of chestnut from eye to side of neck*; lores black; lower side of head light gray. Adult female, immature: upperparts similar to those of adult male, but top of head grayish brown, back without definite reddish or black, and a *light superciliary line present*; no chestnut on side of head, and no black underparts. (Immatures are paler and buffier than adult females.) Compared with other drab-colored sparrows (Fringillidae), the female House Sparrow shows less color variation on the back, no streaking below, a relatively short tail, and a light area near the base of the outer primaries.

DISTRIBUTION AND VARIATION: Resident, in many cases by introduction, in most major land areas of the world, but not in northern South America, Central America, and doubtless in some major portions of the Old World. The Holarctic race, *P. d. domesticus*, occupies all settled parts of Florida.

729. Java Sparrow, *Padda oryzivora* (L.)[1]

IDENTIFICATION: Similar to the House Sparrow in size and profile, averaging slightly smaller (total length about 140 mm). *Head and throat black*, except for prominent *white cheeks*; upperparts and breast lavender gray; underparts otherwise pinkish gray; *bill and feet pinkish, the bill very large with the nostrils widely separated*.

DISTRIBUTION: Native to Java and Sumatra; widely introduced in warmer parts of Asia; also apparently established at Miami, Florida (Ogden and Stimson, personal communication).

ORDER PASSERIFORMES: CXLI. FAMILY FRINGILLIDAE, Finches, Grosbeaks, Buntings, and Sparrows.[2] Small to medium-small birds with a rather heavy cone-shaped bill and, usually, a high tarsus:wing ratio. Tail usually proportionately longer than in Thraupidae or Ploceidae, or the rectrices pointed at the tips. Tomium angled or abruptly decurved toward base. Brownish plumage predominant, especially in females.

730. Evening Grosbeak, *Hesperiphona vespertina* (Cooper)

IDENTIFICATION: Similar in color to American Goldfinch, but much larger (total length 190 to 215 mm). Male: body mostly brownish yellow, but browner on breast and upper back; top of head blackish, in striking contrast to the bright yellow forehead and superciliary stripe; wings black with a large white patch; tail black. Female: similar in pattern, but much grayer and without an obvious superciliary stripe. Immature: similar to adult female, but more brownish above and buffy below. In all plumages the *bill is very thick and light colored*.

DISTRIBUTION AND VARIATION: Breeds from British Columbia, Alberta, central Saskatchewan, southern Manitoba, central Ontario, west-central Quebec, and northern New Brunswick southward to central California, northern Nevada, central Arizona (and in the mountains to southern Mexico), the northern parts of Minnesota, Michigan, and New York, and to Massachusetts. Winters southward to the southern edge of the United States, including north Florida (mid-December, 1968,

---

1. Often placed in the Estrildidae.
2. A Black-throated Sparrow (*Amphispiza bilineata*) was collected at Tallahassee in February 1976 (TT 3468).

to early May, 1969; Stevenson, 1969a). The eastern race is *H. v. vespertina*.

731. Purple Finch, *Carpodacus purpureus* (Gmelin)

IDENTIFICATION: About size of House Sparrow (total length 125 to 165 mm). Male: *upperparts largely rose red*, mixed with blackish on back; wings and tail dusky, with reddish or brown feather edgings; *chin, throat, breast, and sides of head rose red*, shading to white on belly and under tail coverts. ("Purple" Finch is a misnomer.) Female: upperparts dark grayish brown, with lighter streaks; *white superciliary stripe conspicuous*; underparts white with dark brown spots in longitudinal rows (rarely pinkish on head and neck). Sparrows similar in color to the female Purple Finch usually do not have such a *heavy bill* or a *notched tail*.

DISTRIBUTION AND VARIATION: Breeds from northern British Columbia and Alberta, central Saskatchewan and Manitoba, northern Ontario, central Quebec, and Newfoundland southward to northern Baja California, south-central British Columbia, central Alberta, southern Saskatchewan, North Dakota, central Minnesota, Wisconsin, and Michigan, northeastern Ohio, West Virginia, northeastern Pennsylvania, and southeastern New York. Winters from southwestern British Columbia to central Baja California and southern Arizona; also from the southern parts of Manitoba, Ontario, and Quebec southward to southeastern Texas, the Gulf Coast, and northern Florida (rarely central); only *C. p. purpureus* has been identified in Florida; November to late April.

732. Pine Siskin, *Spinus pinus* (Wilson)

IDENTIFICATION: About size of American Goldfinch (105 to 135 mm). Upperparts streaked with dark brown and buffy; wings and tail dusky, the *bases of the secondaries and rectrices more or less yellowish*; underparts whitish, streaked (except in middle of belly) with dark brown; *tail notched*; bill comparatively slender.

DISTRIBUTION AND VARIATION: Breeds from southern Alaska, central Yukon, southern Mackenzie, central Saskatchewan, southern Manitoba, northern Ontario, central and southeastern Quebec, southern Labrador, and Newfoundland southward to northern Baja California, in the mountains to Guatemala, east to the mountains to south-central and northeastern Kansas, northwestern Iowa, central Minnesota, northern Wisconsin, central Michigan, southern Ontario, northern Pennsylvania, New York, Connecticut, and, in the Appalachians, to eastern Tennessee and western North Carolina. Winters in most of breeding range and also southward to lower elevations in Mexico and western United

States, to the northern Gulf Coast, and (rarely) southern Florida. Only *S. p. pinus* can be expected in Florida; November-May.

733. American Goldfinch, *Spinus tristis* (L.)

IDENTIFICATION: Total length about 105 to 135 mm. Adult male in breeding plumage: *forehead, crown, wings, and tail mostly black*, but wing coverts, remiges, and rectrices with white edges or tips; tail coverts white; *remainder of plumage bright yellow*. Adult male in winter: wings and tail similar, but without black on head; upperparts olive brown, shading to grayish on rump and tail coverts; underparts whitish to light gray, tinged with yellow on throat and breast. Adult female: similar to male in winter plumage, but wings and tail more dusky.

DISTRIBUTION AND VARIATION: Breeds from southern British Columbia, central Alberta and Saskatchewan, southern Manitoba, central Ontario, southern Quebec, and northern Nova Scotia southward to northern Baja California, central Utah, southern Colorado, central Oklahoma, northeastern Texas, northern Louisiana, northern Mississippi, southern Alabama, southwestern Georgia, and South Carolina. Winters from southern British Columbia, Montana, South Dakota, Minnesota, Michigan, southern Ontario, New Brunswick, and Nova Scotia southward to northern Mexico, the northern Gulf Coast, and southern Florida, including the Keys. The eastern race, *S. t. tristis*, is the only one likely to occur in Florida; early November (rarely earlier) to late May.

734. Red Crossbill, *Loxia curvirostra* L.

IDENTIFICATION: About size of House Sparrow (145 to 160 mm). Male: *crown, throat, breast, and part of abdomen dull brownish red*; rump brighter red (this shade often more extensive); remainder of body plumage dark brown; wings and tail dusky. Female: *dull olive*, brighter on the rump, and tinged with brownish. In both sexes the *mandibles are crossed*.

DISTRIBUTION AND VARIATION: Breeds over most of Europe, all but southeastern parts of Asia, in northwestern Africa, and from southeastern Alaska, southern Yukon, Saskatchewan, Manitoba, central Ontario, southern Quebec, and Newfoundland southward to northern Baja California, northern Nicaragua, western Texas, northern Wisconsin, eastern Tennessee, northern Georgia, western North Carolina, Long Island, and eastern Massachusetts. Winters within breeding range and also southward to southeastern Texas, southern Louisiana, central Alabama, and northern Florida. All Florida specimens have been referred to *L. c. minor*; December to February.

735. Cardinal, *Cardinalis cardinalis* (L.)

IDENTIFICATION: Total length 190 to 230 mm. Adult male: *region around bill black*; remainder of plumage red, duller above than below. Adult female: much duller red on crest, wings, and tail; remainder of upperparts olive brown; underparts (except tail) grayish buffy. Immature male: between color of male and female adults, except bill largely brownish. In all plumages the *head is crested* and the *bill very thick*.

DISTRIBUTION AND VARIATION: Resident from southeastern California, central Arizona, southern New Mexico, northern Texas, southeastern South Dakota, central Minnesota, western and southern Ontario, west-central New York, and southwestern Connecticut southward to British Honduras, the northern Gulf Coast, and the Florida Keys; introduced into Hawaii, southwestern California, and the Bermudas. The breeding races in Florida are *C. c. cardinalis*, west of the Apalachicola River, and *C. c. floridanus*, remainder of state.

736. Rose-breasted Grosbeak, *Pheucticus ludovicianus* (L.)

IDENTIFICATION: About size of Cardinal, but tail shorter (total length 175 to 205 mm). Adult male: entire head, neck, and upperparts black, except for white patches on wings and tail; *breast and wing lining pinkish red*; remainder of underparts white; in fall the upperparts are brownish, with a light superciliary line, and the underparts spotted with brownish. *Bill largely whitish, very thick at base*. Adult female: upperparts brownish gray (darker on head), streaked on head and back; wing coverts with one or 2 white bars; underparts white, streaked with dusky (devoid of yellow); *lining of wing yellow*. Immature male: similar to adult female, but wings and tail more blackish, underparts tinged with pale brown, and *lining of wing* (to some extent, breast also) *pinkish red*. *No other common Florida bird of similar size and streaked coloration (as the female and immature) has so heavy a bill*.

DISTRIBUTION: Breeds from northeastern British Columbia, northern Alberta, central Saskatchewan, southern Manitoba, western and southern Ontario, southwestern Quebec, northern New Brunswick, and Nova Scotia southward to central and southeastern Alberta, southern Saskatchewan, eastern South Dakota, Nebraska, and Kansas, southwestern and central Missouri, southern Illinois, central Indiana, northern Ohio, southeastern Pennsylvania, southwestern and central New Jersey, and southeastern New York; also in mountains to eastern Tennessee, western North Carolina, and northern Georgia. Winters from central Mexico southward to northern Ecuador, central Colombia, and southwestern and north-central Venezuela; also casually in Cuba, Florida, and Louisiana. Florida: April-May; September-early November (winter).

737. Black-headed Grosbeak, *Pheucticus melanocephalus* (Swainson)[1]

IDENTIFICATION: Size of Rose-breasted Grosbeak (total length 175 to 215 mm). Adult male: *head black*, sometimes with a median stripe of chestnut; nape also chestnut; back black, usually varied with cinnamon; wings and tail black, with patches and spots of white; rump chestnut; sides of head and chin black; *underparts bright buffy brown, shading to yellow on belly*. Adult female: top of head and neck blackish, divided medially by a light stripe, and bordered laterally by white superciliary lines; feathers of back blackish, with buffy brown edges; wings and tail dusky, the former with rounded white spots, some tending to form wing bars on coverts; sides of head blackish, bordered by light gray below; *underparts largely buffy brown* (the feathers of sides with darker shafts), becoming whitish to yellowish on belly; under tail coverts buffy; *some yellow present on lower breast*. Immature male: similar to adult female, but more buffy. Females and immatures can be distinguished from other Florida birds by a *combination of size, shape of bill, and color of underparts*.

DISTRIBUTION AND VARIATION: Breeds from southern British Columbia, Alberta, and Saskatchewan, northwestern North Dakota, and central Nebraska southward to northern Baja California and southern Mexico (in the mountains). Winters from northern Mexico to southeastern Mexico; also casually eastward near the Gulf Coast to southeastern Florida (presumably *P. m. melanocephalus*; Sprunt, 1963); October-April.

738. Blue Grosbeak, *Guiraca caerulea* (L.)

IDENTIFICATION: Slightly larger than a House Sparrow (total length 165 to 190 mm). Adult male: *plumage mostly dark purplish blue*; lores, remiges, and rectrices blackish; *2 bars of chestnut on wing coverts*; bill whitish. Adult female and immature: upperparts brownish gray; underparts buffy brown; *wing bars buffy. Bill very heavy*.

DISTRIBUTION AND VARIATION: Breeds from central California, southern Nevada, southern and eastern Utah, southern Colorado, central South Dakota and Missouri, southern Illinois, southwestern Kentucky, northeastern Tennessee, western North Carolina, southeastern Pennsylvania, and southern New Jersey southward to Costa Rica, the northern Gulf Coast, and north-central Florida (Kissimmee). Winters from southern Baja California, southern Sonora, and central Veracruz southward to western Panama; also, casually, in Louisiana, Florida, the Bahamas, and Cuba. The Florida race is *G. c. caerulea*; late March-late May (summer) and August-early November (winter).

---

1. Sometimes considered conspecific with the Rose-breasted Grosbeak.

739. Indigo Bunting, *Passerina cyanea* (L.)[1]

IDENTIFICATION: Smaller than a House Sparrow (total length 120 to 145 mm). Adult male: *plumage mostly brilliant violet blue to greenish blue* (lighter than in Blue Grosbeak); wings and tail blackish, *without wing bars*; bill whitish. Adult female and immature: upperparts mostly buffy brown (including wing bars); wings and tail otherwise blackish; belly whitish; remainder of underparts pale buffy to buffy brown, *faintly streaked with dusky on throat and breast*.

DISTRIBUTION: Breeds from southwestern South Dakota, southern Manitoba, northern Minnesota, western and southern Ontario, and the southern parts of Quebec, Maine, and New Brunswick southward (east of the Rockies) to south-central and southeastern Texas, the Gulf Coast, and central Florida. Winters from central Mexico and southern Florida southward to central Panama; casually south to northern South America and north to the northern Gulf Coast. Florida: late March-late May (summer) and August-November (winter).

740. Painted Bunting, *Passerina ciris* (L.)

IDENTIFICATION: About size of Indigo Bunting. Adult male: top and sides of head and neck purplish blue; upper back yellowish green; lower back and rump brick red; remiges mostly blackish; greater coverts bluish green; tail dark reddish brown; underparts vermilion. *No other Florida bird is similar*. Adult female: *upperparts dark bluish green*, including greater coverts; remiges dusky, more or less edged with olive green; underparts greenish yellow. Immature: duller above and buffier below than adult female. In both Florida species of buntings and the Blue Grosbeak, immature males in intermediate plumages are often seen.

DISTRIBUTION AND VARIATION: Breeds from southern New Mexico, central Oklahoma, east-central Kansas, southern Missouri, and western Tennessee southward to northern Mexico, southern Texas, southern Mississippi, and southern Alabama; also near the Atlantic Coast from southeastern North Carolina southward to central Florida. Casual in summer outside breeding range in Florida, but nested at Apalachicola in 1966 (Ogden and Chapman, 1967) and 1970. Winters from northern Mexico and central (rarely northern) Florida southward to western Panama and Cuba, casually in southern Louisiana. Florida: April-May (summer); July-November (winter); *P. c. ciris*.

741. Black-faced Grassquit, *Tiaris bicolor* (L.)

IDENTIFICATION: One of the world's smallest finches (total length 100 to 115 mm). Male: *head, throat, and breast blackish*, gradually

---

1. Sometimes considered conspecific with the Lazuli Bunting (*Passerina amoena*).

merging to dull greenish above and white in center of abdomen; remiges and rectrices mostly dusky; *no yellow present*. Female: head and underparts dull olive gray; otherwise like male.

DISTRIBUTION AND VARIATION: Resident in the West Indies (except Cuba and the Isle of Pines); also in northwestern South America. *T. b. bicolor* is accidental in southern Florida (Mason, 1961b; Austin, 1963; Browning, 1964; Robertson, 1968).

742. Dickcissel, *Spiza americana* (Gmelin)

IDENTIFICATION: Size of a House Sparrow (total length 150 to 165 mm). Adult male: top of head and nape grayish; remainder of upperparts grayish brown (quite similar to those of female House Sparrow); superciliary stripe yellow to white; sides of head gray; *a black patch on throat*; *breast more or less yellowish*; remainder of underparts whitish. Adult female: similar to male, but black and yellow almost lacking. Immature: similar to adult female, but more-or-less tinged with buffy.

DISTRIBUTION: Breeds from eastern Montana, northwestern North Dakota, southern Manitoba, northwestern and central Minnesota, northern Wisconsin, southern Michigan and Ontario, central New York, and Massachusetts southward to central Colorado, western Oklahoma, west-central Texas, the southern parts of Louisiana, Mississippi, and Alabama, and central Georgia and South Carolina. Winters from southwestern Mexico southward to central Colombia, southern Venezuela, British Guiana, and French Guiana; also, casually, in the southeastern United States. Rare in Florida, September to May.

743. Rufous-sided Towhee, *Pipilo erythrophthalmus* (L.)

IDENTIFICATION: Slightly larger than a House Sparrow, but tail much longer (total length 190 to 215 mm). Adult male: *entire upperparts, chin, throat, and upper breast black*, but for white or buffy spots on wings and lateral rectrices; lower breast and belly white; *sides, flanks, and under tail coverts reddish brown*. Adult female: color pattern as in male, but black replaced by some shade of brown. Juvenal male: wings and tail as in adult male, but remainder of upperparts blackish with tawny feather edgings; underparts whitish to buffy, heavily streaked with blackish on throat, breast, and sides. Juvenal female: similar to juvenal male, but upperparts more brownish, and streaks of underparts also browner. Iris color varies with subspecies from white through yellow to red, but may be brown or gray in immatures.

DISTRIBUTION AND VARIATION: Breeds from southern British Columbia, central Alberta and Saskatchewan, southern Manitoba, northern Minnesota and Michigan, southern Ontario, northern New York and Vermont, central New Hampshire, and southwestern Maine southward to southern Baja California, Guatemala, western Texas, northern Oklahoma, northern Arkansas, south-central Louisiana, the coast of Mississippi and Alabama, and southern Florida. Winters north to southern British Columbia, Utah, Colorado, Nebraska, Iowa, northern Illinois and Ohio, New York, and Massachusetts; casually southward to the Florida Keys. The following races breed in Florida: *P. e. canaster*, extreme western Panhandle; *P. e. rileyi*, Okaloosa County eastward to Madison County; *P. e. alleni*, remainder of Peninsula, westward along coast to Bay County; in winter *P. e. erythrophthalmus* ranges southward to the Kissimmee Prairie (Dickinson, 1952) and probably farther, and *P. e. arcticus* has been collected in Franklin County (Stevenson and Baker, 1970).

744. Lark Bunting, *Calamospiza melanocorys* Stejneger

IDENTIFICATION: Total length 155 to 190 mm. Adult male: *mostly slaty black, but with wing coverts largely white* and edges of inner remiges margined with white; in fall, similar to female. Adult female: brownish gray above, streaked with dusky; *wing patch present*, but less conspicuous; underparts white, streaked on breast and sides with dusky. Immature: similar to female, but more buffy.

DISTRIBUTION: Breeds from southern Alberta and Saskatchewan, southwestern Manitoba, southeastern North Dakota, and southwestern Minnesota southward to south-central Montana, southeastern New Mexico, northern Texas, western Oklahoma, and south-central and eastern Kansas; also irregularly west to Utah. Winters from southern California and Nevada, central Arizona, southern New Mexico, and northeastern Texas southward to southern Baja California, Jalisco, Hidalgo, and northern Tamaulipas; casually eastward in fall to the Atlantic Coast, including Florida (Stevenson, 1964b; Gaither and Gaither, 1968; Robertson and Ogden, 1969).

745. Savannah Sparrow, *Passerculus sandwichensis* (Gmelin)

IDENTIFICATION: Smaller than a House Sparrow (total length 125 to 155 mm). Upperparts brownish, heavily streaked with dusky or blackish; superciliary stripe white, usually shading to yellow in front of eye; underparts white, streaked with dark brown or blackish; tarsus and toes pinkish in life. (Compare with Song Sparrow.)

DISTRIBUTION AND VARIATION: Breeds from northern part of Alaska and Canada southward at higher elevations to Guatemala, Missouri, Illinois, Indiana, Ohio, West Virginia, western Maryland, and the southeastern parts of Pennsylvania and New York. Winters from southern British Columbia, southern Nevada and Utah, central New Mexico, Oklahoma, Tennessee, and Massachusetts southward to El Salvador, Swan Islands, Grand Cayman, Cuba, and the Bahamas (sight record in Ecuador). The following races have been collected in Florida: *P. s.*

labradorius, *savanna*, *oblitus*, and *nevadensis* (Sprunt, 1963; Stevenson and Baker, 1970); late September to mid-May.

746. Grasshopper Sparrow, *Ammodramus savannarum* (Gmelin)

IDENTIFICATION: A small, short-tailed sparrow (total length 110 to 125 mm). Top of head mostly blackish, with a median light stripe; remainder of upperparts buffy grayish, *streaked with reddish brown on nape and upper back*, and heavily marked with blackish otherwise; *rectrices pointed*; *underparts whitish, tinged with buffy on throat and breast, but unstreaked* (except in juvenals); superciliary line light gray to ochraceous; yellow on bend of wing. *Bill very thick at base*; feet flesh-colored.

DISTRIBUTION AND VARIATION: Breeds from northern California, eastern Washington, southeastern British Columbia, the southern parts of Alberta, Saskatchewan, and Manitoba, northern Minnesota, southern Ontario, southwestern Quebec, northern Vermont, central New Hampshire, and Maine southward to southern California, central Nevada, northern Utah, central Colorado and Texas, northern Mississippi, the central parts of Alabama, Georgia, and South Carolina, and eastern North Carolina; also (*A. s. floridanus*) in south-central Florida, where apparently disappearing. The species winters from central California, southern Arizona, Oklahoma, Arkansas, Tennessee, and North Carolina southward to El Salvador, the Isle of Pines, and the Bahamas; *A. s. pratensis* occurs throughout Florida at that season (late September to mid-May); *A. s. perpallidus* has been collected twice (Stevenson and Baker, 1970).

747. Henslow's Sparrow, *Ammodramus henslowii* (Audubon)

IDENTIFICATION: Size of Grasshopper Sparrow. Sides of crown mostly blackish; median crown stripe, superciliary stripe, and *nape buffy olive*, the nape *speckled with black*; back and *much of wings reddish brown*, mixed with black; bend of wing pale yellow; *rectrices reddish brown, pointed*; chin and throat buffy; breast and sides buffy, streaked with black; abdomen white; tarsus and toes straw-colored.

DISTRIBUTION AND VARIATION: Breeds from eastern South Dakota, central Minnesota, Wisconsin, central Michigan, and from southern Ontario, Vermont, and New Hampshire southward to eastern Kansas (casually southeastern Texas), central Missouri, southern Illinois, northern Kentucky, West Virginia, and North Carolina. Winters in lower Coastal Plain from South Carolina to southeastern Texas, including all but southern tip of Florida; *P. h. henslowii* and *P. h. susurrans*, October to April.

748. Le Conte's Sparrow, *Ammospiza leconteii* (Audubon)

IDENTIFICATION: About size of Grasshopper Sparrow. Upperparts as in that species, but median crown stripe and superciliary stripe (especially the latter) more buffy; throat and breast also buffy; sides of breast and abdomen streaked with blackish; *nape reddish brown*; *rectrices pointed*; feet straw-colored. *Bill decidedly more slender* than in Grasshopper or Henslow's Sparrow.

DISTRIBUTION: Breeds from southern Mackenzie, northeastern Alberta, central Saskatchewan and Manitoba, and northern Ontario southward to north-central Montana, southeastern Alberta, southern Saskatchewan, northern North Dakota, northwestern and eastern Minnesota, northeastern Wisconsin, and northern Michigan (casually to South Dakota, Illinois, and southern Ontario). Winters from west-central Kansas, southern Missouri and Illinois, northern Alabama, and South Carolina southward to southern Texas, the Gulf Coast, and southern Florida (October to early May).

749. Sharp-tailed Sparrow, *Ammospiza caudacuta* (Gmelin)

IDENTIFICATION: Smaller than a House Sparrow (total length 125 to 140 mm). Median crown stripe dark slate, bordered laterally by stripes of blackish mixed with dark brown; remainder of upperparts varying shades of brown or grayish, streaked with white or buffy on back; *sides of head grayish, surrounded by buffy or tawny*; chin whitish or light buff; *throat, upper breast, and sides buffy or tawny, streaked with dark brown*; under tail coverts buffy or tawny; abdomen white; *rectrices pointed*.

DISTRIBUTION AND VARIATION: Breeds from northeastern British Columbia, southern Mackenzie, and central Saskatchewan and Manitoba southward to southern Alberta and North Dakota; also at James Bay and along Atlantic Coast from the lower St. Lawrence River to North Carolina. Winters along Atlantic and Gulf Coasts from New York to southern Texas except for the lower east coast and Keys in Florida. The following races have been collected in Florida: *A. c. subvirgata*, *caudacuta*, *nelsoni*, and *altera* (Sprunt, 1963); late September to mid-May.

750. Seaside Sparrow, *Ammospiza maritima* (Wilson)

IDENTIFICATION: About size of House Sparrow (total length 125 to 155 mm). Adult: *upperparts dark brown* (olive brown on nape) to blackish, streaked on crown (upperparts grayer in some races); chin whitish; remainder of underparts mostly light gray (sometimes buffy on breast), often streaked obscurely on throat, breast, and sides with blackish or dusky; *lores and bend of wing yellow*; *rectrices pointed*. Juvenal: upperparts similar to those of adult; superciliary stripe buffy; *underparts whitish to buffy with bold streaking*.

DISTRIBUTION AND VARIATION: Breeds in salt marshes from Massachusetts to northeastern Florida and from Pasco County, Florida, to southern Texas; reaches southern tip of Florida mainland in winter (Map 58). *A. m. pelonota* breeds along the northeast coast southward to New Smyrna (at least formerly), *A. m. peninsulae* on the Gulf Coast from Pasco County (formerly northern Pinellas County) northward to Dixie County, and *A. m. juncicola* from Taylor County westward (rare west of Wakulla County), being replaced by *A. m. fisheri* in the Pensacola area. *A. m. maritima* winters on the east coast south at least to Fort Pierce and has been collected on the west coast (Cape Sable; Robertson and Ogden, 1968).

Map 58. Distribution of Seaside Sparrow (1), Dusky
Seaside Sparrow (2), Cape Sable Sparrow (3)

751. Dusky Seaside Sparrow, *Ammospiza nigrescens* (Ridgway)[1]

IDENTIFICATION: About size of *Ammospiza maritima*. Adult: *upperparts almost entirely black*, the feather edges lighter and the

1. This population has recently been included in *Ammospiza maritima* (A.O.U., 1973). Thus the description and distribution of that species should be modified accordingly.

primaries more brownish; lores and bend of wing yellow; *underparts* whitish, *heavily streaked* (except in center of belly) *with black*; rectrices pointed. Juvenal: more *brownish* black above, and more narrowly streaked with *dusky* below.

DISTRIBUTION: Brevard County, Florida; chiefly in the St. Johns River marshes, but still present on Merritt Island (Sharp, 1969; Map 58).

752. Cape Sable Sparrow, *Ammospiza mirabilis* (Howell)[1]

IDENTIFICATION: Slightly smaller than other seaside sparrows (total length about 130 to 140 mm). Adult: similar to *Ammospiza maritima*, but *upperparts more greenish* and ground color of underparts more whitish. Juvenal: upperparts not greenish and lores not yellow; underparts suffused with pale yellow, finely streaked on breast with brown (Stimson, *in* Austin, 1968).

DISTRIBUTION: Resident in marshes just inland from mangrove fringe from Ochopee, Florida, southeastward to Taylor Slough (eastern edge of Everglades National Park; Stimson, 1956; Ogden and Stimson, personal communication; Map 58).

753. Vesper Sparrow, *Pooecetes gramineus* (Gmelin)

IDENTIFICATION: About size of House Sparrow (total length 140 to 165 mm). Upperparts dark brown mixed with dusky, with buffy to light gray feather edgings; *bend of wing reddish brown*; *lateral rectrices mostly white*; underparts whitish, streaked on throat, breast, and sides with dark brown; *eye-ring white*.

DISTRIBUTION AND VARIATION: Breeds from central British Columbia, southwestern Mackenzie, central Saskatchewan, southern Manitoba, central and northeastern Ontario, southern Quebec, and Nova Scotia southward to western Oregon, east-central California, central Nevada, southwestern Utah, the central parts of Arizona and New Mexico, western Nebraska, central Missouri, northeastern Tennessee, western North Carolina, and southeastern Maryland. Winters from central California, southern Nevada, central and southeastern Arizona, west-central and north-central New Mexico, eastern Colorado, Arkansas, southern Illinois, Kentucky, West Virginia, southern Pennsylvania, and Connecticut southward to southern Baja California, southeastern Mexico, the northern Gulf Coast, and southern Florida. *P. g. gramineus* is common in Florida, and *P. g. confinis* occurs rarely (Sprunt, 1963; October-April).

---

1. This population has recently been included in *Ammospiza maritima* (A.O.U., 1973). Thus the description and distribution of that species should be modified accordingly.

754. Lark Sparrow, *Chondestes grammacus* (Say)

IDENTIFICATION: About size of House Sparrow (total length 155 to 165 mm). *Crown dark chestnut, divided along midline by a stripe of grayish buff*; remainder of upperparts mostly brownish gray, streaked with dusky on back; wings dusky, with prominent white edges on coverts; *rectrices dusky, all but central ones tipped with white*; *auriculars chestnut*; a narrow black streak from bill to eye, bordered above (superciliary stripe) and below by white; chin and throat white, bordered laterally by a black stripe; underparts essentially white, with a *black spot in center of breast*.

DISTRIBUTION AND VARIATION: Breeds from western Oregon, southern British Columbia, Alberta, Saskatchewan, Manitoba, northwestern and central Minnesota, north-central Wisconsin, southern Michigan and Ontario, western New York, and central Pennsylvania southward to southern California, north-central Mexico, southern Texas, Louisiana, west-central and northwestern Alabama, Tennessee, and central North Carolina (rarely). Winters from central California, southern Arizona, central Texas, and Florida (especially peninsular) southward to El Salvador. Most Florida specimens have been referred to *C. g. grammacus*, but *C. g. strigatus* has been collected once; late July to early May.

755. Bachman's Sparrow, *Aimophila aestivalis* (Lichtenstein)

IDENTIFICATION: About size of House Sparrow (total length 140 to 155 mm). Adult: *top of head, nape, and back streaked with rusty and gray*; remiges brown, sometimes with rusty or buffy edgings; tail dusky; bend of wing yellow; *chin, throat, and breast buffy*; *abdomen whitish to buffy*. Juvenal: similar, but underparts finely streaked. Feet light brown.

DISTRIBUTION AND VARIATION: Breeds from southern Missouri, northern Illinois, central Indiana and Ohio, southwestern Pennsylvania, and central Maryland southward to southeastern Texas, the Gulf Coast, and southern Florida (northern Collier and Martin Counties; formerly northern Palm Beach County). Winters as far north as northeastern Texas, central Mississippi, north-central Alabama, northern Georgia, and North Carolina. The breeding form in the Panhandle is *A. a. bachmani*, which reaches southern Florida in winter (UM 1363 and 4115); in the Peninsula it is replaced as a breeder by *A. a. aestivalis*; *A. a. illinoensis* has been collected in spring (NMNS 478375).

756. Dark-eyed Junco, *Junco hyemalis* (L.)[1]

IDENTIFICATION: Slightly smaller than House Sparrow, but tail relatively longer (total length 140 to 155 mm). *Upperparts, entire head,*

---

1. With the recent inclusion of western populations under the name of Dark-eyed Junco, the plumage may be brownish above and the breeding range in the West extends south into Mexico (A.O.U., 1973).

throat, upper breast, and sides slate gray, sometimes with a brownish cast on back; secondaries edged with grayish buff; *lateral rectrices white*, the central ones blackish; remainder of underparts white; female somewhat browner above and paler on throat and breast than male.

DISTRIBUTION AND VARIATION: Breeds from the northern parts of Alaska and Canada southward to northern British Columbia, central Alberta and Saskatchewan, southern Manitoba, central Minnesota, southeastern Wisconsin, central Michigan, southern Ontario, northeastern Ohio, northern and western Pennsylvania, and Connecticut; also southward in the Appalachians to eastern Tennessee, northern Georgia, and northwestern South Carolina. Winters from southeastern Alaska, southern British Columbia, northwestern Montana, southern Saskatchewan and Manitoba, northern Minnesota, western Ontario, northern Michigan, central Ontario, southern Quebec, and Newfoundland southward to northern Baja California, northwestern Mexico, southern Texas, the Gulf Coast, and northern (rarely to southern) Florida. Both *J. h. hyemalis* and *J. h. cismontanus* have been found in Florida (Stevenson and Baker, 1970); late October to mid-April.

757. Chipping Sparrow, *Spizella passerina* (Bechstein)

IDENTIFICATION: Smaller than the House Sparrow (total length 125 to 140 mm). Adult: crown chestnut, mixed with blackish (streaked with dusky in winter); feathers of nape and upper back dark brown with light rufous edges; *rump grayish*; wings and tail dusky, the former with 2 buffy bars across coverts; underparts whitish. Juvenal: similar to adult, but less reddish brown above, crown streaked with black and buffy gray, and lightly spotted or streaked with dusky on breast and sides.

DISTRIBUTION AND VARIATION: Breeds from east-central Alaska, central Yukon, southern Mackenzie, the northern parts of Saskatchewan, Manitoba, and Ontario, southern Quebec, and southwestern Newfoundland southward to northern Baja California, northern Nicaragua, south-central Oklahoma, eastern Texas, the southern parts of Louisiana, Mississippi, and Alabama, northwestern Florida, and the southeastern parts of Georgia and South Carolina. Winters from central California, southern Nevada, central Arizona and New Mexico, Oklahoma, Arkansas, Tennessee, Virginia, and Maryland southward throughout the breeding range and to the northern Gulf Coast and southern Florida. Only the nominate form is known to Florida; October to early May (summer).

758. Clay-colored Sparrow, *Spizella pallida* (Swainson)

IDENTIFICATION: Similar to Chipping Sparrow except in the following respects: upperparts more buffy; *a clear, buff brown patch on side of head* (indistinct in Chipping Sparrow); *a light median streak through the otherwise brown-to-black crown* (except in immature); a dark streak on each side of throat; some buffy on breast.

DISTRIBUTION: Breeds from northeastern British Columbia, south-central Mackenzie, central Saskatchewan and Manitoba, western Ontario, and northern Michigan southward to southwestern Alberta, south-central Montana, southeastern Wyoming and Colorado, southern Nebraska, northern Iowa, southern Wisconsin, central Michigan, and southern Ontario; possibly farther southward and eastward. Winters from northern Mexico and southern Texas to southeastern Mexico, and rarely eastward to Florida (late September to early May).

759. Field Sparrow, *Spizella pusilla* (Wilson)

IDENTIFICATION: Smaller than House Sparrow (total length 125 to 150 mm). Adult: upperparts more or less reddish brown, the back streaked with dark brown, the crown with a poorly defined median stripe of grayish; 2 rather indistinct whitish bars on wing coverts; sides of head gray, interrupted by a reddish brown stripe running through eye to side of neck, there turning abruptly downward; underparts light gray, generally tinged with buff, but with rufous on sides of breast. Juvenal: similar to adults, but upperparts paler, sides of head more whitish, and underparts lightly streaked. *Bill pinkish orange*; feet flesh-colored.

DISTRIBUTION AND VARIATION: Breeds from northwestern Montana, northern North Dakota, central Minnesota, northern Wisconsin, north-central Michigan, southern Ontario, southwestern Quebec, and southern Maine southward to central Texas, northern and southeastern Louisiana, southern Mississippi, southern Alabama, northwestern Florida, and southern Georgia. Winters from Kansas, Missouri, Ohio, West Virginia, southern Pennsylvania, and Massachusetts southward to northern Mexico, southern Texas, the Gulf Coast, and southern Florida. Only *S. p. pusilla* is known to Florida; October to April (summer).

760. Harris' Sparrow, *Zonotrichia querula* (Nuttall)

IDENTIFICATION: The largest Florida sparrow (total length 175 to 190 mm). Adult: *a black patch extending from crown through forehead to chin and throat*, abruptly contrasting with whitish sides of head and the *pinkish bill*; remainder of upperparts grayish brown, the upper back streaked with blackish; 2 whitish wing bars present; remainder of underparts whitish, streaked along sides with dusky. Immature: differs from adult in having sides of head and neck buffy, *back of crown mixed with buff*, and black of underparts much reduced. Feet flesh-colored.

DISTRIBUTION: Breeds from northwestern and east-central Mackenzie and southern Keewatin southward to northeastern Saskatchewan and northern Manitoba. Winters from southern British Columbia and Idaho, northern Utah, Colorado, and Nebraska, and central Iowa southward to southern California and Nevada, central Arizona, south-central Texas, northern Louisiana, and central Tennessee; casually eastward to the Atlantic Coast, including northern and central Florida (photograph, Seeler, 1958; Young, 1964); November to late March.

761. White-crowned Sparrow, *Zonotrichia leucophrys* (Forster)

IDENTIFICATION: About size of White-throated Sparrow (total length 150 to 190 mm). Adult: *top of head marked with 4 longitudinal black stripes* (on sides of crown and through eyes), connected at the lores, and *separated by 3 white stripes* (a *broad* one in center of crown and a narrow one above each eye); remainder of upperparts grayish brown, with darker brown streaks on upper back, shading to dusky on wings and tail; 2 white bars on wing coverts; underparts largely white, but light gray on throat and breast (also sides of head and neck), and buffy on flanks and under tail coverts. Immature: similar to adult, but *top of head reddish brown, with a grayish brown crown stripe*, a light brownish patch on side of head, and a grayish superciliary stripe. In all plumages the *bill is pinkish* in life, and the feet are flesh-colored to light brown.

DISTRIBUTION AND VARIATION: Breeds from northern Alaska and Canada southward to south-central California, Nevada, central Arizona, northern New Mexico, central Manitoba, southeastern Quebec, and northern Newfoundland. Winters from southern British Columbia, southeastern Washington, southern Idaho, central Wyoming, northeastern Kansas, central Missouri and Kentucky, West Virginia, and western North Carolina southward to central Mexico, the northern Gulf Coast, and northern Florida (rarely to southern Florida and the northern Antilles). The nominate form is the most frequent in Florida, but *Z. l. gambelii* has been collected also (Johnston, 1969); October to May.

762. White-throated Sparrow, *Zonotrichia albicollis* (Gmelin)

IDENTIFICATION: About size of House Sparrow, but tail longer (total length 150 to 185 mm). Adult: head with 4 black and 3 white longitudinal stripes, but the *white crown stripe narrower than the black stripes*; *lores yellow*; remainder of upperparts grayish to reddish brown, with darker brown streaks in middle of back, shading to dusky on remiges and rectrices; 2 whitish bars on wing coverts; abdomen white, breast smoky gray, *sharply contrasting with white throat patch*; flanks buffy; yellow on bend of wing; *bill brownish* in life. Immature: similar to adult, but the throat patch less distinct, crown stripe buffy, bill lighter (but not pinkish), and with faint streaking or mottling on breast and sides. Feet light brown. Some retain this plumage as adults (*Evolution*, 29:611–621).

DISTRIBUTION: Breeds from southern Yukon, central Mackenzie, northern Manitoba and Ontario, west-central and southeastern Quebec, southern Labrador, and northern Newfoundland southward to central British Columbia and Alberta, southern Saskatchewan, north-central North Dakota, central Minnesota, northern Wisconsin, central Michigan, northern Ohio and West Virginia, northeastern Pennsylvania, southeastern New York, northwestern Connecticut, and Massachusetts. Winters from western California, southern Arizona and New Mexico, eastern Kansas, southern Illinois, northern Kentucky, central New York, and Massachusetts southward to southern Texas, the Gulf Coast, and southern Florida; also westward through New Mexico and Arizona to California; casually north of stated limits. Florida: mid-October to mid-May.

763. Fox Sparrow, *Passerella iliaca* (Merram)

IDENTIFICATION: Larger than any common Florida sparrow (total length 165 to 190 mm). Top of head, nape, and back reddish brown, mottled with grayish; *rump and tail brighter brown* (chestnut); *underparts white, heavily spotted with reddish brown* on throat and breast, and with dark brown along sides; *a reddish brown patch on side of head, contrasting with gray above and behind*; base of lower mandible yellowish.

DISTRIBUTION AND VARIATION: Breeds from northern Alaska, northwestern and east-central Mackenzie, and the northern parts of Manitoba, Ontario, Quebec, and Labrador southward to the coast of Washington, the mountains of central California, Nevada, Utah, and Colorado, and (at lower elevations) to central Alberta and Saskatchewan, southern Manitoba, central Ontario, southern Quebec, and Newfoundland. Winters from southwestern British Columbia, northeastern California, southern Utah, central Colorado, eastern Kansas, the southern parts of Iowa, Wisconsin, Michigan, and Ontario, and to New Brunswick southward to northern Baja California, southern Arizona, western and southern Texas, the Gulf Coast (rarely), and central Florida (casually); only the nominate form is likely to reach Florida (mid-November to mid-March).

764. Lincoln's Sparrow, *Melospiza lincolnii* (Audubon)

IDENTIFICATION: Slightly smaller than House Sparrow, but tail longer (total length 130 to 150 mm). *Crown reddish, streaked with black and divided by a gray median stripe*; superciliary stripe, nape, and sides of neck smoky gray; remainder of upperparts reddish brown, with large patches or stripes of blackish; throat white; *breastband buffy, overlaid by longitudinal rows of dark spots that continue onto sides of throat and body*.

DISTRIBUTION AND VARIATION: Breeds from northwestern Alaska and the middle latitudes of Canada southward to southern California, central Arizona, northern New Mexico, southern Manitoba, the northern parts of Minnesota, Wisconsin, and Michigan, southern Ontario, northern New York, central Maine, and Nova Scotia. Winters from northern California, southern Nevada, northern Arizona, New Mexico, and Oklahoma, eastern Kansas, central Missouri, south-central Kentucky, and northern Georgia (but generally rare in winter east of Mississippi River) southward to El Salvador, the northern Gulf Coast, and casually to Florida (more frequent in southern parts) and the northern Antilles; only *M. l. lincolnii* is to be expected in Florida (October-April).

765. Swamp Sparrow, *Melospiza georgiana* (Latham)

IDENTIFICATION: Slightly smaller than House Sparrow (total length 125 to 160 mm). Adult in breeding plumage: forehead black; *crown chestnut*; superciliary stripe, nape, and *sides of neck smoky gray*; *remainder of upperparts reddish brown*, with large patches or stripes of dark brown; *throat white*; breast gray; remainder of underparts whitish to light buffy. Adult in winter, immature: similar, but crown streaked with black and gray, throat light gray, and underparts faintly streaked with dusky.

DISTRIBUTION AND VARIATION: Breeds from southwestern and south-central Mackenzie, the northern parts of Saskatchewan, Manitoba, and Ontario, central Quebec, and Newfoundland southward to northeastern British Columbia, central Alberta, southern Saskatchewan and Manitoba, eastern Nebraska, northern Missouri, northern Illinois and Indiana, south-central Ohio and West Virginia, Maryland, and Delaware. Winters from eastern Nebraska, central Iowa, southern Wisconsin, Michigan, and Ontario, central New York, and Massachusetts southward to southwestern and northeastern Mexico, the northern Gulf Coast, and southern Florida. Both *M. g. georgiana* and *M. g. ericrypta* have been collected in the state (Sprunt, 1963); early October to early May.

766. Song Sparrow, *Melospiza melodia* (Wilson)

IDENTIFICATION: Near size of House Sparrow, but tail longer (total length 135 to 165 mm). Top of head reddish brown, streaked with blackish, and medially divided by a gray stripe (sometimes indistinct); remainder of upperparts grayish brown to reddish brown, heavily marked with dark brown streaks on back; superciliary stripe light gray; 2 brownish stripes below each eye; underparts mostly whitish, streaked with blackish on breast and on sides of throat and body, the streaks often converging to form a spot in center of breast. The *long, reddish brown tail*, lack of yellow lores, and darker bill and feet help to distinguish it from the Savannah Sparrow.

DISTRIBUTION AND VARIATION: Breeds from southwestern Alaska, southern Yukon and Mackenzie, the northern parts of Saskatchewan, Manitoba, and Ontario, central Quebec, and Newfoundland southward to south-central Baja California, south-central Mexico, northern New Mexico, northeastern Kansas, northern Arkansas, southwestern Kentucky, central Tennessee, northeastern Alabama, northern (casually central) Georgia, northwestern South Carolina, western North Carolina, central Virginia, and along Atlantic Coast southward to North Carolina. Winters from southern Alaska, southern British Columbia, southeastern Montana, South Dakota, the southern parts of Minnesota, Wisconsin, Michigan, Ontario, and Quebec, central New Brunswick, and Novia Scotia southward to southern limit of western breeding range, southern Texas, the northern Gulf Coast, and the mainland of southern Florida (casually to Keys). Subspecies collected in Florida are *M. m. melodia*, *juddi*, and *euphonia* (Stevenson and Baker, 1970); October to late April.

767. Lapland Longspur, *Calcarius lapponicus* (L.)

IDENTIFICATION: Slightly larger than a House Sparrow (total length 150 to 180 mm). Winter plumage: crown blackish, with light median stripe and bordered below by whitish superciliary stripe; remainder of upperparts light grayish brown, streaked with blackish; *nape more or less rufous*; *a blackish patch behind and below eye and another on breast*, from which a dark streak runs up each side of throat; remainder of underparts white, streaked on sides with dusky; bill yellowish brown with a dark tip. *Hind claw very long.*

DISTRIBUTION AND VARIATION: Breeds from northern Alaska, northern Canada, central Greenland, and northern Eurasia southward to southwestern Alaska, northwestern and middle latitudes of Canada, southern Greenland, southern Scandinavia, and southeastern Siberia. Winters from southern British Columbia, Montana, South Dakota, central Minnesota and Michigan, southern Ontario and Quebec, and New Brunswick southward to northeastern California, northern Arizona and New Mexico, northeastern Texas, southern Louisiana, West Virginia, and northern Virginia, casually southeastward to Florida (presumably *C. l. lapponicus*; Monroe, 1958b; Stevenson, 1963b), late October to January; also southward in the Old World to southern Europe and central Asia.

768. Chestnut-collared Longspur, *Calcarius ornatus* (Townsend)

IDENTIFICATION: Size of a small sparrow (total length 135 to 155 mm). Adult male: *nape rufous*; *underparts mostly black* (much reduced in winter); chin and throat often buffy; crown and stripe from eye and ear onto throat black, the former streaked in winter; superciliary stripe white; back brownish gray and blackish; outermost rectrices white, except for distal ends; lesser wing coverts black. Adult female: upperparts grayish brown, streaked with blackish; underparts buffy white, the breast faintly streaked with dusky; outermost rectrices similar to those of male; amount of white on rectrices in both sexes decreases toward middle pair; bill fleshy brown with dark tip. *Hind claw very long.* Immature: similar to adult female.

DISTRIBUTION: Breeds from southern Alberta, southern Saskatchewan, and southern Manitoba southward to northeastern Colorado, north-central Nebraska, and southwestern Minnesota (formerly western Kansas). Winters from northern Arizona, central New Mexico, northeastern Colorado, and central Kansas (rarely east to Ontario) southward to northern Mexico, southern Texas, and northern Louisiana, casually to Florida (2 records at Tallahassee; Stevenson, 1964a; Robertson, 1967).

769. Snow Bunting, *Plectrophenax nivalis* (L.)

IDENTIFICATION: Larger than the House Sparrow (total length 160 to 190 mm). Winter: *head, throat, and breast white, overlaid partly with rusty*; upperparts otherwise a mixture of black, rust, and whitish; *large white patches in secondaries*; *outer rectrices largely white*; remainder of remiges and rectrices mostly black; remainder of underparts white; bill flesh-colored with dusky tip. *Hind claw very long.*

DISTRIBUTION AND VARIATION: Breeds in Arctic Zone of North America, Europe, and Asia. Winters southward to Oregon, Utah, Colorado, Kansas, Indiana, Georgia (coast), northern France, Russia, northern China, and northern Japan; casual in Bermuda and northern Africa; accidental in the Bahama Islands and in Florida (2 records, probably *P. n. nivalis*; Stevenson, 1970a).

769H. HYPOTHETICAL LIST

PROBABLY EXTIRPATED FROM FLORIDA BEFORE 1900:
Passenger Pigeon (*Ectopistes migratorius*). The record of "several thousand" in Florida in 1907, only a few years before the species became extinct (Howell, 1932), is very unlikely. No other Florida records fell within the present century.

FLORIDA OCCURRENCE NOT WELL SUBSTANTIATED:
Least Grebe (*Podiceps dominicus*). Several sight records of one bird in Polk County. Said to have been photographed, but evidence not seen. (See Agey, 1967.) Also a sight record at Miami (Robertson, 1971).

Wandering Albatross (*Diomedea exulans*). One old sight record off Jacksonville Beach (Howell, 1932).

Yellow-nosed Albatross (*Diomedea chlororhynchos*). Sight record off Cocoa Beach (Stevenson, 1958).

White-bellied Storm-Petrel (*Fregetta grallaria*). None of the specimens of the Black-bellied Storm-Petrel (*Fregetta tropica*) mentioned by Howell (1932) have been located. According to W.R.P. Bourne (E.M. Reilly, Jr., personal communication), the original measurements of the specimens indicated *Fregetta grallaria* rather than *F. tropica*. In the absence of a specimen, neither species seems entitled to a place on the Florida list.

Red-billed Tropicbird (*Phaethon aethereus*). One sight record off Cocoa Beach (Stevenson, 1964c).[1]

Jaçana (*Jaçana spinosa*). The specimen or specimens referred to by Howell (1932) have never been located. There was a specimen in the Bailey-Law collection (no. 4992) at the Virginia Polytechnic Institute, Blacksburg, Virginia, said to have been taken near Belle Glade, Florida, by H.H. Bailey on September 10, 1932. It seems strange, however, that this record was not published and that Bailey's associates who were questioned about the record in 1968 had no knowledge of it. These facts suggest the possibility of error regarding the provenience of the specimen. There has also been a recent sight record by Ludlow Griscom (Stevenson, 1957).

Eurasian Lapwing (*Vanellus vanellus*). A recent sight record near Fort Lauderdale was not well qualified (Swem, 1969).

Eskimo Curlew (*Numenius borealis*). A recent sight record of this nearly extinct species was made at Port Canaveral. See Sprunt, 1963.

Curlew Sandpiper (*Calidris ferruginea*). Two recent sight records were made in the central part of the Peninsula (Edscorn, 1968; Mason, 1968a and 1968b). Also see Sprunt, 1954.

Skua (*Catharacta skua*). A few records near the Keys (Howard, 1961; *American Birds*, 28:630).

Arctic Tern (*Sterna paradisaea*). A sight record was reported by Hebard (1952), but the species is notoriously difficult to identify.[2]

Common Murre (*Uria aalge*). One was found dead on a beach near Fort Pierce, December 28, 1971, but the specimen was not saved (D.C. Scott, personal communication).

Blue-headed Quail-Dove (*Starnoenas cyanocephala*). See Howell (1932) for early records. A specimen in the Bailey-Law collection (no. 3943) was obtained in Miami on October 12, 1928. The record was not published, and the possibility exists that the specimen was a captive obtained from a local zoo.

Snowy Owl (*Nyctea scandiaca*). One sight record by Tucker (1950).

---

1. A specimen is now at the Florida State Museum (Gainesville).
2. A specimen is now at the University of Maimi museum.

Lesser Nighthawk (*Chordeiles acutipennis*). Seen by several observers on the Dry Tortugas (Stimson, 1966).

Broad-billed Hummingbird (*Cynanthus latirostris*). One sight record at Pensacola. See Sprunt, 1954.

Black-backed Three-toed Woodpecker (*Picoides arcticus*). One old specimen of doubtful provenience (Howell, 1932).

Tropical Kingbird (*Tyrannus melancholicus*). Several sight records in south Florida, but this species is difficult to distinguish in the field from the Western Kingbird. See Stimson, 1942, and Stevenson, 1957.

Loggerhead Kingbird (*Tyrannus caudifasciatus*). Seen at Islamorado by several experienced observers in the winter of 1971–72. The photographs, however, were referred by James Bond to the Giant Kingbird (*Tyrannus cubensis*). (Stevenson, 1972a).

Fork-tailed Flycatcher (*Muscivora tyrannus*). Two sight records (Sprunt, 1954; J. Edscorn, personal communication).

Great Kiskadee (*Pitangus sulphuratus*). Sight records and photographs of one bird near Fort Lauderdale, but there is no evidence that the possibility of a Lesser Kiskadee (*P. lictus*) was considered, and the photographs could hardly show the difference in the two species. See Inwood, 1961.

Black Phoebe (*Sayornis nigricans*). Two sight records, Key Largo and Cocoa Beach (Sprunt, 1954 and 1963).

Say's Phoebe (*Sayornis saya*). Two sight records: Sanibel Island and Seahorse Key (Robertson and Ogden, 1968; *American Birds*, 29:46).[1]

Violet-green Swallow (*Tachycineta thalassina*). Two sight records, but the species is difficult to distinguish from the Tree Swallow (Eifrig, 1946; Ogden, 1964).

Bahama Mockingbird (*Mimus gundlachii*). One carefully studied at Fort Jefferson (Dry Tortugas) on May 3, 1973 (P.A. Buckley, personal communication).

Long-billed Thrasher (*Toxostoma longirostre*). One published sight record (Cobb, 1946), but difficult to distinguish from the Brown Thrasher which occasionally has an abnormally long bill.

Thick-billed Vireo (*Vireo crassirostris*). Three sight records, but difficult to distinguish from the White-eyed Vireo (Stevenson, 1961d; Cunningham, 1964; Christmas Count, West Palm Beach, 1969).

Sutton's Warbler ("*Dendroica potomac*"). Three sight records, near Tampa and on Key West (Brewer, 1954; Sprunt, 1954; Stevenson, 1955). Even if these identifications were correct, it should be mentioned that this bird is probably not a full species, but a hybrid between the Yellow-throated Warbler and the Northern Parula.

Audubon's Warbler (*Dendroica auduboni*).[2] Two sight records, Miami and Sarasota areas (Smith, 1943; Stevenson, 1960b).

---

1. Photographed near Orlando in late 1975.
2. Now merged with the Myrtle Warbler under the common name of Yellow-rumped Warbler (A.O.U., 1973).

Townsend's Warbler (*Dendroica townsendi*). Sight records in Corkscrew Swamp, near Immokalee (Stevenson, 1969a); and at Panama City (Imhof, 1972).

Melodious Grassquit (*Tiaris canora*). Several sight records and photographs. There is no evidence, however, that the photographs were seen by competent ornithologists, and their present whereabouts is unknown. See Sprunt (1954), Abramson and Stevenson (1961), and Austin (1963).

White-winged Junco (*Junco aikeni*).[1] One sight record, but the species is difficult to identify in the field. (See Sprunt, 1963).

Tree Sparrow (*Spizella arborea*). One sight record, but the species may be confused with certain other sparrows under most field conditions (Farrar, 1950).

White-winged Crossbill (*Loxia leucoptera*). One sight record. See Stevenson, 1950.

ESCAPED OR RECENTLY INTRODUCED; NOT CONSIDERED ESTABLISHED IN FLORIDA:

Mute Swan (*Cygnus olor*). The common, semi-domestic swan, now escaped and established in New England. One apparently feral bird was seen on Merritt Island in 1967 (Robertson and Ogden, 1968).

Black-bellied Tree Duck (*Dendrocygna autumnalis*). The only known semi-feral population is in the vicinity of Crandon Park Zoo, near Miami. See Sprunt, 1963, and Stevenson, 1968b.

Ring-necked Pheasant (*Phasianus colchicus* ssp.). Attempts by the Florida Game and Freshwater Fish Commission to introduce the strain known as the Iranian Pheasant in eastern Jackson County have not been convincingly successful.

Black Francolin (*Francolinus francolinus*). The Florida Game and Freshwater Fish Commission has introduced this species on protected land northeast of Avon Park. It is too early to know whether the species can maintain itself under natural conditions and hunting pressure in Florida.

Gray-necked Wood Rail (*Aramides cajanea*). Small populations are said to exist under protected conditions near Miami and Vero Beach (Carter Bundy and Herbert Kale, personal communication).

Green Parakeet (*Aratinga holochlora*). The sight records and specimen mentioned by Howell (1932) probably represented escaped birds.

Orange-fronted Parakeet (*Aratinga canicularis*). Apparently becoming established in Miami with one record near Delray Beach in 1972 (Ogden, 1972).

Monk Parakeet (*Myiopsitta monachus*). Nesting records on Key Biscayne and Key Largo by 1972 (Ogden, 1972).

---

1. Now merged with *Junco hyemalis* under the name of Dark-eyed Junco (A.O.U., 1973).

Orange-chinned Parakeet (*Brotogeris jugularis*). Several records in Miami by 1972 (Ogden, 1972).

Ring-necked Parakeet (*Psittacula krameri*). Present in the Miami area since 1968 (Ogden, 1972; Owre, personal communication).

Black-billed Magpie (*Pica pica*). Accounts of 2 birds seen repeatedly and photographed at Hobe Sound in the winter of 1947–48 (Dawes, 1948; Sprunt, 1954) indicated that they were not escaped individuals, but there was no indication that a thorough check was made of the numerous zoos and aviaries along the lower east coast. Because of the difficulty of determining the provenience of such a bird, and the fact that magpies are often kept in captivity, the likelihood that these birds had escaped seems more probable than that they had wandered so far from their normal far-western range.

Hill Mynah (*Gracula religiosa*). Seen frequently in small numbers near the lower east coast since about 1970 and possibly becoming established there.

More recently Owre (1973) has pointed out that perhaps the Red-crowned Parrot (*Amazona viridigenalis*), Brown-throated Parakeet (*Aratinga pertinax*), Brazilian Cardinal (*Paroaria coronata*), and possibly a few other avian species should be added to the above list.

CLASS MAMMALIA: MAMMALS

ORDER MARSUPIALIA: CXLII. FAMILY DIDELPHIDAE, Opossums. These are the only North American mammals in which the female is provided with an external, abdominal pouch. In most other obvious respects, the various species of marsupials vary widely, but only one species lives a feral existence in Florida.

770. Opossum, *Didelphis marsupialis* L.

IDENTIFICATION: Florida's only marsupial bears a distant resemblance to a large rat, but is larger (adults 50 to 90 cm in total length), has a more pointed snout, and a tail almost completely devoid of hair. It also differs from all rodents in having the *first toe* (hallux) *of the hind foot clawless and opposable to the other toes*. The typical color of the pelage is mostly grayish with white hairs interspersed, but black individuals may be seen in some parts of Florida.

DISTRIBUTION AND VARIATION: Eastern Colorado, all but the northwestern part of Nebraska, southeastern Iowa, northern Minnesota, Wisconsin, Michigan, southeastern Ontario, Vermont (except northeastern), and southwestern New Hampshire southward through western Texas and most of Mexico (except higher elevations) into South

America, and to the northern Gulf Coast and southern Florida; also, west of the Sierra Nevada mountains, from Washington to extreme northwestern Baja California, and in southeastern Arizona. *D. m. pigra* occurs throughout Florida.

ORDER INSECTIVORA: CXLIII. FAMILY SORICIDAE, Shrews. Insectivores resemble mice, but have a *more tapered snout and vestigial eyes*. Shrews are generally smaller than moles, do not have the front feet enlarged, and have an *incomplete zygomatic arch*. Both have dense, slate gray pelage.

771. Southeastern Shrew, *Sorex longirostris* Bachman

IDENTIFICATION: This shrew differs from the other 2 Florida species in having a *proportionately longer tail* (more than 25 percent of total length) and a *reduced third unicuspid* (5 unicuspids are present). Pelage: dark *brown* above, paler underneath. Other Florida shrews are more gray than brown. Total length 75 to 100 mm.

DISTRIBUTION AND VARIATION: Illinois (except northwestern), southwestern Indiana, Kentucky (except northernmost part), southern West Virginia, and southwestern Maryland southward to northern Louisiana, southern Mississippi and Alabama, and northeastern and central Florida (to Polk County; *S. l. longirostris*). *S. l. eionis* has been described from Homosassa Springs (Davis, 1957).

772. Short-tailed Shrew, *Blarina brevicauda* (Say)

IDENTIFICATION: Resembles the Southeastern Shrew in having 5 upper unicuspids, but the *fifth is not readily visible*, and the third is about the same size as the fourth. The *tail is relatively shorter* (less than 25 percent of total length), but the body is larger than in that species. Compared with the Least Shrew, it is larger, shows more upper unicuspids, and is more of a *slate* gray color. Total length 90 to 130 mm.

DISTRIBUTION AND IDENTIFICATION: Southeastern Saskatchewan across southern Canada to Newfoundland and southward to southeastern Texas, the northern Gulf Coast, and southern Florida. Northern and central Florida are inhabited by *B. b. carolinensis*, the Fort Myers area by *B. b. shermani*, and the remainder of the Peninsula by *B. b. peninsulae* (limits of each subspecies uncertain).

773. Least Shrew, *Cryptotis parva* (Say)

IDENTIFICATION: Smaller than *Blarina* and with a proportionately shorter tail than that of *Sorex*, this shrew may be further distinguished from others in Florida by its unicuspid teeth. Only 3 are readily visible in the upper row, the *fourth being much smaller than the first 3*. Four or 5 unicuspids are readily discernible in the other shrews. The light *brownish* gray color of this species is a further distinction. Total length 75 to 90 mm.

DISTRIBUTION AND VARIATION: From about 40 degrees North latitude and coastal Connecticut southward into northeastern Mexico, to the northern Gulf Coast, and the southern tip of Florida; westward to western South Dakota and Nebraska, northeastern Colorado, western Kansas and Oklahoma, and northeastern, central, and the Big Bend region of Texas. Most of Florida is occupied by *C. p. floridana*, but *C. p. parva* occurs west of the Apalachicola River.

ORDER INSECTIVORA: CXLIV. FAMILY TALPIDAE, Moles. Generally larger than shrews, with *greatly widened front feet, unpigmented teeth*, and a complete zygomatic arch. Snout more tapering and tail shorter than in rodents. *Eyes and pinnae lacking.*

774. Eastern Mole, *Scalopus aquaticus* (L.)

IDENTIFICATION: A velvety, burrowing mammal with *enlarged front feet and no semblance of eyes or ear flaps*. Snout tapering and naked; tail short and sparsely haired. Total length about 150 to 180 mm (larger than Florida shrews). The teeth lack the dark pigmentation of the shrew's.

DISTRIBUTION AND VARIATION: Southeastern Colorado, southern South Dakota and Minnesota, southern and western Wisconsin, lower Michigan, extreme southwestern Ontario, Ohio (except eastern), southern and eastern West Virginia, southeastern Pennsylvania, southeastern New York and Massachusetts (except northern) southward to the mouth of the Rio Grande, the Gulf Coast, and the southern tip of the Florida mainland. The following races occur in the state: *S. a. howelli*, west of Apalachicola River; *S. a. australis*, remainder of mainland except as follows: *S. a. anastasae*, Anastasia Island; *S. a. parvus*, Pasco, Pinellas, and Hillsborough Counties, probably only near coast; *S. a. bassi*, vicinity of Englewood; and *S. a. porteri*, vicinity of Miami.

ORDER CHIROPTERA: CXLV. FAMILY PHYLLOSTOMATIDAE, Leaf-nosed Bats. Bats are easily distinguished from other mammals by their *wings*, supported by one series of metacarpals and one or more series of phalanges. Their skulls bear a distant resemblance to a monkey's, but with important differences in dentition. Leaf-nosed bats—not more than accidental in Florida—are readily distinguished by the *leaflike projection from the nose*, the lack of an obvious tail and of a palatal notch in the skull, and by having 3 bony phalanges in the third finger.

775. Jamaican Fruit-eating Bat, *Artibeus jamaicensis* Leach

IDENTIFICATION: Differs from all other Florida bats in the *presence of a nose leaf* and the *lack of a visible tail*. (Also see description of family.) Head-body length about 80 to 90 mm—larger than most Florida bats.

DISTRIBUTION AND VARIATION: Widely distributed through Central America, South America, and the West Indies; *A. j. parvipes* has been reported from Key West, Florida (Miller and Kellogg, 1955).

ORDER CHIROPTERA: CXLVI. FAMILY VESPERTILIONIDAE, Twilight Bats. These are the common bats of Florida. They differ from the Molossidae in the fact that the *tail does not extend beyond the interfemoral membrane*, and from the Phyllostomatidae (chiefly tropical) in the absence of a nose leaf. Like the Molossidae, they show *only 2 bony phalanges* distal to the third metacarpal.

776. Southeastern Myotis, *Myotis austroriparius* (Rhoads)

IDENTIFICATION: A small bat similar to the Gray Myotis in its evenly grayish brown upperparts and the sagittal crest on the skull. It differs, however, in that the *wing membrane attaches low on the side of the foot*. The dorsal hairs are not blackish basally as they are in *M. sodalis*. Total length 80–95 mm (averaging slightly larger than most others in its genus).

DISTRIBUTION: Locally distributed from the southern parts of Indiana and Illinois to southeastern Oklahoma, northeastern Texas, and the Gulf Coast from Louisiana to Florida (Map 59); there appears to be a hiatus from the Aucilla to the Suwannee Rivers (Rice, 1957).

777. Gray Myotis, *Myotis grisescens* Howell

IDENTIFICATION: A small grayish brown bat with the corner of the *wing membrane attached well above the toes*. The *dorsal hairs are of the same color throughout their length* (tawny buff or dark brown) and are comparatively short (about 6 mm), and the *forearm is longer* than in most members of the genus (about 41 to 46 mm). The skull shows a sagittal crest and is longer than in other small bats (greatest length at least 15.5 mm). Total length about 80 to 95 mm.

DISTRIBUTION: Northeastern Oklahoma, northern Arkansas, most of Missouri, southern Illinois and Indiana, and Kentucky (except northeastern) southward through Tennessee (except western and northeastern), Alabama (except western), and western Georgia to Marianna, Florida, where resident and breeding in caves (Rice, 1955a and 1955b; Jennings and Layne, 1957).

NUMBERS AND KINDS OF TEETH IN THE FLORIDA GENERA OF VESPERTILIONID BATS

Genus	Incisors		Canines		Premolars		Molars		Totals
	UPPER	LOWER	UPPER	LOWER	UPPER	LOWER	UPPER	LOWER	
Myotis	4	6	2	2	6	6	6	6	38
Pipistrellus	4	6	2	2	4	4	6	6	34
Eptesicus	4	6	2	2	2	4	6	6	32
Lasiurus	2	6	2	2	2 or 4	4	6	6	30 or 32
Nycticeius	2	6	2	2	2	4	6	6	30
Plecotus	4	6	2	2	4	6	6	6	36

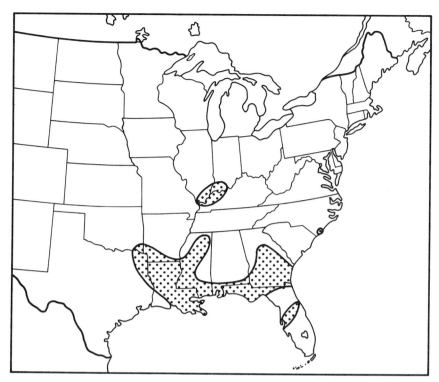

Map 59. Distribution of the Southeastern Myotis
(*Myotis austroriparius*)

778. Keen's Myotis, *Myotis keeni* (Merriam)

IDENTIFICATION: A small, dark brown bat very similar to *M. austroriparius*, but differing in having *slightly longer ears* (more than 15 mm). It differs from *M. grisescens* in having the wing membrane attached to the base of the first toe. The second metacarpal is shorter than the third, and the skull is narrow (least width of interorbital constriction less than 4 mm). It differs from other small bats in its dull *grayish* brown upperparts and *lack of a complete sagittal crest* on the skull. Total length about 80 to 90 mm.

DISTRIBUTION AND VARIATION: Central Saskatchewan and southern Manitoba eastward through southern Canada to Newfoundland and southward through eastern Montana, northeastern Wyoming, northern and eastern Nebraska, and eastern Kansas and Oklahoma to southern Arkansas, southeastern Missouri, western Kentucky and Tennessee, northeastern Mississippi, northern and eastern Alabama, northern and western Georgia, western North Carolina, and Virginia; casually (?) to Marianna, Florida (*M. k. septentrionalis*, October 30, 1954; Rice,

1955b). A disjunct population (*M. k. keeni*) lives near the Pacific Coast from southeastern Alaska to Washington.

779. Indiana Myotis, *Myotis sodalis* Miller and Allen

IDENTIFICATION: A small reddish brown bat whose *dorsal hairs are blackish basally and long* (about 10 mm). Otherwise it may be confused with *M. austroriparius* in coloration, although less reddish than that species. It differs, however, in having a keel on the calcar, shorter hairs on the toes, and a shorter foot. The brownish tips of the dorsal hairs may contrast with a grayish middle portion as well as the blackish base, thus giving a tricolored effect, but not so noticeably as in *Pipistrellus*. The skull has a sagittal crest. Total length about 70 to 90 mm.

DISTRIBUTION: Southern parts of Wisconsin and Michigan, most of Ohio and Pennsylvania, northern New York, and western New England southwestward to eastern Iowa, central Missouri, eastern Oklahoma, central Arkansas, western Kentucky and Tennessee, Alabama (except southwestern), northwestern Georgia, western North Carolina, western and northern Virginia, western Maryland, and northern New Jersey, and (casually?) Marianna, Florida (October 14, 1955; Jennings and Layne, 1957).

780. Eastern Pipistrelle, *Pipistrellus subflavus* (Cuvier)

IDENTIFICATION: *A small bat with tricolored hairs*. General coloration usually yellowish brown above and below (with little contrast), but body hairs dark at base, light at middle, and medium dark at tip. Near the joints the *wing membrane is reddish in life, appearing translucent in the preserved skin*. It also differs from other small bats in having *34 teeth*. Total length about 70 to 90 mm.

DISTRIBUTION AND VARIATION: Southeastern Minnesota, northern Wisconsin, Upper Michigan, southern Ontario, Quebec, and Nova Scotia southward (east of the Great Plains) to Honduras, the northern Gulf of Mexico, and central Florida; accidental on Sugarloaf Key. *P. s. subflavus* ranges from Tallahassee westward and *P. s. floridanus* in the Peninsula southward to Lake Okeechobee (Davis, 1959).

781. Big Brown Bat, *Eptesicus fuscus* (Beauvois)

IDENTIFICATION: A large bat with dark brown upperparts and lighter underparts. Bases of hairs even darker. Ear moderately long (about 15 mm). Interfemoral membrane naked. From other bats of similar size, it differs in its *dark coloration* and in having *4 upper incisors*. The greatest width of the tragus is more than half its length. Total length about 100 to 120 mm.

DISTRIBUTION AND VARIATION: Southern Canada, the United States (except central and southern Texas and southern Florida), highlands of Mexico and Central America, the Bahama Islands, and the Greater Antilles. The precise ranges of *E. f. fuscus* (northwestern Florida) and *E. f. osceola* (peninsular Florida) are imperfectly known.

782. Red Bat, *Lasiurus borealis* (Müller)

IDENTIFICATION: Bats of this genus are medium-sized to large, have short, rounded ears, and some have white-tipped hairs. They have only 2 upper incisors and a densely furred interfemoral membrane (at least anteriad). In the Red Bat the dorsal fur is yellowish toward the base, reddish farther out, and usually has whitish tips; females are darker than males (chestnut buff), but both sexes may have much of the fur frosted with white and have a *yellowish face*. Total length about 90 to 110 mm.

DISTRIBUTION AND VARIATION: Southern Canada and most of the United States (except Rocky Mountains) southward, at moderately high elevations, to Panama; also to the northern Gulf Coast and northern Florida (*L. b. borealis*). Winters northward to West Virginia, Indiana, Illinois, and Missouri.

783. Seminole Bat, *Lasiurus seminolus* (Rhoads)

IDENTIFICATION: Distinguishable from the Red Bat only by its coloration of mahogany brown, with or without silver-tipped dorsal hairs, and its lack of yellowish on the face. Total length about 95 to 105 mm.

DISTRIBUTION: Southern Arkansas, northern parts of Mississippi, Alabama, and Georgia, northwestern South Carolina, central North Carolina, and southeastern Virginia southward to northeastern Mexico, the northern Gulf Coast, and southern Florida; wanders farther north in late summer.

784. Hoary Bat, *Lasiurus cinereus* (Beauvois)

IDENTIFICATION: Although somewhat resembling the Seminole Bat in its white-tipped dorsal pelage, these *hairs are brown basally, and the white tips very conspicuous*. The underparts are lighter, especially on the belly, and the *ears are edged with black*. The skull may also be distinguished by its size (more than 15.5 mm long). Total length 125 to 145 mm (the largest Florida bat).

DISTRIBUTION: Breeds from north-central, southwestern, and southeastern Canada southward to at least the middle latitudes of the United States; in winter southward to South America (resident at high elevations?), the Gulf Coast, and central Florida, and northward to southern parts of breeding range. Florida records range from late September to April and as far south as Orange County (Sherman, 1956; Cooley, 1954).

785. Yellow Bat, *Lasiurus intermedius* H. Allen[1]

IDENTIFICATION: A large bat with a decidedly *yellowish* brown dorsal pelage, often with darker tips. Unlike other bats in this genus, the Yellow Bat lacks the white frosting on its hairs, and it differs markedly from the Hoary and Evening Bats in color pattern, skull size, and tooth characters (see Key). Total length about 115 to 130 mm.

DISTRIBUTION AND VARIATION: Cuba; southeastern Virginia (casually northern New Jersey) southwestward through Atlantic and Gulf Coastal Plain to Texas, continuing southward on Atlantic slope of Mexico to Chiapas; *L. i. floridanus* occurs nearly or quite throughout the mainland of Florida (Map 60).

786. Evening Bat, *Nycticeius humeralis* (Rafinesque)

IDENTIFICATION: Similar in size and dark brownish color to certain bats of the genus *Myotis*, but differing in its *more rounded ear, wider and more curved tragus* (width at least half its length), and *smaller number of teeth* (30). The dorsal hairs are shorter than in *M. sodalis* (about 6 mm). Total length about 80 to 95 mm.

DISTRIBUTION AND VARIATION: Southeastern Nebraska, central Iowa, northern Illinois, southern Michigan, western Ohio, Kentucky (except eastern), Tennessee (except eastern), northern Georgia, South Carolina (except northwestern), the eastern half of North Carolina and Virginia, central Maryland, and south-central Pennsylvania southward to northeastern Mexico (through eastern and southern Texas), the Gulf Coast, and the southern tip of the Florida mainland. Most of Florida is occupied by *N. h. humeralis*, but *N. h. subtropicalis* has been described from the extreme southern part.

787. Eastern Lump-nosed Bat, *Plecotus rafinesquei* Lesson[2]

IDENTIFICATION: A medium-sized bat with *ears about 25 mm long*. The common name derives from *2 distinctive lumps in front of the eyes*. The color is dark gray brown, and the interfemoral membrane is naked. Total length about 90 to 105 mm.

DISTRIBUTION: Western, southern, and eastern Arkansas, southern Illinois and Indiana, western and northeastern Kentucky, Tennessee (except northeastern), and most of North Carolina southward to eastern

1. See Hall and Jones, 1961.

2. See Handley, 1959.

Map 60. Distribution of the Yellow Bat (*Lasiurus intermedius*)

Texas, the Gulf Coast, and Orange County, Florida (Neill, 1953; Pearson, 1954).

ORDER CHIROPTERA: CXLVII. FAMILY MOLOSSIDAE, Free-tailed Bats. These bats differ from all others in Florida in having a *part of the tail free of the interfemoral membrane*, the tragus virtually lacking, and the palatal notch oval-shaped or lacking.

788. Brazilian Free-tailed Bat, *Tadarida brasiliensis* (St. Hilaire)

IDENTIFICATION: A medium-sized bat with a *protruding tail* and with the *bases of the ears not connected to each other*. The color is usually *blackish*, though partial albinism is frequent. Also the palatal notch is narrowed anteriad, making it oval-shaped, and the skull is less than 20 mm long. Total length about 90 to 105 mm.

DISTRIBUTION AND VARIATION: Southern Oregon, Nevada and Utah (except northern parts), western Colorado, most of New Mexico and Texas, southeastern Nebraska, most of Oklahoma, southern Arkansas, central parts of Mississippi, Alabama, and Georgia, and southern South Carolina southward to southern Florida, the Antilles, northern Baja California, and through Central America to much of South America. The Florida race is *T. b. cynocephalus*. (See Schwartz, 1955.)

789. Wagner's Mastiff Bat, *Eumops glaucinus* (Wagner)

IDENTIFICATION: Along with *Tadarida*, the only Florida bat whose *tail extends beyond the interfemoral membrane*, but differing from that species in its larger size (total length about 130 mm) and the fact that the *bases of the ears are conjoined*. General coloration cinnamon brown to yellowish brown.

DISTRIBUTION AND VARIATION: Much of South and Central America (to southeastern Mexico), as well as Cuba, the Isle of Pines, and Jamaica; *E. g. floridanus* is frequent and perhaps established at Miami, Florida (Koopman, 1971).

ORDER PRIMATES: CXLIX. FAMILY CERCOPITHECIDAE, Old-World Monkeys and Apes. Medium-sized to relatively large primates, with the tail long or short, but not prehensile (absent in one species). Body almost completely covered with hair, but face mostly bare. Muzzle protruding. Forelimbs shorter than hind limbs. Teeth 32. Locomotion bipedal or quadripedal.

790. Rhesus Monkey, *Macaca mulatta* Zimmerman

IDENTIFICATION: A rather large, heavy-set monkey with strong limbs; face mostly bare (except forehead); body mostly covered with rather long hair, ranging in color from grayish brown to rufous yellow, but lighter ventrally. Head-body length 40 to 55 cm; tail length 17 to 25 cm (males larger than females).

DISTRIBUTION: Northern India; introduced and established in Puerto Rico and near Silver Springs, Florida (Neill, 1957, p. 206).

ORDER PRIMATES: CL. FAMILY HOMINIDAE, Man. Covering of hair very incomplete. External tail lacking. Anterior aspect of skull flattened, with incisors below nose. Stance upright. Height extremely variable, but usually 150 to 190 cm.

791. Man, *Homo sapiens* L.

IDENTIFICATION: See description of family.
DISTRIBUTION: Cosmopolitan.

ORDER EDENTATA: CLI. FAMILY DASYPODIDAE, Armadillos. *The only Florida mammal with an exoskeleton.* Head small, narrowed anteriad into a snout. Teeth homodont, none toward front of mouth.

792. Nine-banded Armadillo, *Dasypus novemcinctus* L.

IDENTIFICATION: See description of family. This species has 9 narrow rings around the middle of the body and many on the tail. Only 4 toes on each front foot. Total length about 60 to 75 cm.

DISTRIBUTION AND VARIATION: Southeastern Kansas, southwestern Missouri, Arkansas (except northeastern), Mississippi (except northern), central and southern Alabama, the southern parts of Georgia and South Carolina southward and westward through Texas, eastern, southern, and southwestern Mexico, and Central America into South America (Cleveland, 1970; Map 61). *D. n. mexicanus* occupies most of the Florida Peninsula (absent southwestward) and the Panhandle west of the Apalachicola River (Hollister, 1952; Wolfe, 1968); no evidence of occurrence in the Tallahassee area until July, 1972 (J. Wiese, personal communication; *contra* Neill, 1952).[1]

ORDER LAGOMORPHA: CLII. FAMILY LEPORIDAE, Rabbits and Hares. Jumping mammals with *short tails, long ears, and exceptionally long hind legs.* The incisors are like those of rodents, except that there is a small second pair behind the larger upper pair. They also differ from that group in having canine teeth.

793. Marsh Rabbit, *Sylvilagus palustris* (Bachman)

IDENTIFICATION: A small, *short-tailed, reddish brown* rabbit of the marshes. It differs from the only other common rabbit in the state, the Cottontail, in having the tail shorter (usually less than 39 mm, maximum 50 mm), the ears shorter (averaging less than 50 mm), and the *under side of the tail and body gray.* The feet are also much darker than in the Cottontail. The skull differs in that the backward extension of the supraorbital process is largely fused to the cranium. Total length 36 to 45 cm.

DISTRIBUTION AND VARIATION: Atlantic and Gulf Coastal Plain from southeastern Virginia through Florida and southern Georgia to

---

1. Also see *Fla. Field Nat.*, 2:8–10.

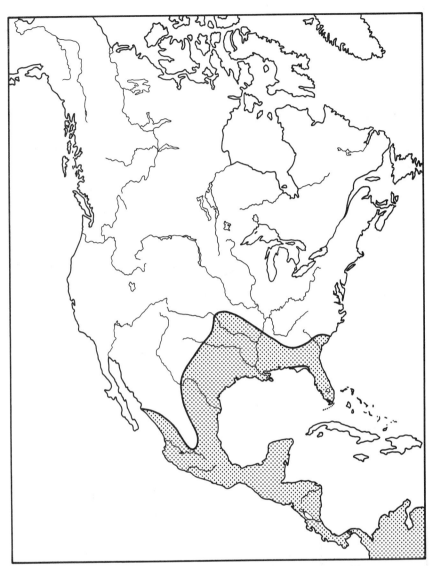

Map 61. Distribution of the Nine-banded Armadillo

Mobile Bay, Alabama (Map 62). Largely, but not entirely, allopatric with the Swamp Rabbit (*S. aquaticus*). The Peninsula of Florida (except near Georgia) is populated by *S. p. paludicola*, and northern Florida by *S. p. palustris*. (See Shanholtzer, 1967.)

794. Eastern Cottontail, *Sylvilagus floridanus* (Allen)

IDENTIFICATION: Slightly larger than the Marsh Rabbit, from which it differs in its *grayish* brown coloration (reddish on nape) and the *white under surface of the tail and upper surface of the feet*. The larger size is evident in *average* measurements of total length, tail length, and ear length, but there is overlap in all cases. Any rabbit seen in dry, upland situations is likely to be this species. The Cottontail's skull can be distinguished from the Marsh Rabbit's by the *partially free posterior extension of the supraorbital process*. Total length 33 to 53 cm.

DISTRIBUTION AND VARIATION: Southern Canada (locally) and west-central Vermont southward, east of the Sierra Nevada Mountains, to Central America (locally), the northern Gulf Coast, and the tip of the Florida mainland, except southwestward. The range of *S. f. mallurus* extends across north Florida, *S.f. floridanus* occupies most of the Peninsula, and *S.f. paulsoni* is limited to extreme southeast Florida (Schwartz, 1956c). *S.f. ammophilus*, known only from the type specimen near Micco, may be an aberrant example of *S. f. floridanus*.

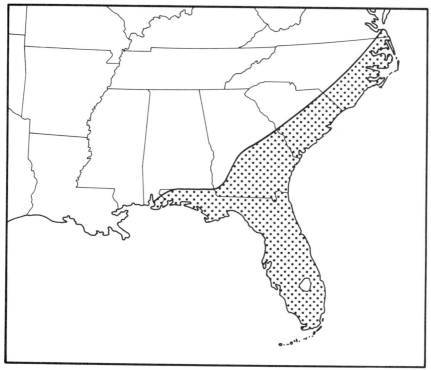

Map 62. Distribution of the Marsh Rabbit (*Sylvilagus palustris*)

795. Black-tailed Jack Rabbit, *Lepus californicus* Bachman

IDENTIFICATION: A large rabbit with *very long ears* (100 mm or more). The general coloration is grayer above and lighter on the sides than in other Florida rabbits, and the *tips of the ears and upper surface of the tail are black.* Its skull is distinguished by the presence of an *interparietal bone.* Total length 48 to 58 cm.

DISTRIBUTION AND VARIATION: Southwestern Oregon, south-central Washington, southern Idaho, Utah (except northeastern), Colorado (except northwestern and central), southeastern Wyoming, southern South Dakota, Nebraska, and western Missouri southward throughout Baja California and to northern Mexico and the Gulf Coast of Texas; also introduced in Kentucky (unsuccessfully?) and Florida (southern Broward County to Miami, subspecies?; Layne, 1965b).

ORDER RODENTIA: CLIII. FAMILY SCIURIDAE, Squirrels. Rodents are most readily distinguished by their teeth. The *long, curved, bevel-edged incisors* (2 pairs) *are separated by a wide gap from the molariform teeth.* The only similar teeth are those of rabbits, which have *small canine teeth* and a *second pair of small upper incisors* immediately behind the first. The animals themselves differ markedly from rabbits, *with comparatively long tails and short ears.* The squirrels (Sciuridae) may be distinguished by head shape and individual color patterns from smaller carnivores, and from other large rodents by their bushier tails. Their skulls have more teeth (20 or 22) than those of any other Florida rodents except the Beaver and Pocket Gopher (20 each).

796. Eastern Chipmunk, *Tamias striatus* (L.)

IDENTIFICATION: The combination of *reddish brown pelage and 2 dorsal light stripes bounded by dark brown* should distinguish this ground squirrel from any other eastern mammal. The tail is bushy like that of the tree squirrels. In having only 4 upper molars its skull is like that of the Fox Squirrel, but the Chipmunk's skull is much smaller (less than 50 mm long). Total length 23 to 30 cm.

DISTRIBUTION AND VARIATION: Southern Manitoba and the central parts of Ontario and Quebec southward to southeastern Louisiana, southern Mississippi and Alabama, northwestern Florida (Okaloosa County, presumably *T. s. pipilans*; Stevenson, 1962b), central Georgia, northwestern South Carolina, and central North Carolina. The western, southern, and southeastern limits are irregular, being dependent on predominantly hardwood forests (Map 63).

797. Gray Squirrel, *Sciurus carolinensis* Gmelin

IDENTIFICATION: A medium-sized, bushy-tailed squirrel, typically *gray dorsally, reddish brown laterally, and light gray to whitish below.* The smaller Flying Squirrel differs in having loose skin on each side of the body, and the larger Fox Squirrel (highly variable in color) by having a white nose. The presence of a *vestigial fifth upper molar* separates the

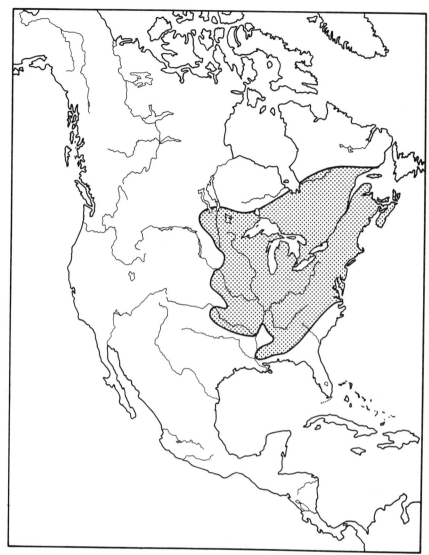

Map 63. Distribution of the Eastern Chipmunk (*Tamias striatus*)

skull of this squirrel from that of any other common in Florida. Total length 40 to 45 cm.

DISTRIBUTION AND VARIATION: Southern Manitoba and the eastern half of North Dakota eastward through northern Minnesota and Michigan, southeastern Ontario, and southern Quebec to Maine (except northern and eastern) and southward to the Gulf Coast from Aransas Bay, Texas, to the Florida Keys; ranges west to western Iowa, southeastern Nebraska, eastern Kansas and Oklahoma, and central Texas. Introduced to the British Isles. The distribution of subspecies in Florida has not been clarified, but it appears that *S. c. carolinensis* ranges southward to the central Peninsula and *S. c. extimus* from there to upper Matecumbe Key. An albino population (with a dark middorsal streak) from Gadsden County has been introduced at the lower Ochlockonee River, Franklin County. (See Loftin, 1970.)

798. Red-bellied Squirrel, *Sciurus aureogaster* Cuvier

IDENTIFICATION: Slightly larger than the Gray Squirrel. Most Florida individuals are *completely melanistic*; the remainder are *grizzled above and chestnut red below*, the latter color extending dorsally behind the shoulders.

DISTRIBUTION AND VARIATION: Lowlands and mountains of the Gulf slope in southeastern Mexico; introduced on Elliott Key, Florida (Brown, 1969). The subspecies in Florida is uncertain (Map 64).

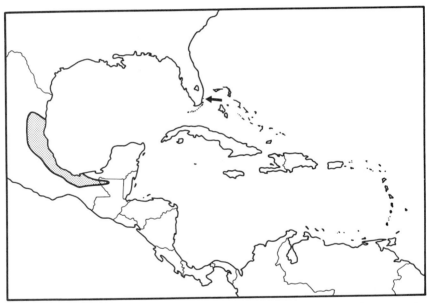

Map 64. Distribution of the Red-bellied Squirrel (*Sciurus aureogaster*)

799. Fox Squirrel, *Sciurus niger* L.

IDENTIFICATION: The largest Florida squirrel, its *total length ranging up to 63 cm or more*. The coloration is highly variable, but over most of Florida the pelage is chiefly silvery gray with black on the head and *white on the snout*; some individuals are almost entirely black. Individuals from west Florida may show considerable buff or reddish brown. The skull may be distinguished from that of the Gray Squirrel by the *absence of a vestigial fifth upper molar*, and from that of other Florida squirrels by its larger size (length more than 50 mm).

DISTRIBUTION AND VARIATION: Most of North Dakota, Minnesota, and Michigan, Ohio, extreme western New York, western and southern Pennsylvania, and southwestern New Jersey southward to northeastern Mexico, the Gulf Coast, and the southern tip of the Florida mainland. *S. n. avicennia* is restricted to the southern part of the Peninsula, northward to Lake Okeechobee and Tampa Bay; the remainder of the Peninsula is occupied by *S. n. shermani*; *S. n. niger* ranges from the Aucilla River over most of the Panhandle, but *S. n. bachmani* may occur in the extreme western portion (Moore, 1956).

800. Southern Flying Squirrel, *Glaucomys volans* (L.)

IDENTIFICATION: A small brownish gray squirrel with *large eyes and a fold of loose skin along each side*; underparts white to creamy. The skull can be distinguished from those of other squirrels by its combination of *small size* (less than 40 mm long) and *5 well-developed upper molars*. Total length about 20 to 25 cm.

DISTRIBUTION AND VARIATION: Central Minnesota and Wisconsin, Upper Michigan, southeastern Ontario, northern Vermont and New Hampshire, and southwestern Maine southward to central Texas and the Gulf Coast from Aransas Bay, Texas, to south-central Florida. The race in the Panhandle is *G. v. saturatus*; from the Tallahassee area eastward and southward to the tip of the mainland *G. v. querceti* is found.

ORDER RODENTIA: CLIV. FAMILY GEOMYIDAE, Pocket Gophers. These burrowing rodents are large-headed, heavy-set, and have *enlarged front claws*, small ears and eyes, *cheek pouches*, and a short tail. Distinctive features of the skull include the presence of *20 teeth* and the *grooved incisors*.

801. Southeastern Pocket Gopher, *Geomys pinetis* Rafinesque

IDENTIFICATION: Florida's only species of pocket gopher has a sandy or reddish brown coloration and a total length of about 25 to 30 cm. (See description of family.)

DISTRIBUTION AND VARIATION: Central Alabama and central Georgia southward to or near the Gulf Coast from Baldwin County, Alabama, to Estero Bay, Florida. The race west of the Apalachicola River is *G. p. mobilensis*; from there eastward to the Atlantic Coast, and southward to Orlando and Merritt Island, *G. p. floridanus* is found; *G. p. austrinus* ranges from Levy County southward to Estero Bay and Arcadia; and *G. p. goffi* is restricted to the vicinity of Eau Gallie.

ORDER RODENTIA: CLV. FAMILY CASTORIDAE, Beavers. Enormous rodents with *broad, scaly, dorsoventrally flattened tails*. The skull may be distinguished by its large size, the *orange incisors*, and the *transverse ridges on the grinding surface of the molars*. Pelage dark glossy brown; feet webbed. Total length about 75 to 125 cm. Skull length about 130 to 155 mm.

802. American Beaver, *Castor canadensis* (Kuhl)

IDENTIFICATION: See description of family.

DISTRIBUTION AND VARIATION: Alaska (except western and northern) and Canada (except northern) southward to central California, northern Nevada, most of Utah, Baja California (via Colorado River), southern Arizona, northern Mexico, the Gulf Coast (Texas to Alabama), and the Florida Panhandle (*C. c. carolinensis*). Introduced in Russia.

ORDER RODENTIA: CLVI. FAMILY CRICETIDAE, New-World Rats and Mice. Difficult to separate from Muridae, but tail relatively short (often shorter than head plus body), obviously hairy above, and underparts often more whitish. The 2 families are best separated by the tooth characters in the Key.

803. Rice Rat, *Oryzomys palustris* (Harlan)

IDENTIFICATION: Rats differ from mice in their larger size (total length usually 225 mm or more), more sparsely furred tail, and relatively smaller ears. The present species differs from other native rats (Cricetidae) in its shorter hair and smaller skull (see Key) and particularly from *Sigmodon* in its even coloration and whitish feet. Tooth characters separating the 2 families will distinguish it from the larger Norway and Black Rats. The general coloration is medium gray, lighter below, with a wash of brownish on the sides. Maximum length about 230 to 300 mm.

DISTRIBUTION AND VARIATION: Southeastern Kansas, northern Arkansas, southeastern Missouri, southern Illinois and Kentucky, Tennessee (except northeastern), North Carolina (except northwestern), eastern Virginia and Maryland, southeastern Pennsylvania, and southern New Jersey southward to the Gulf Coast from southern Texas to the tip of the Florida mainland; also through Mexico (except north-central) and extreme southern Baja California to Costa Rica (Hall, 1960; Map

65). This rat is represented across northern Florida by *O. p. palustris*; in the northern and central Peninsula by *O. p. natator*; and from about Lake Okeechobee southward by *O. p. coloratus*. More restricted populations are *O. p. planirostris*, on the mainland near Fort Myers, and *O. p. sanibeli*, known only from Sanibel Island. Recently found on Cudjoe Key (J. Layne, unpubl. mss.).

804. Eastern Harvest Mouse, *Reithrodontomys humulis* (Audubon and Bachman)

IDENTIFICATION: This mouse can be differentiated from all others in its family by its *smaller size* (head-body length less than 70 mm in adults). The grayer House Mouse (Family Muridae) is of the same size range except for its longer tail (more than 90 percent of head-body length). Both of these species differ from immatures of some larger mice in having grayer underparts. The upperparts in *Reithrodontomys* are dark gray with a brownish tinge. In most other measurements it is smaller than most Florida mice (total length, 107 to 128 mm; tail, 45 to 60 mm; hind foot, 15 to 17 mm; ear, 8 to 9 mm). The skull is easier to recognize than the skin, the *upper incisors being grooved*.

DISTRIBUTION AND VARIATION: Ohio River, southern Ohio, and western Maryland southward (east of the Mississippi River) to the Gulf Coast and Lake Okeechobee (*R. h. humulis*); also extending west of the Mississippi River through most of Louisiana to eastern Texas and across northern Arkansas (Map 66).

805. Old-field Mouse, *Peromyscus polionotus* (Wagner)

IDENTIFICATION: Unlike most other small mice, members of this genus have *white underparts*. The Old-field Mouse is of *paler, often sandy, coloration dorsally* than other species, and has a *proportionately short tail* (usually less than 65 percent of head-body length). Compared with other members of the genus, it is small in all measurements (total length, 122 to 153 mm; tail, 40 to 60 mm; hind foot, 15 to 19 mm). Even the skull can be distinguished by its small size, the greatest length being about 24 to 26 mm; from those of the equally small Harvest Mouse and House Mouse, it is separable by its lack of grooved incisors and the presence of only 2 rows of cusps on the upper molars.

DISTRIBUTION AND VARIATION: Northern Alabama and Georgia and northwestern and central South Carolina southward to the Gulf Coast from Baldwin County, Alabama, to south Florida (Tampa Bay, Highlands and Broward Counties). It is represented in Florida by 13 subspecies: *P. p. polionotus*, near Alabama state line and in Gadsden and northern Liberty Counties; *P. p. leucocephalus*, restricted to Santa Rosa Island; *P. p. peninsularis*, restricted to St. Joseph's Peninsula and the coast of the adjacent mainland north to Port St. Joe; *P. p.*

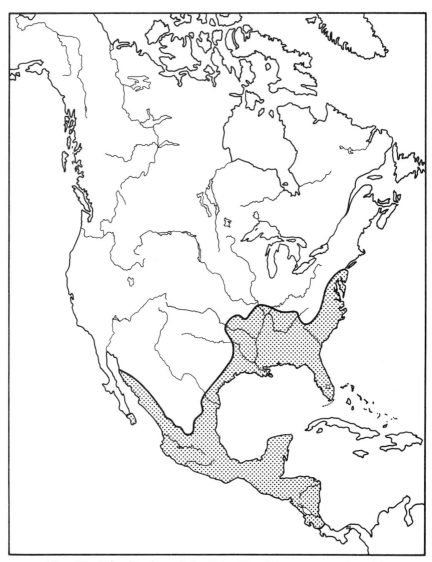

Map 65. Distribution of the Rice Rat (*Oryzomys palustris*)

trissyllepsis, coast of Escambia County (mainland); *P. p. griseobracatus*, coast of Santa Rosa and Okaloosa Counties on mainland east to Fort Walton Beach; *P. p. allophrys*, coastal strip from mouth of Choctawhatchee Bay east to mouth of St. Andrews Bay; *P. p. albifrons*, parts of inland Okaloosa and Walton Counties; *P. p. sumneri*, northern Bay County, southern Washington County, and southwestern Jackson County; *P. p. subgriseus*, eastern Jackson County east to upper Suwannee River and southward in the interior to Levy

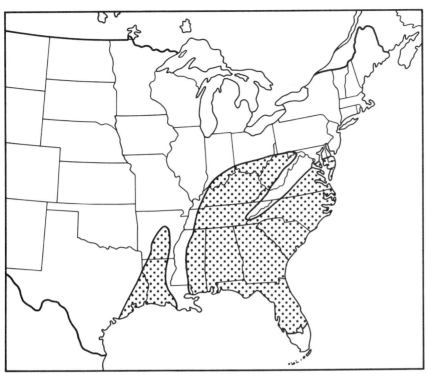

Map 66. Distribution of the Eastern Harvest Mouse
(*Reithrodontomys humulis*)

and Marion Counties; also a disjunct population in Pasco and Hillsborough Counties; *P. p. phasma*, Atlantic Coast north of St. Johns
County; *P. p. decoloratus*, near coast of Volusia County; *P. p.
rhoadsi*, disjunct areas in the Peninsula from Leon and Putnam Counties south to Highlands County; *P. p. niveiventris*, Atlantic Coast
from about Titusville south to Broward County (Bowen, 1968).

806. Cotton Mouse, *Peromyscus gossypinus* (Le Conte)

IDENTIFICATION: A medium-large *Peromyscus* with *dark gray
brown pelage dorsally*, sometimes with reddish brown on sides. No
other Florida mouse in the genus is so dusky above. Immatures may be
as small as *P. polionotus*, but have a proportionately longer tail (usually
more than 65 percent of head-body length). The Golden Mouse and
Florida Deer Mouse are much more brightly colored. The ratio of ear
length to hind-foot length (not more than 75 percent) is lower in this
mouse than in others of its genus. In this species and *P. polionotus* the
posterior palatal foramina are located farther forward (at anterior edge of
second molars) than in others of the genus, and the skull is longer in *P*.

gossypinus (about 28 mm) than in *P. polionotus*. Total length about 160 to 200 mm.

DISTRIBUTION AND VARIATION: Southeastern Oklahoma, central and northeastern Arkansas, southeastern Missouri, southern Illinois, southern Kentucky, Tennessee (except eastern), Georgia and the Carolinas (below about 1000 feet), and southeastern Virginia southward to the Gulf Coast from Galveston Bay to the Florida Keys. *P. g. gossypinus* inhabits north Florida as far south as Gainesville and the mouth of the Suwannee River, *P. g. palmarius* most of the remaining Peninsula, and *P. g. telmaphilus* the extreme southwestern part of the Peninsula, northward to Naples. Other races are more restricted: *P. g. anastasae*, Anastasia Island (and Cumberland Island, Ga.); *P. g. restrictus*, vicinity of Englewood; and *P. g. allapaticola*, upper Key Largo.

807. Florida Deer Mouse, *Peromyscus floridanus* (Chapman)

IDENTIFICATION: This mouse is the largest member of its genus in Florida, is restricted to the Peninsula, and differs from all others in usually having *only 5 plantar pads*. The tail, however, is scarcely longer than that of *P. gossypinus* or *O. nuttalli*, being only about 75 percent of head-body length. The *pale gray brown upperparts* resemble those of the much smaller Old-field Mouse (*P. polionotus*). The basilar length of the skull (more than 20 mm) exceeds that of other Florida mice. Total length, 185 to 220 mm; head-body length, 105 to 125 mm; hind-foot length, 24 to 29 mm; ear length, 22 to 25 mm.

DISTRIBUTION: Peninsular Florida (Gainesville and Anastasia Island southward to Palm Beach County (formerly Miami), Highlands County and Marco Island (J. Layne, unpubl. mss.).

808. Golden Mouse, *Ochrotomys nuttalli* (Harlan)[1]

IDENTIFICATION: The most brightly colored of Florida mice, the upperparts being an ochraceous or golden brown, or (in immatures) the head so colored. The tail is longer than in most mice, being about 85 to 95 percent as long as the head and body. The skull differs from that of *floridanus* in its smaller size (basilar length less than 20 mm), and from that of other Florida mice in the genus *Peromyscus* in that the posterior palatal foramina are at the same level as the *posterior* part of the second molars. Total length, 150 to 190 mm; tail, 75 to 90 mm; hind foot, 17 to 20 mm.

DISTRIBUTION AND VARIATION: From southern Missouri, Kentucky, and Virginia southward to southeastern Oklahoma, northeastern

---

1. See Hooper and Hart, 1962.

Texas, Louisiana (except southwestern corner), the Gulf Coast, and central Florida (*O. n. aureolus*).

809. Hispid Cotton Rat, *Sigmodon hispidus* Say and Ord

IDENTIFICATION: This rat is larger and has a less hairy tail than any Florida mouse, but can be confused with other species of rats. Its *grizzled appearance* contrasts with the more evenly colored pelage of other species, however. It differs from the equally small Rice Rat (*Oryzomys*) in having *dark feet* and a relatively *shorter tail* (less than 85 percent of head-body length). The relative ear length (55 to 60 percent of hind foot) is greater than that of *Neofiber* or *Oryzomys*, but less than that of *Neotoma*. The hind toes are not webbed. The skull may be distinguished from that of *Neofiber* and *Oryzomys* by the curving, often S-shaped pattern of uneven ridges on the molars, and from that of *Neotoma* by the absence of black on the molars. The skull is also shorter than that of *Neotoma* (basilar length less than 30 mm). Total length about 225 to 350 mm; tail length, 85 to 165 mm; hind foot, 28 to 41 mm; ear, 16 to 24 mm.

DISTRIBUTION AND VARIATION: Lower Colorado River through central Arizona to southwestern Arizona, southern New Mexico, northern Texas, southeastern Colorado, western and northeastern Nebraska, western and southern Missouri, western Kentucky, southern and eastern Tennessee, North Carolina (except near coast), and southeastern Virginia southward through Mexico (except northwestern) to Panama, the Florida Keys, and the coast of South Carolina. The race across north Florida is *S. h. hispidus*, and throughout the remainder of the state (with the exceptions listed below), *S. h. littoralis*. Other subspecies are restricted to small areas: *S. h. floridanus*, east side of Lake Okeechobee; *S. h. insulicola*, vicinity of Fort Myers, especially on islands; *S. h. spadicipygus*, southern tip of Peninsula; *S. h. exsputus*, Lower Keys (Schwartz, 1954); Cotton Rats on the Upper Keys have not been subspecifically identified. There is a possibility that *S. h. komareki* occurs in Florida near the Alabama line.

810. Eastern Wood Rat, *Neotoma floridana* (Ord)

IDENTIFICATION: Florida's largest native rat, comparable in size with the introduced Black and Norway Rats. It differs from those 2 species, however, in its *whitish feet and underparts* (including under side of tail), from the Black Rat in its *proportionately shorter tail* (not more than 85 percent of head-body length). (Note also tooth differences in the 2 families.) It does not have the conspicuously grizzled fur of the smaller Cotton Rat nor the long hair (about 14 mm) of the Rice Rat. Furthermore, it lacks the webbing between the hind toes and the dark underparts of *Neofiber*. The skull can be distinguished from all others except that of *Neofiber* by its *partly black molars*, and from that skull by the lack of *alternating* triangular ridges on the molars. Total length about 310 to 430 mm; tail, 130 to 200 mm; hind foot, 35 to 45 mm; ear, 24 to 29 mm.

DISTRIBUTION AND VARIATION: Southwestern South Dakota, western Nebraska, eastern Colorado, Kansas, southern and western Missouri, southern Illinois, Indiana, and Ohio, Pennsylvania (except extreme northwestern), southeastern New York, and western Connecticut southward through New Mexico into northeastern Mexico and (west of the Appalachian Mountains) to the Gulf Coast and central Florida; also eastward through Georgia to northwestern and southern South Carolina and southeastern North Carolina (includes ranges of *Neotoma "magister"* and *N. "micropus"*; see Schwartz and Odum, 1959). In Florida, *N. f. illinoensis* is found west of the Apalachicola River, and *N. f. floridana* from that river southeastward to Sebastian and near Fort Myers; an isolated race (*N. f. smalli*) has been described from Key Largo (Sherman, 1955) and is introduced on Lignum Vitae Key (J. Layne, unpubl. mss.).

811. Pine Vole, *Microtus pinetorum* (Le Conte)

IDENTIFICATION: The extremely short tail (about 20 percent of head-body length) will readily separate animals in this genus from all other Florida mice. This species also differs from other mice in its *rusty brown* dorsal pelage; the *underparts are gray*. Total length 105 to 145 mm.

DISTRIBUTION AND VARIATION: Southeastern Nebraska, Iowa (except northwestern), southeastern Minnesota, western, central, and southern Wisconsin, lower Michigan, extreme southern Ontario, northern New York, central Vermont and New Hampshire, and extreme southwestern Maine southward to central Texas, southern Louisiana, Mississippi, Alabama, and Georgia, the northern Florida Peninsula, and the coast of South Carolina (Map 67). *M. p. pinetorum* may occur in northwestern Florida, but most specimens there represent intergrades with *M. p. parvulus*, which is chiefly restricted to Alachua and Marion Counties. (See Hooper and Hart, 1962, and Arata, 1965.)

812. Florida Water Rat, *Neofiber alleni* True

IDENTIFICATION: A medium-sized rat with *webbing between the second and third hind toes*. The fur is dense, smoky or dark brownish gray above, and whitish below. It may be further distinguished from *Oryzomys* by its shorter tail (less than 80 percent of head-body length), and from *Neotoma* and *Sigmodon* by its shorter ears (less than 50 percent of hind foot). It most resembles the Muskrat (not known to occur in

Florida), but even adults are much smaller, and the tail shape (round in cross section) further differentiates the Florida animal. The skull differs from those of *Sigmodon* and *Oryzomys* in the *partly black molars*, from that of the Muskrat in its smaller size (basilar length less than 52 mm), and from that of *Neotoma* in that the triangular ridges on the molars *alternate in position*. Total length about 285 to 380 mm.

DISTRIBUTION AND VARIATION: Peninsular Florida and southeastern Georgia: *N. a. apalachicolae* from Lake Miccosukee and Madison southwestward to Wakulla County and Apalachicola Bay; *N. a. alleni* from Lake City through the interior to Winter Haven and to the Atlantic Coast (Flagler Beach to Stuart); *N. a. nigrescens*, from southern limit of *alleni* to the Gulf Coast (Tampa to south of Naples) and south on Atlantic Coast to Delray Beach; *N. a. struix*, remainder of the Peninsula (southern tip).

ORDER RODENTIA: CLVII. FAMILY MURIDAE, Old-World Rats and Mice. External features of these introduced species differ only slightly from those of certain native rats and mice, but as a group they have a

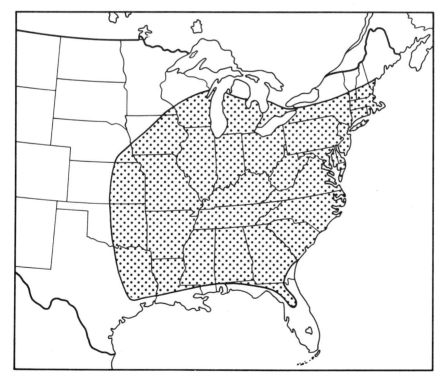

Map 67. Distribution of the Pine Vole (*Microtus pinetorum*)

relatively longer tail (about equal to or longer than the head and body in 2 of the 3 species). The tail is also *more nearly naked* than in most native forms. Their skulls can be recognized by the *3 rows of tubercles on the molars.*

813. Black Rat, *Rattus rattus* (L.)

IDENTIFICATION: Any Florida rat with a tail at least as long as its combined head-body length must be this species or *Oryzomys*. The Black Rat is usually much darker (especially underneath) and has a much longer ear (more than 60 percent of hind foot length). The skull is most like that of the Norway Rat, but has *no notches on the outer edge of the anterior cusps of the first molar.* Dorsal coloration light gray brown; whitish underneath. Total length about 325 to 450 mm; tail length, 165 to 230 mm.

DISTRIBUTION: Pacific Coast from southern British Columbia to Central America, throughout most of Mexico, around Gulf Coast to Atlantic Coast as far north as Massachusetts and inland to the Appalachian Mountains, northern Alabama, western Tennessee, and eastern Arkansas. Also widespread in the Old World.

814. Norway Rat, *Rattus norvegicus* (Berkenhout)

IDENTIFICATION: Similar in appearance to the usually darker Black Rat, but differing in its *shorter tail* (shorter than head-body length) and shorter ears (less than 60 percent of hind foot length). The relatively short tail makes this species confusible with certain native rats. The relatively short ear will separate it from *Neotoma* and *Sigmodon*, and the dense fur of *Neofiber* is in contrast with the sparser hair of *Rattus*. Total length about 320 to 450 mm; tail length, 125 to 215 mm.

DISTRIBUTION: Widespread in the Old World. In the New World ranges north to northern Alaska and Newfoundland near the coast, and to southern Canada in the interior. Said to be more restricted to human habitations than the Black Rat.

815. House Mouse, *Mus musculus* L.

IDENTIFICATION: The external differences between this small rodent and the native mice are not striking. Perhaps the best is the *absence of true white on the underparts.* The tail tends to be longer (about equal to the head-body length), more scaly, and not bicolored. The Harvest Mouse is of similar color, but smaller (total length less than 130 mm) and has a relatively shorter tail (less than 90 percent of head-body length). The skull can always be distinguished by the *3 rows of cusps on the molars.* Total length 130 to 195 mm; tail length, 63 to 100 mm.

DISTRIBUTION: Widespread in the Old World. In the New World, ranges north to northern Alaska and Newfoundland near the coast and in the interior to southern Canada.

ORDER RODENTIA: CLVIII. FAMILY CAPROMYIDAE, Hutias and Nutrias. Medium-sized to large rodents with relatively large head and short to moderately long tails (shorter than head-body length) that may be either scaly or hairy. A peculiarity of the skull is that the *infraorbital openings are larger than the foramen magnum*. All are native to South America, but one species is widely introduced into the United States.

816. Nutria, *Myocaster coypus* (Molina)

IDENTIFICATION: Adults are much larger than any other Florida rodents except the Beaver, which differs from all in its flattened tail. Immatures, however, can be recognized by their *webbed hind feet*. The skull is easily distinguished by the *size of the infraorbital openings* (see family description) and the *orange incisors*; the skull is about 130 mm long in adults. Total length about 50 to 65 cm, the tail making up about one-third of this.

DISTRIBUTION AND VARIATION: Most of temperate South America; introduced and established locally from North Carolina to eastern Texas; also in Ohio, Montana, Oregon, and Washington, as well as in Russia. Florida records of *M. c. bonariensis* range from Walton County westward (near the coast) and from Duval and Jefferson Counties southward to Hillsborough and Brevard Counties (Griffo, 1957; Florida Game and Fresh Water Fish Comm., 1963). I noted a colony in Hamilton County near the Georgia line in 1969.

ORDER CARNIVORA: CLIX. FAMILY CANIDAE, Foxes, Coyotes, and Wolves. Members of this order range from medium to large size and differ from certain other orders in having the tail both long and well covered with hair. Skull characters are more distinctive than the skin: *canines present and well developed*; upper incisors in 3 pairs and not especially large. The family Canidae consists of larger carnivores than some families, with long, bushy tails and *42 teeth (3 pairs of lower molars)*.

817. Coyote, *Canis latrans* Say

IDENTIFICATION: A doglike carnivore of *gray or reddish gray pelage and rust-colored legs and ears*. (The color pattern is not strikingly different from those of the Red Wolf and Gray Wolf, presently unknown in Florida.) In addition to color differences, it may also be distinguished from foxes by its greater size and *relatively shorter tail* (less than half length of head and body). The skull has *convex postorbital processes* and its basilar length is in excess of 145 mm. Total length about 100 to 130 cm; tail about 30 to 38 cm.

DISTRIBUTION AND VARIATION: Alaska (except west-central and southwestern), Yukon, western and southern Mackenzie, western, central, and southern Saskatchewan, southern Manitoba, Ontario (except northern), and extreme southwestern Quebec southward through Mexico (except Yucatan Peninsula) to western Costa Rica, and eastward to northwestern Indiana, southern Illinois, southwestern Missouri, central Arkansas, and extreme western Louisiana. Because of eastward range expansion and frequent introductions, records occur in most other states, including northern and central Florida (Cunningham and Dunford, 1970). There is considerable doubt which of the 19 described subspecies of the Coyote are represented by such introductions.

818. Red Fox, *Vulpes vulpes* L.[1]

IDENTIFICATION: Foxes differ from the Coyote and from wolves in their smaller size, in coloration, and in having a *relatively longer tail* (more than 50 percent of head-body length). The Red Fox is *sandy red* dorsally, whitish below, *with blackish feet and ears*. There is much black in the tail, which is *white-tipped*. (Black and intermediate phases occur elsewhere.) The skull may be distinguished from that of a Coyote by its smaller size (basilar length less than 145 mm) and the postorbital processes, which are concave dorsally; compared with that of the Gray Fox, the lower jaw is not incised posteriad. Total length about 90 to 100 cm; tail length, 35 to 40 cm.

DISTRIBUTION AND VARIATION: Holarctic, ranging southward in North America to the southern United States (except most of the Pacific Coast, southern Texas, the southeastern Coastal Plain, and the most arid portions of the West). *V. v. fulva* is established in Florida west of the Apalachicola River and sporadically southward to Polk County (Lee and Bostelman, 1969) and Highlands County (R. Sanderson, personal communication); others seen in the Peninsula probably represent recent liberations.

819. Gray Fox, *Urocyon cinereoargenteus* (Schreber)

IDENTIFICATION: About the size of the Red Fox, with which it is sometimes confused because of its *reddish brown pelage ventrally*. However, the more prevalent *gray of the upperparts* (hairs tipped with white) earns this animal its vernacular name. It lacks the black feet and white tail tip of the Red Fox, but has a *black streak down the dorsal*

---

1. See Churcher, 1959.

midline of the tail. The small size of the skull, along with dorsally-concave postorbital processes, will separate it from that of a Coyote; compared with the skull of a Red Fox, the *lower jaw is notched posteriad.* Total length about 80 to 100 cm; tail length, 28 to 40 cm.

DISTRIBUTION AND VARIATION: The western half of Oregon, northeastern California, central Nevada, central and northeastern Utah, northern and southeastern Colorado, northern Texas, the eastern half of the Great Plains states, extreme southwestern and southeastern Ontario, Upper Michigan, southern Quebec and southwestern Maine southward through Mexico and Central America into South America; absent from Rhode Island, eastern Connecticut, and southeastern Massachusetts. The southeastern race, *U. c. floridanus*, occurs throughout the Florida mainland.

ORDER CARNIVORA: CLX. FAMILY URSIDAE, Bears. Because of their huge size, extremely *short tails*, and *plantigrade feet*, bears can hardly be confused with any other animals. The skull may be known by its 42 teeth (3 pairs of lower molars) and position of the nasal bone (ending halfway between orbits and upper incisors).

820. Black Bear, *Ursus americanus* (Pallas)

IDENTIFICATION: A huge, black, short-tailed, flat-footed carnivore, quite unlike any other wild animal in Florida. The color is occasionally brownish, and the snout usually so. Adults may reach a length of 150 to 180 cm and a weight of 90 to 180 kg. (For skull, see description of family.)

DISTRIBUTION AND VARIATION: Formerly from Alaska (except northern and western) and Canada (except extreme northern) southward to central California (or southern in mountains), northern Mexico, Texas (except southern tip), the Gulf Coast, and the southern tip of Florida (*U. a. floridanus*). Now doubtless extirpated from many areas in the East, but still occurs throughout most of the Florida mainland (J. Layne, unpubl. MS).

ORDER CARNIVORA: CLXI. FAMILY PROCYONIDAE, Raccoons. Medium-sized carnivores with long tails and pentadactyl, plantigrade feet. The teeth total 40 or more, including 2 pairs of upper molars. In several species the tail is long-haired and alternately ringed with light and dark.

821. Raccoon, *Procyon lotor* (L.)

IDENTIFICATION: A medium-sized, flat-footed carnivore with a *long, bushy, ringed tail.* The typical pelage is a brownish gray, darker above, lighter below—but lighter, even sandy-colored, animals are frequent in Florida. In all phases, the *dark mask over each eye, bounded by white*, will distinguish the species. The skull resembles that of a dog, the canines being comparatively dull. Total length about 60 to 90 cm; tail length, 20 to 40 cm.

DISTRIBUTION AND VARIATION: Southern British Columbia, northern, eastern, and southern Alberta, western and southern Saskatchewan, southwestern Manitoba, northern Minnesota, and southern Ontario and Quebec southward (except in Rocky Mountains and most arid parts of the West) throughout the United States, Mexico, and Central America to Panama; introduced in eastern Europe, Asia, and the Bahama Islands. Florida subspecies are: *P. l. varius*, west of the Apalachicola River; *P. l. elucus*, most of remaining mainland; *P. l. marinus*, southwestern tip of mainland, northeastward to Lake Okeechobee; *P. l. inexperatus*, upper Florida Keys; *P. l. auspicatus*, Key Vaca; *P. l. incautus*, Big Pine Key to Key West.

ORDER CARNIVORA: CLXII. FAMILY MUSTELIDAE, Mustelids. Small to medium-sized carnivores, with long tails, never banded and sometimes not bushy. The skull differs from that of the Procyonidae in having only 34 to 36 teeth, with only one pair of upper molars. Except for the 2 species of skunks, Florida species have long, powerful necks and short-haired tails. Feet plantigrade.

822. River Otter, *Lutra canadensis* (Schreber)

IDENTIFICATION: Body form similar to that of the Mink, but *tail much thicker, toes webbed, and size much greater*. Except for the more extensive light area on the face and throat, the color is also that of the Mink. In addition to its size, the skull differs from those of other mustelids in having 5 (rather than 4) *pairs of upper molariform teeth*. Total length, about 90 to 125 cm; tail length, 20 to 50 cm; width of tail at base, about 55 mm.

DISTRIBUTION AND VARIATION: Alaska and Canada (except northernmost parts) southward through the United States to central California, western and northern Nevada, northeastern and southeastern Utah, extreme southern Nevada and southeastern California, southern Arizona, southwestern and central New Mexico, northern and eastern Texas, the Gulf Coast, and the southern tip of Florida (remarkably similar to the distribution of the Mink). *L. c. vaga* is the only race known to Florida, though typical *canadensis* may occur in the northwestern corner.

823. Long-tailed Weasel, *Mustela frenata* Lichtenstein

IDENTIFICATION: Members of this genus have conspicuously long, thick necks and long tails with relatively short hair. The Long-tailed

Weasel is much smaller than the other Florida species, the Mink. It differs further in its color pattern of rich, dark brown above, *white below, with a black tail tip*. The skull is also smaller, the basilar length being less than 55 mm. Total length about 30 to 50 cm; tail length, 10 to 15 cm. (The population in the northern United States and Canada becomes almost entirely white in winter.)

DISTRIBUTION AND VARIATION: Central and southern British Columbia (except on coast), central Alberta, southern Saskatchewan, southwestern and south-central Manitoba, northwestern Minnesota, Upper Michigan, southern Ontario and Quebec, northern Maine, and southwestern Newfoundland southward through the United States (except for southwestern California, southern Nevada, and most of Arizona, Mexico (except northwestern), and Central America into South America. The 2 subspecies in Florida are *M. f. olivacea* across the northern portion, and *M. f. peninsulae* south of Ocala on the mainland at least to Collier County (Larry Brown, personal communication).

824. Mink, *Mustela vison* Schreber

IDENTIFICATION: Very similar to the weasel in body form, but differing in its larger size and *entirely blackish brown pelage*, except for the white chin. The larger size (of adults) is illustrated by the basilar skull length of more than 55 mm, the total length of 50 to 70 cm, and the tail length of 15 to 20 cm. In all of these dimensions it is much smaller than the River Otter, although very similar to that species in color.

DISTRIBUTION AND VARIATION: Alaska, Canada (except some of northern portion), and the United States southward to central California, western and northeastern Nevada, northern and central Utah, Colorado, northern New Mexico, Oklahoma, eastern and central Texas, the Gulf Coast, and the southern tip of the Florida mainland. Introduced in Scandinavia and Iceland. *M. v. mink* ranges across the Panhandle eastward to Taylor and Columbia Counties; *M. v. lutensis* occupies most of the remaining Peninsula; and *M. v. evergladensis* occurs in the southwestern tip of the Peninsula (see Schwartz, 1949). Localized and largely coastal in Florida (J. Layne, unpubl. MS).

825. Spotted Skunk, *Spilogale putorius* (L.)

IDENTIFICATION: A small skunk with *interrupted stripes, slashes, and spots of white* on a black background. The tail is relatively shorter and less bushy than in the Striped Skunk. Skull differences in the 2 species include the more flattened interorbital region of *Spilogale*, and the fact that its *last upper molar is wider than long*. Total length about 40 to 50 cm; tail length, 15 to 23 cm.

DISTRIBUTION AND VARIATION: Southwestern British Columbia, Washington, Idaho, southwestern Montana, central and northeastern

Wyoming, southeastern Montana, southern and eastern North Dakota, northern Minnesota, western Wisconsin, Iowa, Missouri, southern Illinois and Indiana (not near Mississippi River north of Louisiana), northern Kentucky, central West Virginia, south-central Pennsylvania, western Maryland, central Virginia and the Carolinas, and Georgia (except near coast) southward to Baja California, central and southeastern Mexico, Yucatan Peninsula, Costa Rica, the Gulf Coast of the United States, and the Florida mainland (except for Panhandle, northeastern corner, and southwestern tip; *S. p. ambarvalis*, intergrading with *S. p. putorius* near the Georgia line; see Van Gelder, 1959).

826. Striped Skunk, *Mephitis mephitis* (Schreber)

IDENTIFICATION: Larger than the Spotted Skunk and typically with a different color pattern. Usually black with *2 broad white or buffy stripes down the back*. However, the amount of white may vary from none to almost the entire upperparts. None of these various color patterns could be confused with that in *Spilogale*. The skull differs in the *more convex interorbital region* and the *shape of the last upper molar* (square in cross section). Total length about 60 to 70 cm; tail length, 22 to 35 cm.

DISTRIBUTION AND VARIATION: British Columbia (except near coast), southern Mackenzie, northern Saskatchewan, central Manitoba and Ontario, southern Quebec, and western Newfoundland southward throughout the United States (except parts of southeastern California, southern Nevada, and southwestern Utah) to extreme northern Mexico. Introduced in Russia. *M. m. elongata* is found throughout most of the Florida mainland.

ORDER CARNIVORA: CLXIII. FAMILY FELIDAE, Cats. Medium to large carnivores with *digitigrade feet* (5 toes on front feet, 4 on hind). The shorter, more rounded head (skull) will distinguish them from the Canidae at a glance. The tail is sometimes short, but when long (*Felis*) it is shorter-haired than in the Canidae. The skull contains only 28 or 30 teeth, with only 3 pairs of molariform teeth in the lower jaw.

827. Panther, *Felis concolor* L.

IDENTIFICATION: A huge, golden brown to brownish gray cat with whitish underparts and a long tail. The Jaguarundi, also rare in Florida, is similar in one color phase, but hardly larger than the *spotted young* of the Panther. Both of these species have *4 pairs of upper molariform teeth* (unlike *Lynx*), but the Jaguarundi's skull never exceeds a length of 116 mm, compared with a minimum of 158 mm in an adult Panther. Total length about 180 to 270 cm; tail length about 50 to 80 cm; maximum weight about 90 kg.

DISTRIBUTION AND VARIATION: Originally from northern British Columbia, southern Alberta, southwestern Saskatchewan, northwestern and central North Dakota, northern Minnesota, southern Ontario and Quebec, Newfoundland, and Nova Scotia southward throughout the United States, Mexico, Central America, and South America to Patagonia; now extirpated from most of the eastern and central United States and Canada, and from parts of the West. Undoubtedly a few individuals of *F. c. coryi* still persist in the wilder sections of the state, but the unreliability of some reports leads to uncertainty regarding the present distribution. (See Hamilton, 1943, and Pearson, 1954.)

828. Jaguarundi, *Felis yagouaroundi* (Berlandier)

IDENTIFICATION: A medium-sized cat with a long neck and a long, rather bushy tail. Because of great differences in size, it could hardly be confused with the Panther. (Both species are rare in Florida.) An immature, especially if seen in poor light, might be mistaken for a large house cat. One color phase is *bright reddish brown* except for the white throat and under side of the head; the other is *uniformly slate gray*. Total length about 90 to 140 cm; tail length, 33 to 60 cm. The skull differs from the Panther's in its smaller size (less than 116 mm long) and from that of a house cat in having 2 teeth posterior to the largest upper molariform tooth.

DISTRIBUTION AND VARIATION: South America, Central America, and the lowlands of Mexico, reaching the United States in southeastern Arizona and extreme southern Texas. Reports of this animal have come from various parts of the Florida Peninsula (photo examined). The nearest geographic race is *F. y. cacomitli* in southern Texas, but it is problematic whether Florida specimens have invaded undetected along the Gulf Coast or represent introductions. In the latter case their subspecific identity would be conjectural. "Specimens identified as Jaguarundis" from Jena and near Lake Placid (de Vos et al., 1956) probably have not been studied critically.

829. Bobcat, *Lynx rufus* (Schreber)

IDENTIFICATION: The *very short tail* of this medium-sized cat makes it almost unmistakable. Its color pattern of *black spots* on a reddish, buffy, or gray ground color separates it from all other cats except an immature (and *long*-tailed) Panther. If the tail cannot be seen in the field, the *triangular ears* will aid in identification. The skull differs from that of *Felis* cats in having *only 3 pairs of upper molariform teeth*. Total length about 75 to 125 cm; tail length, 13 to 18 cm.

DISTRIBUTION AND VARIATION: Southern Canada southward throughout the United States to southern Baja California and, at higher elevations, southeastern Mexico. The Florida race is *L. r. floridanus* which occurs as far south as Lower Matecumbe Key.

ORDER PINNIPEDIA: CLXIV. FAMILY PHOCIDAE, Hair Seals. With their *flipperlike limbs*, seals can hardly be mistaken for any other Florida mammals, as whales, dolphins, and sea cows lack the hind limbs. *The skull* resembles that of the Carnivora, but *has only 2 or 4 lower incisors*. Seals of the family Phocidae are not able to rotate their hind limbs forward and *lack the pinnae of land mammals*.

830. Harbor Seal, *Phoca vitulina* L.

IDENTIFICATION: The smallest of the 3 seals recorded in Florida (maximum length 120 to 180 cm). As in the Hooded Seal, the *claws are well developed* on both front and hind limbs, but males lack the rostral pouch of that seal. It is the only seal in Florida with *3 pairs of upper incisors*. As in the West Indian Seal, the premaxilla makes contact with the nasal. Color variable, but usually spotted or mottled.

DISTRIBUTION AND VARIATION: Arctic, Pacific, and Atlantic coasts southward to Baja California, South Carolina, and Florida (one record of *P. v. concolor* near Daytona Beach; Caldwell and Caldwell, 1969a); also up St. Lawrence River to Lake Ontario.

831. West Indian Seal, *Monachus tropicalis* (Gray)

IDENTIFICATION: *Upperparts plain brownish gray*, underparts whitish. *Claws of hind foot vestigial*. Skull more slender than that of Hooded Seal, with *2 pairs of lower incisors*, and the *premaxilla in contact with the nasal*. Total length about 150 to 230 cm.

DISTRIBUTION: Bahama Islands to Honduras; recorded in Florida prior to 1932 (Keys and Cape Florida; Ray, 1961). Probably extinct (Caldwell, personal communication).

832. Hooded Seal, *Cystophora cristata* (Erxleben)

IDENTIFICATION: Variable in coloration: gray to bluish or blackish above, sometimes whitish below. Males have an *inflatable pouch on the snout* and usually *darker or white blotches on the sides*. Females are virtually without the pouch and may lack spots on the sides. In any case, this seal may be distinguished from the West Indian Seal by the *well-developed claws on its hind toes*. The skull also differs, having *only one pair of lower incisors*, and the *premaxilla not in contact with the nasal*. Maximum length of males more than 300 cm; of females, 240 cm.

DISTRIBUTION: North Atlantic from Canada to Iceland, ranging southward in winter, once as far as Brevard County, Florida (Moore, 1953).

ORDER SIRENIA: CLXV. FAMILY TRICHECHIDAE, Manatees. Large marine animals with flat, *paddlelike tails*, flipperlike forelimbs, and *no hind limbs.* Muzzle broad and covered with bristles. Pinnae lacking and hair vestigial.

833. Manatee, *Trichechus manatus* L.

IDENTIFICATION: (See description of family.) The Manatee is easily separated from seals by the *absence of hind limbs,* and from whales and porpoises by the *paddle-shaped tail,* the forward-directed eyes, the shape of the head, and the presence of vestigial nails. Color evenly grayish black. Total length 240 to 450 cm; weight to 680 kg. The skull can be distinguished from most others by its large diastema, and from those of cetaceans by its shape and by the presence of small incisors in young.

DISTRIBUTION AND VARIATION: Atlantic and Gulf Coasts from North Carolina (Beaufort) and the mouth of the Rio Grande southeastward through the West Indies and Central America into northern South America. *T. m. latirostris* inhabits the southeastern United States and enters the lower parts of Florida rivers, ascending farther up the St. Johns, Caloosahatchee, and Suwannee Rivers (Map 68). The Suwannee River apparently marks its normal northward limit on the Gulf Coast (Layne, 1965a), but strays reach the Panhandle (Moore, 1951). Restricted to south Florida in winter.

ORDER CETACEA: CLXVI. FAMILY DELPHINIDAE, Dolphins. The *generally fishlike shape and absence of hind limbs* distinguish members of this order from all other mammals. The skull is recognized by its *homodont teeth* (or *absence* of teeth) and *upward-directed nares.* The Delphinidae are small to medium-sized animals with upper and lower jaws of *about equal length* and each with *numerous teeth* (in most species). The dorsal fin is well developed and about midway on the back. Rostrum long (equal to or longer than remainder of skull).

834. Rough-toothed Dolphin, *Steno bredanensis* (Lesson)

IDENTIFICATION: A small, dark blue black dolphin with white underparts and a *gradually sloping profile.* The lower jaws are united for a greater distance than in other members of the family (25 percent length of ramus). The common name derives from the fact that the *crowns of the teeth are rugose.* The teeth are larger and fewer than in some dolphins (less than 30 per upper row). Maximum length about 245 cm.

DISTRIBUTION: Almost cosmomarine. Florida records come from Taylor County (16 stranded) and Tampa (Layne, 1965a).

835. Cuvier's Dolphin, *Stenella frontalis* Cuvier

IDENTIFICATION: A small, blackish, beaked dolphin; underparts whitish. The profile is indented at the base of the beak, which is sepa-

Map 68. Distribution of the Manatee (*Trichechus manatus latirostris*)

rated by a groove from the forehead. With the Spotted Dolphin it shares a shorter rostrum (less than 60 percent of skull length) and fewer teeth (less than 45 per upper row). However, it *lacks the numerous small, white spots* of that porpoise, has shorter flippers (less than 20 percent of total length), and has a skull width of less than 192 mm. Maximum length about 185 cm.

DISTRIBUTION: Western Atlantic southward to Florida; south temperate and tropical eastern Atlantic; Japan. Florida records are near Miami and Sebastian (Moore, 1953).

836. Long-beaked Dolphin, *Stenella longirostris* (Gray)

IDENTIFICATION: A small, beaked dolphin, dark gray dorsally, lighter below, but mottled above and below. The profile is noticeably indented and grooved at the base of the beak. Unlike other members of the genus in Florida, its *rostrum makes up more than 60 percent of the total skull length*; each upper jaw contains more than 45 teeth. The flipper is relatively small (less than 20 percent of total length). Maximum length about 215 cm.

DISTRIBUTION: Locally in the Pacific, Atlantic, and Indian Oceans. Two published records in Florida: one 73 km off Miami Beach (Moore, 1953) and 36 stranded on Dog Island, Franklin County, September, 1961 (Layne, 1965a); also unpublished records near St. Petersburg and Fort Walton Beach (Caldwell, personal communication).

837. Spotted Dolphin, *Stenella plagiodon* (Cope)

IDENTIFICATION: A small, dark blue gray, beaked dolphin *with light spots*; grayish on the sides. Like several other dolphins, its profile is sharply indented and grooved at the base of the beak. Diagnostic skull characters are the shorter rostrum (less than 60 percent of the total skull length) and the *number of teeth in each upper row* (about 35). The *flipper length is about 20 percent of the total length*, which reaches 215 cm.

DISTRIBUTION: North Atlantic Ocean to the Gulf of Mexico and Caribbean Sea; also coast of Brazil. Recorded at various points along the Florida coast, both Atlantic and Gulf, especially during the warmer months.

838. Gray's Dolphin, *Stenella coeruleoalba* (Gray)

IDENTIFICATION: General form and size as in other members of its genus, but easily distinguished by dark lines running from eye to anus and from eye to flipper. Color otherwise blackish above and white below, except black posterior to vent. The flipper is about 15 percent of total length. There are about 45 to 50 teeth in each upper row and about 75 vertebrae. Maximum length about 245 cm.

DISTRIBUTION: Atlantic Ocean from Greenland to Jamaica; Mediterranean Sea; Pacific Ocean from Bering Sea to New Zealand; 2 records at Indian Rocks Beach, Florida (Caldwell and Caldwell, 1969b; Caldwell, personal communication).

839. Common Dolphin, *Delphinus delphis* L.

IDENTIFICATION: A small cetacean with a definite beak set apart from the forehead by a groove, which leads to a *black ring around the eye*. Blackish above, grayish laterally and ventrally, alternating on sides with bands of yellow and white. The skull may be recognized by the *deep palatal grooves* and the *numerous teeth* (47 or more per row). Maximum length about 250 cm.

DISTRIBUTION AND VARIATION: Temperate and tropical oceans. *D. d. delphis* has been recorded in Florida only in St. Johns County in January and February (Moore, 1953; Layne, 1965a).

840. Atlantic Bottle-nosed Dolphin, *Tursiops truncatus* Montagu

IDENTIFICATION: A dark slate dolphin with white underparts and a *short beak* (not more than 3 percent of total length). *No groove between beak and forehead.* Each tooth row contains about 20 to 25 teeth. Maximum length about 370 cm.

DISTRIBUTION AND VARIATION: Virtually cosmomarine. *T. t. truncatus* is the common "porpoise" of Florida waters. Enters fresh water in large rivers such as the St. Johns (Caldwell, personal communication).

841. False Killer Whale, *Pseudorca crassidens* (Owen)

IDENTIFICATION: Easily distinguished from the Killer Whale by its entirely blackish color pattern, more falcate dorsal and pectoral fins, and the fact that the pterygoids contact one another. Its head profile, diameter of the teeth, and width of the rostral base distinguish it from the Short-finned Pilot Whale. The Pygmy Killer Whale (*Feresa*) is very similar in form, but much smaller. Maximum length about 550 cm.

DISTRIBUTION: Temperate and tropical oceans. Florida records come from Indian River, St. Lucie, Palm Beach, Broward, and Dade Counties (Moore, 1953; Caldwell, Caldwell, and Walker, 1970).

842. Killer Whale, *Orcinus orca* (L.)

IDENTIFICATION: This species, Risso's Dolphin, and the Short-finned Pilot Whale lack the beak of other dolphins. The Killer Whale has a triangular, erect dorsal fin that is longer than wide. The color pattern is a contrasting black dorsally and white ventrally, with a prominent white slash over each eye. The basal width of the rostrum is not more than 70 percent of its length, and the teeth are more than 13 mm in diameter. The pectoral fin is rounded and paddle-shaped. Maximum length about 9 m.

DISTRIBUTION: Cosmomarine.[1] Florida records range from Marineland through the Keys to Collier County, December to March (Moore, 1953; Layne, 1965a), and near Destin (Caldwell, Layne, and Siebenaler, 1956).

843. Risso's Dolphin, *Grampus griseus* (Cuvier)

IDENTIFICATION: Body form, especially the *truncate snout*, as in the Short-finned Pilot Whale. It differs from that species in having a *largely white snout with a vertical crease in front* and a shorter flipper. Moreover, it differs from all other members of the family in usually lacking teeth on the upper jaw and in never having more than 7 pairs on

---

1. Includes *Grampus (=Orcinus) rectipinna* of Hall and Kelson, 1959.

the lower jaw. Coloration dark above (frequently with lighter scratch-marks), mostly whitish below. Maximum length about 400 cm.

DISTRIBUTION: Widespread in temperate and tropical seas; in the western Atlantic, from Massachusetts to Texas, the Lesser Antilles, and Florida (near St. Augustine and Tarpon Springs; Paul, 1968; Caldwell, personal communication).

844. Short-finned Pilot Whale, *Globicephala macrorhyncha* Gray

IDENTIFICATION: This cetacean differs from most others in the Delphinidae in its *truncate snout*, and from Risso's Dolphin in having the *entire body largely black*. The flipper is long and slender (about 20 percent of total length). It differs from most non-beaked dolphins in having a wider rostral base (at least 80 percent length of rostrum) and smaller teeth (less than 13 mm in diameter). Maximum length about 8.5 m.

DISTRIBUTION: Temperate and tropical oceans of the world. There are many records of strandings all around the coast of Florida, representing collectively nearly all months of the year.

ORDER CETACEA: CLXVII. FAMILY PHYSETERIDAE, Sperm Whales. Whales of varying size and a *truncate snout*. The *lower jaw is comparatively small*, being shorter than the upper. The skull features *at least 9 teeth in each lower jaw, but none in the upper jaw*.

845. Sperm Whale, *Physeter catodon* L.

IDENTIFICATION: (See description of family.) The Sperm Whale is much larger than the Pygmy Sperm Whale (males up to 18 or 19 m long), has a *median row of bumps* replacing the dorsal fin, and the *lower jaw is comparatively small* and much shorter than the upper. Coloration dark bluish, sometimes lighter below. The lower jaw contains 20 or more pairs of teeth, and the *zygomatic arch is complete*.

DISTRIBUTION: Cosmomarine. Florida records have come from Gulf, Franklin, and Broward Counties, Englewood, Naples, Marco Island, and Delray Beach, representing most months of the year.

846. Pygmy Sperm Whale, *Kogia breviceps* (Blainville)

IDENTIFICATION: (See description of family.) This small whale differs from the Sperm Whale by having a *normally developed dorsal fin*, and the upper jaw projects only slightly beyond the lower. The flipper is relatively longer than in the Sperm Whale. Color blackish above, fading to whitish below. *Zygomatic arch incomplete*; lower jaw with 9 to 15 pairs of teeth. Maximum length about 4 m.

DISTRIBUTION: Oceans of temperate (rarely tropical) zones. Numerous Florida records on Atlantic Coast and one each at St. Petersburg and Destin; no records for May through July (Caldwell and Siebenaler, 1960; Layne, 1965a).

ORDER CETACEA: CLXVIII. FAMILY HYPEROODONTIDAE, Beaked Whales. Distinguished from other whales by having *not more than one or 2 teeth in evidence on each lower jaw*, upper and lower jaws of about equal length, and a *grooved throat*. They have no baleen plates and only one blowhole.

847. Antillean Beaked Whale, *Mesoplodon europaeus* Gervais

IDENTIFICATION: In this genus the angle of the jaws is almost directly below the eye. Color pattern similar to that of *M. mirus*, but *without spotting. No teeth at tip of lower jaw*. Maximum length about 450 cm.

DISTRIBUTION: Western Atlantic Ocean from Long Island southward to Gulf of Mexico and Trinidad; also English Channel (Moore, 1960).

848. True's Beaked Whale, *Mesoplodon mirus* True

IDENTIFICATION: Differs from *M. europaeus* in having a pair of teeth *at the tip* of the lower jaw. There seems to be no consistent color difference in the 2 species. Both are slate black above and lighter below, but in some individuals of *M. mirus* the sides are flecked with light spots, the ventral midline is darkened, and there is a gray anterior to the vent. Maximum length about 500 cm.

DISTRIBUTION: North Atlantic Ocean southward to Flagler Beach, Florida, and the British Isles; also in south Atlantic off Africa. See Moore, 1957 and 1966.

849. Goose-beaked Whale, *Ziphius cavirostris* Cuvier

IDENTIFICATION: Coloration not distinctive; grayish, darker above than below. Differs from other Florida members of this family in that the *angle of the jaws is far anterior to the eye*. The *nasal bone is widened and almost obscures the narial openings* from the dorsal aspect. Maximum length about 8 or 9 m.

DISTRIBUTION: Northern and southern oceans, some from the North Atlantic reaching Gulf of Mexico and many West Indian islands; also in Pacific Ocean. Recorded in Florida at St. Augustine, Daytona Beach, Eau Gallie, and in Pasco County (Moore, 1953; Hansen and Weaver, 1963; Layne, 1965a; Caldwell, personal communication).

ORDER CETACEA: CLXIX. FAMILY BALAENOPTERIDAE, Fin-backed Whales. Very large whales with *baleen plates* (whalebone) *replacing teeth* in the upper jaw. *Two blowholes* and a dorsal fin are present, the latter small and far back on the body. *The outer surface of the throat is conspicuously grooved.* The baleen plates are shorter and wider than in the Balaenidae, and the rostrum wider. The family includes the largest living animals.

850. Little Piked Whale, *Balaenoptera acutorostrata* Lacépède

IDENTIFICATION: A smaller baleen whale whose throat grooves extend to its chin. The dorsal coloration is gray black, with white ventrally and white markings on the flipper. The baleen plates are yellowish white and shorter than in those of other members of the genus—less than 30 cm long. The dorsal fin is about two-thirds of the distance from snout to fluke. Maximum length about 10 m.

DISTRIBUTION: Cold and temperate oceans. Recorded twice on the Florida Keys and once near the mouth of the Aucilla River (Moore, 1953; Moore and Palmer, 1955).

851. Sei Whale, *Balaenoptera borealis* Lesson

IDENTIFICATION: A very large baleen whale with small pectoral and dorsal fins, the dorsal about two-thirds of the distance from snout to fluke. The *pectoral fin lacks the white spot* of the Little Piked Whale, and it is less than 15 percent of the animal's total length. The dorsal coloration is blue gray, most of the underparts being whitish. The *throat grooves do not reach the chin*, unlike those of *B. acutorostrata*. The *baleen plates are chiefly blackish* and range in length up to 74 cm. Maximum length about 18 or 19 m.

DISTRIBUTION: Cold and temperate oceans. One record off Duval County, Florida, May, 1919 (Moore, 1953).

852. Bryde's Whale, *Balaenoptera edeni* Anderson

IDENTIFICATION: Very similar to the Sei Whale, but with the following differences: dorsolateral ridges sometimes present from tip of snout to blowholes; ventral grooves extending back to region of umbilicus; bristles of baleen plates coarse and gray; baleen plates usually shorter; several differences in body and skull proportions (Omura, *in* Norris, 1966). Maximum length about 15 or 16 m.

DISTRIBUTION: Widespread in Pacific Ocean; also eastern and tropical parts of Atlantic; one record at Panacea, Florida, April 2–4, 1965 (Rice, 1965).

853. Finback Whale, *Balaenoptera physalus* (L.)

IDENTIFICATION: An enormous whale with a conspicuously grooved throat, a protruding lower jaw, and small pectoral and *dorsal fins*, the latter *placed more than three-fourths of the way back*. Coloration gray black above and white below. The *flipper is shorter* than that of *Megaptera* (less than 15 percent of total length) and lacks the white patch of the Little Piked Whale. The *length of the baleen plates ranges up to 90 cm*, and they are *not yellowish* as in the Little Piked Whale. Maximum length about 25 m.

DISTRIBUTION: Cosmomarine. One record at Ormond Beach, Florida, May, 1950 (Moore, 1953).

854. Humpback Whale, *Megaptera novaeangliae* (Borowski)

IDENTIFICATION: Easily distinguished from other baleen whales by its *longer pectoral fin* (about one-third of total length). It is further distinguished by *fleshy knobs around the mouth*. The *throat grooves are widely spaced* and do not extend to the chin, and the baleen plates are blackish. The skull may be recognized (in addition to the lack of teeth) by the *convex outline of the relatively wide rostrum*. Coloration black above, whitish below. Maximum length about 15 or 16 m.

DISTRIBUTION: Cosmomarine. Florida records consist of one photographed off Egmont Key, April, 1962; a skull from Delray Beach (Layne, 1965a); and many seen 65 km off Miami (Moore, 1953).

ORDER CETACEA: CLXX. FAMILY BALAENIDAE, Whalebone Whales. Large cetaceans similar to those of the Balaenopteridae in the *absence of teeth* and the *presence of baleen plates and 2 blowholes*. They differ, however, in the *absence of a dorsal fin and of throat grooves*. The skull can be recognized by its *more narrow rostrum* (width of unflared base only 25 percent length of rostrum). Size very large.

855. Atlantic Right Whale, *Eubalaena glacialis* (Borowski)

IDENTIFICATION: This large whale can be separated from those of similar appearance by its *strongly arched lower jaw line*, a character that should be visible at some distance. (Also see description of family.) The color is chiefly blackish, as is that of the baleen plates. Maximum length about 17 m.

DISTRIBUTION AND VARIATION: Northern and southern oceans, *E. g. glacialis* wintering south to the Bermuda Islands and Florida. Individuals have been seen off the east coast of Florida from January to March (Duval, Flagler, St. Johns, Brevard, and Indian River Counties; Layne, 1965a; Caldwell and Caldwell, 1971), and in the Gulf of Mexico (Moore and Clark, 1963).

ORDER ARTIODACTYLA: CLXXI. FAMILY SUIDAE, Pigs.[1] This order contains the only *wild* hoofed animals in Florida. They differ from such domestic hoofed animals as horses in the fact that the foot has *2 functional hoofs*. Pigs (Suidae) differ from deer in their *sparse, stiff hairs* and *3 pairs of upper incisors*. The *upper canines project laterally to form tusks*, and no horns or antlers are present in either sex.

856. Pig, *Sus scrofa* L.[2]

IDENTIFICATION: (See description of family.) The Pig, with its sparse, bristly hair, truncate, naked snout, short legs, and spiral tail, is too well known to require a detailed description. Feral animals are smaller and rangier than some domestic ones.

DISTRIBUTION: Although domestic pigs occur throughout Florida, and many are not contained in pens, it is not always obvious which ones are fully self-supporting. Truly feral individuals are not easily observed, but many exist throughout most or all of the Florida mainland. Belonging to the same species is the Wild Boar, of Europe, introduced into the mountains of the Southeast and on islands off California.

ORDER ARTIODACTYLA: CLXXII. FAMILY CERVIDAE, Deer. Long-legged hoofed mammals with 2 functional toes (hoofs) and antlers (at least in males of North American species). Body hair neither unusually sparse nor stiff. *Upper incisors lacking.* Upper canines rudimentary or lacking in North American species.

857. Sambar Deer, *Cervus unicolor* (Cuvier)

IDENTIFICATION: A large, uniformly brownish deer with a *shaggy mane on the throat, 3-tined antlers in the male*, and unspotted young. There is a basal tine of the antler projecting forward, and the 2 distal tines result from a *dichotomous fork* posteriad; they may be of equal length (adults) or unequal (immatures). Both sexes have *rudimentary upper canine teeth*. The posterior portion of the nasal cavity, unlike that of *Odocoileus*, is *not divided by the vomer*. Length, 185 to 215 cm; weight, 160 to 340 kg (males larger than females). (Brooke, 1878; Schaller, 1967.)

DISTRIBUTION: Native to India, Ceylon, Malay Peninsula, East Indies, and Philippine Islands. Introduced and evidently established on islands of the southwest Pacific, in eastern Australia, and on St. Vincent Island, Florida (Allen, 1952; de Vos et al., 1956).

---

1. Feral goats (Family Bovidae) are established on Little St. George Island (Franklin County).
2. The East Indian Pig (*Sus vittatus*) may also be in the ancestry of the Florida animal.

858. Axis Deer, *Axis axis* (Erxleben)

IDENTIFICATION: A small, reddish brown deer with *white spots at least part of the year*. The under side of the tail is white, as in *Odocoileus*. It resembles the Sambar Deer, but differs from the White-tailed, in having the posterior portion of the nasal cavity not divided by the vomer and the *antlers 3-pronged*. There are no upper canines. Unlike the Sambar Deer, it has the *middle tine of the antler longer than the others*. Total length about 155 cm; weight about 32 to 36 kg (Brooke, 1878; Walker, 1964).

DISTRIBUTION: Native to Ceylon and India. Introduced and evidently established in New Zealand, Hawaii, Brazil, Argentina, and in Florida in an area between the St. Johns River and the Atlantic Coast (Allen, 1952; de Vos et al., 1956); possibly established in Australia.

859. White-tailed Deer, *Odocoileus virginianus* Zimmermann

IDENTIFICATION: A medium-large, reddish brown (grayer in winter?) native Florida deer; fawns white-spotted. Its habit of raising the tail in flight makes the *white under side of the tail* an excellent field mark, but it is the *only* species of deer over most of the state. The relative *tail length* (longer than the ear) *is greater than in the other 2 Florida species*. The antlers of this deer can be distinguished by the nondichotomous branching, and the skull by the fact that the *vomer separates the posterior portion of the nasal cavity* medially. Upper canines are rarely present. Total length about 140 to 180 cm; tail length, 15 to 33 cm; ear length, 14 to 23 cm; weight, 23 to 135 kg. Deer on the lower Keys are much smaller than those on the mainland.

DISTRIBUTION AND VARIATION: Central British Columbia (except near coast), southern parts of Alberta, Saskatchewan, and Manitoba, northern Ontario, southern Quebec, and southern Newfoundland southward to southern Oregon, extreme northeastern California, northwestern and central Colorado, New Mexico (except northwestern), central and southern Arizona, through Mexico and Central America into South America; also to the Gulf Coast and Florida Keys. Introduced in Finland, New Zealand, Cuba, "and other islands" (de Vos et al., 1956). Two subspecies inhabit the Florida mainland: *O. v. osceolus*, Panhandle eastward to about Madison County and southward near the Gulf to Tampa Bay; *O. v. seminolus*, remainder of the Peninsula. The much smaller Key Deer, *O. v. clavius*, is restricted to a range of about 27 by 24 km on the lower Keys, centered around Big Pine Key (Allen, 1952).

859H. HYPOTHETICAL LIST

Three species of mammals are known to occur just outside Florida, and their occurrence within the state seems possible. Two of these, the

Swamp Rabbit (*Sylvilagus aquaticus*) and the Muskrat (*Ondatra zibethicus*), occur in the Mobile Bay area of Alabama, and either species might occur in the northwestern corner of Florida. A rabbit that I collected near Century, Florida, was considered a hybrid between the Swamp Rabbit and the Marsh Rabbit (*Sylvilagus palustris*) (Shanholtzer, 1967). At least one specimen of the Star-nosed Mole (*Condylura cristata*) has been collected in the Okefenokee Swamp in Georgia. Extensions of this swamp cross the Florida line, suggesting the possibility that this burrowing mammal may be there, as proved to be the case with the Carpenter Frog (*Rana virgatipes*).

Squirrel Monkeys (*Saimiri* sp.) are living a feral existence near Silver Springs, Florida (King, 1968), and may eventually be accorded a place on this state's faunal lists.

ADDENDA

Based on more recent information, 21 species may be added to the foregoing list, 10 of which are exotic fishes intentionally or inadvertently introduced. One species is an amphibian, 7 are birds, and 3 are mammals. I am deeply indebted to Walter R. Courtenay for information regarding the Florida status of all fishes in this section.

860. Armored Catfish, *Hypostomus* sp.

IDENTIFICATION: The members of this family (Loricariidae) differ from most Florida fishes in their *covering of enlarged plates*, arranged in 3 or 4 longitudinal rows on each side. The mouth is ventral and the lips thick and fleshy. In *Hypostomus* the ventral surface is covered with splinters of bone. Also the first dorsal fin is long-rayed, the peduncle rather slender, and the *ventral lobe of the caudal fin longer than the dorsal lobe*. There are barbels around the mouth. Some species attain total lengths of 380 mm or more.

DISTRIBUTION: The family is restricted to northern and central South America. At least one species of *Hypostomus*, thus far unidentified, is established in Hillsborough and Dade Counties, Florida.

861. Black Molly, *Poecilia latipinna* x *velifera* (?)

IDENTIFICATION: This fish and the four species immediately following in this list belong to the Family Poeciliidae (no. XXV), for which a description may be found in the text. The hybrid known as the Black Molly differs from typical *Poecilia latipinna* (Sailfin Molly) in its larger dorsal fin, with backward-curving rays, and its color pattern—*lustrous black with reddish spots along the edge of the dorsal fin*. Maximum length about 115 mm (Sterba, 1962).

DISTRIBUTION: Established in Hillsborough, Palm Beach, Broward, and Dade Counties, Florida.

862. Liberty Molly, *Poecilia sphenops* (Valenciennes)

IDENTIFICATION: Differs from *P. latipinna* in its lower counts of rays in the dorsal (8 to 11) and anal (8 to 10) fins. Males typically have much *brilliant orange red in the dorsal and caudal fins*, with black bars and spots, but the coloration varies geographically. The body is iridescent blue above, whiter below, with longitudinal rows of light spots. Females are plainer, but may show longitudinal rows of reddish spots. Length to 115 mm (Sterba, 1962).

DISTRIBUTION: Mexico to Colombia; euryhaline. Established in Hillsborough County, Florida.

863. Green Swordtail, *Xiphophorus helleri* Heckel

IDENTIFICATION: Males are readily recognized by the *long sword-like process from the ventral lobe of the caudal fin*. In both sexes the number of rays in the dorsal fin (11 to 14) is somewhat high for the genus, but this fin is always smaller than in *Poecilia*. Males are mostly greenish, with longitudinal stripes of reddish, blue, or duller colors, and there may be rows of reddish spots in the dorsal fin. Females are similar in color, but lack the sword and have a normally expanded anal fin. Maximum length about 130 mm (Sterba, 1962).

DISTRIBUTION: Southern Mexico to Guatemala near the Gulf Coast. Established in Palm Beach and Hillsborough Counties, Florida.

864. Platy, *Xiphophorus maculatus* (Günther)

IDENTIFICATION: Both sexes resemble *Gambusia* in body form, but not in color. The wild type has a background of various pale colors, with *large dark blotches* frequent, especially *on the peduncle and dorsal fin*. The rays in the dorsal and anal fins may also be partly darkened. In females the scales are large (greatest diameter about equal to that of eye), and in both sexes *some scales may be outlined with black*. Brightly colored varieties occur, at least in aquaria. There are 10 rays in the dorsal fin. Maximum length about 65 mm (Sterba, 1962).

DISTRIBUTION: Near Gulf of Mexico in Mexico and Guatemala. Introduced in Florida and established in Hillsborough and Palm Beach Counties.

865. Variegated Platy, *Xiphophorus variatus* (Meek)

IDENTIFICATION: As the name implies, highly variable in color, but dorsal fin often with black-tipped rays and caudal fin often with a greenish tip. The relative development of the dorsal fin is almost equal to that in some mollies (*Poecilia*). Counts of fin rays and oblique scale rows are similar to those in *X. maculatus*. Maximum length 65 to 75 mm (Sterba, 1962).

DISTRIBUTION: Southern Mexico. Established in Hillsborough and Palm Beach Counties, Florida.

866. Jack Dempsey, *Cichlasoma biocellatum* Regan

IDENTIFICATION: This fish and the three that follow in this list are members of the Family Cichlidae (XXXIX), distinguished by *2 incomplete lateral lines*, the posterior portion of each more ventral than the anterior one. Of the several Florida species in this family, the present one is distinguished by its *large number of spines*, about 19 in the dorsal fin and 8 in the anal. It is heavy-set, with thick lips and a protruding lower jaw. Ground color dark to medium dark, *heavily spotted with blue* on the median fins, body, and head. Caudal fin somewhat rounded. Young have several dark vertical bars. Maximum length about 180 mm (Sterba, 1962).

DISTRIBUTION: Much of the Amazon Basin. Established in Palm Beach and Hillsborough Counties, Florida.

867. Rio Grande Perch, *Cichlasoma cyanoguttatum* (Baird and Girard)

IDENTIFICATION: Adults are easily distinguished by their mostly *sky blue coloration*, but the young, with dark vertical bars, resemble those of *C. biocellatum*. There are from 15 to 18 spines in the dorsal fin and 5 in the anal. Females are less colorful than males. Maximum length about 300 mm (Sterba, 1962).

DISTRIBUTION: Southern Texas and northern Mexico. Introduced and established in Polk County, Florida.

868. Jewelfish, *Hemichromis bimaculatus* Gill

IDENTIFICATION: The 3 anal spines are the smallest number in any Florida cichlids except the two species of *Tilapia*; there are 13 to 15 spines in the dorsal. It is also more slender than most others and has a more rounded caudal fin. Ground color greenish or yellowish to reddish, with *3 large, dark blotches along sides*—one on the operculum, one at midbody, and one at the caudal base. These blotches may be connected by a longitudinal dark band, and there may be fainter vertical dark bands. Maximum length about 150 mm (Sterba, 1962).

DISTRIBUTION: Northern and central Africa. Introduced and established in Dade County, Florida.

869. Blue Tilapia, *Tilapia aurea* (Steindachner)[1]

IDENTIFICATION: With 3 anal spines, as in the Jewelfish, but 15 or 16 dorsal spines; also each fin has slightly more rays than in that species (10–14 and 9–11, respectively). It differs from *Tilapia melanotheron* in having *fewer gill rakers* (8–12) *on the first gill arch*, but the color pattern of this species is not distinctive. Maximum length about 85 mm (Boulenger, 1915).

DISTRIBUTION: Native to West Africa, but introduced and established in Florida from Pasco and Orange Counties southward to De Soto and Manatee (possibly Sarasota) Counties.

870. Pine Barrens Tree Frog, *Hyla andersoni* Baird

IDENTIFICATION: Resembles the Green Tree Frog in dorsal aspect, but *additional white lines are present* on the head, the limbs, and running around the vent, and the *underparts are dark* (to lavender in life); also much concealed orange on hind legs. Average length of adults about 40 mm.

DISTRIBUTION AND VARIATION: Until 1970, known in only three disjunct areas—the southern parts of New Jersey and North Carolina and near the Savannah River in Georgia. A colony found recently in the northern half of Okaloosa and Walton Counties, Florida (Christman, 1970), is thought to be subspecifically distinct (Means, personal communication).

871. Bar-tailed Godwit, *Limosa lapponica* (L.)

IDENTIFICATION: Total length about 380 to 450 mm. Similar in color to the Marbled Godwit, but ground color of head, neck, and underparts more grayish, and *upper tail coverts not barred*. The tarsus is relatively shorter than in that species (see Key), and the tail is whiter between the dark bars.

DISTRIBUTION: Breeds in the Arctic Zone of Eurasia and Alaska. Winters chiefly along the coasts (except Arctic) of the Old World, but not south of the Equator in Africa. Accidental in North America, including a Florida record near Cocoa in the winter of 1970–71 (photograph in *Fla. Nat.*, 44:62).

872. Lesser Black-backed Gull, *Larus fuscus* L.

IDENTIFICATION: After the first year, very similar in plumage to the Great Black-backed Gull (*L. marinus*), but much smaller (near size of the Herring Gull) and with *yellowish legs* in adults. First-year plumage as in Herring Gull.

---

1. *Tilapia melanopleura* of Boulenger (1915).

DISTRIBUTION: Breeds along the coast of northern Europe. Winters southward to northern Africa, also casually to eastern North America. Several Florida records, including two specimens (USF, TT).

873. Black-headed Gull, *Larus ridibundus* L.

IDENTIFICATION: Adults in winter are similar in size and color pattern to those of the Laughing and Franklin's Gulls, but the *legs and bill are more reddish* and the *outermost primaries almost entirely white*; the vestige of blackish on the head may be even more restricted. It is also similar to these two species in the immature plumage, but at that stage the *legs and feet are yellowish*. Adults in breeding plumage have the *head brown* rather than black.

DISTRIBUTION: Breeds in the northernmost parts of Europe, central Russia, and in northern and central Asia. Winters southward to northern Africa, southern Asia, the Philippines, and rarely to the eastern United States and Puerto Rico; recognizably photographed at Cocoa, Florida, winter of 1971–72 (John Edscorn, personal communication); also 2 previous sight records.

874. Canary-winged Parakeet, *Brotogeris versicolurus* (Müller)

IDENTIFICATION: Total length about 230 mm. A green parakeet with a large yellow patch in each wing and a long, graduated tail.

DISTRIBUTION AND VARIATION: Tropical portions of South America from French Guiana and southeastern Colombia southward to Bolivia, Paraguay, and northern Argentina; recently introduced at Miami, Florida, and apparently established by 1971; also seen at Fort Lauderdale and Islamorado (*B . v . versicolurus*; Owre, 1973).

875. Antillean Palm Swift, *Tachornis phoenicobia* Gosse

IDENTIFICATION: Smaller than the Chimney Swift or Bank Swallow (total length 95 to 105 mm). Black above except for a white rump; underparts white *with a dark collar*. Tail slightly forked, the rectrices not spinose.

DISTRIBUTION: Cuba, Isle of Pines, Jamaica, Hispaniola, and a few smaller islands; accidental at Key West, Florida, in the summer of 1972 (Ogden, 1972; photographs by J. Edscorn and P. Sykes).

876. Bahama Woodstar, *Calliphlox evelynae* (Bourcier)

IDENTIFICATION: About size of Ruby-throated Hummingbird, but with the *tail deeply forked in the male*; gorget purplish. Similar to the Rufous Hummingbird in that both sexes have much brownish on the underparts and tail, but the *upperparts are mostly greenish*.

DISTRIBUTION: Bahama Islands; two records near lower East Coast (Fisk, *American Birds*, 28:855).

877. Ipswich Sparrow, *Passerculus princeps* Maynard[1]

IDENTIFICATION: Very similar to the paler races of the Savannah Sparrow, but larger (total length 160 to 170 mm, wing length 75 to 82 mm). Little or no yellow on lores.

DISTRIBUTION: Breeds almost entirely on Sable Island (Nova Scotia); winters on Atlantic Coast from Massachusetts to Georgia and northeastern Florida (several sight records and one specimen). The only photograph that has come to my attention surely did not represent this species.

878. Pygmy Killer Whale, *Feresa attenuata* Gray

IDENTIFICATION: Form similar to that of *Pseudorca*, but much smaller (maximum length about 300 cm). Dark gray to black, with large white patches ventrally and smaller white areas around mouth. Dorsal fin about 12 percent of total length. There are about 70 vertebrae and 10 to 13 teeth in each upper row (Nishiwaki, in Ridgway, 1972).

DISTRIBUTION: Known from scattered localities in the Atlantic, Pacific, and Indian Oceans (Best, 1970); there are 4 or 5 Florida records through February 1976 (Caldwell, personal communication).

879. Dwarf Sperm Whale, *Kogia simus* Owen

IDENTIFICATION: Strongly similar to the Pygmy Sperm Whale in form, size, and coloration, but averaging slightly smaller (maximum length about 270 cm). Probably the best external distinction between the two is the position of the dorsal fin, halfway between the anterior and posterior tips in *K. simus*, more than halfway in *K. breviceps*; this fin is also relatively larger in *K. simus*. Other distinctive features of *K. simus* are the presence of one to 3 pairs of maxillary teeth (with exceptions?), less than 12 teeth per mandibular row, and no ventral keel on the mandibular symphysis (Handley, *in* Norris, 1966; Nishiwaki, *in* Ridgway, 1972).

DISTRIBUTION: Widespread in warmer oceans, with records as far north as Japan and south to southern Australia; Florida records come from St. Augustine, Cape Canaveral, Destin, and the Keys (Caldwell, personal communication).

---

1. Now reduced to a subspecies of the Savannah Sparrow, *P. sandwichensis* (A.O.U., 1973).

880. Blainville's Beaked Whale, *Mesoplodon densirostris* (Blainville)

IDENTIFICATION: A rather small, dark beaked whale similar in form and size to *M. europaeus* and *M. mirus* (maximum length about 450 cm). The position of the dorsal fin is somewhat more anterior than in those species (about 60 percent of distance from anterior to posterior tips). The teeth are farther back on the lower jaw than in the other 2 species (posterior to the mandibular symphysis) (Nishiwaki, *in* Ridgway, 1972).

DISTRIBUTION: Atlantic and Indian Oceans; one Florida record at Crescent Beach, June, 1969 (Caldwell and Caldwell, 1971).

Chapter Four

Collecting and Preserving
Vertebrates

LOWER VERTEBRATES

COLLECTING: Most lower vertebrates may be freely handled without fear of harm to the captor. Poisonous snakes are obvious exceptions. The most important principle to observe in capturing the latter is to keep the head securely pinned down until the neck may be grasped and firmly held *immediately* behind the head. Because large nonvenomous snakes may also inflict painful bites, they are best handled in the same way. Caution should also be exercised in handling several other lower vertebrates. The Giant Toad (*Bufo marinus*), now established in parts of southern Florida, and the Cuban Tree Frog (*Hyla septentrionalis*), may cause some distress to their handlers if secretions from their parotoid glands or skin are inadvertently applied to eyes, lips, or other sensitive areas. Spines of some fishes may inflict painful wounds, especially those of certain catfishes (Siluriformes).

TRANSPORTING: Fishes should be carried in an ample quantity of water that is exposed to air. Amphibians may become too dry if carried for an hour or more, a situation that can easily be prevented by keeping them in jars with perforated lids, or in plastic bags, along with some plant matter and moist soil. Small lizards can be treated in like fashion, but reptiles generally (and large or venomous snakes particularly) are best carried in cloth bags. A drawstring at the top is unnecessary, the best procedure being to tie a knot in the top of the bag itself. There is little danger of snake bites through such a bag if normal precautions are exercised.

KILLING: Three basic methods of killing reptiles are drowning, injecting, or etherizing. Full-strength alcohol or a 10 percent formalin solution drowns them much more quickly than water. (As it is commonly sold, formalin is at full strength, even though the information on the label may seem to indicate otherwise. Therefore it must be diluted with 9 parts of water to obtain a 10 percent solution.) Full-strength alcohol or 10 percent formalin may also be injected with a hypodermic syringe into the body cavity to kill reptiles and large amphibians, although small fishes and amphibians may simply be dropped into a less concentrated solution of either fluid. A 10 percent solution of veterinary pentobarbital is excellent for injections. If time is not too important, even the largest reptiles may be killed by enclosing them in sealed jars with a wad of cotton soaked in ether. Use of chloroform is *not* recommended. Also, reptiles and amphibians are easily killed by freezing.

A more desirable procedure for killing salamanders, and perhaps for small frogs and fishes, is by anesthetizing them in a solution of Chloretone—chloral hydrate, 28 grams (1 ounce) per gallon of water. (These crystals will dissolve more readily if a small amount of alcohol is added.) This leaves the specimens in a much more relaxed condition than do the preserving fluids. However, as it is not a preserving fluid itself, they should be transferred to 10 percent formalin or full-strength alcohol within a few hours after death.

Alternatively, these animals may be killed with a 50 percent solution of isopropyl alcohol, arranged in the desired position between cloths or paper towels, and then hardened with 10 percent formalin.

HARDENING: For proper preservation, all parts of the body must be penetrated by the preservative—preferably 10 percent formalin. (In an emergency, full-strength alcohol may be used for reptiles, but *not* for amphibians.) Small amphibians may simply be dropped into such a solution and allowed to remain for a few hours, but penetration will not be effective for large specimens unless other steps are taken. For larger amphibians and reptiles formalin may be injected into the body, head, limbs, and tail, or cuts may be made in these places to permit ready entrance of the preservative. In fishes more than 125 mm long, a cut is usually made in the right side. If alcohol is used, the specimen may be transferred after a few days directly to alcohol of a lower concentration. A 40 percent solution of isopropyl alcohol, or 70 percent for ethyl alcohol, is ideal. If the hardening agent becomes discolored during the few days the specimen is kept there, it should be changed at least once before transferring the specimen to the final preservative. Whenever formalin is used for hardening, the specimen should be soaked in water afterward for an approximately equal period of time if it is to be stored in alcohol. Formalin itself, at a strength of 5 to 10 percent, may be used as the final preservative, but it has the disadvantage of being irritating to the user, affecting sensitive parts of the body, such as the eyes. The larger the specimen the more care must be taken to see that the hardening agent penetrates thoroughly (especially in turtles), and several cuts must always be made in very large specimens. The mouths of salamanders and turtles must be propped open during the hardening process. Care should be taken when using syringes not to insert them into skeletal parts, as they may be broken or become plugged. The amount of the hardening agent should be three or four times the mass of all specimens to be hardened in it. All equipment used for hardening should be thoroughly rinsed with water after use, especially if formalin is used as the hardening agent.

PRESERVATION: The preferred preservative for most animals is 40 percent isopropyl alcohol or 70 percent ethyl alcohol, but 10 percent formalin is often used for turtles. Fading of brightly colored animals seems inevitable, but it may be reduced by storage of specimens in a cool, dark place. Color may be preserved better by adding a small amount of Ionol (about 1 percent by volume) to a solution of 10 percent formalin.

LABELS: For scientific purposes a specimen without data is of no value. The minimum information required is the date and place of capture and the name of the collector. Notes on ecology, color, and other data are also desirable. These data should be recorded on waterproofed paper or parchment with Higgins India or Eternal ink. If these are not available, a hard-lead pencil (3H or 4H) is acceptable. The label should be kept in the jar with the specimen(s), though not necessarily attached to it unless other specimens in the same jar have a different set of data. Individually tagged specimens facilitate the correlation of data with specimens when studied.

BIRDS

No responsible ornithologist advocates the indiscriminate shooting of birds, even for museum purposes. Perhaps unfortunately, a view of the opposite extreme is often encountered—that is, against collecting birds under any circumstances. Without a doubt, the truth regarding the advisability of collecting lies somewhere between these two extremes.

Among the reasons that may be advanced for occasional collecting of birds are these:

Certain institutions need to build representative collections of museum skins to enable students of general ornithology to study the characteristics of various species of common birds, as well as those of larger taxonomic groups.

A few larger collections of museum skins are necessary to allow advanced students of ornithology to study variations *within* species. Such a study may lead to descriptions of new subspecies, the validation or revision of those previously described, or to determining the subspecific identity of an unknown specimen.

Field observations of a very unusual nature should be supported, whenever feasible, by a preserved specimen. Only the observer himself can be certain of the record's accuracy without a specimen, and there is abundant evidence that he is sometimes in error. However strong his faith in the observer, no reviewer of field records can ever be *positive* of their accuracy without the substantiating evidence of a specimen (or recognizable photograph). With such evidence he can be sure that, at least, the identification of the species was correct.

In order to collect most birds in the United States one must obtain a permit issued by the U.S. Fish and Wildlife Service. This permit currently prohibits the collecting of any bird on a federal refuge or other sanctuary or the collecting of such rare or endangered species as the Great White Heron, Trumpeter Swan, Whooping and Sandhill Cranes, Eskimo Curlew, Bald Eagle, Ivory-billed Woodpecker, and others.

In addition to the federal permit, most states require one of their own. The Florida permit now in use does not authorize collecting of the Reddish Egret, Glossy Ibis, Roseate Spoonbill, American Flamingo, most species of kites, Caracara, Short-tailed Hawk, Limpkin, or certain other rare species. Special, more restrictive permits are sometimes issued.

Certain species of birds are not protected by law and may be shot in Florida without a permit, although the wisdom of extensively shooting even these is open to question. Such species are the House (or English) Sparrow, Starling, Common Crow, and Fish Crow. It should be emphasized that *all hawks and owls* are now protected by law.

Suitable methods of collecting may vary somewhat according to the purpose of the collection. It is always desirable to obtain the specimen with the least possible damage to it, and unless collecting it is imperative this philosophy may determine the method employed. Generally speaking, birds caught in traps or Japanese mist nets are in better condition than birds that have been shot. Netting and trapping are indiscriminate, but healthy birds may always be released. Numbers of the unprotected House Sparrow may often be caught in sparrow traps baited with chicken grain and used for practice by beginning students. For any particular bird, a shotgun is a more dependable method of collecting. For the purpose of substantiating an abnormal record, collecting the bird may be more important than turning out a perfect museum skin. Sometimes the observer has only one shot and cannot afford to miss. Even in such situations, however, he should keep in mind that some kind of recognizable skin must be prepared, therefore the distance, load, and other factors of the collecting attempt should be considered. Wounded or captive birds may be killed quickly by compressing the sides of the breast near the wings.

Once the bird has been collected it must be preserved until it can be skinned. Blood should be rinsed off with cold water or a weak solution of ammonia in water. If soap is used, it may be removed with white gasoline. In any case, the feathers should then be dried with fine sawdust and "fluffed" for best results. Cotton should be stuffed into the throat and vent, or the throat only in the case of small birds. The bird should be transported in such a way as to avoid ruffling of the feathers. The skin should be made up within a few hours of the death of the specimen, longer in periods of cool weather; otherwise refrigeration will be necessary. In the field, decay may be postponed by injecting the specimen with a saturated alum solution. Small birds may be kept for a few days and larger birds for a week or more at a temperature slightly above freezing. For longer periods, specimens should be kept frozen, in which condition they will be preserved almost indefinitely. In some cases, however, the skin on the head may become too dry within one year unless they are kept in air-tight containers. Thawing requires about two hours for a small bird, but at least twice as long for a larger specimen. It may be hastened by the use of goose-neck lamps or other sources of heat in small quantities. Fine sawdust or corn meal may be used to absorb the water given off as it thaws.

Before the actual skinning is begun certain data must be obtained and recorded. Invariably, these should include the bird's total length—a straight-line measurement taken from the tip of its bill to the tip of its longest tail feather as the bird lies on its back, neck extended but not stretched. Specimens still in *rigor mortis* must first be flexed to measure length or wing spread accurately. Recording the weight at this time is also desirable. Various other measurements may be reliably taken after the skin has been prepared, but the total length of the finished skin often varies from that of the original specimen by 10 percent or more. Either before or shortly after skinning, any bright colors of such unfeathered areas as the bill, cere, lores, eye (iris), lower legs, and feet should be recorded since such colors tend to fade in the museum specimen. (Rarely they may fade within an hour after collecting.)

With the bird lying on its back, begin a cut at the posterior end of the sternum to extend to the vent (Fig. 11). The utmost caution should be employed to insure that this cut be only through the thin skin for if it penetrates the body cavity the skin may be ruined. As the cut is extended posteriad the skin should be separated from the underlying tissues in a lateral direction until the leg is encountered. During this procedure and subsequently, fine sawdust (or corn meal) should be used copiously to absorb blood or other fluids. As the skin is worked over the knee, that joint may be grasped with forceps, pulled upward, and ampu-

Fig. 11. Preparing a bird skin—Opening cut

tated. Following this, the rest of the leg should be skinned almost to the joint below (heel). Now cut the tendons and remove the flesh from this segment of the leg before pushing the skin back over the leg. Follow the same procedure for the other leg. The skin must now be separated from the ventral body wall backward to the vent and downward to the backbone. When the latter has been accomplished, carefully push a dull probe, followed by one point of the scissors, under the vertebral column so as to cut it along with the rest of the body just anterior to the vent (Fig. 12). In making this cut exercise extreme care not to cut through the skin of the back or the bases of the tail feathers.

The skin may now be pushed down the back toward the wings. Since these are eventually to be cut at the shoulder, they may be broken from the body there whenever they interfere with the skinning process. Each wing should then be skinned and fleshed in turn as far out as possible onto the second segment without loosening the secondaries. Remove muscles on this segment by cutting their tendons opposite the elbow and pulling them down toward the wrist, near which they should be cut.

After cleaning the wing bones, work the skin down the neck to the head. It can be forced over the head carefully without tearing except in birds with a relatively large head (ducks, crows, hawks, owls, woodpeckers, and so forth). When it is apparent that it cannot be forced over the head, a longitudinal cut should be made in the neck (about 2 or 3 cm long, depending on the size of the bird) at the base of the skull (Fig. 13).

Fig. 12. Cut at base of tail

Fig. 13. Cut in skin of neck

This may be done on the dorsal, ventral, or lateral surface. In any case it will have to be sewn together before the skin is completed.

As the skin is pushed over the head it will slip out of the ear of small birds, at least if pressure is applied, but will have to be cut out close to the bone on larger ones. It should be forced completely past the eyes, then loosened by cutting the connective tissue attaching it to the skull. This will make it possible to remove the eyes by carefully cutting around them and finally cutting or tearing the optic nerve under each. Be careful not to puncture the eyeball, else the surrounding feathers may be soiled.

When all flesh and tissue have been removed from the eye socket the skull may be cut across vertically at its base (through occipital bone). The base of this cut may then be extended forward on the right and left sides of the roof of the mouth (Fig. 14).

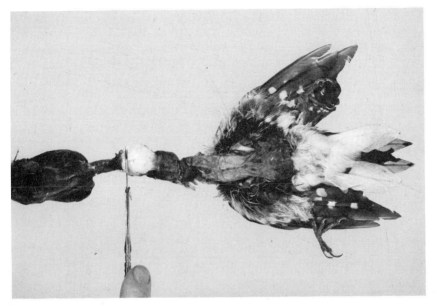

Fig. 14. Cut at base of skull

The brain may sometimes be removed intact by separating its meninges from the inner surface of the skull; in any case all soft tissue must be thoroughly removed and the skull cleaned out with a cotton swab. Small muscles and other bits of tissue around the bill and jaws must be removed, especially in large birds.

With a dull blade, all fat and other loose tissue should be scraped from the skin, scraping toward the head. In this process meal or sawdust must be used to absorb grease. Remove the flesh from the base, and the oil gland (if present) from the upper side, of the tail. Sew the scapular feather tracts of the back closer together, approximating the distance between the scapulae on the bird's body. Turn legs and wings inside out. The skin is now ready for poisoning. Dust borax (or arsenic, or a mixture of the two) over the inner surface of the skin, skull, wing and leg bones, and base of the tail. The heads of the humeri should be tied closer together so that these bones lie approximately parallel. Insert a small, rounded cotton ball through the skull into each eye socket, holding it with the forceps and smoothing its outer surface with the fingers. (Alternatively, this may be done later through the mouth.) The skull may now be pushed back through the skin, which may have to be moistened if it has become too dry. A dull probe may then be used to arrange the feathers on the head and neck by slipping it under the skin at the edges of the eye sockets and gently massaging the inside of the skin. Feathers may be dried and sawdust removed by blowing on them, or by the use of compressed air.

A thin layer of cotton should be wrapped around each leg bone in place of the muscles. Then the body is filled out by wrapping cotton around a stick to simulate the original body. This stick should be just the right length to reach the base of the tail from the base of the upper mandible without stretching the skin, unless it is to project from the body (see below). The sharpened end of the stick should be *pushed up into* the base of the upper mandible. If the stick is not properly lodged there, the specimen will have a wobbly head and the dried skin of the neck will eventually split open. The opposite end may finally be inserted into the base of the tail, but care should be exercised not to throw any of the tail feathers out of line in doing so. This end of the stick may instead project well beyond the body. When the cotton body is in place additional pieces of cotton may be added as needed. It is desirable to insert a small piece of cotton through the mouth to fill out the throat.

The cut in the skin may now be sewed together. The seam should begin at the anterior end of the cut and proceed posteriad, making each stitch from the inside out. The bill should be tied shut by running the thread through the nostrils and around the lower mandible, then tying. The lower legs are tied together in a crossed position, or so tied to the stick if it projects. The label must be tied securely to them, but not directly against them (Fig. 15). Pertinent data to be recorded on the label include the locality and date of collection, name of collector, name of preparator, sex of the specimen, and its total length. The scientific name of the bird is often recorded also, and a museum number will be added later.

The sex may be determined at this time (if it has not been determined previously) by dissection of the body. The ovary (usually on the left side only) will be relatively large and granular (tapioca-like) in appearance, the testes smaller, more compact, and bean-shaped. Both lie in the lower back near the anterior end of the kidneys.

The final appearance of the specimen will depend to a large extent on how it is arranged for drying. It should be symmetrical, with the wings folded practically on its back. Anatomical parts should be arranged so as to show special features (webbed or lobed toes, white outer rectrices, and so forth). If it is wrapped in cotton, pinned to cardboard, or placed in a close-fitting, flat-bottomed metal trough, in the correct position and permitted to dry for several days, its permanent appearance will be assured. Care should also be taken that the feathers on the crown are not ruffled before drying.

The beginner should not be discouraged by the results of his first few attempts at making museum skins. Experience is the most important factor in his ultimate success.

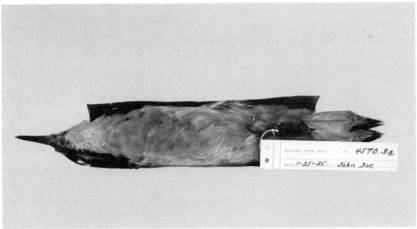

Fig. 15. Completed skin ready for drying

MAMMALS

The basic procedures to follow in preserving mammal skins are similar to those already described for the preparation of bird skins, so the emphasis here is on the differences, rather than the similarities, between the two processes. Also the reasons for collecting mammals are almost identical to those for collecting birds. In Florida, however, no special permits are required in order to collect most kinds of nongame mammals.

The methods employed by mammal collectors are more varied than those used by ornithologists. Guns are widely used for various species, but the kind of gun and the load should be determined by the size of the mammal. Also the same Japanese mist nets used for birds are effective in the capture of bats. Unlike birds, however, many small mammals are caught in specially devised traps. Snap traps of various sizes may be used for rats, mice, and shrews. (Other types of traps are available if live specimens are desired.) Steel traps are used for larger mammals, but present some disadvantages, as well as dangers to domestic animals and children. Special traps are available for moles and gophers.

Four standard measurements are made of mammals *before* the skin is prepared (see page 6). These are recorded in millimeters on the label in the following order: total length, length of tail, length of hind foot, length of ear. These numbers are separated by dashes. The following data taken from a Cotton Rat (*Sigmodon hispidus*) will serve for illustration: 285–122–32–18. An additional measurement is taken in bats, the length of the tragus (see Glossary). As these parts are subject to shrinkage when dried, it is imperative to take the measurements from the specimen before skinning. Also the skin (total length) may be stretched or shortened in the skinning process. The sex should also be recorded on the label at this time.

Skinning is begun in the mammal, as in birds, on the ventral side, beginning near the forelegs. The hind leg is cut through the knee joint in both instances. An important difference is that the skin must be removed from the tail of a mammal. With small mammals this can be done best by first separating the skin from the posterior end of the body around the base of the tail. Then the base of the tail may be grasped (inside the skin) with the thumb and 2 fingers, nails gripping the tail vertebrae and pushing against the skin of the tail. In this way the skin may be *pushed* off the tail and not inverted, thus saving much time and trouble. In any case, if the skin cannot be removed in this way, a longitudinal cut must be made in order to remove the tail vertebrae, and this must later be sewed up.

Once past the tail, push the skin down over the body and skin out each foreleg, cutting at the elbow joint. All flesh should be removed from both front and hind legs. Carefully cut around the base of the ears as they are encountered. As the skull will not be left in the completed skin, the eyes may be carefully cut loose from the skin but left, for the present, in the skull. When the skin has been inverted to the tip of the snout, cut it loose from the skull anterior to the nasal bone. In like manner, separate it from the tip of the lower jaw. The lips may be sewed together with three stitches forming a triangle, one perforation near the tip of the lower lip, one on each side of the tip of the upper lip.

Although borax is commonly used as a preservative for both bird and mammal skins, some claim that it causes certain colors of the feathers or hairs to fade. Although I know of no experimental evidence of this, the possibility may be avoided by the use of arsenic. This, of course, presents other dangers. If used, it should be applied by the use of forceps holding a wad of cotton, and the hands should be washed afterward as a double precaution. All fat on the inner surface of the skin must be removed before any preservative is applied.

The cotton body of the mammal skin (especially small ones) is not usually supported by a stick or wire, as is that of a bird skin. Cut or tear off a square of cotton somewhat wider than the length of the animal's body (including the head). This should be rolled until the circumference, when slightly compressed, is equal to that of the body. Now cut off the excess length from what will be the ventral side, then fold the remaining dorsal flap over to produce a smooth, rounded posterior end for the body. At this time the anterior end should be attenuated so as to fit the head and nose. This calls for compressing it from the dorsal and lateral sides.

The inverted skin may now be turned over the cotton body. Begin by holding the nose end (with forceps) against the inner surface of the nose, then turn the skin of the head and body, successively, over the cotton body, insuring the symmetrical arrangement of the eyes and ears as you pass them. (Before the body skin is turned back you will need to remove the forceps.) As in birds, the bones of the upper legs must also be wrapped with cotton. Some preparators, prior to this process, insert a piece of wire alongside the bones of each leg. In any case, the legs should be arranged ultimately parallel to the plane of the body, soles down, with the forelegs pointing anteriad and the hind legs posteriad.

It is now necessary to prepare a support for the tail. Usually this is a piece of wire of smaller diameter than that of the tail. In fact, it may have to be filed on one end to reach the tip of the tail. It should also be long enough to project for an inch or so into the body cavity, placed ventrally there. Depending on the relative diameter of the wire and the tail, it may be advisable to wrap firmly a thin layer of cotton around the wire, dusted with the preservative you are using. If not, the preservative should be sprinkled into the open end of the tail, the tip pointed down. The wire cannot be pushed to the tip of the tail if the diameter is too great, or if the cotton is loosely wrapped.

Sewing of the ventral cut does not differ materially from the same process in birds. When this has been done, the sex should be determined and recorded on the tag (if not earlier). In the case of females, the presence or absence of embryos, including the number and their length if present, should be written on the label. The specimen is then laid on a large enough piece of cardboard, ventral side down, and arranged in the proper position for drying by pins stuck alongside the legs and tail. If the ears are long, as in rabbits, they should be laid back against the head and held by pins or tied down. With rabbits it is also advisable to secure the long hind legs by tying them to a thin piece of wood which may run under, or into, the body.

Attention may now be given to the skull (or more of the skeleton), which must bear a label showing either the same data as that for the skin or otherwise identifying it with the skin. The brain, larger muscles, and other soft parts should be removed. Ideally, the skeletal parts should then be dried and placed in a colony of dermestid beetles or mealworm beetles, the larvae of which feed on the dried flesh. Occasionally some damage may also be done to the cartilage if the specimens remain in the colony too long. After the beetles have done their work, some preparators prefer to whiten the skulls by soaking them in ammonia (full strength) for at least 30 minutes. They may then be stored, with their labels, in cardboard boxes or in glass vials plugged with cotton.

A quicker method that dissolves the flesh is that of soaking the skull in a bleach. If this method is used, the skull of small mammals must be carefully watched at frequent intervals and removed before any damage is done to thin bones.

When beetles are not available, skeletons may be macerated (cleaned by bacterial action in water). A tight lid on the container will prevent the escape of the odor. The brain and most of the flesh should be removed first. (See Hall and Kelson, 1959.)

Glossary

The terms selected for this glossary are chiefly those that are necessary for the successful use of this key, excluding those defined in the key itself. Generally speaking, terms that should be in the student's general vocabulary do not appear here unless their meaning in connection with some group of vertebrates is somewhat specialized.

FISHES

ADIPOSE FIN   A fatty or fleshy median fin that lacks rays and spines. Located dorsally and posteriad on catfishes and other species.

ADNATE FIN   A median fin (usually dorsal) that is attached for its full length to the back or fused to the caudal fin.

ANADROMOUS   Migrating into fresh water to breed.

ANAL FIN   The median fin that lies immediately posterior to the anus (Fig. 1).

APPRESSED   Pressed against the body.

BARBEL   A fleshy projection on the head or jaw region of certain fishes (chubs and catfishes, for example).

BASICAUDAL   At the base of the caudal fin.

BRANCHIOSTEGAL (—RAY)   One of the supporting skeletal elements in the *branchiostegal membrane*, at the ventral edge of the operculum.

CATADROMOUS   Migrating into salt water to breed.

CAUDAL FIN   The tail fin (Fig. 1).

DEPTH   The maximum depth of a fish's body, or other part. The body depth is measured exclusive of fins.

DISK   The body of any greatly flattened fish of rounded contour.

DORSAL FIN   Any of the median fins on the upper surface of a fish (Fig. 1).

DORSUM   The upper surface.

EMARGINATE   Indented or very slightly forked. Generally used with reference to the caudal fin of fishes.

EURYHALINE   Inhabiting, to some degree, both fresh and salt water.

FRENUM   A ridge connecting the snout with the upper margin of the mouth.

GAPE   The distance across the mouth.

GENITAL PAPILLA   A fleshy projection just anterior to the anal fin in certain fishes.

GILL ARCH   The skeletal support of the gill. Usually curved or angled, thus resulting in *upper* and *lower limbs*.

GILL FILAMENT   One of the many slender, vascularized projections from the posterior edge of a gill arch.

GILL MEMBRANE   The membrane that connects the lower edge of the operculum with the throat, or with the corresponding membrane of the other operculum.

GILL RAKER   One of the fleshy projections from the anterior edge of a gill arch. The shape in various species may vary from short and thick to long and slender.

GONOPODIUM   The modified anal fin found in males of the Poeciliidae; used as an intromittent organ.

GULAR PLATE   A hardened (sometimes bony) plate located ventrally between the opercula of certain primitive fishes.

HETEROCERCAL   That type of caudal fin in which the vertebrae turn up toward or into the upper lobe, which is accordingly longer than the lower lobe; an asymmetrical caudal fin.

HOMOCERCAL   That type of caudal fin in which the vertebrae do not turn up toward the dorsal lobe, with the result that the two lobes are of about equal length; an externally symmetrical caudal fin.

HUMERAL SCALE   A scale (usually enlarged) located above the base of the pectoral fin.

INFERIOR MOUTH   One that opens downward, with the upper jaw projecting beyond the lower.

INSERTION   The posterior or ventral connection of a fin.

ISTHMUS   The narrowed portion of the throat region that projects anteriad between the opercula.

LABIAL GROOVE   A groove extending from some part of the edge of the mouth.

LATERAL LINE   A system of tubes (one per scale) extending from the upper edge of the operculum to, or toward, the caudal fin. May be either complete, incomplete, or absent in the various species. (Not to be confused with the dark lateral *stripe* of many species.) Also see page 5 and Fig. 1.

LENGTH   See page 5.

MANDIBULAR PORES   Small paired openings along the lower jaw that connect with the mandibular canal of the lateral-line system.

MAXILLA   The chief bone of the upper jaw, lying posterior to the premaxilla, or dorsal to its posterior end (Fig. 1).

MELANOPHORE   A cell that contains dark pigment granules.

OPERCULUM   The bony gill cover (Fig. 1); also *opercle*.

ORBIT   The bony socket that contains the eye.

ORIGIN   The anterior or dorsal connection of a fin.

PECTORAL FINS   The pair of fins attached to the sides of the body immediately behind the head (Fig. 1).

PEDUNCLE   The narrow posterior portion of the body to which the caudal fin is attached; extends from the anus to the base of the caudal fin (Fig. 1).

PELVIC FINS   The ventrally located pair of fins that lie posterior to (or more ventral than) the pectoral fins. May be closer to anterior or posterior end in various fishes (Fig. 1).

PERITONEUM   The shiny inner lining of the body wall.

PREDORSAL   Anterior to the (first) dorsal fin.

PREMAXILLARY   The bone that is located at the tip of the upper jaw (Fig. 1).

PREOPERCLE   An L-shaped bone forming the anterior part of the operculum.

PROTRACTILE   Capable of being thrust out.

RAY   One of the flexible skeletal elements of a fin. They also differ from spines in being segmented and usually branched (Fig. 1).

RETICULATE   In the form of a network or chainlike arrangement.

SCALLOPED   Having an undulating margin.

SNOUT   The part of the head lying between the eye and the tip of the upper jaw.

SPINE   One of the stiffened skeletal elements of a fin. It differs from a ray in being unbranched and unsegmented (Fig. 1).

STANDARD LENGTH   See page 5.

SUBABDOMINAL   Almost abdominal (between thoracic and abdominal).

SUPERIOR MOUTH   One that opens upward and has the lower jaw projecting beyond the upper.

SUPRAMAXILLA   A small bone lying above the posterior part of the maxilla in certain fishes.

TERMINAL MOUTH   That type of mouth that opens directly in front and has jaws of about equal length.

TRUNCATE   Square-tipped; generally used with reference to the caudal fin.

VILLIFORM TEETH   Slender, crowded teeth, all of which are about equal in size.

ABDOMINAL PLATES   A pair of large plates located at about the middle of the plastron in turtles (Fig. 4).

ALVEOLAR SURFACE   The masticating surface in a turtle's mouth, immediately to the inside of the cutting edge. Usually represented in the Emydidae by a pair of low, rounded ridges.

ANAL PLATE[1]   An enlarged scale (or scales, if divided) immediately anterior to the vent of a snake; also a pair of plates in a similar position on the plastron of turtles (Fig. 4). (In lizards, the scales at this location are called *preanals*.)

AURICULARS   Modified scales at the anterior edge of a lizard's ear.

CARAPACE   The upper part of a turtle's shell.

CAUDAL PLATES   See SUBCAUDAL PLATES.

CHIN SHIELDS   Enlarged scales on the under side of the head in lizards and snakes (Figs. 5 and 6). In snakes there are 2 pairs typically elongate and arranged longitudinally.

CIRRUS (plural, CIRRI)   A fleshy projection from the front of the upper lip in certain salamanders.

COSTAL GROOVES   The downward-curving grooves that alternate with the positions of the ribs in certain salamanders. They are counted as described on page 5.

COSTAL (LATERAL) PLATES   On the carapace of a turtle, the row of plates lying immediately lateral to the medial row, that is, the vertebrals (Fig. 4).

CRANIAL CRESTS   Hardened ridges on the head of a toad.

DEWLAP   An erectile and skeletally supported fold of skin on the throat of anoles (*Anolis*), where it is brightly or contrastingly colored.

DORSAL FIN   The membranous upper portion of the tail in the larvae of frogs, toads, and certain salamanders (Fig. 2).

DORSAL SCALES   The small, unmodified scales covering the back and sides of a snake or lizard. The number of longitudinal rows ranges from about 13 to 30 in Florida snakes.

DORSOLATERAL RIDGE   A ridge of skin running along the side of the back in certain frogs.

FEMORAL PLATES   A pair of plates on the plastron of turtles just anterior to the anal plates (Fig. 4).

FRONTAL PLATE   The enlarged median scale located between the eyes of reptiles; smaller and more numerous in some lizards (Figs. 5 and 6).

---

1. The terms "plate," "scale," and "scute" are often used interchangeably. It seems appropriate to restrict the use of "plate" to the larger, more specialized scales such as those on the heads of lizards and snakes and covering the shells of turtles.

FRONTONASAL PLATES One or more scales lying anterior to the prefrontals, posterior to the internasals, and dorsal to the loreals in lizards (Fig. 5); also present in turtles.

FRONTOPARIETAL PLATES A pair of plates lying between the frontal and parietal plates in lizards (Fig. 5).

GULAR PLATES A pair of plates at the anterior tip of the plastron in turtles; sometimes single (Fig. 4).

GULAR FOLD See DEWLAP.

HORNY BEAK The hardened, partly pigmented upper and lower mandibles of the larval anuran (Fig. 3).

HUMERAL PLATES The second pair of plates on the plastron of a turtle, lying immediately behind the gulars (Fig. 4).

INFRALABIAL PLATES See LABIAL PLATES.

INFRAMARGINAL PLATES Small scales located between the bridge and the marginal plates of certain turtles.

INTERNASAL PLATES A pair (usually) of scales lying between the nasals. In snakes and most lizards they are thus immediately posterior to the rostral (Figs. 5 and 6). Also called SUPRANASALS in lizards.

INTERPARIETAL PLATE A median scale lying between the parietals in lizards and containing the parietal eye, if present (Fig. 5).

LABIAL PLATES The enlarged scales bordering the mouth. Those on the upper jaw are termed UPPER LABIALS (SUPRALABIALS), those on the lower jaw LOWER LABIALS (INFRALABIALS). Scales at the tip of the lower jaw and behind the corner of the mouth are not considered labials (Figs. 5 and 6).

LABIAL TEETH Small horny processes arranged parallel to one another like the teeth of a comb in rows above and below the horny beak of tadpoles. Commonly there are one or 2 rows above and 2 or 3 rows below (Fig. 3).

LATERAL PLATES See COSTAL PLATES.

LOREAL SCALE An elongate scale located between the eye and the nostril in snakes, but usually not bordering either. Lacking in certain species of snakes, and one or 2 rows in certain lizards (Figs. 5 and 6).

MARGINAL PLATES The relatively small plates arranged around the edge of the carapace (except the anterior tip) in turtles (Fig. 4).

MENTAL PLATE The median scale at the tip of the lower jaw in lizards and snakes (Figs. 5 and 6).

NASAL PLATE The scale that surrounds the nostril; in snakes it is sometimes divided into a PRENASAL and a POSTNASAL. The latter type is sometimes numerous in lizards (Figs. 5 and 6).

NASOLABIAL GROOVE A small furrow running from the nostril to the upper lip of certain salamanders; may appear to be only an unpigmented streak.

NUCHAL PLATE   A single plate anterior to the vertebrals of turtles (Fig. 4); also enlarged scales at the back of the head in lizards (Fig. 5).

PAPILLAE   The fleshy protuberances surrounding much of the mouth of a tadpole (Fig. 3).

PARAMEDIAN STRIPES   A pair of light stripes lying just on each side of the midline of the head and neck of certain turtles.

PARASPHENOID TEETH   Small teeth located on the roof of the mouth behind the vomerine teeth of certain salamanders. They overlie the parasphenoid bone.

PARIETAL PLATES   A pair of plates on the sides of the head and just behind the frontal (or frontoparietal) in lizards and snakes (Figs. 5 and 6).

PAROTOID GLANDS   A pair of raised skin glands located behind the eyes of toads and certain other anurans.

PECTORAL PLATES   The third pair of plastral plates in turtles, lying immediately behind the humerals (Fig. 4).

PLASTRON   The lower part of a turtle's shell.

POSTLABIAL PLATES   Enlarged scales immediately posterior to the upper labials (Figs. 5 and 6).

POSTMENTAL PLATES   One or 2 scales on the midventral line immediately posterior to the mental scale in lizards (Fig. 5).

POSTNASAL PLATE   See NASAL PLATE.

POSTORBITAL RIDGE   A lateral extension of the cranial crest.

PREFRONTAL PLATES   The enlarged scales immediately anterior (often extending somewhat lateral) to the frontal. Typically one or 2 pairs, but rarely single (Figs. 5 and 6).

PRENASAL PLATE   See NASAL PLATE.

PREOCULAR SCALES   One or more scales at the anterior edge of the eye; absent in some species. Usually elongate vertically, especially if single.

RETICULATE   See Fish section of Glossary.

ROSTRAL PLATE   The single, more or less enlarged scale at the tip of the snout (Figs. 5 and 6).

SPIRACLE   The excurrent tube on the left side (usually) of a tadpole's body (Fig. 2).

SUBCAUDAL PLATES (CAUDAL PLATES)   The scales on the under side of the tail; typically enlarged and in 2 rows in snakes, but less specialized and smaller in other reptiles.

SUBLABIAL PLATES   Enlarged scales below and medial to the lower labials in lizards.

SUBOCULAR SCALES   One or more relatively small scales lying between the eye and the upper labials; absent in most snakes.

SUPERCILIARY SCALES   Small scales on the upper eyelid of lizards (Fig. 5).

SUPRALABIAL PLATES   See LABIAL PLATES.

SUPRANASAL PLATES   See INTERNASAL PLATES.

SUPRAOCULAR PLATES   One or more enlarged scales lying above the orbit in snakes and lizards (Figs. 5 and 6).

TAIL CRESTS   The finlike portions of a tadpole's tail, especially the dorsal half (DORSAL FIN) (Fig. 2).

TAIL MUSCULATURE   The thickened, axial portion of a tadpole's tail (Fig. 2).

TEMPORAL PLATES   Enlarged scales located on the sides of the head, behind the postoculars, in snakes and lizards. The 2 or more descending rows, beginning anteriad, are sometimes called the PRIMARY, SECONDARY, and TERTIARY temporals (Figs. 5 and 6).

TRANSVERSE LAMELLAE   Transverse rows of soft, overlapping scales under the toes of certain lizards.

TYMPANUM   The eardrum; rounded, smooth, and conspicuous in most anurans.

VENTRAL FIN   The membranous lower portion of the tail in the larvae of frogs, toads, and certain salamanders (Fig. 2).

VENTRAL PLATES   The scales on the ventral side of the body; large and extending across the body in all Florida snakes; somewhat enlarged in some lizards (for example, *Ophisaurus*) but not in others.

VERTEBRAL PLATES   The median row of large plates on the carapace of turtles (excepting the nuchal at the anterior edge and the marginals at the posterior tip); also known as CENTRALS (Fig. 4).

VOMERINE TEETH   Small teeth located near the anterior tip of the palate of certain amphibians.

BIRDS

AURICULARS   Somewhat modified feathers covering the ear.

AXILLARS   Feathers on the under side of the wing next to the body.

BILAMINATE   Having 2 lines of junction between the scales on the tarsus, one lateral and one medial.

CARUNCLE   A fleshy, unfeathered outgrowth about the head.

CERE   A soft, sometimes swollen area at the base of the upper mandible in doves, hawks, and certain other birds.

COMPRESSED   Laterally flattened.

CONTOUR FEATHERS   The small, unmodified feathers that cover most of the body, head, neck, wings, and upper legs; therefore, those that determine the bird's contour.

COVERTS   Small feathers that cover the bases of the large flight feathers of the wing and tail, or other specialized feathers (Fig. 7).

CREST   Elongate, often erectile feathers on the crown.

CULMEN   The middorsal ridge of the bill (Fig. 7).

DECURVED   Curving downward; used with reference to the bill.

DERTRUM   The enlarged tip of the upper mandible in plovers and certain other birds; homologous to the nail of Anatidae.

DORSUM   See Fish section of Glossary.

ENTIRE   Having the margin continuous and without indentations; used with reference to the webbing between the toes of certain water birds.

FACIAL DISK   The area around the eyes of owls, in which the feathers are radially arranged.

FRONTAL SHIELD   An expanded horny area covering the forehead in certain birds (gallinules and coots).

GAPE   See Fish section of Glossary.

GONYS   The ventral outline of the bill.

GORGET   The iridescent throat patch of male hummingbirds.

GRADUATED   Tapered; used to describe the tail of birds in which the outermost feather (rectrix) is the shortest and each succeeding feather considerably longer up to the middle pair.

GULAR POUCH   A fold of loose, bare skin at the base of the lower mandible in pelicans and their allies.

HALLUX   The hind toe (Fig. 7).

IMPERFORATE   Having the nostrils separated by a partition.

INCISED   Having the margin notched or indented.

IRIDESCENCE   The production of rainbowlike colors caused by the diffraction of light from the feathers of certain birds (for example, grackles).

LAMELLAE (singular, LAMELLA)   The thin, platelike ridges along the margins of the bill in most ducks, geese, and swans. This type of bill is said to be LAMELLATE.

LAMINIPLANTAR   That type of tarsus that is sharply keeled behind and largely invested by one large scale.

LOBED (LOBATE)   Having flanges on the side of the toes, as in the case of grebes, coots, and phalaropes.

LORE   The area between the eye and the bill.

LOWER MANDIBLE   The lower of the 2 elements of the bill; homologous to the mandible of mammals (Fig. 7).

MALAR   Pertaining to the side of the head, below the eye (Fig. 7).

MANTLE   The region of the upper back and folded wings, especially if these are unicolored and contrasting with the rest of the upperparts.

NAIL   Used with reference to the flattened dertrum of ducks, geese, and swans; also called the ROSTRAL NAIL. (The flattened claws of grebes are also called nails.)

NAPE   The back of the neck (Fig. 7).

NASAL GROOVES   The grooves in which the nostrils of some birds are located.

OCCIPUT   The back of the head (region of the occipital bone) (Fig. 7).

OPERCULATE   With a membranous lid covering the dorsal part of the nostril.

ORBITAL SKIN   Bare skin around the eye, as in pigeons and doves.

PECTINATE   Having comblike serrations, as on the middle claw of herons and egrets.

PERFORATE   Lacking a partition between the nostrils.

PLUME   A narrow, elongate feather like those appearing on the head and upper back of herons and egrets during the breeding season.

PRIMARY   One of the large flight feathers inserted on the third (outermost) segment of the wing (Fig. 7).

RAMI   The 2 basal branches of the lower mandible.

RAPTORIAL   Adapted for capturing live prey, therefore with a strongly hooked bill, large feet, and sharp, curved claws.

RECTRICES (singular, RECTRIX)   The large tail feathers; 12 in most species of Florida birds (Fig. 7).

REMIGES (singular, REMEX)   A collective term embracing the large flight feathers of the wing (primaries, secondaries, and tertiaries).

RETICULATE   See Fish section of Glossary.

RICTAL BRISTLES   Stiff, hairlike bristles (modified feathers) around the corners of the mouth (Fig. 7).

RUDIMENTARY   Imperfectly developed and of small size, as the outermost primary of many passerines. (A more appropriate term in this instance is VESTIGIAL.)

RUMP   The area just anterior to the bases of the rectrices and their coverts, and just posterior to the lower back (Fig. 7).

SCANSORIAL   Adapted for climbing, with long toes and tarsi.

SCUTELLATE   A type of tarsal scalation in which the scales are square or rectangular and arranged in rows like shingles (Plate VIII-4).

SECONDARIES   The flight feathers that are inserted into the second segment of the wing; largely hidden by primaries when the wing is folded (Fig. 7).

SEMIZYGODACTYL   Having the outer (lateral) toe reversible (Fig. 8).

SERRATE   With projections resembling the teeth of a saw, as on the edges of the bill in mergansers.

SPECULUM   A patch of contrasting color on the distal ends of the secondaries, usually visible when the wing is folded.

SQUARE   The condition of a tail tip in which all rectrices are approximately equal in length; also a wing tip with the 3 outermost primaries of nearly equal length.

SUPERCILIARY STRIPE   A white or light longitudinal stripe just above the eye (Fig. 7).

SYNDACTYL   Having the front toes (or any 2 of them) fused together at their bases.

TARSUS   The third segment of the leg (actually a part of the foot); covered with scales in most species, but feathered in some (Fig. 7).

TERTIARIES   Large feathers inserting on the basal segment of the wing; much more flexible than primaries and secondaries.

TIBIA   The second segment of the leg; completely feathered in most land birds, but with the lower part usually scaly in water birds (Fig. 7).

TOMIUM   The cutting edge of the bill (Fig. 7).

TOOTH   A pointed projection on the edge of the biil.

TOTAL LENGTH   See page 6.

UPPER MANDIBLE   The upper half of the bill; homologous to the maxilla (and premaxilla) of other vertebrates (Fig. 7).

ZYGODACTYL   Having the outer (lateral) toe permanently reversed, so that 2 toes are directed forward and 2 backward.

MAMMALS

AUDITORY BULLA   The bony capsule that encloses the middle ear. Usually appears as a rounded prominence on the skull below and posterior to the orbit (Fig. 9).

BALEEN (WHALEBONE)   Horny plates attached to the upper jaw of certain whales.

CALCAR   A hardened process of one of the tarsal bones (the calcaneum, or fibulare) that helps to support the interfemoral membrane.

CANINE TEETH   The unicuspid teeth posterior or lateral to the incisors (Fig. 9).

CUSP (TUBERCLE)   A projection on the grinding surface of a tooth. Cusps are distinguished from ridges in being *isolated* projections.

DIASTEMA   A gap in the teeth on the jawline, as in rodents and rabbits.

DIGITIGRADE   A type of foot in which only the toes contact the ground and no hoof is present.

FLIPPER   An appendage that is highly modified for swimming, such as that of a seal, manatee, whale, or dolphin.

FLUKE   The expanded tail of cetaceans.

FORAMEN (plural, FORAMINA)   An opening in a bone through which a nerve or blood vessel passes, except that the *foramen magnum* accommodates the spinal cord (Fig. 9).

FRONTAL BONE   A large bone that lies dorsally between the orbits (Fig. 9).

GUARD HAIRS   Longer hairs interspersed among other body hairs.

HETERODONT   Having teeth of differing sizes and shapes (incisors, canines, and so forth).

HOMODONT   Having teeth of similar size and shape, such as those of whales and armadillos.

INCISOR   The teeth borne at the front of the jaws, usually adapted for biting (Fig. 9).

INFRAORBITAL FORAMEN   An opening in the maxilla at the front of the orbit, and beside the rostrum; usually small.

INTERFEMORAL MEMBRANE (UROPATAGIUM)   A membrane extending between the tail and the hind leg in bats and continuous with the wing.

INTERORBITAL CONSTRICTION   The narrowed part of the skull between the orbits as viewed from the dorsal aspect.

INTERPARIETAL BONE   A bone found in certain mammals between the parietals and immediately above the occipital bone (Fig. 9).

MANDIBULAR RAMUS   The right or left side of the lower jaw.

MAXILLA   A bone forming most of the upper jaw; also called the maxillary (Fig. 9).

METACARPAL   One of the long bones in the palm of the hand (or the corresponding part of another mammal's forefoot); care must be taken to distinguish metacarpals from phalanges in bats' wings.

MIDSAGITTAL CREST   A bony ridge along the dorsal midline of the skull, for example, in bats.

MOLARS   The large teeth toward the back of each jaw, often modified for grinding (Fig. 9). (The term MOLARIFORM includes the premolars.)

NASAL BONE   A bone located around and posterior to the external nares (Fig. 9).

ORBIT   See Fish section of Glossary.

PALATAL FORAMEN   An opening on either side of the bony palate in the region of the molars.

PALATAL NOTCH   An indentation at the anterior end of the palate of most North American bats, separating the right and left upper incisors.

PALATINE BONE   A bone located in the posterior part of the roof of the mouth (Fig. 9).

PARIETAL BONES   A pair of large bones surrounding the brain laterally and dorsally (Fig. 9).

PECTORAL FIN   See FLIPPER (or look under "Fishes").

PELAGE   The covering of hair.

PHALANGES   The bones of the fingers or toes. (See also META-CARPALS.)

PLANTAR PAD   A thickened, often naked protuberance on the sole of the hind foot.

PLANTIGRADE   A type of foot in which the sole comes in contact with the ground.

POSTORBITAL PROCESS   A projection from the skull posterior to the eye.

PREMAXILLA   A bone at the anterior end of the upper jaw in certain mammals (Fig. 9).

PREMOLAR   Those teeth located between the canines (or, in their absence, a diastema) and the molars.

PTERYGOIDS   A pair of bones posterior to the palatines and usually separated by the basioccipital.

ROSTRUM   The elongate portion of the skull that projects in front of the eyes of most mammals.

SUPRAORBITAL PROCESS   A flattened, lateral projection of the skull above the eye of certain mammals (for example, rabbits).

SUPRAORBITAL RIDGE   A ridge located on the skull above the orbit.

TOTAL LENGTH   See page 6.

TRAGUS   A leaflike projection from the base of a bat's ear, anterior to the auditory opening.

TUBERCLE   See CUSP.

UNICUSPIDS   Several teeth in shrews located between the incisors and the molars, each having only one cusp.

UROPATAGIUM   See INTERFEMORAL MEMBRANE.

ZYGOMATIC ARCH   A bony bridge lateral to or just below the orbit, formed of the maxilla, jugal, and squamosal bones; sometimes incomplete (Fig. 9).

Bibliography

The following references are concerned chiefly with the taxonomy of living vertebrates and their distribution in Florida. Most publications prior to 1954 had been utilized in several major works that dealt with the respective classes of vertebrates, such as Carr and Goin (1955), Blair et al. (1957), Conant (1958), Sprunt (1954), American Ornithologists' Union (1957), and Hall and Kelson (1959). For this reason the great majority of references listed here appeared after 1954, thus bringing the accounts up to date. Sources are not cited for most individual records outside Florida.

Abramson, Ira Joel, and Henry M. Stevenson. 1964. Regional reports: Florida region. *Aud. Field Notes* 15:402–404.

Ager, Lothian A. 1971. The fishes of Lake Okeechobee, Florida. *Quart. Jour. Fla. Acad. Sci.* 34:53–62.

Agey, H. Norton. 1967. Notes on sightings of what appears to be the Least Grebe (*Colymbus dominicus*). *Fla. Nat.* 40:101.

Aldrich, John W., and Allen J. Duvall. 1958. Distribution and migration of races of the Mourning Dove. *Condor* 60:108–128.

Allen, E. Ross, and Wilfred T. Neill. 1954. The Florida deer. *Fla. Wildlife* 7:21, 37.

—————. 1955. Establishment of the Texas Horned Toad, *Phrynosoma cornutum*, in Florida. *Copeia* 1955:63–64.

Allen, Robert P. 1952. The Key Deer: a challenge from the past. *Aud. Mag.* 54:76–81.

American Fisheries Society. 1970. *A list of common and scientific names of fishes from the United States and Canada*. 3d ed. Ann Arbor.

American Ornithologists' Union. 1957. *Check-list of North American birds*. 5th ed. Baltimore: The Lord Baltimore Press, Inc.

—————. 1973. Thirty-second supplement to the American Ornithologists' Union check-list of North American birds. *Auk* 90:411–419.

American Society of Ichthyologists and Herpetologists. 1963 et seq. *Catalogue of American amphibians and reptiles*. William J. Riemer and Herndon G. Dowling, eds. New York.

Anderson, Paul K. 1961. Variation in populations of brown snakes, genus *Storeria*, bordering the Gulf of Mexico. *Amer. Midland Nat.* 66:235–249.

Anthony, H.E. 1928. *Field book of North American mammals*. New York: G.P. Putnam's Sons.

Arata, Andrew A. 1965. Taxonomic status of the Pine Vole in Florida. *Jour. Mamm*. 46:87–94.

Arnold, Ida V. 1961. Blue-gray Tanagers (*Thraupis virens*). *Fla. Nat*. 34:44–45.

Auffenberg, Walter. 1954. A reconsideration of the Racer *Coluber constrictor* in eastern United States. *Tulane Studies in Zool*. 2:89–155.

Austin, Oliver L., Jr. 1963. On the American status of *Tiaris canora* and *Carduelis carduelis*. *Auk* 80:73–74.

———. 1965. Louisiana Waterthrush (*Seiurus motacilla*) nesting in Gainesville. *Fla. Nat*. 38:144.

———. 1968. *Life histories of North American cardinals, grosbeaks, buntings, towhees, finches, sparrows, and allies*. U.S. Nat'l Mus. Bull. 237, part 2. Washington, D.C.

Bailey, Reeve, and Robert H. Gibbs, Jr. 1956. *Notropis callitaenia, a new Cyprinid fish from Alabama, Florida, and Georgia*. Occ. Papers Mus. Zool. Univ. Mich., no. 576. Ann Arbor.

Bailey, Reeve M., Howard Elliott Winn, and Lavett Smith. 1954. Fishes from the Escambia River, Alabama and Florida, with ecologic and taxonomic notes. *Proc. Acad. Nat. Sci*. (Phila.) 106:109–164.

Bancroft, Griffing. 1969. Black-headed Gull, *Larus ridibundus*, on Sanibel Island. *Fla. Nat*. 42:134.

Banks, Richard C., and Roxie C. Layborne. 1968. The Red-whiskered Bulbul in Florida. *Auk* 85:141.

Banta, Benjamin H. 1961. On the original description of *Sceloporus undulatus*. *Herpetologica* 17:136.

Barbour, Roger W., and Wayne H. Davis. 1969. *Bats of America*. Lexington: University of Kentucky Press.

Barbour, Thomas, and Charles T. Ramsden. 1919. The herpetology of Cuba. *Mem. Mus. of Comp. Zool*. 47 (no. 2):71–213.

Baxter, Mrs. Gordon P. 1970. Harcourt's Petrel, *Oceanodroma castro*. *Fla. Nat*. 43:68.

Beebe, William, and John Tee-Van. 1933. *Field book of the shore fishes of Bermuda*. New York: G.P. Putnam's Sons.

Below, Lilla. 1965. White-winged Doves (*Zenaida asiatica*) and Lark Sparrows (*Chondestes grammacus*) near Naples. *Fla. Nat*. 38:65.

Berg, L.S. 1947. *Classification of fishes both recent and fossil*. Ann Arbor: J.A. Edwards, Inc.

Best, Peter B. 1970. Records of the Pygmy Killer Whale, *Feresa attenuata*, from southern Africa, with notes on behaviour in captivity. *Ann. South African Mus*. 57:1–14.

Bigelow, Henry B., and William C. Schroeder. 1948. *Fishes of the western North Atlantic*. Part I. New Haven: Sears Foundation, Yale University.

———. 1953. *Fishes of the western North Atlantic*. Part II. New Haven: Sears Foundation, Yale University.

Bigelow, Henry B. et al. 1963. *Fishes of the western North Atlantic*. Part III. New Haven: Sears Foundation, Yale University.

Birdsong, Ray S., and Ralph W. Yerger. 1967. A natural population of hybrid sunfishes: *Lepomis macrochirus* X *Chaenobryttus gulosus*. *Copeia* 1967: 62–71.

Bishop, Sherman C. 1943. *Handbook of salamanders*. Ithaca: Comstock Publishing Co.

Blair, W. Frank et al. 1968. *Vertebrates of the United States*. 2d ed. New York: McGraw-Hill Book Co.

Blake, Emmet Reid. 1953. *Birds of Mexico*. Chicago: University of Chicago Press.

Blaney, Richard M. 1971. An annotated check list and biogeographic analysis of the insular herpetofauna of the Apalachicola Region, Florida. *Herpetologica* 27:406–430.

Blaney, Richard M., and Kenneth Relyea. 1967. The Zigzag Salamander, *Plethodon dorsalis* Cope, in southern Alabama. *Herpetologica* 23:246–247.

Böhlke, James. 1956. A new pygmy sunfish from southern Georgia. *Notulas Naturae*, no. 294, pp. 1–11.

Bond, Gorman, 1963. Geographic variation in the Thrush *Hylocichla ustulata*. *Proc. U. S. Nat'l Mus*. 114:373–385.

Bond, James. 1961. *Birds of the West Indies*. Boston: Houghton Mifflin Co.

Bonney, Christine, and Leo Johnston. 1964. The nesting of Roseate Terns at Cocoa Plum Beach—Crawl Key, Florida. *Fla. Nat.* 37:57–58.

Booth, Ernest S. 1949. *How to know the mammals*. Dubuque: Wm. C. Brown Co.

Boulenger, G.A. 1915. *Catalogue of the fresh-water fishes of Africa in the British Museum*. Vol. III. London: Brit. Mus. Nat. History.

Bowen, W. Wedgwood. 1968. Variation and evolution of Gulf Coast populations of Beach Mice, *Peromyscus polionotus*. *Bull. Fla. State Mus*. 12:1–91.

Boyd, Claude E. 1964. The distribution of cricket frogs in Mississippi. *Herpetologica* 20:201–202.

Boyd, Claude E., and David H. Vickers. 1963. Distribution of some Mississippi amphibians and reptiles. *Herpetologica* 19:202–205.

Brame, Arden H., Jr. 1967. A list of the world's recent and fossil salamanders. *Herpeton* 2:1–26.

Breder, Charles M., Jr. 1948. *Field book of marine fishes of the Atlantic coast*. New York: G.P. Putnam's Sons.

Brewer, Mrs. Talbot. 1954. Sutton's Warbler (*Dendroica potomac*) at Anna Maria. *Fla. Nat.* 27:96.

Briggs, John C. 1958. A list of Florida fishes and their distribution. *Bull. Fla. State Mus*. 2:223–318.

Brooke, Victor. 1878. On the classification of the Cervidae with a synopsis of the existing species. *Proc. Zool. Soc. of London*, pp. 883–927.

Brown, Jack S., and Herbert T. Boschung, Jr. 1954. *Rana palustris* in Alabama. *Copeia* 1954:226.

Brown, Larry N. 1969. Exotic Squirrel in Florida. *Fla. Wildlife* 23(6):4–5.

——. 1972. Presence of the Knight Anole (*Anolis equestris*) on Elliott Key, Florida. *Fla. Nat.* 45:130.

Brown, L.N., and G.C. Hickman. 1970. Occurrence of the Mediterranean Gecko in the Tampa, Florida, area. *Fla. Nat.* 43:68.

Browning, M. Ralph. 1964. Third United States record of the Black-faced Grassquit (*Tiaris bicolor*). *Auk* 81:233.

Bundy, Carter. 1962. The Scarlet Ibis in Florida. *Fla. Nat.* 35:87.

——. 1965. A new Floridian: the Scarlet Ibis. *Aud. Mag.* 67:84–85.

Burt, W.H. 1952. *A field guide to the mammals*. R.P. Grossenheider, illus. Boston: Houghton Mifflin Co.

Caldwell, David K., and Melba C. Caldwell. 1966. *Observations on the distribution, coloration, behavior and audible sound production of the Spotted Dolphin*, Stenella plagiodon (*Cope*). Los Angeles County Mus. Contributions in Science, no. 104.

————. 1969a. The Harbor Seal, *Phoca vitulina concolor*, in Florida. *Jour. Mamm.* 50:379–380.

————. 1969b. Gray's Dolphin, *Stenella styx*, in the Gulf of Mexico. *Jour. Mamm.* 50:612–614.

————. 1971. Sounds produced by two rare cetaceans stranded in Florida. *Cetology*, no. 4, pp. 1–6.

Caldwell, David K., and J.B. Siebenaler. 1960. Sperm and Pigmy Sperm Whales stranded in the Gulf of Mexico. *Jour. Mamm.* 41:136–138.

Caldwell, David K., Melba C. Caldwell, and Cecil M. Walker, Jr. 1970. Mass and individual strandings of False Killer Whales, *Pseudorca crassidens*, in Florida. *Jour. Mamm.* 51:634–636.

Caldwell, David K., James N. Layne, and J.B. Siebenaler. 1956. Notes on a Killer Whale (*Orcinus orca*) from the northeastern Gulf of Mexico. *Quart. Jour. Fla. Acad. Sci.* 19:189–196.

Campbell, Howard W. 1962. An extension of the range of *Haldea valeriae* in Florida. *Copeia* 1962:438–439.

Carr, Archie. 1952. *Handbook of turtles of the United States, Canada, and Baja California*. Ithaca: Comstock Publ. Co.

————. 1967. *So excellent a fishe*. Garden City, N.Y.: Natural History Press.

Carr, Archie, and John W. Crenshaw, Jr. 1957. A taxonomic reappraisal of the turtle *Pseudemys alabamensis*. *Bull. Fla. State Mus.* 2:25–42.

Carr, Archie, and Coleman J. Goin. 1955. *Guide to the reptiles, amphibians, and fresh-water fishes of Florida*. Gainesville: University of Florida Press.

Carr, Archie, and Robert M. Ingle. 1959. The Green Turtle (*Chelonia mydas mydas*) in Florida. *Bull. Marine Sci. Gulf and Caribbean* 9:315–320.

Chapman, Frank L. 1967. The subspecies of *Cassidix mexicanus* (Gmelin) in the eastern United States. Master's thesis, Florida State University, Tallahassee, Fla.

Chapman, Frank M. 1934. *Handbook of birds of eastern North America*. New York: Appleton-Century-Crofts.

Chermok, Ralph L. 1952. *A key to the amphibians and reptiles of Alabama*. Geol. Survey of Ala., mus. paper no. 33.

Christensen, R.F. 1965. An ichthyological survey of Jupiter Inlet and Loxahatchee River, Florida. Master's thesis, Florida State University, Tallahassee, Fla.

Christman, Steven P. 1970. *Hyla andersoni* in Florida. *Quart. Jour. Fla. Acad. Sci.* 33:80.

Churcher, Charles S. 1959. The specific status of the New World Red Fox. *Jour. Mamm.* 40:513–520.

Cleveland, Arthur G. 1970. The current geographic distribution of the armadillo in the United States. *Texas Jour. Sci.* 22:90–92.

Cobb, Marie P. 1946. Report of spring migration. *Fla. Nat.* 19:67–68.

Collette, Bruce B., and Ralph W. Yerger. 1962. The American Percid fishes of the sub-genus *Villora*. *Tulane Studies in Zool.* 9:213–230.

Committee of American Society of Ichthyologists and Herpetologists. 1956. Common names for North American amphibians and reptiles. *Copeia* 1956:172–185.

Conant, Roger. 1958. *A field guide to reptiles and amphibians*. Boston: Houghton Mifflin Co.

————. 1963. Evidence for the specific status of the Water Snake *Natrix fasciata*. *Amer. Mus. Novit.* no. 2122.

Cooley, R. 1954. Second record of the Hoary Bat in Florida. *Jour. Mamm.* 35:116–117.

Crawford, Ronald W. 1956. A study of the distribution and taxonomy of the Percid fish *Percina nigrofasciata* (Agassiz). *Tulane Studies in Zool.* 4:1–55.

Crenshaw, John W., Jr., and W. Frank Blair. 1959. Relationships in the *Pseudacris nigrita* complex in southwestern Georgia. *Copeia* 1959:215–222.

Crenshaw, John W., Jr., and Milton N. Hopkins, Jr. 1955. The relationships of the Soft-shelled Turtles *Trionyx ferox ferox* and *Trionyx ferox aspera*. *Copeia* 1955:13–23.

Crossman, E.J. 1966. A taxonomic study of *Esox americanus* and its subspecies in eastern North America. *Copeia* 1966:1–20.

Croulet, C.H. 1965. Evidence of conspecificity of the Western Ringneck Snake (genus *Diadophis*). *Herpetologica* 21:80.

Cruickshank, Allan D. 1949. Black-headed Gull in Florida. *Auk* 66:205–206.

———. 1967. First razorbill for Florida. *Fla. Nat.* 40:48–49.

Cunningham, Richard L. 1964. Regional reports: Florida region. *Aud. Field Notes* 18:442–446.

Cunningham, Vernon D., and Robert D. Dunford. 1970. Recent coyote record from Florida. *Quart. Jour. Fla. Acad. Sci.* 33:279–280.

Dahlberg, Michael D., and Donald C. Scott. 1971. The freshwater fishes of Georgia. *Bull. Ga. Acad. Sci.* 29:1–64.

Davis, David E. 1960. The spread of the Cattle Egret in the United States. *Auk* 77:421–424.

Davis, Joseph A., Jr. 1957. A new shrew (*Sorex*) from Florida. *Amer. Mus. Novit.* no. 1844.

Davis, Wayne H. 1959. Taxonomy of the Eastern Pipistrel. *Jour. Mamm.* 40:521–531.

Dawes, Mrs. B.G. 1948. Mag and Maggie on Jupiter Island. *Fla. Nat.* 21:60.

de Vos, A., R.H. Manville, and R.G. Van Gelder. 1956. Introduced mammals and their influence on the native biota. *Zoologica* 41:163–194.

Dickerman, Robert W., and Kenneth C. Parkes. 1968. Notes on the plumages and generic status of the Little Blue Heron. *Auk* 85:437–440.

Dickie, Eva S. 1965. Miami nesting of Killdeer. *Fla. Nat.* 38:31.

Dickinson, J.C., Jr. 1952. Geographic variation in the Red-eyed Towhee of the eastern United States. *Bull. Mus. Comp. Zool.* 107:271–352.

Dilger, William C. 1956. Relationships of the thrush genera *Catharus* and *Hylocichla*. *Syst. Zool.* 5:174–182.

Dobie, James L. 1972. Correction of distributional records for *Graptemys barbouri* and *Graptemys pulchra*. *Herpetological Review* 4:23.

Duellman, William E. 1955. Systematic status of the Key West Spadefoot Toad, *Scaphiopus holbrooki albus*. *Copeia* 1955:141–143.

Duellman, William E., and Albert Schwartz. 1958. Amphibians and reptiles of southern Florida. *Bull. Fla. State Mus.* 3:181–324.

Eddy, Samuel. 1957. *How to know the freshwater fishes*. Dubuque: Wm. C. Brown Co.

Edscorn, John B. 1968. Curlew Sandpiper (*Erolia ferruginea*) found near Lakeland. *Fla. Nat.* 41:126.

Eifrig, C.W.G. 1946. More western birds seen in Florida. *Fla. Nat.* 19:63.

Eisenmann, Eugene. 1962. Notes on nighthawks of the genus *Chordeiles* in southern Middle America, with a description of a new race of *Chordeiles minor* breeding in Panama. *Amer. Mus. Novit.* no. 2094.

Fanning, Sheryl A. 1967. A synopsis and key to the tadpoles of Florida. Master's thesis, Florida State University, Tallahassee, Fla.

Farrar, Merritt C. 1950. A sight record of the Eastern Tree Sparrow in Florida. *Fla. Nat.* 23:46.

Feuer, Robert C. 1971. Intergradation of the Snapping Turtles *Chelydra serpentina serpentina* (Linnaeus, 1758) and *Chelydra serpentina osceola* (Stejneger, 1918). *Herpetologica* 27:379–384.

Fisk, Erma J. 1966. A happy newcomer in a fruitful land. *Fla. Nat.* 39:10–11.

———. 1968. White-winged Doves breeding in Florida. *Fla. Nat.* 41:126.

Florida Game and Fresh Water Fish Commission. 1963. 1963 Nutria survey. Mimeographed letter, Feb. 11, 1963.

Folkerts, George W. 1967. A Spotted Turtle, *Clemmys guttata* (Schneider), from southeastern Georgia. *Herpetologica* 23:63.

France, Mrs. Ralph L. 1969. Red-breasted Nuthatch in Venice. *Fla. Nat.* 42:134.

Freeman, John R. 1959. A record-size (*sic*) Dwarf Siren. *Herpetologica* 15:16.

Fugler, Charles M., and George W. Folkerts. 1967. A second record of *Hemidactylium* from Florida. *Herpetologica* 23:60.

Funderburg, John B. 1962. Masked Duck (*Oxyura dominica*) at Lakeland. *Fla. Nat.* 35:92.

Funderburg, John B., and David S. Lee. 1967. Distribution of the Lesser Siren, *Siren intermedia*, in central Florida. *Herpetologica* 23:65.

Funk, Richard S. 1964. The size attained by *Desmognathus auriculatus* (Holbrook). *Herpetologica* 20:204.

Gaither, Agnes, and Harold Gaither. 1966. Records from northwest Florida. *Fla. Nat.* 39:30.

———. 1968. Lark Bunting (*Calamospiza melanocorys*) at Destin. *Fla. Nat.* 41:82–83.

Gaither, H.E. 1964. Prairie Warbler (*Dendroica discolor*) nesting in northwest Florida. *Fla. Nat.* 37:91.

Gibbs, Robert H., Jr. 1957. Cyprinid fishes of the subgenus *Cyprinella* of *Notropis*. III. Variation and subspecies of *Notropis venustus* (Girard). *Tulane Studies in Zool.* 5:173–203.

Gilbert, Carter R. 1964. The American Cyprinid fishes of the subgenus *Luxilus* (genus *Notropis*). *Bull. Fla. State Mus.* 8:95–194.

Ginsburg, Isaac. 1932. A revision of the genus *Gobionellus* (Family Gobiidae). *Bull. Bingham Ocean. Coll.* 4 (art. II):1–51.

———. 1933. A revision of the genus *Gobiosoma* (Family Gobiidae) with an account of the genus *Garmannia*. *Bull. Bingham Ocean. Coll.* 4 (art. V):1–59.

———. 1951. Western Atlantic tonguefishes with descriptions of six new species. *Zoologica* 36:185–201.

Glass, Bryan P. 1951. *A key to the skulls of North American mammals*. Minneapolis: Burgess Publ. Co.

Gloyd, Howard K. 1969. Two additional subspecies of North American Crotalid snakes, genus *Agkistrodon*. *Proc. Biol. Soc. Wash.* 82:219–232.

Goin, Coleman J., and Olive B. Goin. 1969. The wood thrush breeding in Alachua County. *Fla. Nat.* 42:172.

Golley, Frank B. 1962. *Mammals of Georgia: a study of their distribution and functional role in the ecosystem*. Athens: University of Georgia Press.

Goodwin, Derek. 1967. *Pigeons and doves of the world*. London: Trustees of the British Museum.

Green, Mrs. Dwight. 1961. Manx Shearwater found on beach at Juno. *Fla. Nat.* 34:93.

Greene, Earl R. 1946. Birds of the lower Florida Keys. *Quart. Jour. Fla. Acad. Sci.* 8:199–265.

Greenwood, P. Humphry et al. 1966. Phyletic studies of Teleostean fishes, with a provisional classification of living forms. *Bull. Amer. Mus. Nat. Hist.* 131(4):341–455.

Griffo, James V., Jr. 1957. The status of the Nutria in Florida. *Quart. Jour. Fla. Acad. Sci.* 20:209–215.

Gunter, Gordon. 1968. The status of seals in the Gulf of Mexico with a record of feral Otariid seals off the United States Gulf Coast. *Gulf Res. Reports* 2:301–311.

Gunter, Gordon, and Gordon E. Hall. 1963. Additions to the list of euryhaline fishes of North America. *Copeia* 1963:596–597.

Hall, E. Raymond. 1960. *Oryzomys couesi* only subspecifically different from the Marsh Rice Rat, *Oryzomys palustris*. *Southwestern Nat.* 5:171–173.

Hall, E. Raymond, and J. Knox Jones, Jr. 1961. North American yellow bats, "*Dasypterus*," and a list of the named kinds of the genus *Lasiurus*. *Publ. of Kans. Mus. Nat. Hist.* 14:73–98.

Hall, E. Raymond, and Keith R. Kelson. 1959. *The mammals of North America*. 2 vols. New York: Ronald Press.

Hallman, Roy C. 1961. Common Tern (*Sterna hirundo*) nesting in northwest Florida. *Fla. Nat.* 34:221–222.

———. 1962. American Flamingo captured in Bay County. *Fla. Nat.* 35:92.

———. 1963. Black-chinned Hummingbird, a new bird for Florida. *Fla. Nat.* 36:89.

———. 1965. Record of Whooping Crane (*Grus americana*) killed in St. Johns County, Florida. *Fla. Nat.* 38:23.

———. 1966a. American Coot (*Fulica americana*) nestings in Bay County, Florida. *Fla. Nat.* 39:117.

———. 1966b. Harcourt's Petrel (*Oceanodroma castro castro*) in north Florida. *Fla. Nat.* 39:30.

Hames, Frances. 1956. Masked Duck in Florida. *Auk* 73:291.

———. 1959. Harcourt's Petrel (*Oceanodroma castro castro*). *Fla. Nat.* 32:145–146.

———. 1960. Bahama Honeycreeper at Key West. *Fla. Nat.* 33:170.

Hamilton, William J., Jr. 1943. The mammals of eastern United States. Ithaca: Comstock Publ. Co.

Handley, Charles O., Jr. 1959. A revision of the American bats of the genera *Euderma* and *Plecotus*. *Proc. U.S. Nat'l Mus.* 110:95–246.

Hansen, Keith L., and Harold F. Weaver. 1963. Another Florida record of the Goose-beaked Whale. *Jour. Mamm.* 44:575.

Hastings, Robert W. 1967. Ecology, life history, and geographic variation of the Diamond Killifish, *Adinia xenica* (Jordan and Gilbert). Master's thesis, Florida State Univ., Tallahassee, Fla.

Hebard, Frederick V. 1952. Sight record of Arctic Tern in Florida. *Fla. Nat.* 25:126.

Heinzman, George, and Dorotha Heinzman. 1963. White-tailed Kite (*Elanus leucurus*) in Polk County. *Fla. Nat.* 36:126.

———. 1965. Ringed Turtle Doves (*Streptopelia risoria*) at Homestead. *Fla. Nat.* 38:31.

Hellier, Thomas R., Jr. 1967. The fishes of the Santa Fe River system. *Bull. Fla. State Mus.* 11:1–46.

Herald, E.S., and R.R. Strickland. 1948. An annotated list of fishes of Homosassa Springs, Florida. *Quart. Jour. Fla. Acad. Sci.* 4:99–109.

Hershkovitz, Philip. 1961. On the nomenclature of certain whales. *Fieldiana: Zoology* 39:547–565.

————. 1966. *Catalog of living whales*. U.S. Nat'l Mus. Bull. 246. Washington, D.C.

Highton, Richard. 1956. Systematics and variation of the endemic Florida snake genus *Stilosoma*. *Bull. Fla. State Mus.* 1:73–96.

————. 1961. A new genus of lungless salamander from the coastal plain of Alabama. *Copeia* 1961:65–68.

————. 1962. Revision of North American salamanders of the genus *Plethodon*. *Bull. Fla. State Mus.* 6:235–367.

Hollister, J. Murray. 1952. The armadillo in Florida. *Fla. Nat.* 25:21–23.

Hooper, Emmet T. 1958. *The male phallus in mice of the genus* Peromyscus. Misc. Publ. Mus. Zool., Univ. Mich., no. 105. Ann Arbor.

Hooper, Emmet T., and Barbara S. Hart. 1962. *A synopsis of recent North American Microtine rodents*. Misc. Publ. Mus. Zool., Univ. Mich., no. 120. Ann Arbor.

Howard, Mrs. John K. 1961. Skuas seen near Florida Keys. *Fla. Nat.* 34:221.

Howell, Arthur H. 1932. *Florida bird life*. New York: Coward-McCann, Inc.

Hubbard, Lyle S. 1965. White-crowned Pigeon (*Columba leucocephala*) near Fort Pierce. *Fla. Nat.* 38:60, 65.

Hubbs, Carl L., and E. Ross Allen. 1943. Fishes of Silver Springs, Florida. *Proc. Fla. Acad. Sci.* 6:110–130.

Hubbs, Carl L., and Robert Rush Miller. 1965. *Studies of Cyprinodont fishes, XXII. Variation in* Lucania parva, *its establishment in western United States, and description of a new species from an interior basin in Coahuila, Mexico*. Misc. Publ. Mus. Zool., Univ. Mich., no. 127. Ann Arbor.

Huhey, James E. 1959. Distribution and variation in the Glossy Water Snake, *Natrix rigida* (Say). *Copeia* 1959:303–311.

Hundley, Margaret H., and Frances Hames. 1960–62. Bird life of the lower Florida Keys. *Fla. Nat.* 33:15–24, 56, 91–94, 149–155, 209–214; 34:25–34, 74–80, 129–135, 164, 203–207; 35:17–19, 30, 55–56, 78–81, 123–128.

Imhof, Thomas A. 1972. The changing seasons: central southern region. *Amer. Birds* 26:769–774.

Inwood, Arthur. 1961. Kiskadee Flycatcher (*Pitangus sulphuratus*) on Christmas bird count. *Fla. Nat.* 34:95.

————. 1965. Key West Quail-Dove (*Geotrygon chrysia*) at Hillsboro Beach. *Fla. Nat.* 38:65.

Jackson, James F. 1973. Distribution and population phenetics of the Florida Scrub Lizard, *Sceloporus woodi*. *Copeia* 1973:746–761.

James, Frances C. 1968. Regional reports: central southern region. *Aud. Field Notes* 22:445–448.

Jehl, Joseph R. 1968. The systematic position of the Surfbird, *Aphriza virgata*. *Condor* 70:206–210.

Jennings, William L., and James N. Layne. 1957. *Myotis sodalis* in Florida. *Jour. Mamm.* 38:259–260.

Johnston, David W. 1961. *The biosystematics of American crows*. Seattle: University of Washington Press.

————. 1969. Sage Thrasher and other unusual birds in north-central Florida. *Auk* 86:754–755.

Jordan, David Starr. 1929. *Manual of the vertebrate animals of the northeastern United States*. New York: World Book Co.

Jordan, David Starr, and Barton Warren Evermann. 1896. *The fishes of North and Middle America*. Vols. 1–4. U.S. Nat'l Mus. Bull. 47. Washington, D.C.

Kilby, John D., Edward Crittenden, and Lovett E. Williams. 1959. Several fishes new to Florida freshwaters. *Copeia* 1959:77–78.

King, Wayne. 1958. Observations on the ecology of a new population of the Mediterranean Gecko, *Hemidactylus turcicus*, in Florida. *Quart. Jour. Fla. Acad. Sci.* 21:317–318.

————. 1968. As a consequence many will die. *Fla. Nat.* 41:99–103, 120.

King, Wayne, and Thomas Krakauer. 1966. The exotic herpetofauna of southeast Florida. *Quart. Jour. Fla. Acad. Sci.* 29:144–154.

Koopman, K.F. 1971. The systematic and historical status of the Florida *Eumops*. *Amer. Mus. Novit.* no. 2478.

Kortright, Francis H. 1942. *The ducks, geese, and swans of North America.* Harrisburg, Pa.: The Stackpole Co., and Washington: Wildlife Management Inst.

Krakauer, Thomas. 1968. The ecology of the Neotropical Toad, *Bufo marinus*, in south Florida. *Herpetologica* 24:214–221.

Langridge, H.P. 1959. Cory's Shearwater at Palm Beach, Florida. *Auk* 76:241.

————. 1960. Arctic Loon at Palm Beach. *Auk* 77:351.

————. 1962. Masked Duck (*Oxyura dominica*). *Fla. Nat.* 35:62.

————. 1963. Stripe-headed Tanager in Palm Beach. *Fla. Nat.* 36:89.

————. 1964. Ruff near West Palm Beach. *Fla. Nat.* 37:57.

Layne, James N. 1965a. Observations on marine mammals in Florida waters. *Bull. Fla. State Mus.* 9:131–181.

————. 1965b. Occurrence of Black-tailed Jack Rabbits in Florida. *Jour. Mamm.* 46:502.

————. 1969. Strange mammals in Florida (in "Letters"). *Fla. Nat.* 42:50.

LeBuff, Charles R., Jr. 1957. The range of *Crocodylus acutus* along the Florida Gulf Coast. *Herpetologica* 13:188.

Lee, David S. 1969. The Spring Redeye Chub in Lithia Springs. *Fla. Nat.* 42:39.

Lee, David S., and Ernest Bostelman. 1969. The Red Fox in central Florida. *Jour. Mamm.* 50:161.

Lesser, Frederick H., and Allen R. Stickley. 1967. Occurrence of the Saw-whet Owl in Florida. *Auk* 84:425.

Letson, Orrin W. 1968. Band-tailed Pigeon (*Columba fasciata*). *Fla. Nat.* 41:126.

Ligon, David J. 1963. Breeding range expansion of the Burrowing Owl in Florida. *Auk* 80:367–368.

Little, Frank J., Jr. 1957. Key to the fishes of Alligator Harbor, St. George's Sound—Apalachee Bay region, Florida Gulf Coast. Mimeographed, Dept. of Oceanography, Florida State University, Tallahassee, Fla.

Loftin, Horace, and Storrs Olson. 1960. *Arenaria interpres interpres* in Florida. Auk 77:352.

Loftin, Robert W. 1970. The white squirrels of north Florida. *Fla. Nat.* 43:53.

Lynch, John D. 1963. Additional evidence for the recognition of *Limnaoedus* (Amphibia: Hylidae). *Copeia* 1963:566–568.

McConkey, Edwin H. 1954. A systematic study of the North American lizards of the genus *Ophisaurus*. *Amer. Midland Nat.* 51:133–171.

McDowell, Samuel B. 1964. Partition of the genus *Clemmys* and related problems in the taxonomy of the aquatic Testudinidae. *Proc. Zool. Soc. London* 143:239–279.

McGowan, Albert F. 1969. Common Terns, *Sterna hirundo*, nesting on roof in Pompano Beach. *Fla. Nat.* 42:172.

Martof, Bernard S., and H. Carl Gerhardt. 1965. Observations on the geographic variation in *Ambystoma cingulatum*. *Copeia* 1965:342–346.

Mason, C.R. 1959. Rufous Hummingbird (*Selasphorus rufus*). *Fla. Nat.* 32:96.
————. 1960. Bahama Honeycreeper (*Coereba bahamensis*). *Fla. Nat.* 33:38.
————. 1961a. New Yellowthroat for Florida. *Fla. Nat.* 34:44.
————. 1961b. Black-faced Grassquit (*Tiaris bicolor*) in Everglades National Park. *Fla. Nat.* 34:45–46.
————. 1964a. Cory's Shearwater (*Puffinus diomedea*). *Fla. Nat.* 37:24.
————. 1964b. Blue-gray Tanager (*Thraupis virens*) in South Miami. *Fla. Nat.* 37:122.
————. 1968a. Curlew Sandpiper (*Erolia testacea*) sighted at Zellwood. *Fla. Nat.* 41:34.
————. 1968b. Curlew Sandpiper date. *Fla. Nat.* 41:82.
Mayr, Ernst, and Lester L. Short. 1970. *Species taxa of North American birds*. Publ. no. 9. Cambridge, Mass.: Nuttall Ornith. Club.
Means, D. Bruce, and Clive J. Longden. 1970. *Desmognathus monticola* in Florida. *Herpetologica* 26:396–399.
Mecham, John S. 1959. The genetic relationships of two chorus frogs of the genus *Pseudacris*. *Jour. Ala. Acad. Sci.* 31:100.
Merrill, I.L. 1969. Red-breasted Nuthatch. *Fla. Nat.* 42:94.
Meyer de Schauensee, R. 1964. *The birds of Colombia*. Narberth, Pa.: Livingston Publ. Co.
Meyerriecks, Andrew J. 1963. Caspian Tern breeds in Florida. *Auk* 80:365–366.
Miller, Gerrit S., Jr., and Remington Kellogg. 1955. *List of North American recent mammals*. U.S. Nat'l Mus. Bull. 205. Washington, D.C.
Mittelman, M.B. 1950a. The generic status of *Scincus lateralis* Say, 1823. *Herpetologica* 6:17–20.
————. 1950b. Miscellaneous notes on some amphibians and reptiles from the southeastern United States. *Herpetologica* 6:20–24.
Mittelman, M.B., and J.C. List. 1953. The generic differentiation of the swamp treefrogs. *Copeia* 1953:80–83.
Monroe, Burt L., Jr. 1957. A breeding record of the cowbird for Florida. *Fla. Nat.* 30:124.
————. 1958a. Western Meadowlark (*Sturnella neglecta*). *Fla. Nat.* 31:28.
————. 1958b. Lapland Longspur (*Calcarius lapponicus*). *Fla. Nat.* 31:28.
————. 1958c. First record of the Bronzed Grackle in Florida. *Fla. Nat.* 31:137.
————. 1959a. Occurrence of the Yellow-green Vireo in Florida. *Auk* 76:95–96.
————. 1959b. Little Gull. *Fla. Nat.* 32:96.
Moore, Joseph C. 1951. The range of the Florida Manatee. *Quart. Jour. Fla. Acad. Sci.* 14:1–19.
————. 1953. Distribution of marine mammals to Florida waters. *Amer. Midland Nat.* 49:117–118.
————. 1956. Variation in the Fox Squirrel in Florida. *Amer. Midland Nat.* 55:41–65.
————. 1960. New records of the Gulf-Stream Beaked Whale, *Mesoplodon gervaisi*, and some taxonomic considerations. *Amer. Mus. Novit.* no. 1993.
————. 1966. Diagnoses and distributions of beaked whales of the genus *Mesoplodon* known from North American waters. In Norris, Kenneth S., ed.: *Whales, Dolphins and Porpoises*. International Symposium on Cetacean Research. Berkeley: University of California Press.
Moore, Joseph C., and Eugenie Clark. 1963. Discovery of Right Whales in the Gulf of Mexico. *Science* 141:269.

Moore, Joseph C., and Ralph S. Palmer. 1955. More Piked Whales from southern North Atlantic. *Jour. Mamm.* 36:429–433.

Moore, Joseph C., and F.G. Wood, Jr. 1957. Differences between the beaked whales *Mesoplodon mirus* and *Mesoplodon gervaisi*. *Amer. Mus. Novit.* no. 1831.

Mount, Robert H. 1964. New locality records for Alabama anurans. *Herpetologica* 20:127–128.

———. 1965. Variation and systematics of the scincoid lizard, *Eumeces egregius* (Baird). *Bull. Fla. State Mus.* 9:183–213.

Mount, Robert H., and George W. Folkerts. 1968. Distribution of some Alabama reptiles and amphibians. *Herpetologica* 24:259–262.

Mumford, Russell E., and James B. Cope. 1964. Distribution and status of the Chiroptera of Indiana. *Amer. Midland Nat.* 72:473–489.

Murphy, Robert Cushman. 1967. *Serial atlas of the marine environment.* (Folio 14: Distribution of North Atlantic pelagic birds). New York: Amer. Geogr. Soc.

Murray, Bertram G., Jr. 1968. The relationships of sparrows in the genera *Ammodramus*, *Passerherbulus*, and *Ammospiza* with a description of a hybrid Le Conte's X Sharp-tailed Sparrow. *Auk* 85:586–593.

Myers, Charles W. 1958. A possible introduction of the snake *Typhlops* in the United States. *Copeia* 1958:338.

———. 1967. The Pine Wood Snake, *Rhadinaea flavilata* (Cope). *Bull. Fla. State Mus.* 11:47–97.

Neill, Wilfred T. 1952. The spread of the armadillo in Florida. *Ecology* 33:282–284.

———. 1953. A Florida specimen of Le Conte's Lump-nosed Bat. *Jour. Mamm.* 34:382–383.

———. 1954. Ranges and taxonomic allocations of amphibians and reptiles in the southeastern United States. *Publ. Res. Div. Ross Allen's Reptile Inst.* 1:75–96.

———. 1957a. The status of *Rana capito stertens* Schwartz and Harrison. *Herpetologica* 13:47–52.

———. 1957b. Historical biogeography of present-day Florida. *Bull. Fla. State Mus.* 2:175–220.

———. 1961. *How to preserve reptiles and amphibians for scientific study.* Silver Springs, Fla: Ross Allen's Reptile Inst., Inc.

———. 1964a. A new species of salamander, genus *Amphiuma*, from Florida. *Herpetologica* 20:62–66.

———. 1964b. Taxonomy, natural history, and zoogeography of the Rainbow Snake, *Farancia erytrogramma* (Palisot de Beauvois). *Amer. Midland Nat.* 71:257–295.

Nelson, Milton G. 1966. Key West Quail-Dove (*Geotrygon chrysia*) at Lake Worth. *Fla. Nat.* 39:154.

Nicholson, Donald J. 1955. Whistling Swan, *Cygnus columbianus*, killed in Osceola County, Florida. *Fla. Nat.* 28:59.

———. 1960. Ringed Turtle Dove (*Streptopelia risoria*) (an introduced species) found near Orlando, Fla. *Fla. Nat.* 33:224.

Norman, J.R. 1934. *A systematic monograph of the flatfishes (Heterosomata).* Vol. I. London: Oxford University Press.

Ogden, John. 1964. Violet-green Swallow in Florida. *Fla. Nat.* 37:121.

———. 1968. Sharp-tailed Sandpiper collected in Florida. *Auk* 85:692.

———. 1972. The nesting season: Florida region. *Amer. Birds* 26:847–852.

Ogden, John C., and Frank L. Chapman. 1967. Extralimital breeding of Painted Buntings in Florida. *Wilson Bull*. 79:347.

Olson, Clark, and Louis A. Stimson. 1966. Eared Grebe (*Podiceps caspicus*) near Miami. *Fla. Nat*. 39:57.

Olson, Mary Ann. 1965. Ruff (*Philomachus pugnax*) found in Bay County. *Fla. Nat*. 38:106.

Olson, Storrs L. 1961. A second breeding record of the Robin in Florida. *Fla. Nat*. 34:222.

Owre, Oscar T. 1962. The first record of the King Eider, *Somateria spectabilis* (Linnaeus), and the occurrence of other Anseriformes in Florida. *Auk* 79:270–271.

————. 1973. A consideration of the exotic avifauna of southeast Florida. *Wilson Bull*. 85:491–500.

Palmer, Ralph S. 1954. *The mammal guide*. Garden City, N.Y.: Doubleday and Co., Inc.

————. 1962. *Handbook of North American birds*. Vol. I (Loons through Flamingos). New Haven: Yale University Press.

Parkes, Kenneth C. 1958. Specific relationships in the genus *Elanus*. *Condor* 60:139–140.

Paul, John R. 1968. Risso's Dolphin, *Grampus griseus*, in the Gulf of Mexico. *Jour. Mamm*. 49:746–748.

Paulson, Dennis R. 1966. Variation in some snakes from the Florida Keys. *Quart. Jour. Fla. Acad. Sci*. 29:295–308.

Pearson, Paul G. 1954. Mammals of Gulf Hammock, Levy County, Florida. *Amer. Midland Nat*. 51:468–480.

Peters, James A. 1964. *Dictionary of herpetology*. New York: Hafner Publ. Co.

Peters, James L. 1960. *Check-list of birds of the world*. Vol. 9. Cambridge, Mass.: Mus. Comp. Zool.

Peterson, Roger Tory. 1947. *A field guide to the birds*. Boston: Houghton Mifflin Co.

Pylka, Joseph M., and Richard D. Warren. 1958. A population of *Haideotrition* in Florida. *Copeia* 1958:334–336.

Rand, A.L. 1957. *Lanius ludovicianus miamensis* Bishop, a valid race from southern Florida. *Auk* 74:503–505.

————. 1960. Races of the Short-tailed Hawk, *Buteo brachyurus*. *Auk* 77:448–459.

Ray, Clayton E. 1961. The Monk Seal in Florida. *Jour. Mamm*. 42:113.

Regan, C. Tate. 1905. A revision of the fishes of the South American cichlid genera *Acara*, *Nannacara*, *Acaropsis*, and *Astronotus*. *Ann. and Mag. of Nat. Hist*. (ser. 7), 15:329–347.

Reichard, Sherwood M., and Henry M. Stevenson. 1964. Records of *Eleutherodactylus ricordi* at Tallahassee. *Fla. Nat*. 37:97.

Relyea, Kenneth G. 1965. Taxonomic studies of the Cyprinodont fishes, *Fundulus confluentus* Goode and Bean and *Fundulus pulvereus* (Evermann). Doctoral dissertation, Florida State University, Tallahassee, Fla.

Rice, Dale W. 1955a. Status of *Myotis grisescens* in Florida. *Jour. Mamm*. 36:289–290.

————. 1955b. *Myotis keenii* in Florida. *Jour. Mamm*. 36:567.

————. 1957. Life history and ecology of *Myotis austroriparius* in Florida. *Jour. Mamm*. 38:15–32.

————. 1965. Bryde's Whale in the Gulf of Mexico. *Norsk Hvalfangst-Tidende* 54:114–115.

Ridgway, Robert. 1887. *A manual of North American birds*. Philadelphia: J.P. Lippincott Co.

Ridgway, Sam (ed.). 1972. *Mammals of the sea*. Springfield, Ill.: Charles C. Thomas.

Riemer, William J. 1958. Giant toads of Florida. *Quart. Jour. Fla. Acad. Sci.* 21:207–211.

Rivas, Luis Rene. 1962. The Florida fishes of the genus *Centropomus*, commonly known as snook. *Quart. Jour. Fla. Acad. Sci.* 25:53–64.

————. 1966. The taxonomic status of the Cyprinodontid fishes *Fundulus notti* and *F. lineolatus. Copeia* 1966:353–354.

————. 1969. A revision of the Poeciliid fishes of the *Gambusia punctata* species group, with a description of two new subspecies. *Copeia* 1969:778–795.

Robertson, William B., Jr. 1964. The terns of the Dry Tortugas. *Bull. Fla. State Mus.* 8:1–94.

————. 1967. Regional reports: Florida region. *Aud. Field Notes* 21:407–413.

————. 1968. Regional reports: Florida region. *Aud. Field Notes* 22:516–520.

————. 1969. Transatlantic migration of juvenile Sooty Terns. *Nature* 223:632–634.

Robertson, William B., Jr., and C. Russell Mason. 1965. Additional bird records from the Dry Tortugas. *Fla. Nat.* 38:131–138.

Robertson, William B., Jr., and John C. Ogden. 1968. Regional reports: Florida region. *Aud. Field Notes* 22:25–31.

————. 1969. Regional reports: Florida region. *Aud. Field Notes* 23:35–41.

————. 1971. Regional reports: Florida region. *Aud. Field Notes* 25:44–49.

Robertson, William B., Jr., Dennis R. Paulson, and C. Russell Mason. 1961. A tern new to the United States. *Auk* 78:423–424.

Rohwer, Sievert A. 1968. Second breeding record of the Caspian Tern in Florida. *Fla. Nat.* 41:35.

Rose, Francis L., and James L. Dobie. 1963. *Desmognathus monticola* in the coastal plain of Alabama. *Copeia* 1963:564–565.

Rosen, Donn Eric. 1964. The relationships and taxonomic position of the half-beaks, killifishes, silversides, and their relatives. *Bull. Amer. Mus. Nat. Hist.* 127:217–268.

Rosen, Donn Eric, and Reeve M. Bailey. 1963. The Poeciliid fishes (Cyprinodontiformes), their structure, zoogeography, and systematics. *Bull. Amer. Mus. Nat. Hist.* 126:1–176.

Rossman, Douglas A. 1959. Ecosystematic relationships of the salamanders *Desmognathus fuscus auriculatus* Holbrook and *Desmognathus fuscus carri* Neill. *Herpetologica* 15:149–155.

————. 1963a. The Colubrid snake genus *Thamnophis*: a revision of the *sauritus* group. *Bull. Fla. State Mus.* 7:99–178.

————. 1963b. *Relationships and taxonomic status of the North American natricine snake genera* Liodytes, Regina, *and* Clonophis. Occ. Papers Mus. Zool. La. State Univ. no. 29. Baton Rouge.

————. 1964. A new subspecies of the Common Garter Snake, *Thamnophis sirtalis*, from the Florida Gulf Coast. *Proc. La. Acad. Sci.* 27:67–73.

Sabath, Michael D., and Laura Elsa Sabath. 1969. Morphological intergradation in Gulf Coastal Brown Snakes, *Storeria dekayi* and *Storeria tropica. Amer. Midland Nat.* 81:148–155.

Savage, Jay M. 1954. Notulae herpetologicae 1–7. *Trans. Kans. Acad. Sci.* 57:326–334.

Schaller, George B. 1967. *The deer and the tiger*. Chicago: University of Chicago Press.

Schmidt, Karl P., and D. Dwight Davis. 1941. *Field book of snakes of the United States and Canada*. New York: G.P. Putnam's Sons.

Schreiber, Ralph W., and James J. Dinsmore. 1972. Caspian Tern nesting records in Florida. *Fla. Nat.* 45:160–161.

Schwartz, Albert. 1949. A second specimen of *Mustela vison evergladensis*. *Jour. Mamm.* 30:315–316.

————. 1954. Observations on the Big Pine Key Cotton Rat. *Jour. Mamm.* 35:260–263.

————. 1955. The status of the species of the *brasiliensis* group of the genus *Tadarida*. *Jour. Mamm.* 36:106–109.

————. 1956a. *The relationships and nomenclature of the soft-shelled turtles (genus* Trionyx) *of the southeastern United States*. Charleston Mus. Leaflet no. 26.

————. 1956b. Geographic variation in the Chicken Turtle. *Fieldiana: Zoology*, Chic. Nat. Hist. Mus. 34:461–503.

————. 1956c. The Cottontail Rabbits (*Sylvilagus floridanus*) of peninsular Florida. *Proc. Biol. Soc. Wash.* 69:145–152.

————. 1957. Chorus Frogs (*Pseudacris nigrita* Le Conte) in South Carolina. *Amer. Mus. Novit.* no. 1838.

————. 1965. Geographic variation in two species of Hispaniolan *Eleutherodactylus*, with notes on Cuban members of the *ricordi* group. *Studies on the Fauna of Curaçao and other Caribbean Islands* 22:98–123.

————. 1966. Geographic variation in *Sphaerodactylus notatus* Baird. *Rev. Biol. Trop.* 13:161–185.

Schwartz, Albert, and Eugene P. Odum. 1957. The woodrats of the eastern United States. *Jour. Mamm.* 38:197–206.

Seeler, Katherine. 1958. Harris' Sparrow appears in Winter Park. *Fla. Nat.* 31:89–90, 93.

Seibert, Henri C. 1964. The Coal Skink in Florida. *Jour. Ohio Herp. Soc.* 4:79.

Selander, Robert K. 1954. A systematic review of the booming nighthawks of western North America. *Condor* 56:57–82.

Shanholtzer, G. Frederick. 1967. The taxonomy and distribution of the Swamp and Marsh Rabbits. Master's thesis, Florida State Univ., Tallahassee, Fla.

Sharp, Brian. 1969. Let's save the Dusky Seaside Sparrow. *Fla. Nat.* 42:68–70.

Sherman, H.B. 1955. Description of a new race of woodrat from Key Largo, Florida. *Jour. Mamm.* 36:113–120.

————. 1956. Third record of the Hoary Bat in Florida. *Jour. Mamm.* 37:281–282.

Short, Lester L., Jr. 1965. Hybridization in the flickers (*Colaptes*) of North America. *Bull. Amer. Mus. Nat. Hist.* 129:307–428.

Sibley, Charles G., and Lester L. Short, Jr. 1964. Hybridization of the orioles of the Great Plains. *Condor* 66:130–150.

Simmons, Robert, and Jerry D. Hardy, Jr. 1959. The River-Swamp Frog, *Rana heckscheri* Wright, in North Carolina. *Herpetologica* 15:36–37.

Smith, Fred E. 1943. [Notes from correspondents.] *Fla. Nat.* 16:41–42.

Smith, Hobart. 1946. *Handbook of lizards*. Ithaca: Comstock Publ. Co.

————. 1953. The generic name of the newts of eastern North America. *Herpetologica* 9:95–99.

Smith, Hobart, John D. Lynch, and B. Gail Puckette Browne. 1965. Proposed suppression of the snake name *Coluber doliatus* Linnaeus, 1776. *Herpetologica* 21:1–5.

Smith, Michael H. 1967. Variation in plantar tubercles in *Peromyscus polionotus*. *Quart. Jour. Fla. Acad. Sci.* 30:108–110.

Smith-Vaniz, William F. 1968. *Freshwater fishes of Alabama*. Auburn, Ala.: Agri. Exp. Sta.

Spangler, James A., and Robert H. Mount. 1969. The taxonomic status of the Natricine snake *Regina septemvittata mabila* (Neill). *Herpetologica* 25:113–119.

Springer, Victor G. 1964. A revision of the Carcharhinid shark genera *Scoliodon*, *Loxodon*, and *Rhizoprionodon*. *Proc. U. S. Nat'l Mus.* 115:559–632.

Sprunt, Alexander, Jr. 1953. The Fork-tailed Flycatcher in Florida. *Fla. Nat.* 26:54.

———. 1954. *Florida bird life*. New York: Coward-McCann, Inc.

———. 1958. Second record for Common Eider in Florida. *Fla. Nat.* 31:53.

———. 1963. Addendum to *Florida bird-life* (1954). 24 pp.

Steffee, Nina Dean, and C. Russell Mason. 1967. Zenaida Dove (*Zenaida aurita*) reported from Osceola County. *Fla. Nat.* 40:103.

Stein, Robert Carrington. 1963. Isolating mechanisms between populations of Traill's Flycatchers. *Proc. Amer. Phil. Soc.* 107:21–50.

Sterba, Günter. 1962. *Freshwater fishes of the world*. London: Vista Books (Longacre Press Ltd.).

Stevenson, Henry M. 1950. Distribution of certain birds in the southeastern United States. *Amer. Midland Nat.* 43:605–626.

———. 1955. Regional reports: Florida region. *Aud. Field Notes* 9:19–22.

———. 1956. Probable breeding of the Northern Prairie Warbler (*Dendroica discolor discolor*) in Florida. *Auk* 73:134.

———. 1957. Regional reports: Florida region. *Aud. Field Notes* 11:257–263.

———. 1958a. A record of *Hemidactylium scutatum* in Florida. *Copeia* 1958:49.

———. 1958b. Regional reports: Florida region. *Aud. Field Notes* 12:405–408.

———. 1960a. *A key to Florida birds*. Tallahassee: Peninsular Publ. Co.

———. 1960b. Regional reports: Florida region. *Aud. Field Notes* 14:379–383.

———. 1961a. Apparent breeding of the American Redstart in Florida. *Fla. Nat.* 34:44.

———. 1961b. Possible breeding of the Worm-eating Warbler in Florida. *Fla. Nat.* 34:222–223.

———. 1961c. Nesting of the Louisiana Waterthrush in Florida. *Fla. Nat.* 34:223.

———. 1961d. Regional reports: Florida region. *Aud. Field Notes* 15:320–325.

———. 1962a. Evidence of the breeding of two new species of warblers in Florida. *Fla. Nat.* 35:134–135.

———. 1962b. Occurrence and habits of the Eastern Chipmunk in Florida. *Jour. Mamm.* 43:110–111.

———. 1963a. A winter specimen of the Tennessee Warbler in Florida. *Fla. Nat.* 36:63–64.

———. 1963b. Another specimen of the Lapland Longspur in Florida. *Fla. Nat.* 36:64.

———. 1964a. A record of the Chestnut-collared Longspur in Florida. *Auk* 81:559.

———. 1964b. A Lark Bunting in Florida. *Fla. Nat.* 37:25.

———. 1964c. Regional reports: Florida region. *Aud. Field Notes* 18:503–505.

———. 1965. On the specific identity of summering scaups in Florida. *Fla. Nat.* 38:82.

————. 1966. The status of Traill's Flycatcher in Florida. *Fla. Nat.* 39:151.

————. 1967. Additional specimens of *Amphiuma pholeter* from Florida. *Herpetologica* 23:134.

————. 1968a. Records of the Coal Skink in Florida. *Quart. Jour. Fla. Acad. Sci.* 31:205–206.

————. 1968b. Regional reports: Florida region. *Aud. Field Notes* 22:599–602.

————. 1969a. Regional reports: Florida region. *Aud. Field Notes* 23:468–473.

————. 1969b. Occurrence of the Carpenter Frog in Florida. *Quart. Jour. Fla. Acad. Sci.* 32:233–235.

————. 1970a. Snow Buntings in Florida. *Fla. Nat.* 43:22.

————. 1970b. An inland record of the Great Cormorant in Florida. *Fla. Nat.* 43:178.

————. 1972a. Regional reports: Florida region. *Amer. Birds* 26:592–596.

————. 1972b. Recent breeding of the Sandwich Tern (*Thalasseus sandvicensis*) in Florida. *Fla. Nat.* 45:94–95.

————. 1972c. The recent history of Bachman's Warbler. *Wilson Bull.* 84:344–347.

————. 1973. An undescribed insular race of the Carolina Wren. *Auk* 90:35–38.

Stevenson, Henry M., and W. Wilson Baker. 1970. Records of new avian subspecies in Florida. *Fla. Nat.* 43:69–70.

Stimson, Louis A. 1942. Sight records of *Tyrannus melancholicus*. *Fla. Nat.* 15:56.

————. 1956. The Cape Sable Seaside Sparrow: its former and present distribution. *Auk* 73:489–502.

————. 1960. Inland record of Melodious Grassquit. *Fla. Nat.* 33:172.

————. 1962. Escaped Red-whiskered Bulbuls (*Pycnonotus jocosus*) increasing in Dade County. *Fla. Nat.* 35:93.

————. 1966. A remarkable ten days at the Dry Tortugas. *Fla. Nat.* 39:149–150.

Stitt, William T. 1964. Ruff (*Philomachus pugnax*) at Clewiston. *Fla. Nat.* 37:57.

Stoddard, Herbert L. 1962. *Bird casualties at a Leon County, Florida, TV tower, 1955–1961*. Tall Timbers Res. Sta., Bull. no. 1, Tallahassee, Fla.

Stoddard, Herbert L., Sr., and Robert A. Norris. 1967. *Bird casualties at a Leon County, Florida, TV tower: an eleven-year study*. Tall Timbers Res. Sta., Bull. no. 8, Tallahassee, Fla.

Suttkus, Royal D. 1958. Distribution of Menhaden in the Gulf of Mexico. *Trans. N. Amer. Wildlife Conf.* 23:401–410.

Sutton, George Miksch. 1946. A baby Florida Sandhill Crane. *Auk* 63:100–101.

Swem, Theodore R. 1969. Lapwing at West Palm Beach. *Fla. Nat.* 42:39.

Swift, Camm C. 1970. A review of the eastern North American cyprinid fishes of the *Notropis texanus* species group (subgenus *Alburnops*), with a definition of the subgenus *Hydrophlox*, and materials for a revision of the subgenus *Alburnops*. Doctoral dissertation, Florida State Univ., Tallahassee, Fla.

Tagatz, Marlin E. 1967. Fishes of the St. Johns River, Florida. *Quart. Jour. Fla. Acad. Sci.* 30:25–50.

Taylor, Edward H. 1935. A taxonomic study of the cosmopolitan Scincoid lizards of the genus *Eumeces*. *Kans. Univ. Sci. Bull.* 23:1–643.

Taylor, William Ralph. 1969. *A revision of the catfish genus* Noturus *Rafinesque, with an analysis of higher groups in the Ictaluridae*. U.S. Nat'l Mus. Bull. 282.

Telford, Sam R., Jr. 1955. Notes on an exceptionally large Worm Lizard, *Rhineura floridana*. *Copeia* 1955:258–259.

————. 1966. Variation among the southeastern crowned snakes, genus *Tantilla*. *Bull. Fla. State Mus.* 10:261–304.

Todd, W.E. Clyde. 1958. The Newfoundland race of the Gray-cheeked Thrush. *Can. Field-Nat.* 72:159–161.

Truchot, Edina K. 1962. Bronzed or Red-eyed Cowbird (*Tangavius aeneus*). *Fla. Nat.* 34:135.

Tucker, Marcia Brady. 1950. Sight record of the Snowy Owl in Florida. *Fla. Nat.* 23:72.

Turner, William R. 1971. Occurrence of *Brevoortia gunteri* in Mississippi Sound. *Quart. Jour. Fla. Acad. Sci.* 33:273–274.

Valentine, Barry. 1963. The salamander genus *Desmognathus* in Mississippi. *Copeia* 1963:130–139.

Van Gelder, Richard G. 1959. A taxonomic revision of the spotted skunks (genus *Spilogale*). *Bull. Amer. Mus. Nat. Hist.* 117:229–392.

Veronee, William R. 1959. Western Grebe (*Aechmophorus occidentalis*). *Fla. Nat.* 32:145.

Volpe, E. Peter. 1957. The early development of *Rana capito sevosa*. *Tulane Studies in Zool.* 5:207–225.

Walker, Ernest P. et al. 1964. *Mammals of the world.* Vol. II. Baltimore: The Johns Hopkins Press.

Webb, Robert G. 1959. Description of a new softshell turtle from the southeastern United States. *Publ. of Kans. Mus. Nat. Hist.* 11:517–525.

————. 1962. North American recent soft-shelled turtles (Family Trionychidae). *Publ. of Kans. Mus. Nat. Hist.* 13:429–611.

Weston, Francis M. 1960. Eared Grebe again at Pensacola. *Fla. Nat.* 33:99.

————. 1965. *A survey of the birdlife of northwestern Florida.* Tall Timbers Res. Sta. Bull. no. 5, Tallahassee, Fla.

Weston, Francis M., and Joyce T. (Mrs. G.P.) Baxter. 1957. Surfbird (*Aphriza virgata*). *Fla. Nat.* 30:86.

Wetherbee, David Kenneth, and Brooke Meanley. 1965. Natal plumage characters in rails. *Auk* 82:500–501.

Wetmore, Alexander. 1964. *A revision of the American vultures of the genus* Cathartes. Smithsonian Misc. Coll. no. 146, Washington, D.C.

White, Clayton M. 1968. Diagnosis and relationships of the North American tundra-inhabiting Peregrine Falcons. *Auk* 85:179–191.

Williams, Kenneth L., and Larry D. Wilson. 1967. A review of the Colubrid snake genus *Cemophora* Cope. *Tulane Stud. in Zool.* 13:103–124.

Williams, Lovett E., Jr. 1963. New specimen records for north Florida. *Fla. Nat.* 36:127.

————. 1968. Specimen of the Harlequin Duck in Florida. *Wilson Bull.* 80:488–489.

Wolfe, James L. 1968. Armadillo distribution in Alabama and northwest Florida. *Quart. Jour. Fla. Acad. Sci.* 31:209–212.

Woolfenden, Glen E. 1962. A range extension and subspecific relations of the Short-tailed Snake, *Stilosoma extenuatum*. *Copeia* 1962:648–649.

————. 1965. A specimen of the Red-footed Booby from Florida. *Auk* 82:102–103.

————. 1967. A specimen of the Golden-cheeked Warbler from Florida. *Auk* 84:115.

Woolfenden, Glen E., and Fred E. Lohrer. 1968. Great Cormorant seen on Old Tampa Bay, Florida. *Fla. Nat.* 41:169.

Wright, Albert H., and Anna A. Wright. 1949. *Handbook of frogs and toads*. Ithaca: Comstock Publ. Co.

Yerger, Ralph, and Kenneth Relyea. 1968. The flat-headed bullheads (Pisces: Ictaluridae) of the southeastern United States, and a new species of *Ictalurus* from the Gulf Coast. *Copeia* 1968:361–384.

Yerger, Ralph W., and Royal D. Suttkus. 1962. Records of freshwater fishes in Florida. *Tulane Studies in Zool.* 9:323–330.

Young, Katherine D. 1964. Harris' Sparrow (*Zonotrichia querula*) at Melbourne Beach. *Fla. Nat.* 37:122.

Zillig, Louise D. 1958. The status of *Haldea* Baird and Girard and *Virginia* Baird and Girard. *Copeia* 1958:152.

Index

to common and scientific names (omitting trinomials)

Bold-face numerals indicate location of species accounts and descriptions of orders and families.

A

Acantharchus 206
    pomotis 30, **206**, 207
Acara 572
Acaropsis 572
Accipiter cooperii 101, **329**
    gentilis 101, **328**, 329
    striatus 101, **329**
Accipitridae 84, 100, **327**
Achirus 225
    lineatus 41, **225**
Acipenser brevirostrum 17, **158**
    oxyrhynchus 17, **157**, 158
Acipenseridae 11, 17, **157**
Acipenseriformes 11, 17, **157**
Acris 242
    crepitans 50, 54, **243**
    gryllus 50, 54, **243**
Actitis macularia 106, **350**
Adinia 185
    xenica 27, **184**, 185, 567
Aechmophorus occidentalis 89, **293**, 577
Aegolius acadicus 113, **382**
Agelaius humeralis 128, **451**
    phoeniceus 128, **450**, 451

Agkistrodon 566
    contortrix 76, **287**
    piscivorus 76, **288**
Agnatha 3, 9, 10, 155
Agonostomus monticola 40, **221**
Aimophila aestivalis 132, **471**
Aix sponsa 99, **318**
Ajaia ajaja 83, 93, **308**
Alaudidae 85, 117, **399**
Albatross, Wandering 90, **479**
    Yellow-nosed 90, **479**
Alca torda 110, **372**
Alcedinidae 84, 114, **387**
Alcidae 80, 110, **372**
Alewife 161
Alle alle 110, **372**
Alligator, American 77, **289**, 290
Alligator mississipiensis 77, **289**
Alosa 163
    aestivalis 18, **162**
    alabamae 18, **162**
    chrysochloris 18, **161**
    mediocris 18, **162**
    pseudoharengus 161
    sapidissima 18, **162**
Amazona viridigenalis 482
Ambloplites rupestris 31, **206**

Brevoortia (continued)
  smithi 18, **163**
  tyrannus 18, **163**
Brotogeris jugularis 112, **482**
  versicolurus 112, **531**
Bubo virginianus 113, **381**
Bubulcus ibis 92, **303**
Bucephala albeola 98, **321**
  clangula 95, **320**
Budgerigar 112, 376, **377**
Bufflehead 98, **321**
Bufo americanus 237
  marinus 48, 53, **238**, 533, 569
  quercicus 48, 53, **238**
  terrestris 48, 53, **237**, 238
  woodhousei 44-45, 48, 53, 237,
    238
Bufonidae 42, 48, 235, **236**
Bulbul, Red-whiskered 119, **409**,
  562, 576
Bullfinch, Cuban 129
Bullfrog 51, **244**
Bullhead, Brown 23, **178**, 179
  Flat 23
  Snail 23, **177**, 178
  Spotted 23, **178**
  Yellow 23, **179**
Bulweria bulwerii 90
Bunting, Indigo 133, **464**
  Lark 130, 132, **466**, 566, 575
  Lazuli 464
  Painted 133, **464**, 572
  Snow 129, **478**, 576
Buteo brachyurus 101, **331**, 572
  jamaicensis 101, **330**
  lagopus 101, **331**, 332
  lineatus 101, **330**
  platypterus 101, **330**, 331
  swainsoni 101, **331**
Butorides virescens 93, **302**, 303

C

Caiman sclerops 76, **290**
Caiman, Spectacled 76, 289, **290**
Calamospiza melanocorys 130,
  132, **466**, 566
Calcarius lapponicus 129, **477**, 570
  ornatus 129, **478**

Calidris acuminata 107, **352**
  alba 106, **355**
  alpina 106, **354**
  bairdii 107, **353**
  canutus 105, **351**, 352
  ferruginea 106, **479**
  fuscicollis 106, **353**
  maritima 106, **352**
  mauri 106, **354**
  melanotos 107, **352**
  minutilla 106, **353**
  pusillus 106, **354**
Callichelidon cyaneoviridis 118,
  399
Calliphlox evelynae 114, **531**
Campephilus principalis 115, **391**
Canidae 140, 149, **509**, 514
Canis familiaris 149
  latrans 149, **509**
  rufus 149
Canvasback 98, **319**, 322
Capella gallinago 105, **348**
Caprimulgidae 84, 113, **382**
Caprimulgiformes 84, 113, **382**
Caprimulgus carolinensis 113, **383**
  vociferus 113, **384**
Capromyidae 141, 149, **509**
Caracara 101, **334**, 538
Caracara cheriway 101, **334**
Carangidae 16, 37, **214**
Caranx hippos 37, **214**
Carassius auratus 19
Carcharhinidae 10, **156**
Carcharhinus leucas 10, **156**
Cardinal 129, 458, **462**
  Brazilian 482
Cardinalis cardinalis 86-87, 129,
  462
Carduelis carduelis 562
Caretta caretta 65, **255**, 256
Carnivora 140, 149, 150, **509**, 511,
  512, 514, 516
Carp 19, **175**
Carphophis amoena 76, **290**
Carpiodes 175
  cyprinus 22, **175**
  velifer 12-13, 22, **175**
Carpodacus purpureus 134, **460**

Flycatcher (*continued*)
  Alder xvi, 396
  Ash-throated 116, **395**
  Fork-tailed 115, **480**, 575
  Great Crested 116, 393, **394**
  Kiskadee 568
  Least 117, 396, **397**
  Olive-sided 116, **398**
  Scissor-tailed 115, **393**
  Traill's xvi, 117, **396**, 397,
    575, 576
  Vermilion 116, **398**
  Wied's Crested 116, **394**, 395
  Willow xvi, 396
  Yellow-bellied 117, **395**
Fox, Gray 149, **510**
  Red 149, **510**, 511, 564, 569
Francolin, Black 102, **481**
Francolinus francolinus 102, **481**
Fregata magnificens 92, **301**
Fregatidae 80, 92, **301**
Fregetta grallaria 91, **479**
  tropica 91, 479
Frigatebird, Magnificent 92, **301**
Fringillidae 88, 129, 458, 459
Frog, Barking Tree 50, **240**
  Bird-voiced Tree 49, **240**
  Bronze 51, 244, **245**
  Carpenter 51, **245**, 527, 575
  Cuban Tree 49, 240, **241**, 535
  Gopher 51, 245, **246**
  Greenhouse 48, **239**
  Green Tree 50, **241**, 530
  Leopard 51, **246**
  Little Grass 50, **241**
  Northern Cricket 50, **243**
  Ornate Chorus 50, **242**
  Pickerel 51, 247
  Pig 52, **244**
  Pine Barrens Tree 49, **530**
  Pine-woods Tree 49, **239**
  River 51, **244**, 245, 574
  Southern Chorus 51, **242**, 574
  Southern Cricket 50, **243**
  Southern Gray Tree 49, **240**
  Squirrel Tree 50, **240**, 242
  Upland Chorus 51, **242**

Fulica americana 102, **342**, 567
  caribaea 102
Fundulus 166
  chrysotus 28, 183, 185, **189**,
    193
  cingulatus 28, **188**, 189
  confluentus 28, **188**, 572
  grandis 12-13, 24-25, 28, 185,
    186
  heteroclitus 28, **185**, 186
  jenkinsi 27, **184**, 188
  lineolatus 28, 188, **189**, 190,
    573
  majalis 27, **188**, 190
  notti 28, **189**, 190, 573
  olivaceus 27, 184, 188, **190**
  pulvereus 572
  seminolis 27, 184, **186**, 187,
    189, 193
  similis 27, **187**, 188

G

Gadwall 99, **314**
Galliformes 84, 102, **335**, 336
Gallinula chloropus 102, **342**
Gallinule, Common 102, **342**
  Purple 102, **341**
Gambusia 166, 183, 194, 528
  affinis 29, 187, 189, **193**, 194
  punctata 194, 573
  rhizophorae 29, **194**
  species 194
Gannet 91, **299**
Gar, Alligator 17, **158**
  Florida 17, 158, **159**, 160
  Long-nosed 17, **159**
  Spotted 17, **158**
Garmannia 566
Gasterosteiformes 11, 29, **196**
Gastrophryne carolinensis 52, **247**
Gavia arctica 89, **291**
  immer 89, **290**
  stellata 86-87, 89, **291**
Gaviidae 80, 89, **290**
Gaviiformes 80, 89, **290**
Gecko, Ashy 66, 262, **263**
  Indo-Pacific 66, **261**

Gull (*continued*)
 Great Black-backed 108, **362**, 530
 Herring 108, 361, **362**, 530
 Iceland 108, **361**
 Laughing 108, 360, **363**, 364, 365, 531
 Lesser Black-backed 108, **530**
 Little 109, **364**, 570
 Ring-billed 108, 294, 360, **362**, 363, 372
 Sabine's 108, **365**

H

Haematopodidae 82, 103, **343**
Haematopus palliatus 103, **343**
Haideotriton 572
 wallacei 46, **232**
Haldea 578
 valeriae 564
Haliaeetus leucocephalus 100, **332**
Harengula pensacolae 18, **164**
Hawk, Broad-winged 101, **330**
 Cooper's 101, **329**
 Harris' 100
 Marsh 101, **333**
 Red-shouldered 101, 327, 328, **330**, 331, 333, 334, 404
 Red-tailed 101, **330**, 331, 333, 334
 Rough-legged 101, **331**
 Sharp-shinned 101, 328, **329**
 Short-tailed 101, **331**, 538, 572
 Swainson's 101, **331**
Hawksbill 64, **255**
Helmitheros vermivorus 127, **429**
Hemichromis bimaculatus **529**
Hemidactylium 566
 scutatum 46, **232**, 575
Hemidactylus 60-61
 garnoti 66, **261**
 turcicus 66, **261**, 262, 569
Heron, Black-crowned Night 93, **304**
 Great Blue xv, 93, **302**, 305, 307, 338
 Great White xv, 92, **302**, 304, 538

Green 93, **302**
 Little Blue 93, **303**, 304, 305, 307, 308, 565
 Louisiana 93, 303, **304**
 Wurdemann's 302
 Yellow-crowned Night 93, **304**
Herring, Blue-backed 18, **162**
 Skipjack 18, **161**, 162
Hesperiphona vespertina 134, **459**, 460
Heterandria 184
 formosa 29, **194**, 195
Heterodon 286
 platyrhinos 70, **285**, 286
 simus 72, **286**, 287
Heterosomata 571
Himantopus mexicanus 107, **358**
Hippocampus 196
Hirundinidae 85, 117, **399**
Hirundo rustica 118, **401**
Histrionicus histrionicus 98, **322**
Hogchoker 42, **225**, 226
Hominidae 137, 144, **492**
Homo sapiens 144, **493**
Honeycreeper, Bahama 567, 570
Hummingbird, Black-chinned 114, **386**, 567
 Broad-billed 114, **480**
 Ruby-throated 114, **386**, 387, 531
 Rufous 114, **387**, 531, 570
Hybognathus hayi 21, **172**
Hybopsis 173
 aestivalis 12-13, 19, **174**
 amblops 19, **173**
 harperi 19, 169, **174**
 winchelli 173
Hydrobatidae 80, 91, **296**
Hydroprogne caspia 109, **369**
Hyla 241
 andersoni 49, 55, **530**, 564
 avivoca 49, 54, **240**, 241
 chrysoscelis 49, 55, **240**
 cinerea 50, 55, **241**
 crucifer 50, 54, **239**
 femoralis 49, 55, **239**, 243
 gratiosa 50, 55, **240**

Micropalama himantopus 106, **356**
Micropogon undulatus 39, **219**
Micropterus 200
    coosae 32, **201**
    notius 32, **200**, 201
    punctulatus 32, **201**, 202
    salmoides 32, 201, **202**
    species 201
Microtus pinetorum 145, **506**, 507
Micrurus fulvius 76, **286**
Mimidae 89, 119, **414**
Mimus gundlachii **480**
    polyglottos 119, **414**
Mink 150, 512, **513**
Minnow, Cypress 21, **172**
    Eastern Mud 19, **166**
    Lake Eustis 27, **191**
    Northern Starhead 28, **189**
    Pug-nosed 20, **175**
    Sheepshead 26, 38, **190**, 191, 192
    Silver-jawed 20, **172**
    Southern Starhead 28, **189**, 190
Minytrema melanops 22, **176**
Mniotilta varia 124, **428**
Mockingbird 119, 392, 403, **414**, 415, 416, 419, 422
    Bahama **480**
Mojarra, Spotfin 38, **215**, 216
    Yellowfin 38, **215**
Mole, Eastern 142, **484**
    Star-nosed 142, 527
Molly, Black **527**
    Liberty **528**
    Sailfin 29, **194**, 527
Molossidae 136, 144, 485, **491**
Molothrus ater 128, **454**, 455
Monachus tropicalis 151, **516**
Monkey, Rhesus 144, **492**
    Squirrel 144, 527
Morone chrysops 30, **199**
    saxatilis 30, **198**
Morus bassanus 91, **299**
Mosquitofish, Common 29, **193**, 194
    Mangrove 29, **194**
Motacillidae 85, 121, **421**

Mouse, Beach 563
    Cotton 149, **503**
    Eastern Harvest 148, **501**, 503, 508
    Florida Deer 148, **503**
    Golden 148, **503**
    House 149, 501, **508**
    Old-field 149, **501**
Moxostoma carinatum 22, 226
    duquesnei 22, 176
    poecilurum 22, **176**
    species 22, **176**
Mugil 220
    cephalus 40, **220**
    curema 40, **220**
    trichodon 40, **220**
Mugilidae 16, 40, **220**
Mullet, Fantail 40, **220**
    Mountain 40, **221**
    Striped 40, **220**
    White 40, **220**
Mummichog 28, **185**, 186
Muridae 141, 149, 500, 501, **507**
Murre, Common 110, 479
Mus musculus 149, **508**
Muscivora forficata 115, **393**
    tyrannus 115, **480**
Muskrat 148, 527
Mustela frenata 150, **512**, 513
    vison 150, **513**, 574
Mustelidae 140, 150, **512**
Mycteria americana 93, **307**
Myiarchus cinerascens 116, **395**
    crinitus 116, **394**
    tyrannulus 116, **394**, 395
Myiopsitta monachus 111, **481**
Mynah, Hill 85, 121, **482**
Myocastor coypus 149, **509**
Myotis 142, 486, 490
    austroriparius 143, **485**, 487, 488, 572
    grisescens 142, **485**, 572
    keeni 143, **487**, 488, 572
    sodalis 138-139, 143, **488**, 490, 568
Myotis, Gray 142, **485**
    Indiana 143, **488**

Salamander (*continued*)
    Long-tailed 48, **233**
    Marbled 46, **231**
    Many-lined 47
    Mole 46, **230**
    Mud 48, **233**
    Northern Dusky 47, **234**
    Red 48, **233**
    Red Hills 47, 247
    Seal 47, **235**, 237
    Slimy 48, **232**, 235
    Spotted 46, 247
    Southern Dusky 47, **235**, 236
    Tiger 46, **231**
    Two-lined 48, **234**
    Zigzag 47, 247, 563
Salamandridae 42, 43, **228**, 231
Salientia 42, 48, 49, 51, 52, **235**,
    238, 239, 244, 246
Salmoniformes 15, 18, 19, **165**, 166
Sanderling 106, **355**, 358, 359
Sandpiper, Baird's 107, **353**
    Buff-breasted 106, **355**
    Curlew 106, **479**, 565, 570
    Least 106, **353**, 354
    Pectoral 107, **352**, 353, 354
    Purple 106, **352**
    Semipalmated 106, 353, **354**
    Sharp-tailed 107, **352**, 571
    Solitary 106, **350**
    Spotted 106, **350**
    Stilt 106, **356**
    Upland 106, **349**, 355
    Western 106, 353, **354**
    White-rumped 106, **353**
Sapsucker, Yellow-bellied 115,
    390
Sardine, Scaled 18, **164**
Sauger 33, **208**
Sawfish, Large-toothed 11, **157**
    Small-toothed 11, **156**
Sayornis nigricans 116, **480**
    phoebe 116, **395**
    saya 116, **480**
Scalopus aquaticus 142, **484**
Scaphiopus holbrooki 48, 52, **236**,
    565
Scardafella inca 111, **375**

Scaup, Greater 98, **319**, 320
    Lesser 98, **320**
Sceloporus undulatus 66, **259**, 260,
    562
    woodi 65, **260**, 568
Schoolmaster 38, **215**
Sciaenidae 16, 39, **217**
Sciaenops ocellata 39, **218**
Scincella 268
    lateralis 67, 68-69, **266**
Scincidae 58, 67, **266**
Scincus lateralis 570
Sciuridae 141, 144, **496**
Sciurus aureogaster 145, **498**
    carolinensis 145, **496**, 498
    niger 145, **498**, 499
Scoliodon 575
Scolopacidae 82, 103, 104, **347**
Scoter, Black 95, **323**, 324
    Surf 95, 322, **323**
    White-winged 95, **323**
Sea Lion, California 151
Seal, Harbor 151, **516**, 564
    Hooded 151, **516**
    Monk 572
    West Indian 151, **516**
Seatrout, Spotted 39, **218**
Seiurus aurocapillus 125, **443**
    motacilla 125, **444**, 562
    noveboracensis 125, **443**, 444
Selasphorus rufus 114, **387**, 570
Seminatrix 274
    pygaea 75, **274**, 275
Semotilus 166, 173
    atromaculatus 19, **174**
Serpentes 58, 70, 76, 270, 286
Setophaga ruticilla 122, **448**
Shad, Alabama 18, **162**
    American 18, **162**
    Gizzard 17, **164**
    Hickory 18, **162**
    Threadfin 17, **164**
    Yellowfin 18, **163**
Shark, Atlantic Sharp-nosed 10,
    156
    Bull 10, **156**
Sharksucker 37, **214**

T

Tachornis phoenicobia 114, **531**
Tachycineta thalassina 118, **480**
Tadarida 492
    brasiliensis 138-139, 144, **492**, 574
Tadpoles 5, 42
Talpidae 137, 142, **484**
Tamias striatus 145, **496**, 497
Tanager, Blue-gray 128, **455**, 456, 562, 570
    Scarlet 129, **457**
    Stripe-headed 129, **455**, 569
    Summer 129, 457, **458**
    Western 129, **457**
Tangavius aeneus 128, **455**, 577
Tantilla 274, 577
    coronata 75, **284**, 285
    oolitica 75
    relicta 75
Tarpon 17, 159, **160**
Teal, Blue-winged 100, **315**, 316, 321
    Cinnamon 100, **316**
    Green-winged 99, **315**
Teeidae 58, 67, **265**
Telmatodytes palustris 119, **413**
Tern, Arctic 110, **479**, 567
    Black 110, **370**
    Bridled 109, **368**
    Caspian 109, **369**, 570, 573, 574
    Common 110, **336**, 567, 569
    Forster's 110, **366**, 367, 369
    Gull-billed 109, **365**, 366
    Least 110, 367, **368**, 370
    Royal 109, **369**, 371
    Roseate 110, 366, **367**, 563
    Sandwich 110, **369**, 370, 576
    Sooty 109, **367**, 573
Terrapene carolina 59, 250, **251**
Terrapin, Diamondback 59, **251**
Testudines 247
Testudinidae 58, 64, **254**, 569
Thalasseus maximus 109, **369**
    sandvicensis 110, **369**, 576

Thamnophis sauritus 70, **276**, 573
    sirtalis 70, **275**, 276, 573
Thrasher, Brown 120, **415**, 416, 480
    Curve-billed 120, **415**
    Long-billed 120, **480**
    Sage 120, **416**, 568
Thraupidae 88, 128, **455**, 459
Thraupis virens 128, **455**, 456, 562, 570
Threskiornithidae 81, 93, **307**
Thrush, Gray-cheeked 121, 417, **418**, 577
    Hermit 120, **417**, 418
    Swainson's 120, **417**, 418
    Wood 120, **416**, 417, 566
Thryomanes bewickii 119, **412**, 413
Thryothorus ludovicianus 119, **413**
Tiaris bicolor 132, **464**, 465, 563, 570
    canora 132, **481**, 562
Tilapia 219, 529
    aurea **530**
    melanopleura 530
    melanotheron 39, **219**, 530
Tilapia, Black-chinned 39, **219**
    Blue **530**
Timucu 26, **183**
Titmouse, Black-crested 118
    Tufted 118, **406**
Toad, American 237
    Eastern Narrow-mouthed 52, **247**
    Fowler's **238**
    Giant 48, **238**, 535
    Key West Spadefoot 565
    Neotropical 569
    Oak 48, **238**
    Southern 48, **237**
    Texas Horned 561
    Woodhouse's 48, **238**
Tonguefish, Black-cheeked 42, **226**
Topminnow, Banded 28, **188**
    Black-spotted 27, **190**
    Golden 28, 183, **189**
    Salt-marsh 27, **184**